Larval Fish Nutrition

Larval Fish Nutrition

Editor

G. Joan Holt

WILEY-BLACKWELL

A John Wiley & Sons, Inc., Publication

Wiley-Blackwell is an imprint of John Wiley & Sons, formed by the merger of Wiley's global Scientific, Technical and Medical business with Blackwell Publishing.

Registered office: John Wiley & Sons Ltd, The Atrium, Southern Gate, Chichester, West Sussex, PO19 8SQ, UK

Editorial offices: 2121 State Avenue, Ames, Iowa 50014-8300, USA
The Atrium, Southern Gate, Chichester, West Sussex, PO19 8SQ, UK
9600 Garsington Road, Oxford, OX4 2DQ, UK

For details of our global editorial offices, for customer services and for information about how to apply for permission to reuse the copyright material in this book please see our website at www.wiley.com/wiley-blackwell.

Library of Congress Cataloging-in-Publication Data

Larval fish nutrition / editor, G. Joan Holt.
p. cm.
Includes bibliographical references and index.
ISBN 978-0-8138-1792-7 (hardcover : alk. paper) 1. Fishes—Larvae—Feeding and feeds. I. Holt, J. (Joan)
SH156.L37 2011
639.3—dc22

2010052427

A catalogue record for this book is available from the British Library.

This book is published in the following electronic formats: ePDF 9780470959831; Wiley Online Library 9780470959862; ePub 9780470959848, and Mobi 9780470959855.

Set in 10 on 12.5 pt Sabon by Toppan Best-set Premedia Limited

1 2011

Contents

Contributors

Cláudia Aragão
Centre of Marine Sciences (CCMAR)
University of Algarve
Faro, Portugal

Rick Barrows
USDA—Agricultural Research Service
Fish Technology Center
Bozeman, MT

Luis Conceição
Centre of Marine Sciences (CCMAR)
University of Algarve
Faro, Portugal

Konrad Dabrowski
School of Environment and Natural Resources
Ohio State University
Columbus, OH

Maria J. Darias
Research and Technology in Food and Agriculture (IRTA)
Sant Carles de la Ràpita, Spain

Ralf Dahm
Department of Biology
University of Padua
Padua, Italy

Alicia Estevez
Research and Technology in Food and Agriculture (IRTA)
Sant Carles de la Ràpita, Spain

Jan Ove Evjemo
Marine Resources Technology
SINTEF
Trondheim, Norway

Hipólito Fernández-Palacios
Grupo de Investigación en Acuicultura (ICCM and ULPGC)
Las Palmas de Gran Canaria, Spain

Trina F. Galloway
SINTEF
Trondheim, Norway

Enric Gisbert
Research and Technology in Food and Agriculture (IRTA)
Sant Carles de la Ràpita, Spain

Kristin Hamre
Institute of Nutrition and Seafood Research (NIFES)
Bergen, Norway

G. Joan Holt
Department of Marine Science
University of Texas at Austin Marine Science Institute
Port Aransas, TX

Ione Hunt von Herbing
Department of Biology
University of North Texas
Denton, TX

Marisol Izquierdo
Grupo de Investigación en Acuicultura (ICCM and ULPGC)
Las Palmas de Gran Canaria, Spain

Marta Jaroszewska (Deceased)
Department of Biology and Earth Science
Nicolaus Copernicus University
Torun, Poland

Elin Kjørsvik
Department of Biology
Norwegian University of Science and Technology (NTNU)
Center of Fisheries and Aquaculture
Trondheim, Norway

William Koven
Israel Oceanographic and Limnological Research
The National Center for Mariculture
Eilat, Israel

Chris Langdon
Coastal Oregon Marine Experiment Station
Hatfield Marine Science Center
Oregon State University
Newport, OR

Juan Pablo Lazo
Centro de Investigación Cientifica y de Educación Superior de Ensenada (CICESE)
Ensenada, Mexico

Mari Moren
Institute of Nutrition and Seafood Research (NIFES)
Bergen, Norway

Birgitta Norberg
Institute of Marine Research
Austevoll, Norway

Gunvor Øie
Marine Resources Technology
SINTEF
Trondheim, Norway

Yngvar Olsen
Department of Biology
Norwegian University of Science and Technology (NTNU)
Trondheim, Norway

Rafael Perez-Dominguez
Institute of Estuarine and Coastal Studies (IECS)
University of Hull
Hull, UK

Kjell Inge Reitan
Marine Resources Technology
SINTEF
Trondheim, Norway

Ivar Rønnestad
Department of Biology
University of Bergen
Bergen, Norway

Michael B. Rust
NOAA Northwest Fisheries Science Center
Seattle, WA

Øystein Sæle
Institute of Nutrition and Seafood Research (NIFES)
Bergen, Norway

Josianne Støttrup
National Institute of Aquatic Resources
Technical University of Denmark
Charlottenlund, Denmark

Meredith Turnbough
University of North Texas
Denton, TX

Rune Waagbø
Institute of Nutrition and Seafood Research (NIFES)
Bergen, Norway

Ken A. Webb
NOAA Northwest Fisheries Science Center
Seattle, WA

Manuel Yúfera
Instituto de Ciencias Marinas de Andalucia (CSIC)
Cadiz, Spain

Preface

Addressing the challenges of larval fish nutrition is complex but necessary in order to make progress in developing diets for optimum growth and development of the early life stages of fish. There are two main approaches to feeding larvae in captivity: the first utilizes live food organisms such as rotifers and brine shrimp, while the second focuses on the development of microparticulate diets. Live foods are difficult to sustain, require considerable space and expense, and may not provide adequate nutrition, while microparticulate diets are easier to maintain and have lower production costs, but have not proven to be successful for raising marine fish larvae. Ingestion, digestion, and assimilation are all critical steps in the utilization of larval diets, and each or all may contribute to the lack of success thus far. Fish larvae reared on microparticulate diets generally have lower growth and survival rates that may be the result of inadequate nutritional quality of the diets, low ingestion rates, or poorly developed digestive systems in the larvae. The lack of nutritional information for fish larvae, and the dependence on live food, has hindered progress in developing larval diets. Since larviculture is a bottleneck to successful production of many preferred, high-quality marine species, it is important to know where we should be headed to successfully develop diets for fish larvae. Development of artificial feeds for fish larvae capable of supporting good survival and growth would be a tremendous benefit to hatcheries that depend on the supply of live feed. This book on larval fish nutrition was developed to address these challenges and support the development of feeding and weaning strategies that will optimize growth and survival of fish for production.

The aim of this book is to provide a comprehensive reference text on larval fish nutrition. It includes sections on digestive development and nutrient requirements, nutrient physiology, and feeds and feeding. Each of these three sections details crucial topics, including ontogeny of the digestive system; current state of knowledge on the requirements for lipids, proteins, and micronutrients; effects of broodstock diet on eggs and larvae; utilization of yolk during endogenous feeding; effects of nutrition on development; metabolism and bioenergetics; regulation of digestion; feeding behavior and live feeds; microparticulate diet technology, testing, and evaluation; and methods for assessing growth, and the potential of the zebrafish as a model organism to study fish nutrition. Reference tables of development events, micronutrients in live prey, fatty acid requirements, vitamin requirements for broodstock, and so on are included. A list of challenges and solutions is provided, along with suggestions for evaluating their success. A strong team of international experts in each of the important areas of larval fish nutrition was assembled to summarize current knowledge and to suggest essential future directions. A total of 34 contributors from Denmark, Europe, Israel, Norway, Mexico, the United Kingdom, and the United States provided summaries of critical information needed to promote rapid progress in this field. It is hoped that this book will be a beneficial reference for researchers advancing knowledge of larval fish nutrition and aquaculture production, as well as a valuable supplementary textbook for advanced courses in nutrition, aquaculture, and larval fish studies.

G. Joan Holt

Larval Fish Nutrition

Section 1

Digestive Development and Nutrient Requirements

Chapter 1

Ontogeny of the digestive tract

Juan Pablo Lazo, Maria J. Darias, and Enric Gisbert

1.1 Introduction

Fishes, like other organisms, require an energy source to fuel their body systems and processes, including growth, metabolism, and reproduction. Different fish species have evolved feeding structures and digestive mechanisms that allow them to exploit a vast array of vegetal and animal food sources; consequently, the digestive tract of fishes has incorporated numerous adaptations for the efficient breakdown and absorption of essential nutrients, including appropriate digestive enzymes and absorptive surface areas (Moyle and Cech 2000). Since the dietary requirements of fish larvae are different from those of juveniles or adults, larval nutrition should always be considered along with the organization and functionality of the digestive system, nutritional needs, and the behavior of larvae at different stages of development. In addition to being the site of nutrient digestion and absorption, the digestive organs provide a barrier to environmental toxins, confer essential immune function, and have important roles in metabolism and salt and water absorption (Wallace et al. 2005).

Knowledge of differentiation of the digestive tract and accessory glands during larval development is essential for understanding the digestive and nutritional physiology of larval fishes, and synchronizing the physiological stage of development with feeding practices and rearing protocols. Thus, one of the main features that determines the end of the transformation from larvae to juvenile stages in teleosts is the development of a complete, functional, fully developed digestive system. Knowledge of the developmental stage will facilitate overcoming one of the major bottlenecks in fish hatcheries, the partial or complete replacement of live prey with a compound inert diet. In this sense, the ontogeny of the digestive tract of fish larvae has been the subject of many studies for the last 25 years, although most of the abovementioned effort has been focused on salmonids and marine finfish species due to their important commercial value for the aquaculture industry.

Larval Fish Nutrition, First Edition. Edited by G. Joan Holt.

Many studies have centered on evaluating the ontogenetic and epigenetic changes in the morphoanatomy and histological organization of the digestive organs by means of microscopy, as well as assessing the activity of different digestive enzymes from the pancreas, stomach, and intestine by means of biochemical quantification. Recently, those approaches have been complemented by molecular biological techniques that provide insight into both temporal and spatial expression patterns of genes involved in the development and functionality of the digestive tract during early ontogeny. Contrary to what was originally claimed, these studies have clearly demonstrated that fish larvae are not challenged with physiological or digestive deficiencies, although they hatch with very immature organs and systems compared with juveniles. In this sense, the digestive system of a fish larva should be considered as a very efficient system that provides the larva with all the nutrients and energy needed for routine maintenance metabolism, swimming, and growth in order to enhance its survival, growth performance, and transformation into a juvenile. Although fishes as a group show a remarkable diversity of structure and function of their nutritional physiology, the basic mechanisms of organ and system development are similar in all teleosts, even though there exist considerable interspecific differences in the relative timing of their differentiation, development, and functionality during early ontogeny. The timing of development of organ and physiological function is affected by the general life history and reproductive strategy of each species and by a variety of abiotic and biotic factors, including water quality, mainly temperature, and food availability and composition.

This chapter reviews the available information on the subject of the digestive physiology of marine and freshwater fish larvae, with special emphasis on the organization and functionality of the digestive tract and accessory organs during early ontogeny. We also focus on the use of different histological and biochemical markers to assess the nutritional condition of fish larvae under different nutritional and rearing conditions. Consequently, the chapter has been divided in three parts: first, a section devoted to describing the histomorphological development of the digestive tract and accessory glands in order to achieve a better understanding of their organization and functionality during early ontogeny, followed by a section focused on the functionality of the digestive organs based on the activity of specific digestive enzymes, and last but not the least, a review of the different biochemical and histological parameters that can be used for assessing and describing the nutritional condition of fish larvae. Throughout this chapter, the focus is on those dietary-induced changes in digestive tract functionality during larval ontogeny in species of aquacultural interest, although information is also included regarding other species of interest for biomedical studies (e.g., zebrafish).

1.2 Organogenesis of the digestive system

In most described species, the alimentary canal at hatching appears histologically as an undifferentiated straight tube lying dorsally to the yolk sac. However, during the lecithotrophic larval stage, the larva undergoes rapid developmental changes leading to the differentiation of several regions and organs of the digestive system, namely buccopharynx, esophagus, intestine, pancreas, and liver, whereas the morphogenesis of the stomach depends on the species. There are important morphological and functional differences between marine and freshwater fish species regarding the developmental events involved in the differentiation and functionality of the digestive tract, even between closely related species. In fact, although both groups of fishes hatch with a simple digestive tract appearing as a straight and undifferentiated tube located dorsal to the yolk sac, closed to the exterior at both extremities (mouth and anus), and lined by a single layer of columnar epithelial cells (future enterocytes) with basal or central nuclei, there exist important morphoanatomical differences at the onset of exogenous feeding. It is generally accepted that in the case of marine fish species, the appearance of gastric glands and the onset of acidic digestion does not take place until metamorphosis far beyond the onset of exogenous feeding (Govoni et al. 1986; Zambonino-Infante and Cahu 2007), whereas in freshwater species with large- and medium-sized eggs, this process takes place during the transition to exogenous feeding, for example, cichlids (Lingling and Qianru 1981; Balon 1985; Fishelson 1995; Morrison et al. 2001; Alvarez-González et al. 2008), acipenserids (Gisbert et al. 1998; Gisbert and Doroshov 2003), salmonids (Sarieyyüpoglu et al. 2000; Rust 2002), or siluriformes (Verreth et al. 1992; Kozarić et al. 2008; de Amorim et al. 2009). However, this rapid development of the digestive system is not a generalized feature among other freshwater fish species, such as coregonids (Loewe and Eckmann 1988; Segner et al. 1993), percids (Ostaszewska 2005), cyprinids (Smallwood and Smallwood 1931; Wallace et al. 2005), or characids (Atencio García et al. 2007), which are characterized by small-sized eggs.

However, generalities normally have their own exceptions; for example, in the case of cichlids there exist species-specific differences in the time of differentiation of different digestive structures depending on the reproductive strategy of each species. Thus, substrate-spawning cichlids, such as the Nile tilapia *Tilapia nilotica* (Lingling and Qianru 1981; Morrison et al. 2001), common bay snook *Petenia splendida* (Alvarez-González et al. 2008), or *Cichlasoma dimerus* (Meijide and Guerrero 2000), tend to develop faster than mouthbrooding cichlids (Balon 1985; Fishelson 1995; Osse and van den Boogart 2004). In this sense, the reproductive strategy (parental care) of different cichlid species has a direct effect on the ontogenetic development of the larva, and consequently, on the sequence of development of its digestive tract. In particular, substrate spawners might need to develop faster in order to achieve full functionality of all their organs and systems, consequently maximizing their chances of survival to respond to environmental constraints. In contrast, those larvae receiving parental protection during their early ontogeny (mouthbrooding) might maximize energy for growth and tissue differentiation and minimize expenditures for other purposes (e.g., avoidance of predators), and consequently, these larvae might develop more slowly.

In the following sections, information about the histological differentiation and histochemical properties of different regions of the digestive tract and accessory organs is considered. Since this chapter is mainly devoted to describing the morphological organization and functionality of the digestive tract in relation to nutrient digestion and assimilation processes, the description of the buccopharynx has been deliberately omitted since this organ is more involved in prey capture than nutrition, and it has been recently reviewed by Zambonino-Infante et al. (2008).

1.3 Histological structure of the digestive tract and accessory glands

The digestive tract of fishes is composed of four basic histological layers, such as the mucosa, submucosa, muscularis, and serosa (Figure 1.1). However, the structure of these different segments of the alimentary canal varies considerably, and some of its parts or their constituents may be lacking depending on the species and/or the stage of development considered (Takashima and Hibiya 1995). In general, the wall of the digestive tract is formed by the serosa, muscularis, submucosa, and mucosa from the esophagus to

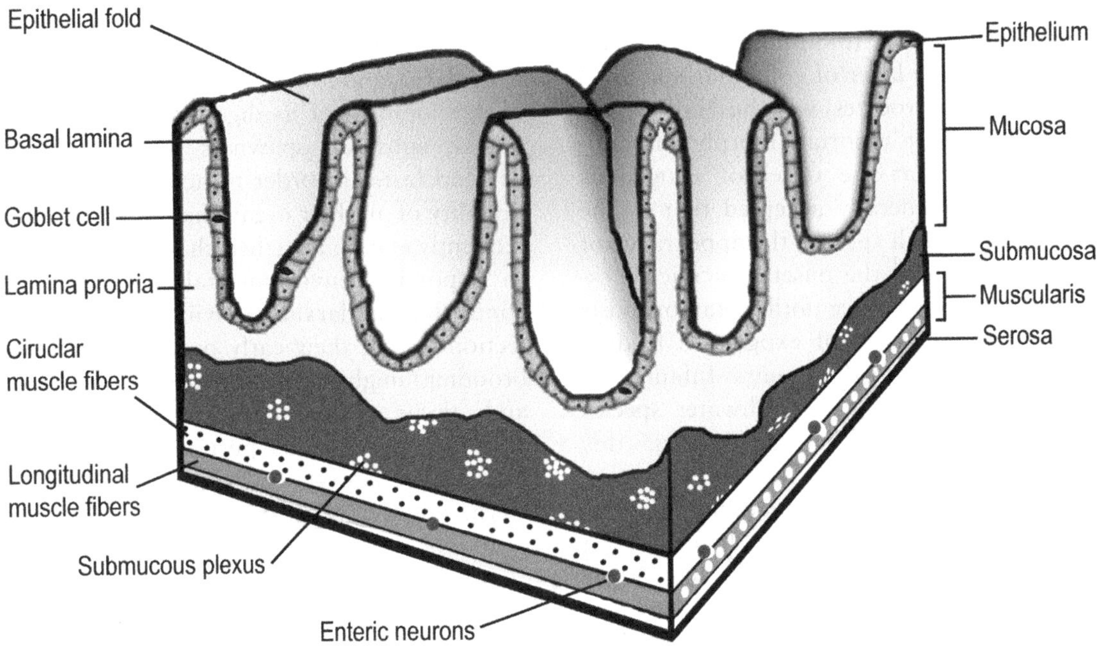

Figure 1.1 General organization of the wall of the digestive tract. Note that this organization in different layers and size of mucosal folds may change depending on the region of the gut considered, esophagus, stomach, or intestine (see details in the text).

the posterior intestine and rectum, whereas the buccal cavity and pharynx lack the serosa (Hossain and Dutta 1996). As the general reader is not normally familiar with the histological characteristics of the above-mentioned layers, and as these may change depending on the species, the stage of development, and the region of the digestive tract considered, we have decided to describe them briefly below (Figure 1.1) since their different histological organization depends on their specific functions.

The **mucosa** is the innermost layer of the digestive tube. Among the rest of layers that compose the digestive tube, the mucosa is the most variable in structure and function, endowing the tube with an ability to perform diverse and specialized digestive tasks along its length. Thus, the mucosal epithelium is regionally differentiated along the digestive tract to pursue multiple specialized functions, such as protection of the inner layers, secretion of the digestive juice, absorption, osmoregulation, and metabolism of nutrients among others. Thus, epithelial mucosal cells may differentiate into absorptive cells such as those of the intestine and pyloric ceca, in gastric and intestinal secretory cells, or in the goblet (mucous) cells found along the entire length of the digestive tract. Other noticeable types of cells that can be observed in the digestive mucosa include the foreign migrant cells (granulocytes, lymphocytes, and macrophages) that are part of the immune system. Under the basal part of the mucosal epithelium, there is the basal lamina, which can be recognized as a thin, dense, and continuous band formed by a complex layer of collagen fibers and polysaccharides. The lamina propria, a thin layer of connective tissue with blood capillaries, underlies the mucosal epithelium. Under the lamina propria, a thin layer of smooth musculature, the muscularis mucosa, may also be found, although it is often poorly developed and commonly absent, in which case, a clear distinction between the connective tissue of the lamina propria and that of the submucosa is not possible.

The **submucosa** of teleosts is composed of one or more layers of connective tissue. This part of the wall of the gut shows some interspecific variability, with some fish species having a single-layered and loose submucosa, and many other species possessing a multilayered submucosa consisting of a stratum compactum, stratum granulosum, and in some cases, an extra layer of loose connective tissue beneath the stratum compactum. The submucosa also contains the submucous plexus, which provides nervous control to the mucosa.

Surrounding the submucosa is the **muscularis** formed by smooth or striated muscle layers depending on the region of the digestive canal. This layer may be organized into concentric inner circular and outer longitudinal muscle fibers. Enteric neurons within plexuses between the two muscle layers endow the digestive tube with an ability to be motile by means of peristaltic movements, which is of special importance in the esophagus and intestine.

The serous membrane, so-called **serosa**, is composed of a simple flat epithelium. This membrane is an extension of the mesentery and covers the external surface of the alimentary canal.

Up to now we have described the general structural characteristics of the wall of the alimentary canal; what follows is the particular histological organization of the different visceral regions of the digestive tract (esophagus, stomach, intestine, and pyloric ceca) and accessory glands (liver, pancreas, and gallbladder). However, due to the intrinsic differences between marine and freshwater species in their phylogenetic status, mode of reproduction, egg size, and larval feeding habits among others, we will describe, for comparative purposes, the ontogeny of the digestive system in freshwater and marine species separately.

1.3.1 Esophagus

At hatching, the esophagus in fish larvae is not a morphoanatomically differentiated region of the digestive tract and its morphogenesis takes place at later stages of development, just before the onset of exogenous feeding. At this stage, the esophagus appears as a short and rudimentary duct that connects the posterior region of the pharynx from the last gill arch with the anterior intestine since the stomach is neither differentiated nor formed in most of the described gastric species during the differentiation of this part of the alimentary canal. Once differentiated, the esophagus in fish is generally short, wide, and straight, and the most distinctive histological features that characterize its differentiation are the histological organization of the esophageal mucosa that varies among species, its longitudinal folding, and the appearance of goblet (mucus-secreting) cells (Figure 1.2a).

The epithelium that lines the esophageal mucosa in fish larvae is similar to that in adults. In general terms, the esophagus of freshwater fish species is lined by a multilayered squamous epithelium with large numbers of goblet cells (Figure 1.2b), whereas that of marine fish species is lined by a columnar epithelium with fewer mucous cells and highly vascularized mucosal folds (Figure 1.2c) (Stevens and Hume 2005). In some species, a ciliated epithelium is found in the esophagus of newly hatched larvae and adult cyclostomes, perch, and some elasmobranches, where this feature may be considered as a plesiomorphic character. However, there exist some exceptions regarding the histological organization of the esophageal epithelium in larvae from different fish species, as pointed out recently by Zambonino-Infante et al. (2008).

Another histological feature that characterizes the differentiation and functionality of the esophagus is the appearance of goblet cells and their histochemical characteristics (Figure 1.2d). The appearance of the first

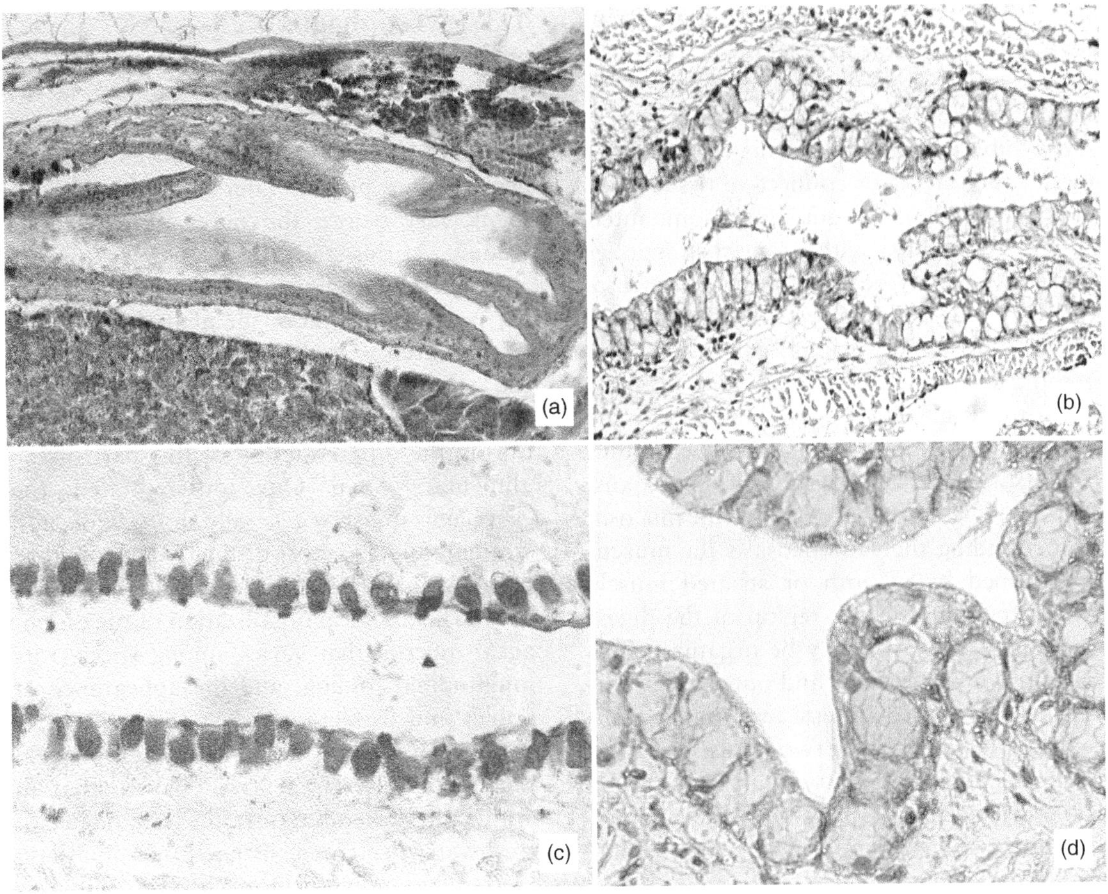

Figure 1.2 Longitudinal paraffin sections of the esophagus from different finfish larvae. (a) Esophagus of a California halibut (*Paralichthys californicus*) larvae aged 15 days posthatch (dph). Note the presence of longitudinal mucosal folds (magnification: 200×; hematoxylin-eosin [H-E] stain). (b) Detail of the esophagus from a bay snook (*Petenia splendida*) larva (24 dph) with abundant and large goblet cells (magnification: 200×; H-E stain). (c) Goblet cells containing neutral glycoconjugates (PAS-positive staining) in the esophageal epithelium of a 17-dph *Paralichthys californicus* larva (magnification: 400×; periodic acid–Schiff [PAS] stain). (d) Detail of goblet cells in the esophageal epithelium of *Petenia splendida* larva aged 13 dph. Note the hyaline content of these cells, indicating the presence of a mixture of secretory mucopolysaccharides (magnification: 400×; H-E stain). (Photographs by E. Gisbert.)

functional goblet cells along the esophageal epithelium varies among species: In Dover sole *Solea solea* (Boulhic and Gabaudan 1992), Senegal sole *Solea senegalensis* (Ribeiro et al. 1999), yellowtail flounder *Limanda ferruginea* (Baglole et al. 1997), large yellow croaker *Pseudosciaena crocea* (Mai et al. 2005), anemonefish *Amphiprion melanopus* (Green and McCormick 2001), Siberian sturgeon *Acipenser baerii* (Gisbert et al. 1999), green sturgeon *Acipenser medirostris* (Gisbert and Doroshov 2003), pike perch *Sander lucioperca* (Ostaszewska 2005), European catfish *Silurus glanis* (Kozarić et al. 2008), and common bay snook (Alvarez-González et al. 2008), goblet cells are detected coinciding with mouth-opening or just before the onset of exogenous feeding, whereas in other species such as the turbot *Scophthalmus maximus* (Segner et al. 1994), brill *Scophthalmus rhombus* (Hachero-Cruzado et al. 2009), California halibut *Paralichthys californicus*

(Gisbert et al. 2004a), gilthead sea bream *Sparus aurata* (Sarasquete et al. 1995), white bream *Diplodus sargus* (Ortiz-Delgado et al. 2003), red porgy *Pagrus pagrus* (Darias et al. 2005), common pandora *Pagellus erythrinus* (Micale et al. 2006), European sea bass *Dicentrarchus labrax* (García-Hernández et al. 2001), cobia *Rachycentron canadum* (Faulk et al. 2007), Atlantic cod *Gadus morhua* (Morrison 1993), and haddock *Melanogrammus aeglefinus* (Hamlin et al. 2000), goblet cells are detected at later stages of development. In haddock, special attention might be required in tanks during the first days of larval rearing, just before the appearance of functional esophageal goblet cells, since the ingestion of live prey may result in desquamation and abrasion of the esophageal epithelium due to the absence of mucus protecting this area. This might lead to significant larval mortality if high water quality is not maintained and bacteria proliferate (Gisbert et al. 2004a).

The species-specific differences in the ontogeny of the goblet cells in the esophageal mucosa of marine fish larvae were recently reviewed by Zambonino-Infante et al. (2008). Goblet cells secrete different types of mucosubstances that differ in their histochemical characteristics. In most of the described species, the mucous cells of the distal esophageal region secrete large amounts of neutral glycoconjugates (Figure 1.2c,d), while those from the anterior region produce, in addition to a minor component of neutral mucosubstances, a major quantity of carboxylated and sulfated acidic glycoconjugates with sialic acid residues. The large amount of mucous secreted by esophageal goblet cells may serve as a lubricant since fish do not have the salivary glands found in higher vertebrates. Mucosubstances produced by goblet cells may have the same functions as mammalian saliva in protecting the mucosa of the entire alimentary canal (Scocco et al. 1998). The presence of sialic acid residues in mucus prevents viruses from recognizing their receptor determinants and also preserves the mucosa from attack by the sialidase produced by bacteria (Zimmer et al. 1992). In addition, the secretion of this combination of mucosubstances has been described by several authors as a mechanism to allow the alimentary canal of young fish to respond to changes in environmental conditions and maintain osmotic balance (Domeneghini et al. 1998; Sarasquete et al. 2001). Ultrastructural studies have revealed that the esophageal epithelium is also involved in iono- and osmoregulation in seawater and freshwater environments. Thus, under hyperosmotic conditions, the ingested water is desalinated in the esophagus by the passive and active removal of Na^+ and Cl^-, although this segment of the digestive tract has low permeability to water and other ions. This results in a reduced osmolality of the water, facilitating its absorption in the intestine (Allen et al. 2009). Furthermore, neutral glycoconjugates secreted by esophageal goblet cells are considered to cooperate in the digestion of food and its transformation into chyme, as well as in the absorption of easily digested substances, such as disaccharides and short-chain fatty acids (Sarasquete et al. 2001).

1.3.2 Intestine

The intestine is the longest portion of the digestive tract, occupying most of the abdominal cavity, and one of the first digestive organs to differentiate. The intestinal mucosa is a very dynamic and active tissue and is the main site of the digestion and absorption of nutrients, as well as being directly involved in the hormonal and nervous activation of enzyme and bile synthesis and their subsequent secretion from the pancreas and liver. At hatching, the intestine is an undifferentiated straight tube with a smooth lumen. During the lecithotrophic (endogenous feeding) phase, the posterior region of the intestine bends and the intestinal valve or ileorectal valve forms as a constriction of the intestinal mucosa, dividing the intestine into two regions: the prevalvular (anterior) (Figure 1.3a) and postvalvular (posterior) intestine (Figure 1.3d). The intestinal mucosa is mostly rectilinear with several short folds. In most species, no histological differences are observed between the pre- and postvalvular intestine; both regions are lined by a simple columnar epithelium with basal nuclei, basophilic cytoplasm, and prominent microvilli (Figure 1.3b).

As the development of the intestine proceeds, the folding of the mucosa increases and the intestine coils, occupying most of the abdominal area. These morphoanatomical changes take place in order to accommodate the increasing length of the digestive tract inside the reduced abdominal cavity, and to increase the digestive and absorptive intestinal surface to cope with the increasing quantities of ingested food. However, not all fish species have a coiled intestine, for example, ayu *Plecoglossus altivelis* (Nakagawa et al. 2002) or zebrafish *Danio rerio* (Wallace et al. 2005); in those cases, the increase in absorptive surface of the intestine is achieved by only incrementing the folding level of the intestinal mucosa (Figure 1.3e). At this point, three different regions can be distinguished along the intestine according to their histological organization and characteristics. The anteromedian segment (prevalvular intestine), which receives the pancreatic and biliary secretions, is histologically characterized by a columnar epithelium with prominent microvilli composed of a high number of goblet cells, especially abundant close to the pyloric sphincter (Figure 1.3a). This region of the intestine is the main site for lipid absorption (Diaz et al. 1997b; Olsen et al. 2000), while proteins are absorbed in the posterior intestine (Deplano et al. 1991). However, other studies have reported that lipid digestion and absorption continue in the posterior and rectal regions of the intestine (García-Hernández et al. 2001; Gisbert et al. 2005). The postvalvular intestine is histologically similar to the anteromedian region except for the number and size of mucosal folds, which are longer, deeper, and more numerous in the prevalvular intestine (Figure 1.3e). In some fish species, such as sturgeons, the distal intestinal region prior to the rectum is modified soon after hatching into the spiral valve, which greatly increases the absorptive area of this region of the posterior intestinal mucosa (Buddington and Doroshov 1986; Gisbert et al. 1998). The intestine terminates in a short rectal zone that, depending on the species, can be lined either by a simple or columnar epithelium with few goblet cells or by a cubical epithelium (Figure 1.3f). Although some differences may be found among species, the organization of the intestinal mucosa throughout its length is quite conserved among teleosts and

Figure 1.3 Longitudinal sections of the intestine from California halibut (*Paralichthys californicus*) larvae. (a) Prevalvular intestine from a 25-dph larva. Note the presence of abundant secretory goblet cells containing neutral mucosubstances (PAS-positive staining) (magnification: 200×; periodic acid–Schiff [PAS] stain). (b) Detail of a mucosal fold of the intestine in a 25-dph larva showing the presence of prominent microvilli and the organization of the columnar epithelium (magnification: 600×; hematoxylin-eosin [H-E] stain). (c) Detail of the posterior intestine in a 15-dph larva showing the presence of eosinophilic supranuclear bodies in the enterocyte cytoplasm and the presence of prominent eosinophilic brush borders (magnification: 400×; H-E stain). (d) Detail of the ileorectal valve forming a constriction of the intestinal mucosa and dividing the intestine into two regions, the prevalvular (anterior) and postvalvular (posterior) intestine. The intestinal lumen appears filled with predigested *Artemia* nauplii (magnification: 400×; H-E stain). (e) General view of the postvalvular (posterior) intestine in a 35-dph larva. Note the reduction in mucosal folding in contrast to Figure 1.3a and the low presence of goblet cells (PAS-positive staining) in the intestinal epithelium (magnification: 100×; PAS stain). (f) Details of the rectum in a 35-dph larva. Note the flattening of columnar cells at the end of the rectum and the presence of incompletely digested *Artemia* nauplii in the intestinal lumen (magnification: 400×; H-E stain). (Photographs by E. Gisbert.)

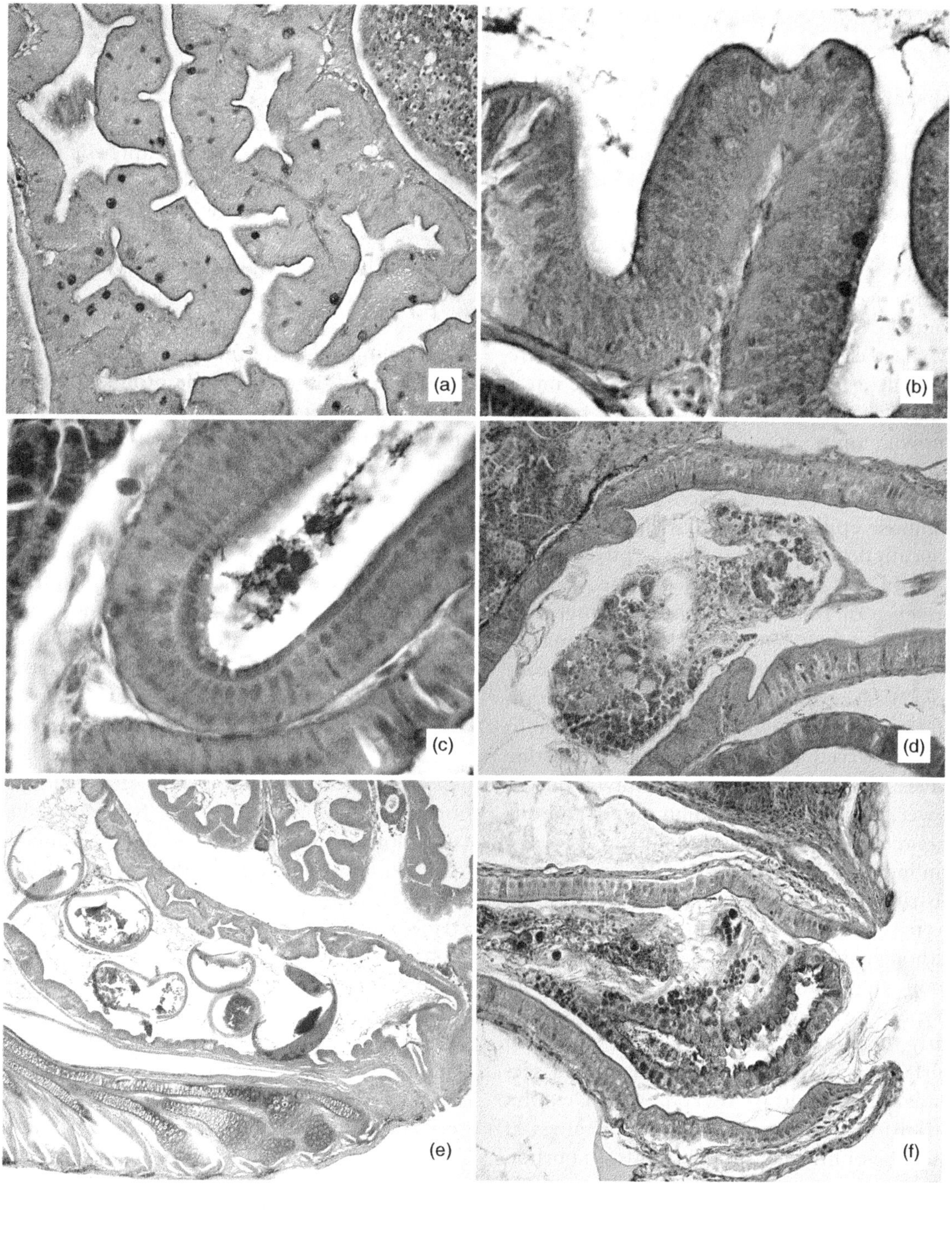
(a)
(b)
(c)
(d)
(e)
(f)

typically characterized into four distinct layers. In particular, zebrafish lack the submucosa layer of the intestine and instead possess an epithelium and lamina propria surrounded by circular and longitudinal smooth muscle layers. Another distinctive feature of the intestinal mucosa of the zebrafish is that the connective tissue found in the mucosa is less complex than in other cyprinid species (Wallace et al. 2005). In the literature, there are only a few reports of crypts in the intestinal mucosa of fish species, and as far as is known, only in cod, common wolffish *Anarhichas lupus*, and burbot *Lota lota* have these structures been described. The crypts comprise epithelial cells that differ morphologically from the general absorptive epithelium of the mucosal surface. Based on the presence of mitotic structures (proliferating cell nuclear antigen [PCNA]-staining nuclei) and less differentiated cells, crypts are considered a place for epithelial cell proliferation and regeneration, which might be similar to that observed in the mammalian intestine where intestinal epithelial cells divide in the lower part of the Lieberkühn crypts and differentiate as they migrate upward to the luminal surface. There exist differences in the distribution and morphology of intestinal crypts among the described fish species, although in all studied species they appear soon after hatching. In common wolffish, crypts are shallower and their openings regularly distributed over the whole mucosal surface of mucosal folds (primary and secondary), in contrast to cod, which exhibit groups of deep crypts with the openings at the base of mucosal primary folds. In burbot, crypts are restricted throughout the proximal intestine, including the pyloric ceca, while in common wolffish, crypts are observed in all parts of the intestine (Hellberg and Bjerkås 2005).

Four different types of cells can be identified along the intestinal epithelium. These include enterocytes, single enteroendocrine cells, rodlet cells, and goblet cells. **Enterocytes** are the most abundant epithelial cell type in the intestine, and they are involved in nutrient absorption, intracellular digestion, and osmoregulation. Enterocytes in the anterior intestine are responsible for the absorption of lipids and amino acids by diffusion, whereas enterocytes in the distal intestine are specialized for the uptake of protein macromolecules by pinocytosis. Intestinal **goblet cells** are the second most abundant cell type and are scattered along the intestinal epithelium. These cells are well known for the production of a physical barrier between the epithelium and the content of the lumen. Goblet-cell-secreted mucins are central to the establishment of this complex mucopolysaccharide barrier. The differentiation of these mucus-secreting cells follows two different patterns according to the larval stage of development at which they differentiate. For example, in wolffish, bay snook, Siberian and green sturgeons, Dover sole, yellowtail flounder, or spotted sand bass, goblet cells appear in the intestinal mucosa, before first feeding and/or coincide with the onset of exogenous feeding, while in other species, such as pike perch, gilthead sea bream, California halibut, Senegal sole, common pandora, kelp grouper, common dentex, cod, or haddock, goblet cells differentiate at latter stages of development (see review in Zambonino-Infante et al. 2008). Intestinal goblet cells contain a mixture of neutral and acid glycoproteins and the histochemical pattern of their content does not change through larval and juvenile periods to adult ages. Mucosubstances produced by rectal and distal postvalvular intestine goblet cells may serve to lubricate the feces, while in other regions of the intestine they protect the digestive mucosa and facilitate the absorption of nutrients. The presence of sulfated acidic glycoproteins produced by goblet cells may regulate the transfer of proteins or protein fragments into enterocytes where these compounds will be digested via pinocytosis (Domeneghini et al. 1998).

Enteroendocrine cells produce and secrete peptide hormones that, in collaboration with the nervous system, control and coordinate the muscular and secretory activities of the gastrointestinal tract. These cells have a pyramidal or spindle shape with a narrow extension to the gut lumen, and their distribution and time of differentiation vary among species. In fish species with a convoluted intestine, enteroendocrine cells are rarely found beyond the anterior intestine, as described for Japanese flounder *Paralichthys olivaceus*, Atlantic halibut *Hippoglossus hippoglossus*, Pacific bluefin tuna *Thunnus orientalis* (Kamisaka et al. 2003), and zebrafish (Wallace et al. 2005), whereas in species with a straight digestive tract, such as the ayu, they are scattered along the entire intestine with the exception of the rectum (Kamisaka et al. 2003). As the former authors suggested, these cells seem to be located in regions where the chyme is retained, and consequently, they can easily receive chemical signals from the food and the digestive process in order to control the release of the digestive hormones. In species that hatch with a well-differentiated digestive system, such as the ayu, enteroendocrine cells are found just after hatching (Kamisaka et al. 2003), whereas in pelagic fish species with a less developed digestive tract, these cells may appear at latter stages of development just before the onset of exogenous feeding (Kurokawa et al. 2000). Intestinal **rodlet cells** have been considered as regulatory elements related to special functions such as osmoregulation, ion transportation, and nonspecific immune response (see review in Manera and Dezfuli 2004).

1.3.3 Pyloric ceca

Pyloric ceca are considered as an adaptation for increasing intestinal surface area without increasing the length or thickness of the intestine itself. These intestinal appendages are also involved in osmoregulatory processes and especially in water uptake in fish exposed to hyperosmotic environments (Allen et al. 2009). Among vertebrates, only teleost fish species have appendages such as the pyloric ceca at the gastrointestinal junction, which are entirely different from those found in birds and mammals, which have fermentation functions (Buddington and Diamond 1987). Pyloric ceca are present in 60% of the known fish species, and their number is highly variable, ranging from none (absent) to numerous (>1,000) depending on the fish group and species. The presence of ceca in some fish and absence in others, as well as variations in the number of ceca between and within species, are regarded as adaptations in the fish digestive system to different feeding habits and morphoanatomical characteristics of their digestive tract (Hossain and Dutta 1996). In addition, the phylogenetic relationships of the pyloric ceca between different fish species and their differences among different teleosts are reviewed in Hossain and Dutta (1996).

These fingerlike projections that form part of the anteriormost region of the intestine increase the surface area for absorption. In addition, pyloric ceca appear to be sites of digestion with the contribution of pancreatic enzymes, and some authors consider them to be accessory food reservoirs and breeding places for gut flora (Buddington and Diamond 1987; Hossain and Dutta 1996). Pyloric ceca have also been shown to neutralize the acid bolus entering the intestine from the stomach, which is supported by the absence of these structures in fishes lacking a stomach (Rust 2002). Hossain and Dutta (1996) reported that the greater the cecum size, the better its functional efficiency; thus, fish species with shorter intestines should have either more or larger pyloric ceca, as described for some detritivorous species. However, this relationship is not consistent for omnivorous species (Albrecht et al. 2001). Drewe et al. (2004) concluded that the relationship between diet and the structure and function of the pyloric ceca is complex and still poorly understood.

The histological organization of pyloric ceca closely resembles that of the intestine;

the four basic histological layers of the intestinal tract (serosa, muscularis, submucosa, and mucosa) are present in all ceca, although the cecal mucosa is more complex compared with that of the intestine. The main differences between these two regions of the digestive tract are the relative thickness of the cecal layers, which are several-fold smaller compared with those of the intestine, and the number of goblet cells secreting a combination of neutral and acidic mucosubstances that are less abundant in the ceca than in the intestine. In addition, the relative thickness of these layers and the lumen space varies within a specimen. In general, a large cecum has more muscle, mucosa, and lumen space than a small cecum. This close resemblance of the histology of ceca and intestine indicates that ceca not only are linked ontogenetically but also are connected functionally with the intestine (Hossain and Dutta 1996).

The development of the pyloric ceca and the stomach are the last steps in the differentiation of the alimentary canal, as well as being the anatomical digestive features that characterize the end of the larval phase and the beginning of the juvenile stage. However, there exist some differences among species. For instance, in some species, such as European sea bass (García-Hernández et al. 2001), spotted sand bass *Paralabrax maculatofasciatus* (Peña et al. 2003), haddock (Hamlin et al. 2000), or pike perch (Ostaszewska 2005), the morphogenesis of the pyloric ceca is contemporaneous with the differentiation of the gastric glands, whereas in cobia (Faulk et al. 2007) or chum salmon *Oncorhynchus keta* (Dabrowski 1984), pyloric ceca develop later than the stomach. The above-mentioned changes in the ontogeny of differentiation of the pyloric ceca might be linked to the species-specific differences in digestive requirements, especially those related to an increasing demand for digestive and absorptive intestinal surfaces coupled with an increase in length of the digestive tract as development proceeds.

1.3.4 Stomach

The main functions of the vertebrate stomach are to store ingested food, to secrete pepsinogen and hydrochloric acid (HCl), and to mix food and the gastric secretions mechanically through the action of muscles that allow for the distension and movement of the organ (Stevens and Hume 2005). The form of the stomach is very diverse, from a single bulge converted to an elongated pouch when it is full, to a well-differentiated sac. It might be bent to form a Y, V, or J shape and is usually separated from the intestine by a pyloric sphincter or valve. It is capable of considerable distension in carnivorous fish species that swallow whole prey, whereas microphagous species have much smaller stomachs. Despite these anatomical variations, all stomachs are relatively homogeneous in their histological structure and differ little, at least in function, from those of higher vertebrates (Guillaume et al. 2001). The stomach is generally divided into three regions, the cardiac (anterior), fundic, and pyloric (posterior) portions, that exhibit specific anatomical and histological adaptations for separate digestive functions. The mucosal epithelium of each region is single layered and folded. The folds of the cardiac portion are normally shallow, but those of the fundic and pyloric portions are deeper. Generally, epithelial cells of the cardiac region are cubical in shape, and those of the rest of the stomach epithelium are columnar. In some fish species, such as sturgeons, the walls (tunica muscularis) of the pyloric region are hypertrophied, especially the dorsal and ventral walls. Because of the presence of such a thickened smooth muscle layer, the pyloric region of the stomach has sometimes been referred to as the gizzard (Buddington and Doroshov 1986; Gisbert et al. 1998) in this group of chondrosteans, and it has a triturative function that may compensate for their lack of dentition (mandibular, oral, or pharyngeal). In some herbivorous species, such as mullets, the pyloric region of the stomach forms a true differentiated organ with toughened walls, surrounded by a very thin circular musculature, following a stomach that has lost its secretory functions. Similarly to sturgeons, this region of the stomach is also called a gizzard and performs a purely grinding function (Guillaume et al. 2001).

The formation of the stomach varies depending on the type of egg cleavage. In fish with holoblastic (complete) cleavage such as sturgeons, where the yolk endoderm participates in the formation of the alimentary canal (Dettlaff et al. 1993), the stomach starts to differentiate in the anterior ventral region of the yolk sac from a fold of stratified epithelium that previously divided the yolk into two compartments; the anterior wall of the furrow becomes the ventral lining of the stomach, while its posterior wall lined with a columnar epithelium becomes the dorsal lining of the intestine. As a consequence of the holoblastic cleavage, epithelial cells lining the alimentary tract and the stomach are filled with eosinophilic yolk granules that gradually disappear as stomach morphogenesis proceeds (Gisbert and Doroshov 2003; Zambonino-Infante et al. 2008). In contrast, in fish with meroblastic (incomplete or partial) egg cleavage, which include the vast majority of finfish species, the digestive system and particularly the stomach differentiate independently of the extraembryonic yolk sac. In the latter fish species, the stomach appears as a slight enlargement, dilatation, or pouch of the esophagus that is accompanied with a notable thickening in the mucosa and narrowing of the lumen demarking the transition to the anterior intestine. This area of the alimentary canal is lined with a cubical epithelium that further differentiates into a columnar one, as has been described in European sea bass (García-Hernández et al. 2001), common pandora (Micale et al. 2006), cobia (Faulk et al. 2007), and Siberian sturgeon (Gisbert et al. 1999) among other species (see review in Zambonino-Infante et al. 2008). Further in development, a sphincter, formed by thickened layers of connective tissue and circular bundles of smooth muscle cells surrounding the digestive epithelium at the level of the constriction, develops at either end of the stomach, separating the gastric region of the alimentary canal from the esophagus and anterior intestine, respectively (Figure 1.4a).

As morphogenesis proceeds, the epithelium of the cardiac and fundic regions of the stomach folds transversally and forms the gastric pits, where the first formed gastric glands open into the gastric lumen. Gastric glands, which are tubular in shape, are generally situated in the lamina propria and surrounded by a loose connective tissue. The secretory cells present in these glands are responsible for the production of both HCl and pepsinogen, and they are called oxynticopeptic cells (Figure 1.4b,c). Ultrastructural characteristics of the secretory cells from undifferentiated and differentiated gastric glands are described in detail in García-Hernández et al. (2001). In short, glandular cells initially have numerous free ribosomes and clear vesicles in the apical zone. As they differentiate and become functional, an apical tubule–vesicular network, a very developed endoplasmatic reticulum, and zymogen granules appear. Histochemically, gastric glands contain glycogen, neutral and especially carboxylated glycoconjugates, and proteins rich in different amino acids, especially proteins rich in arginine, tyrosine, and tryptophan. These amino acids are involved in the synthesis and secretion of enzymatic precursors, that is, pepsinogen (Ortiz-Delgado et al. 2003). Along the epithelial cells of the stomach, goblet cells secreting neutral mucosubstances are found in the lumenally exposed gastric epithelium. The main role of the neutral mucosubstances present in the stomach is to protect the epithelium of the stomach from autodigestion processes caused by HCl and enzymes (e.g., pepsin) produced in gastric glands. In addition, several authors have pointed out that the periodic acid–Schiff (PAS)-positive reaction observed in the gastric epithelial cell

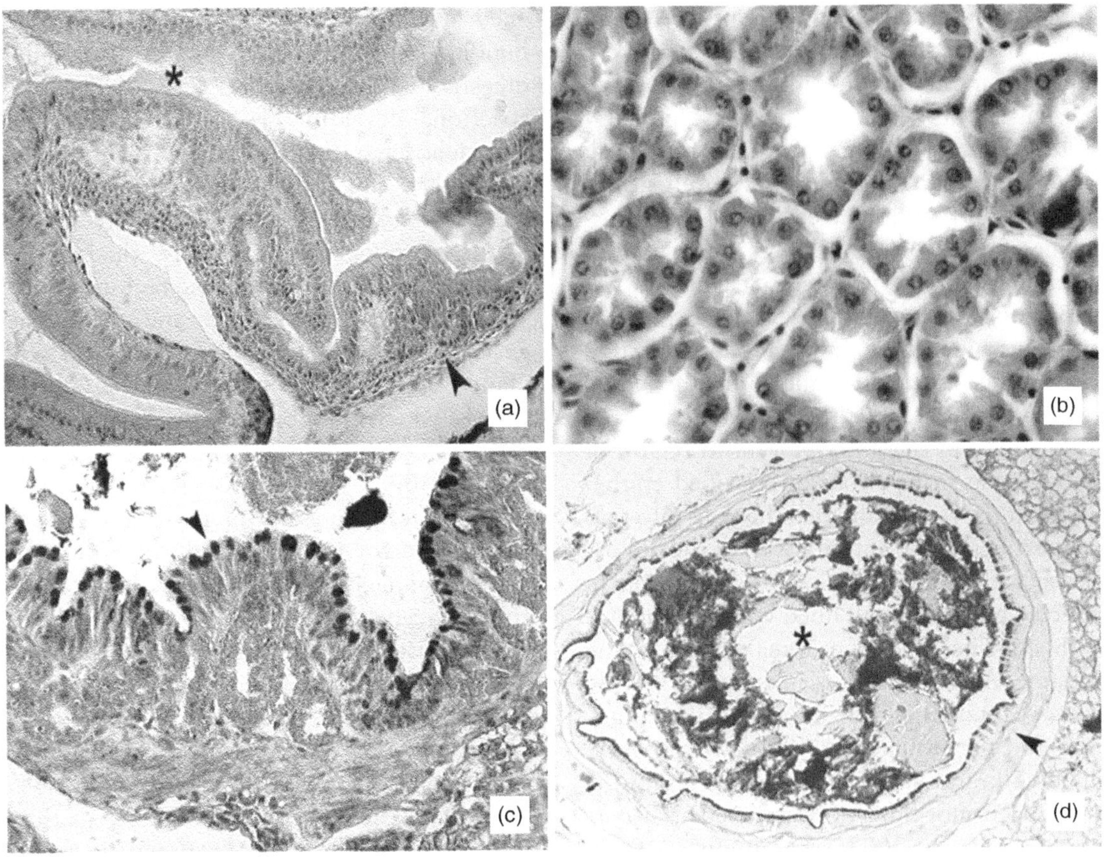

Figure 1.4 Different histological sections of the stomach from larvae of different fish species. (a) Onset of formation of the glandular stomach in a 24-dph California halibut *Paralichthys californicus* larva. Note the transition from the esophagus into the stomach in differentiation (asterisk) and the formation of first gastric glands (arrowhead) (magnification: 200×; hematoxylin-eosin [H-E] stain). (b) Detail of the gastric glands in *Paralichthys californicus* (magnification: 600×; H-E stain). (c) Multicellular tubular gastric glands in Siberian sturgeon *Acipenser baerii* (larva aged 16 dph) composed of a single secretory cell type observed in the cardiac stomach (magnification: 400×; hematoxylin–light green-orange G-acid fuchsin [VOF] stain). (d) Transverse section of the nonglandular stomach in *Acipenser baerii* (larva aged 25 dph). Note the PAS-positive staining of mucous cells lining the epithelial lumen of the stomach and the thick layer of musculature (arrowhead). The asterisk denotes the lumen of the nonglandular stomach filled with chyme (magnification: 400×; PAS stain). (Photographs by E. Gisbert.)

surface resembles that observed in the striated border of intestinal enterocytes, which may indicate nutrient absorption of easily digestible substances such as disaccharides and short-chain fatty acids occurring in this region of the alimentary canal, as previously reported in the esophagus.

The morphogenesis of gastric glands and the achievement of an adultlike digestion, characterized by low pH and gastric proteases, that is, pepsinogen, are achieved at different stages of development depending on the fish species and water rearing temperature. In this sense, there exists a wide variety of ontogenetic stages where gastric glands differentiate (Zambonino-Infante et al. 2008); for instance, salmonids and wolffish possess a functional stomach at the time of first feeding, whereas in others, gastric glands appear later in development, ranging from as early as 10 days after hatching (dah) in turbot (Cousin and Baudin-Laurencin 1985) to as late as 90 dah in Atlantic halibut (Luizi et al. 1999). However, the development of gastric glands is not necessarily accompanied by the onset of stomach activity since morphology does not always mean functionality. An asynchrony between the morphological development of the gastric glands and their functionality has been reported for several species: around a week in summer flounder *Paralichthys dentatus* (Huang et al. 1998), 10 days in red porgy (Darias et al. 2005) and pike perch *Sander lucioperca* (Ostaszewska 2005), and several weeks in common whitefish *Coregonus lavaretus* (Mähr et al. 1983).

In higher vertebrates, gastric glands are mainly located in the fundic region of the stomach (Stevens and Hume 2005), whereas in fishes, several authors have reported the existence of interspecific differences in the localization of gastric glands. In some species, such as turbot (Segner et al. 1994), yellowtail flounder (Baglole et al. 1997), Dover sole (Veggetti et al. 1999), common pandora (Micale et al. 2006), and South American catfish *Rhamdia quelen* (Hernández et al. 2009), gastric glands are only found in the fundic region. In contrast, in gilthead sea bream (Elbal and Agulleiro 1986), white sea bream (Ortiz-Delgado et al. 2003), amberjack *Seriola dumerili* (Grau et al. 1992), European sea bass (García-Hernández et al. 2001), spotted sand bass (Peña et al. 2003), and white, Siberian, and green sturgeons (Gawlicka et al. 1995, Gisbert et al. 1998; Gisbert and Doroshov 2003), gastric glands are observed in the cardiac region of the stomach. In pike perch (Ostaszewska 2005), gastric glands are located in the cardiac and fundic regions, whereas in Senegal sole (Arellano et al. 2001) and cobia (Faulk et al. 2007), gastric glands are observed along the mucosa of fundic and pyloric stomach regions. On the other hand, the stomachs of Atlantic halibut (Murray et al. 1994), tilapia *Tilapia* spp. (Gargiulo et al. 1997), and bay snook (Alvarez-González et al. 2008) are entirely glandular, suggesting that the stomach in these species is mainly involved in chemical rather than a combination of mechanical and chemical digestion of food items, as it is in other fish that show regional differentiation of the stomach (Zambonino-Infante et al. 2008).

1.3.5 Accessory digestive glands

The accessory digestive glands, the liver, pancreas, and gallbladder, are of significant importance for the nutrition and homeostasis of fish (Hoehne-Reitan and Kjørsvik 2004). The liver is the central digestive organ not only for nutrient metabolism, conversion, and transfer to peripheral tissues but also for the production of bile and detoxification of toxins from both endogenous (metabolites) and exogenous sources. The pancreas consists of an exocrine portion that secretes pancreatic juices (digestive enzymes) that are involved in the intestinal digestion of nutrients, and an endocrine portion, the so-called islets of Langerhans, that secretes hormones such as insulin, somatostatin, pancreatic polypeptide, and/or glucagon. A final organ associated with digestion is the gallbladder, which secretes bile produced by the liver and aids in the emulsification of ingested food (i.e., lipids) and increases intestinal pH.

Histologically, the **liver** is composed of the liver *lobuli* (Figure 1.5b,c). In vertebrates, the liver has a primary array based on hepatocytes, bile canaliculi, and sinusoids, and structural differences occur among species in the organization of the stroma and parenchyma. In teleost fish, hepatocytes are arranged in anastomosed laminae around the central vein. Using electron microscopy, different cell types may be identified in the hepatic tissue based on their cellular organelles, stored substances, and cell surface specializations (Takashima and Hibiya 1995). The main cell type in the liver is the parenchymal hepatocyte (often shortened to hepatocyte), while endothelial cells, fat-storing cells, Kupffer cells, mesothelial (serosa) cells,

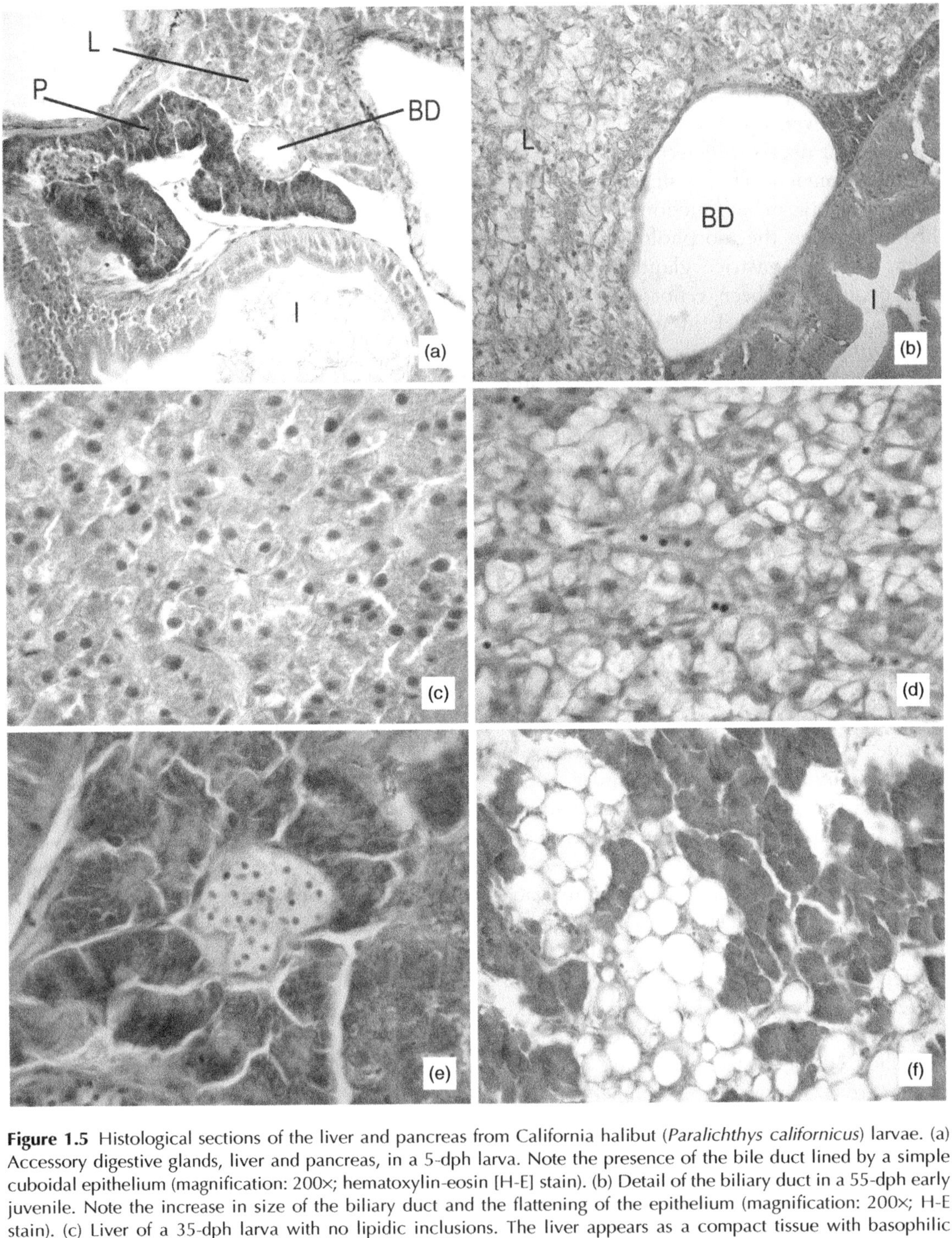

Figure 1.5 Histological sections of the liver and pancreas from California halibut (*Paralichthys californicus*) larvae. (a) Accessory digestive glands, liver and pancreas, in a 5-dph larva. Note the presence of the bile duct lined by a simple cuboidal epithelium (magnification: 200×; hematoxylin-eosin [H-E] stain). (b) Detail of the biliary duct in a 55-dph early juvenile. Note the increase in size of the biliary duct and the flattening of the epithelium (magnification: 200×; H-E stain). (c) Liver of a 35-dph larva with no lipidic inclusions. The liver appears as a compact tissue with basophilic polyhedral hepatocytes with centrally located nuclei (magnification: 400×; H-E stain). (d) Details of the liver of a 35-dph larva showing the large accumulation of lipids (unstained vacuoles) inside the hepatocytes that displaced nuclei to the periphery of the cell (magnification: 400×; H-E stain). (e) Endocrine pancreas (islet of Langerhans) surrounded by exocrine pancreatic tissue (magnification: 400×; H-E stain). (f) Exocrine pancreas with an infiltration of adipose tissue. Note the round shape of the adipocytes (unstained with H-E stain) and the peripheral position of their flattened nuclei (magnification: 400×; H-E stain). BD = biliary duct, I = intestine, L = liver, P = pancreas. (Photographs by E. Gisbert.)

and fibroblasts complement the basic liver architecture. The main stored substances in the fish liver are glycogen and, to a lesser extent, lipids. Eosinophilic and PAS-positive glycogen granules may be found scattered in the cytoplasm or aggregated forming large concentrations, and using electron microscopy they can be identified as rosette-like α particles and single β particles. For a detailed description of the ultrastructural characteristics of the above-mentioned cell types in adult fish, see Takashima and Hibiya (1995).

Timing of liver differentiation varies among species, and is mainly related to their general life history traits (Hoehne-Reitan and Kjørsvik 2004). As these authors stated, the timing of liver development clearly reflects the developmental state at hatching for different species. For example, the liver is already differentiated at hatching in Atlantic cod (Morrison 1993), haddock (Hamlin et al. 2000), common wolffish (Hoehne-Reitan and Kjørsvik 2004), white sea bream (Ortiz-Delgado et al. 2003), percula clownfish *Amphiprion percula* (Önal et al. 2008), tilapia (Morrison et al. 2001), and bay snook (Alvarez-González et al. 2008). In contrast, in some marine species, such as gilthead sea bream (Sarasquete et al. 1995), anemonefish (Green and McCormick 2001), California halibut (Gisbert et al. 2004a), common dentex (Santamaría Rojas et al. 2004), Atlantic halibut (Hoehne-Reitan and Kjørsvik 2004), common pandora (Micale et al. 2006), spotted sand bass (Peña et al. 2003), and the kelp grouper *Epinephelus bruneus* (Kato et al. 2004), and in freshwater species such as green sturgeon (Gisbert and Doroshov 2003), pike perch (Ostaszewska 2005), and European catfish (Kozarić et al. 2008), the liver develops after the larva emerges from the egg envelope during the endogenous feeding phase (Figure 1.5a).

Bile is secreted by the hepatic cells and is discharged into the extracellular bile canaliculi (Figure 1.5b). Bile canaliculi join to form the bile ducts, which subsequently converge into the hepatic duct. The latter leaves the liver and opens into the anterior intestine. In many fishes, the hepatic duct has a branch, the *ductus cysticus*, leading into the gallbladder, which stores bile juice. The walls of the bile ducts consist of a single layer of cuboidal to columnar cells over an underlying layer of connective tissue. The histological organization of the hepatic duct is similar but it includes a layer of smooth muscle (Takashima and Hibiya 1995). The histological development of the liver and the bile transport system develops concomitantly with the gradual maturation of hepatocytes, as well as their functional ability to synthesize, store, and mobilize carbohydrates and lipids (see review in Hoehne-Reitan and Kjørsvik 2004).

The exocrine part of the **pancreas** in teleosts is a diffuse organ, spread throughout the mesentery surrounding the digestive tract and other organs, and interspersed with adipose tissue (Figure 1.5e,f). Portions of it might also be distributed around major blood vessels within the liver of some species forming the hepatopancreas such as in cyprinids, characids, or some siluriformes among others (Takashima and Hibiya 1995; Petcoff et al. 2006), whereas some other species, such as anguillid eels, northern pike, or Japanese catfish, have a distinctive pancreas as is found in higher vertebrates (Hoehne-Reitan and Kjørsvik 2004). A functional exocrine pancreas is characterized by differentiated organ morphology, including developed excretory ducts and the presence of zymogen granulae for all major digestive enzymes.

Histologically, the arrangement of the pancreatic tissue is essentially similar and, in some aspects, resembles the basic architecture of the hepatic chords. Pancreatic secretory cells grouped into acini are deposited around blood vessels and form secretory functional units by the juxtaposition of adjacent cells. The secretory cells generally have a prismatic form with basal nuclei and peripheral heterochromatin and a prominent nucleolus. In light microscopic slides, the cytoplasm of secretory pancreatic cells is strongly basophilic, providing a sharp contrast with the round, intensively acidophilic and eosinophilic zymogen granules. The ultrastructural characteristics of this type of secretory cell are described in detail by Takashima and Hibiya (1995). There is very little information on the histological development of the exocrine pancreas in fish larvae. According to Beccaria et al. (1991), the organogenesis of the exocrine pancreas can be divided into three distinct phases: appearance of a primordium at hatching in the form of a dorsal bud on the digestive tract; differentiation of the exocrine cells and appearance of the excretory ducts and blood vessels before mouth-opening; and growth of the organ during the larval and juvenile period. The quantitative growth after differentiation includes tissue size, an increase in the relative frequency of zymogen granules, and an increase in enzyme synthesis and secretion, while no new structural elements develop. In most of the described species, the exocrine pancreas is histologically differentiated at mouth-opening, as has been reported in Japanese flounder, Atlantic halibut, Atlantic cod, Senegal sole, bay snook, turbot, whitefish, and Siberian sturgeon, among others (Hoehne-Reitan and Kjørsvik 2004; Zambonino-Infante et al. 2008).

The fish **gallbladder** is an accessory digestive organ that stores and secretes concentrated bile. The bile has several functions, such as facilitating several digestive functions, eliminating conjugated metabolites in the liver (including xenobiotics), and participating in the enterohepatic bile circulation. The morphological interrelationship between the liver, the biliary system, and the gallbladder was extensively reviewed by Gilloteaux et al. (1996) and is not covered in the present review. Histologically, three layers are distinguished in the gallbladder of adult fishes: the inner layer, which is composed of a simple epithelium of columnar cells and connective tissue; the intermediate layer, which consists of smooth muscle; and the outer layer, which is the serous membrane (Takashima and Hibiya 1995). These layers are also distinguishable in fish larvae, although there are some ontogenetic differences regarding the type of epithelium lining the inner layer of the gallbladder and the level of development of the smooth muscle fibers that regulate the contraction of the organ. At early stages of development, the inner layer of the gallbladder is lined by a simple squamous epithelium that becomes cubical and columnar with age (Hamlin et al. 2000; Micale et al. 2006; Hachero-Cruzado et al. 2009).

Rodlet cells have been reported in the epithelium of the gallbladder and biliary ducts of some freshwater and marine teleosts. Different studies have shown that the presence of rodlet cells within the teleost gallbladder is species specific and may not necessarily depend on environmental conditions (e.g., pollutants) (see review in Hrubec and Caceci 2001). Considering the function of the liver in detoxification processes, Kramer et al. (2005) hypothesized that the abundance of rodlet cells within the gallbladder epithelium of fish exposed to environmental contamination indicates that this organ could serve as a storage or recruitment site for these cells and provide a portal through which rodlet cell secretions are deposited into the bile and carried away.

1.4 Ontogeny of the digestive enzymes

The development of adequate compound microdiets to replace live foods in the culture of marine fish larvae requires a thorough understanding of the digestion processes occurring during ontogeny (Cahu and Zambonino-Infante 1997; Lazo et al. 2000a). This knowledge is required for reducing the use of live feeds in the rearing of marine fish larvae. The lack of success in completely replacing live foods with compound microdiets from the onset of first feeding has been historically attributed to the presence of an undeveloped digestive system at the time of hatching and consequent low digestive capacity (Lauf and Hoffer 1984; Munilla-Moran et al. 1990; Holt 1993), although most research to date indicates that marine fish larvae have a very defined and specific digestive physiology that merits the development of specific diets and weaning protocols, and that they possess a differentiated and effective digestive system early in development (Sarasquete et al. 1995; Ribeiro et al. 1999; Lazo et al. 2000a; Zambonino-Infante and Cahu 2001).

The conventional approach used for assessing digestive capacity in marine fish larvae has typically involved characterizing the morphological development of the digestive system and associated organs while also quantifying digestive enzyme activities using biochemical, histochemical, and molecular techniques (for an excellent review, see Zambonino-Infante et al. 2008). The morphological and functional development of the digestive system of fish larvae was first reviewed by Tanaka (1973) and Govoni et al. (1986), and more recently by Hoehne-Reitan and Kjørsvik (2004) and Zambonino-Infante et al. (2008). At hatching, the stomach is typically undifferentiated and nonfunctional. Acid digestion and pepsin expression are lacking, and the proton pump used to secrete HCl into the stomach lumen is not functional (Rust 2002; Gawlicka et al. 2001; Darias et al. 2005; Rønnestad et al. 2007). Most species also lack functional mouths and jaws, and the eyes are not yet pigmented. Early larvae typically possess a simple tubelike alimentary canal that is closed at both ends and lined with columnar epithelium. The alimentary canal undergoes rapid transformations during the transition to exogenous feeding. By the onset of first feeding, the alimentary canal has already developed into its different functional regions, but it is still less complex than in juveniles. However, the liver, pancreas, and gallbladder are usually present and functional (Hoehne-Reitan and Kjørsvik 2004). Digestion occurs in the midgut and hindgut, and nutrient absorption takes place through the apical region of the epithelium of each region, which is characterized by columnar cells (i.e., enterocytes). Alkaline proteases play a major role in digestion during the first days of feeding, while acid proteases become increasingly important toward the end of the larval period, concomitant with the appearance of a functional stomach (Lauf and Hoffer 1984; Lazo et al. 2007). As the developmental process progresses, oxynticopeptic cells in the gastric glands become functional, as suggested by the production of HCl through a functional proton pump, the expression of pepsinogen, and its activation to pepsin (Gawlicka et al. 2001). From the perspective of the digestion system, the transformation to the juvenile stage is complete once the stomach is fully differentiated.

High specific activity of digestive enzymes has been observed before the initiation of exogenous feeding in most species studied to date (Zambonino-Infante and Cahu 2001). This suggests the process of enzyme production is initiated by underlying genetic mechanisms (Buddington and Diamond 1989) rather than induced by the diet (Cahu and Zambonino-Infante 1994; Lazo et al. 2000a). While it appears that during the early stages of development digestive enzyme activities are controlled by gene expression rather than by feeding activity, diet composition can influence the maturation of the digestive system by triggering an onset or increase in the activity of some digestive enzymes (Zambonino-Infante and Cahu 2001). Feeding nutritionally unbalanced microdiets to marine fish larvae can disrupt the normal maturation process; the earlier the weaning onto unbalanced microdiets, the more negative the observed effect on maturation (Cahu and Zambonino-Infante 1994; Lazo et al. 2000a). In contrast, some nutrients, such as polyamines, can enhance the maturation and differentiation of the enterocytes involved in nutrient absorption. For example, sea bass (*Dicentrarchus labrax*) larvae fed a diet containing 0.33%

dry weight of the polyamine spermine displayed faster enterocyte maturation compared with fish fed a similar diet lacking in polyamine (Peres et al. 1997). Likewise, Tovar-Ramirez et al. (2002) included the polyamine-producing yeast (*Debaryomyces hansenii* HF1) in the diet of sea bass larvae and observed an increase in digestive enzyme secretion and earlier maturation of the enterocytes that were mediated by spermine and spermidine.

While most species can be effectively weaned onto microdiets before completion of metamorphosis, successful weaning during the early larval stages has proven more challenging (Kolkovski 2001). Only a handful of species can be reared on microdiets from the time of mouth-opening (i.e., red drum *Scieaenops ocellatus* and sea bass). Most species cultured to date require the use of rotifers or *Artemia* at some point during development (Cahu and Zambonino-Infante 2001).

As previously mentioned, early research suggested that problems associated with early weaning were due to low digestive enzyme activity or to the importance of live prey for aiding or triggering the digestive process. In contrast, recent studies indicate that enzymatic activity is high in early larvae, and that the potential contribution of digestive enzymes from the prey is negligible. Typically, enzymes for the luminal digestion of proteins (trypsin, chymotrypsin, and elastase, among others), lipids (lipases and phospholipases), and carbohydrates (amylases and maltases) are present in larvae before exogenous feeding commences or shortly thereafter. Their activity increases with age and length, although there are some exceptions (Alliot et al. 1980; Baragi and Lovell 1986; Cousin et al. 1987; Moyano et al. 1996; Baglole et al. 1998; Izquierdo et al. 2000; Zambonino-Infante and Cahu 2001; Lazo et al. 2007). Intracellular enterocyte digestive enzymes such as tri- and dipeptidases exhibit high levels of activity during the early larval stage and decrease as development progresses (Cahu and Zambonino-Infante 1995; Lazo et al. 2007). In contrast, the activity of intestinal brush border membrane enzymes such as aminopeptidases and alkaline phosphatases are lowest at first feeding and subsequently increase with age. A decrease in intracellular peptidase activity concurrent to an increase in brush border peptidase activity is indicative of the full intestinal maturity of marine fish larvae (Cahu and Zambonino-Infante 1994). The ratio of these two enzymes can be used as an indicator of the maturation of the digestive system in marine fish larvae and will be further described in the last section of this chapter. Thus, although not as complex as the juvenile digestive system, marine fish larvae possess a wide range of digestive enzymes that support the efficient digestion of nutrients if adequate feeds are provided (i.e., larvae can achieve very high growth rates in the wild and under culture conditions).

It has been proposed that exogenous enzymes from live prey could directly aid in larval digestion or activate the zymogens present in larval gut, thus increasing digestion and growth rates (Dabrowski 1979; Lauf and Hoffer 1984). The mechanisms through which exogenous enzymes could aid or stimulate the digestive process are not clearly understood. Moreover, the addition of exogenous enzymes to compound microdiets in the rearing of marine fish larvae has been shown to be beneficial only for sea bass larvae (for a review, see Kolkovski 2001). However, its benefits have not been conclusively demonstrated for other species. Moreover, several authors have reported a lack of significant differences in levels of pancreatic and intestinal enzymes in fish larvae reared with live prey or microdiets (Baragi and Lovell 1986; Cahu and Zambonino-Infante 1997; Lazo et al. 2000b), which indicates the ingestion of live prey does not stimulate enzyme production or secretion into the gut lumen. Kurokawa et al. (1998) estimated the relative contribution of exogenous enzymes to digestion in

Japanese sardine larvae (*Sardinops melanotictus*) and determined that it was only 0.6% of the total protease activity in the intestine, and therefore concluded that the contribution of the prey's enzymes to digestion was minimal. Similarly, Diaz et al. (1997a), using substrate-sodium dodecyl sulfate–polyacrylamide gel electrophoresis (SDS-PAGE) to estimate protease activity in larval sea bream (*Sparus aurata*) and their live prey (rotifers), failed to detect proteases from the prey within the digestive tract. They suggested that the contribution of exogenous enzymes was limited to an autolytic process of the prey in the larval gut. Based on this data, it appears that the contribution of exogenous digestive enzymes to the total digestive capacity of the larvae is negligible in most species.

Since the lack of weaning success at an early date cannot be attributed solely to the absence of a functional stomach and lower digestive enzyme production, other factors have been conjectured to explain the lower performance on microdiets. These include low ingestion rates of the microdiets (Lazo et al. 2002) or the failure of microdiets to effectively stimulate digestive enzyme secretion (Cahu and Zambonino-Infante 2001). The latter would lead to low levels of enzymes in the lumen to digest feed particles. In combination with the relatively fast gut transit time typical of marine fish larvae (Govoni et al. 1986), this would effectively reduce the ability of the larvae to absorb the dietary nutrients necessary for meeting the requirements for normal growth. Recent research has begun to shape a more comprehensive understanding of the development of the digestive system by focusing on the study of the hormonal mechanisms controlling the expression and secretion of digestive enzymes and their modulation through dietary nutrients (recently reviewed by Rønnestad et al. 2007). For example, many compounds present in live feeds have the potential for influencing digestive enzyme activity in fish larvae. Polyamides, algal growth regulators that play multiple roles in stabilizing the intracellular conformation of nucleic acids and membranes (Mathews and van Holde 1990; García-Jimenez et al. 1998), have been shown to stimulate gut hormone (cholecystokinin [CCK]) release in rats, which in turn mediates the release of pancreatic enzymes (Fioramonti et al. 1994). Most formulated diets designed for marine fish larvae contain large amounts of fish meal, which is naturally low in the polyamide spermine (Bardocz et al. 1993). The addition of spermine to microdiets fed to sea bass larvae has been shown to increase pancreatic enzyme secretion and induce earlier intestinal maturation (Peres et al. 1997). In addition, amino acids may increase the secretion of certain hormones, such as somatostatin and bombasin, which also stimulate the secretion of pancreatic enzymes (Chey 1993; Kolkovski et al. 1997). Live feeds contain large amounts of free amino acids, which may stimulate the secretion of trypsin (Dortch 1987; Fyhn 1993). For example, Cahu and Zambonino-Infante (1995) reported increased trypsin secretion in sea bass larvae fed a mixture of free amino acids in their diets.

Both neural and hormonal processes are involved in regulating the secretion of pancreatic enzymes (Fange and Grove 1979) and is discussed in detail in Chapter 9, but a brief description is presented here. The sight, smell, or presence of food triggers a nervous control mediated by the vagus nerve that results in the induction of pancreatic secretion. Hjelmeland et al. (1988) induced secretion of trypsinogen from pancreatic tissue into the intestine of herring larvae (*Clupea harengus*) by feeding polystyrene spheres with no nutritional value. Similarly, Pedersen and Andersen (1992) were able to enhance the secretion of pancreatic enzymes by increasing the size of the inert particles fed to herring larvae. Additionally, gastrointestinal hormones, such as CCK, play an important role not only in the stimulation of pancreatic enzyme secretion but also in gallbladder contraction, intestinal peristalsis, and gut transit time in fish larvae (Rønnestad et al. 2007), all of which are important factors regulating the digestion process. In first feeding larvae, CCK production seems to be genetically hardwired, but in older larvae it can also be regulated by dietary factors such as protein levels and chain length (Cahu et al. 2004). However, distension of the gut wall is not a factor that triggers CCK production (Koven et al. 2002). This indicates that the secretion of pancreatic enzymes is regulated by mechanisms in addition to CCK production and requires further research.

1.5 Expression of digestive enzyme genes

Even though numerous studies have characterized the ontogeny of the digestive system of marine and freshwater fish species, knowledge of the nutritional requirements during the larval period still needs to be improved in order to formulate adequate microdiets for optimal larval rearing. In the last decade, new molecular tools have been used to complement the morphological, histological, histochemical, and biochemical approaches commonly used, allowing researchers to expand the knowledge of the mechanisms underlying the digestive physiology of fish larvae.

The ontogeny of the digestive system is a species-specific and genetically programmed process where digestive enzymes follow a spatiotemporal pattern of gene expression during the larval development. These processes can be influenced by the diet and directly impact nutrient digestion and absorption, and consequently, larval performance and growth. For these reasons, the first studies on digestive enzyme gene expression in larvae were made on species reared under standard conditions using live prey in order to provide the reference gene expression patterns of digestive enzymes for future nutritional experiments.

The expression pattern of digestive enzyme precursors is intimately associated with the degree of development of the organs that produce them. For instance, in the red porgy (*Pagrus pagrus*), the first signs of amylase, lipase, and trypsinogen gene expression were detected in newly hatched larvae, indicating that the enzymatic equipment of the exocrine pancreas is ready to produce the required enzymes for food digestion at the beginning of the exogenous feeding period (Darias et al. 2006, 2007a, 2007b, 2007c). In the next section, we review the most studied digestive enzyme genes during development of marine fish larvae.

1.5.1 Amylase

Alpha (α) amylase (4-α-D-glucan glucanohydrolase, EC 3.2.1.1) is classified as family 13 of the glycosyl hydrolases and is responsible for the hydrolysis of the α-1,4 glycoside bonds in glycogen and starch, and related polysaccharides and oligosaccharides containing three or more (1→4)-α-linked D-glucose units. The main function of the digested and assimilated products is to supply energy to the organism. Alpha amylase is produced as a zymogen granule by pancreatic acinar cells, and their presence is considered to be an indicator of the exocrine pancreas maturation in fish larvae (Cahu and Zambonino-Infante 1994; Cahu et al. 2004). The appearance of acidophilic zymogen granules has been detected after the first exogenous feeding in some species (*Sparus aurata*, Sarasquete et al. 1995; *Solea senegalensis*, Sarasquete et al. 1996; *Paralichthys dentatus*, Bisbal and Bengtson 1995; *Paralichthys olivaceus*, Kurokawa and Suzuki 1996; *Pagrus pagrus*, Darias et al. 2007a), before mouth-opening in others (*Dicentrarchus labrax*, Beccaria et al. 1991; *Sciaenops ocellatus*, Lazo et al. 2000a), or even from hatching (*Melanogrammus aeglefinus*, Hamlin et al. 2000). Zymogen granule detection coincides with the first amylase activity detected in *Solea senegalensis* (Ribeiro et al. 1999), *Diplodus sargus* (Cara et al. 2003), *Sparus aurata* (Moyano et al. 1996), and *Dicentrarchus labrax* (Zambonino-Infante and Cahu 1994a). Additionally, *in situ* hybridization technique has revealed that α-amylase gene expression occurs in the exocrine pancreas (Darias et al. 2006). Although only a few studies have evaluated amylase expression during fish larvae development, it is possible to appreciate differences in the pattern of gene expression among species. For example, in sea bass larvae (*Lates calcarifer*), amylase expression increases early in development (i.e., 5 days posthatch [dph]) to subsequently decrease in later stages (Péres et al. 1996; Ma et al. 2004), while in winter flounder (*Pseudopleuronectes americanus*), amylase expression did not decrease until after metamorphosis (Douglas et al. 2000), and in the red porgy (*Pagrus pagrus*), a constant level of amylase gene expression was observed until 30 dph and subsequently decreased thereafter (Darias et al. 2006). In any case, a relatively elevated level of amylase expression during the first stages of larval development has been observed in most species studied to date, but the main physiological function of this activity has not been completely elucidated.

It has been suggested that the expression of amylase is genetically programmed and regulated at a transcriptional level during the early developmental stages (Péres et al. 1996; Ma et al. 2001; Zambonino-Infante and Cahu 2001). The detection of amylase expression from hatching in *Lates calcarifer* and *Pagrus pagrus* (Ma et al. 2001; Darias et al. 2006) supports the existence of a hereditary component in this process and also indicates that the predisposition to synthesize amylase before the commencement of the exogenous feeding phase is independent of the external diet. This could be a programmed mechanism to ensure sufficient levels of this enzyme to be ready for digestion at the beginning of exogenous feeding.

However, once exogenous feeding commences, amylase expression can be modulated by the quantity and quality of the food (Péres et al. 1996). The different patterns of amylase expression observed in most fish species studied (Péres et al. 1996; Douglas et al. 2000; Ma et al. 2001; Darias et al. 2006) suggest that the variations are mainly due to the rearing conditions, including diet composition, quantity of diet offered, and sampling time during development.

1.5.2 Bile salt–activated lipase (BAL)

BAL is considered one of the most important lipases in fish (Patton et al. 1977; Murray et al. 2003) since it acts on a wide range of substrates of wax esters and triacylglycerols rich in polyunsaturated fatty acids (PUFAs). These substrates are more resistant to hydrolysis by other pancreatic lipases (Chen et al. 1990).

For lipid hydrolysis, pancreatic BAL is secreted to the intestinal lumen and activated by bile salts. Subsequently, the intestine can absorb the resulting substances. Diaz et al. (2002) observed adequate levels of lipase activity, bile function, and intestinal absorption at the beginning of exogenous feeding in three fish species. BAL activity was detected at hatching in several fish species, suggesting adequate enzymatic equipment for lipid digestion at first feeding (Hoehne-Reitan et al. 2001a; Murray et al. 2003; Pérez-Casanova et al. 2004). However, differences in patterns and activity levels have been observed among species.

Several studies evaluated the activity of different lipases in fish, including phospholipases, pancreatic lipases, nonspecific lipases, and BAL (Izquierdo et al. 2000; Hoehne-Reitan et al. 2001a, 2001b; Cahu et al. 2003). Gjellesvik et al. (1992) and Iijima et al. (1998) purified and characterized the BAL of *Gadus morhua* and *Pagrus major*, respectively, and four other studies describe the ontogeny of BAL expression during larval development (Hoehne-Reitan et al. 2001a; Murray et al. 2003; Pérez-Casanova et al. 2004; Darias et al. 2007a).

Similar to trypsinogen and α-amylase gene expression, BAL expression is specifically localized in the exocrine pancreas (Figure 1.6b). BAL expression was detected from hatching in *Melanogrammus aeglefinus* and *Pagrus pagrus* (Pérez-Casanova et al. 2004; Darias et al. 2007a), while from mouth-opening in *Pleuronectes americanus*, progressively increasing during larval development (Murray et al. 2003). In other fish species, BAL expression was shown to be sensitive to changes in diet composition (Pérez-Casanova et al. 2004; Darias et al. 2007a). Variation of BAL activity during larval development of *Psetta maxima,* evaluated by enzyme-linked immunosorbent assay (ELISA), showed a similar pattern to the expression pattern of BAL in *Pagrus pagrus* (Hoehne-Reitan et al. 2001a; Darias et al. 2007a).

The effect of dietary lipid quantity on BAL activity has been demonstrated by Hoehne-Reitan et al. (2001b). They demonstrated that increasing prey density increased ingestion rates and stimulated digestive enzyme synthesis. However, the researchers did not observe an increase in larval growth associated with the increasing ingestion rates. Additionally, they did not detect any positive effects on development associated with the lipid content of rotifers during the first stages of development. In contrast, Zambonino-Infante and Cahu (1999) showed that a diet with high lipid content improved larval development in European sea bass *Dicentrarchus labrax.* Morais et al. (2004) demonstrated for the same species that the use of different neutral lipid sources in the diet did not affect lipase at transcriptional level, in contrast to that observed by Zambonino-Infante et al. (1996) and Cahu et al. (2003). Péres et al. (1996) suggested that diet composition can affect expression of digestive enzymes at the transcriptional and translational levels. In this sense, more experiments using formulated microdiets with different nutritional composition need to be performed to increase the existent knowledge of the digestive physiology of fish larvae.

1.5.3 Trypsinogen

In fish, as in other vertebrate species, trypsinogen is the inactive form of trypsin, an important proteolytic digestive enzyme present early in larval development, when the gastric glands have not yet developed and there is no pepsin activity for acid digestion. The expression of trypsinogen is specifically localized in the exocrine pancreas (Murray et al. 2004; Darias et al. 2007a; Figure 1.6c,d). Douglas and Gallant (1998) described three different trypsinogens in *Pleuronectes americanus* with apparently diverse functions throughout the larval period. Trypsinogen 2 seems to be most important during digestive system development since it is the first detected and the most highly expressed (Murray et al. 2004).

There are few studies concerning trypsinogen during the larval development (Srivastava et al. 2002; Murray et al. 2004; Darias et al. 2007a). In *Paralichthys olivaceus*, trypsinogen 1 expression occurred at 1 dph (Srivastava et al. 2002), while in *Pleuronectes americanus*, the expression of trypsinogen 2 was observed 5 dph and displayed a maximum peak of expression during metamorphosis (Murray et al. 2004). In *Pagrus pagrus*, trypsinogen gene expression was detected from hatching and the maximum levels of expression occurred after first feeding and subsequently remained constant during the first month of development (Darias et al. 2007a). These authors observed a decrease in trypsinogen expression from 50 dph, suggesting a reduction of the importance of trypsin in the digestive process after metamorphosis to the juvenile stage.

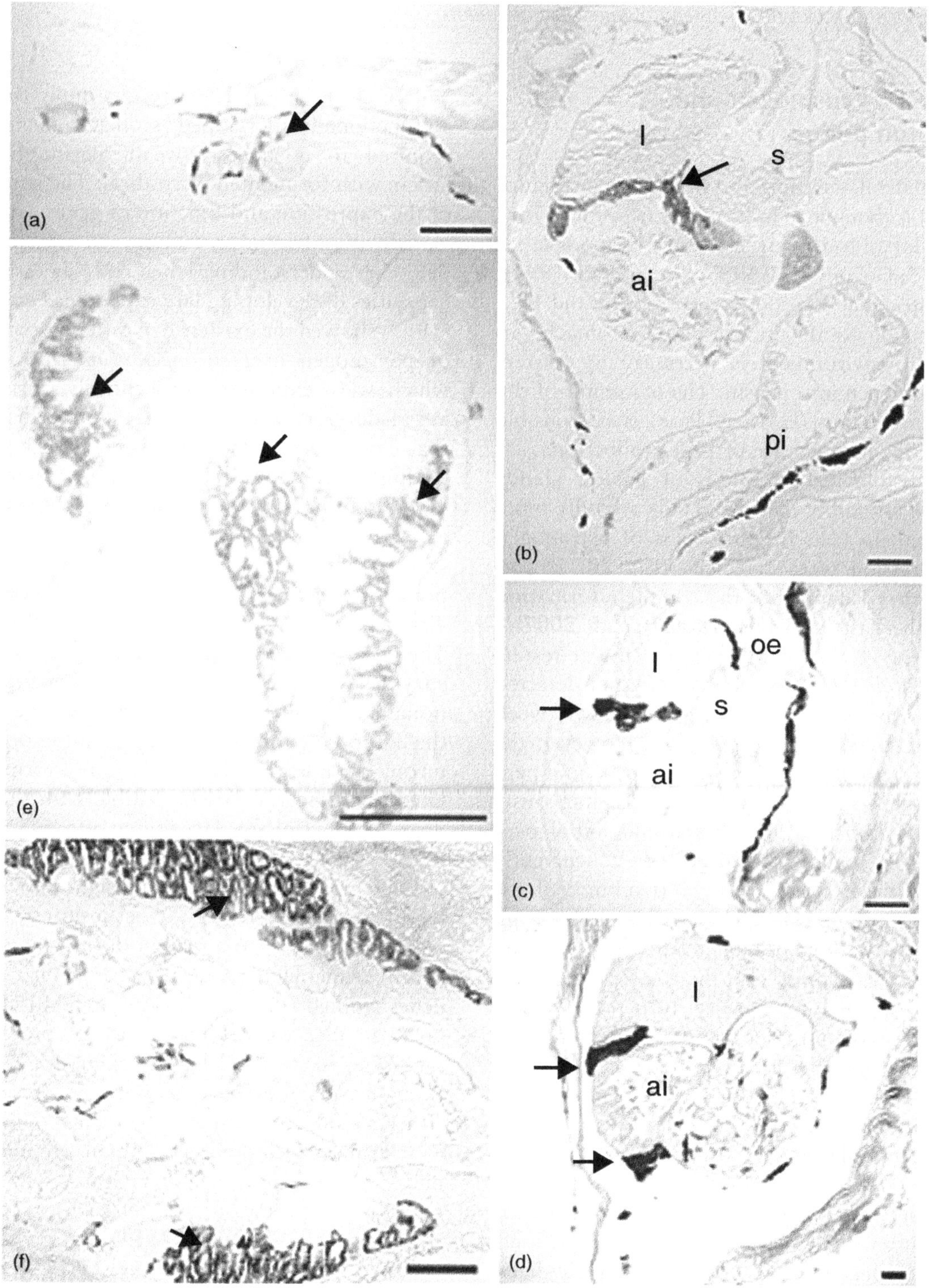

Figure 1.6 Localization of the gene expression of different digestive enzymes in developing red porgy (*Pagrus pagrus*) larvae by *in situ* hybridization. (a) Amylase gene expression at 2 dph in the exocrine pancreas. (b) Bile salt–activated lipase gene expression at 30 dph in the exocrine pancreas. (c, d) Trypsinogen expression at 9 and 35 dph, respectively, in the exocrine pancreas. (e) Proton pump expression at 40 dph in the gastric glands of the stomach. (f) Pepsinogen expression at 50 dph in the gastric glands of the stomach. ai = anterior intestine, oe = esophagus, l = liver, pi = posterior intestine, s = stomach. Scale bar = 100 μm.

1.5.4 Pepsinogen and proton pumps

In general, gastric gland development (Figure 1.4) is thought to indicate the transition from the larval to the juvenile stage (Kapoor et al. 1975; Govoni et al. 1986; Segner et al. 1994). Gastric glands produce pepsinogen and HCl secretion to the lumen of the stomach; an acidic environment is necessary to convert pepsinogen into pepsin. The α subunit of the proton pump (H^+/K^+-ATPase) is responsible for the maintenance of HCl production.

Histological detection of gastric glands does not imply that the glands are fully functional. In fact, the first signs of pepsinogen expression were detected 30 dph in *Pagrus pagrus*, 4 days after the complete formation of the gastric glands (Darias et al. 2007b). Huang et al. (1998) obtained similar results in *Paralichthys dentatus*, where they detected the expression of pepsinogen around 1 week after gastric gland formation. However, the first signs of gastric gland formation and pepsinogen expression occurred simultaneously in *Pleuronectes americanus* and initial pepsinogen expression occurred concurrently with the expression of the α-subunit of the proton pump (H^+/K^+-ATPase) (Douglas et al. 1999). Both pepsinogen and proton pump genes are expressed in the gastric glands and their expression progressively increases during maturation of the gastric glands and stabilizes at the juvenile stage (Douglas et al. 1999; Gawlicka et al. 2001; Darias et al. 2007b; Figure 1.6e,f).

The replacement of live prey by formulated microdiets for larval feeding is of fundamental interest for the marine fish larvae rearing industry. The development of a functional stomach is necessary to reach complete digestive capacity. However, it is important to mention that *Sparus aurata* and *Sciaenops ocellatus* are completely weaned before gastric glands are developed and become functional and are typically reared using a standard protocol (Lazo et al. 2000b; Elbal et al. 2004; Yúfera et al. 2004). These results imply that a functionally developed stomach is not required to adequately wean marine fish larvae with formulated microdiets. The study of the expression and function of genes associated with different proteolytic activities will help in understanding the real digestive capacities of developing larvae. Douglas et al. (1999) showed the existence of different types of pepsinogen in *Pleuronectes americanus*, which were expressed consecutively during larval development. It is interesting to note that pepsinogen IIa was expressed as early as 13 dph, before the gastric glands were formed. Several authors have found acid protease activity before stomach development even in species that never develop a true and functional stomach, such as the puffer fish *Takifugu rubripes* (Kurokawa et al. 2005). Therefore, a better understanding of the enzymes implicit in digestion and their functionality, in combination with knowledge of the natural diet of the larvae, will aid in optimizing the nutritional composition of formulated microdiets to be compatible with their digestive physiology. The timing and quantities of gene expression for the digestive enzyme precursors provide an insight into the larval digestive strategy. For instance, in *Pagrus pagrus*, the expression of trypsinogen was five times higher than all the other enzyme genes studied (such as amylase or BAL) and gives an idea of the importance of protein digestion in early developing larvae even though pepsin activity is not yet present (Darias et al. 2005).

In fish, as well as in amphibians, reptiles, and birds, pepsinogen and HCl are synthesized by one type of gastric cell called oxynticopeptic cells (Helander 1981) that possess characteristics of the oxyntic cell (HCl secretor cells) and zymogen cells (enzyme secretor cells) of mammals (Murray et al. 1994). Some authors suggest the existence of a unique cell type that has different morphologies depending on whether they produce HCl (light cells) or pepsinogen (dark cells) under a transmission electron microscope (Elbal and Agulleiro 1986; Arellano et al. 1999). Gawlicka et al. (2001) confirmed that only one type of cell, the oxynticopeptic cell, is responsible for the synthesis and expression of the different types of pepsinogens and proton pump. They also showed that the mucous neck cells of the stomach epithelium of *Pleuronectes americanus* had HCl, but not pepsinogen, secretory function. This may be due to the need of fish, in an aqueous medium where they continuously drink water, to increase the HCl concentration to counteract the dilution and neutralization of gastric juices (Kapoor et al. 1975). However, in *Pagrus pagrus*, the gene expression of the proton pump was exclusively localized in the gastric glands (Darias et al. 2007b), in contrast to that found in *Pleuronectes americanus*. Since the latter species ingests bigger prey, it might need a higher concentration of HCl and pepsin activity to adequately digest its food. Nevertheless, more studies of the neck cells of the mucosa are needed to test whether there are structural and functional differences among species. Gawlicka et al. (2001) and Darias et al. (2007b) reported that the expression of pepsinogen and proton pump occurred simultaneously in *Pleuronectes americanus* and *Pagrus pagrus*, respectively. The simultaneous secretion of pepsinogen and HCl could be a physiological strategy for promoting fast conversion of pepsinogen into active pepsin (Bal and Ghoshal 1992). In *Pagrus pagrus*, the expression of both genes begins 30 dph and have similar copies of mRNA that increase with larval development. However, in *Pleuronectes americanus*, the expression of pepsinogen is constant from 20 dph onward (Douglas et al. 1999).

Ongoing studies of the molecular mechanisms underlying the gastrointestinal functions of fish larvae reared under different nutritional and rearing conditions will help improve our understanding of the digestive physiology of commercially important fish species.

1.6 Assessing the nutritional condition of fish larvae: histological biomarkers and digestive enzymes

Assessing the nutritional condition of fish larvae is of vital importance in ecological studies since the physical and physiological condition of larval fishes throughout their development influences their growth performance and survival and, ultimately, contributes to recruitment to the adult population. These studies require that accurate, objective, and quantitative criteria be used to characterize the nutritional condition of fish larvae. This approach can also be applied in aquaculture where the development of dependable and sustainable fish larval rearing techniques requires a deep knowledge of the critical aspects of larvae nutrition in relation to the development of the digestive and metabolic systems, as well as establishing the limits for initiating exogenous feeding.

Once exogenous feeding is established, larval development depends on the proper nutrient input provided by the diet, in addition to optimal biotic and abiotic conditions. Periods of food deprivation after the completion of yolk reserves can lead to abnormal behavior and morphological development, degeneration of the alimentary tract and trunk musculature, and reductions in food utilization efficiency and feeding activity. Fish larvae are especially sensitive to nonoptimal feeding conditions or nutritional stressors (dietary imbalances) because most tissues and organs are under progressive and intense differentiation and development, and larvae do not have enough reserves stored to withstand starvation (Ferron and Leggett 1994; Catalan 2003; Gisbert et al. 2008).

The effect of feeding restriction or nutritional imbalance on aquatic organisms is routinely assessed by a number of indicators commonly named "condition indices" used to characterize nutritional condition of fish larvae. Condition indices were extensively reviewed by Ferron and Leggett (1994) and Catalan (2003) in terms of reliability, sensitivity, time response, size and age specificity, field versus laboratory estimates, processing time, costs, and requirements. These authors divided condition indices into three main categories according to the main organization levels: cell, tissue, and organism. In this sense, the physical deterioration of fish larvae resulting from food deprivation or dietary imbalance has been assessed and interpreted by means of morphometric and gravimetric measurements (shape and weight changes), biochemical methods (RNA:DNA ratios, digestive and metabolic enzyme activities), histological criteria, or various combinations of the above-mentioned methods (Ferron and Leggett 1994; Catalan 2003; Gisbert et al. 2008; see also the discussion in Chapter 14). Although there are a wide variety of nutritional condition indices, this section will only cover those related to digestive system organization (histological biomarkers) and function (pancreatic and intestinal enzyme activities).

1.6.1 Histological biomarkers

In vertebrates, different organs of the digestive system have been shown to employ different cellular mechanisms in response to diet quantity and quality. Thus, the use of the intestine and digestive accessory glands as target organs of the nutritional and physiological status in fish is well known and, up to a certain limit, standardized. The use of histological biomarkers for assessing the nutritional condition of fish larvae has been recently reviewed by Gisbert et al. (2008). The histological organization and histochemical properties of the liver, exocrine pancreas, and intestine have been used on a regular basis as targets to elucidate the effects of different dietary regimes or nutrients and starvation levels on larval physiology, nutrition, and early development (Table 1.1).

The histological organization of the intestine, like that of the liver, is particularly sensitive to food deprivation and starvation. Major alterations of the intestinal mucosa include

Table 1.1 Cellular criteria used to grade tissues and assess the nutritional condition in teleost larvae.

	Grade (condition)		
Tissue	**1 (degraded)**	**2 (average)**	**3 (healthy)**
Liver hepatocytes	Nearly all nuclei pycnotic and dark with clumped chromatin; cytoplasm lacks texture; intracellular vacuoles absent; cells small and indistinct	At least 50% of cell nuclei with dark granules and situated medially; nearly 50% of cytoplasm granular; intracellular vacuoles reduced or absent; boundaries of most hepatocytes visible	Nuclei distinct and often displaced laterally; cytoplasm lightly stained with abundant intracellular vacuoles containing lipids and glycogen; boundaries of hepatocytes prominent
Exocrine pancreas	No acinar symmetry remaining; all nuclei dark (pycnotic) and indistinct	Acinar symmetry reduced by 50%; 50% of nuclei dark and indistinct; moderate amounts of zymogen	Cells formed in distinct, circular acini; all nuclei clear and distinct in basal position; abundant zymogen granules
Intestinal epithelium	Mucosal cell height reduced by >50% in height; some loss of striations in bordering microvilli; supranuclear vacuoles reduced or absent	Mucosal cells reduced by 25–50% in height; some loss of striations in bordering microvilli; supranuclear vacuoles reduced or absent	Mucosa deeply convoluted and mosaic; mucosal cells compact, pronounced in height, with distinct nuclei; prominent supranuclear acidophilic inclusions and vacuoles

Data rewritten from Margulies (1993), Catalan (2003), and Gisbert et al. (2004b).

the reduction in the height of the enterocytes and the number and size of epithelial folds. Proteolysis of the intestinal mucosa is a common response to severe starvation, which involves a reduction of the nutrient absorption surface area, and compromises the digestive capabilities of refeeding larvae. For these reasons, the criterion of enterocyte height has been widely used as a valuable histological index of suboptimal feeding or starvation in several fish species (Ferron and Leggett 1994; Catalan 2003; Gisbert et al. 2008). However, Catalan and Olivar (2002) reported that cell heights of the posterior intestine in European sea bass larvae were less useful to distinguish different feeding treatments than other quantitative measurements (e.g., hepatocyte maximum diameter, muscle fiber separation). Consequently, for any selected species, any current or putative nutritional condition index should be tested and validated under laboratory-controlled conditions.

Lipid and protein inclusions in enterocytes may also be used as a biomarker in fish larval nutrition and digestive physiology studies (Gisbert et al. 2008).The presence of acidophilic supranuclear inclusions is a typical feature of the posterior intestine in fish larvae. These inclusions are due to the absorption of protein macromolecules by pinocytosis. In most studied species, supranuclear bodies are observed throughout the larval period, although their number and size decrease as the stomach differentiates and extracellular digestion takes place. Thus, variations in the normal pattern of accumulation of these inclusions may be indicative of changes in the nutritional physiology of the larva and therefore be used in developmental or nutritional studies dealing with larval early stages of development. The presence of lipid inclusions in the enterocytes of fish larvae is a common feature during their early development. The type and size of lipid inclusions vary depending on the fat content of feed and the degree of unsaturation of the lipids ingested. As a result, changes in the size and type of lipid inclusions may be dietary dependent and may be useful for assessing the nutritional condition of a fish larva. Three types of inclusions can be distinguished in fish enterocytes according to their size: particles (20–70 nm in diameter) resembling mammalian very low-density lipoproteins (VLDL); lipoprotein particles (70–500 nm in diameter) considered as chylomicrons; and large inclusions of triglycerides measuring up to 6 µm and described as lipid droplets (Diaz et al. 1997b). In addition, the formation of large lipoproteins and lipid droplets is closely related to an excess of fats in enterocytes caused by the high fatty acid contents of diets. This large accumulation of lipids in the enterocytes may cause some pathological damage since large lipid inclusions produce epithelial abrasion, cellular necrosis, and/or inflammatory reactions along the intestinal mucosa (Deplano et al. 1989) that may affect nutrient absorption and reduce digestive efficiency.

The histological organization of the liver accurately reflects any physiological disorder originated from a nutritionally unbalanced diet or feed deprivation episodes since hepatic energy stores respond sensitively to nutritional changes (Table 1.1). Under food deprivation conditions, liver glycogen and lipids are the first energy sources to be mobilized. As reviewed by Gisbert et al. (2008), large central nuclei are observed in livers containing few lipid inclusions, while peripheral nuclei are detected in livers of larvae showing high levels of lipid deposition. Histopathological changes in food-deprived larvae are similar among different species and include changes in liver organization (shrinkage of the nucleolar volume, swollen and deformed mitochondria, dilated sinusoids, large intercellular spaces, vascularization, increase in lysosomes, cytoplasmatic necrosis, and hypertrophy of the bile canaliculi and the gallbladder) and a decrease in glycogen and lipid deposits stored in the hepatocytes. The liver is also a good biomarker for the nutritional effects of different dietary composition and feeding regimes because the hepatic energy stores respond sensitively and rapidly to nutritional changes in fish larvae. In addition, alterations in fatty acid metabolism derived from unbalanced diets have resulted in modifications of nuclei shape and size, chromatin density, and cytoplasmatic lipid deposition in hepatocytes (Caballero et al. 1999; Mobin et al. 2000). Disorders in glycogen and protein synthesis and/or their utilization may also result in an increased level of basophilia in the cytoplasm of the hepatocytes of larvae fed unbalanced diets (Segner et al. 1994; Mobin et al. 2000).

The earlier differentiation and morphogenesis of the exocrine pancreas in comparison with that of the liver or intestine facilitates its use as a histological index for assessing the condition of the larva as soon as it emerges from the egg envelope. Food deprivation induces degeneration of the exocrine pancreas, which may be summarized as a disruption of the acinar symmetry and organization of the pancreas, a reduction in size of secretory cells, and an increase of pycnotic nuclei (Table 1.1).

Catalan (2003) extensively reviewed the use of histological methods in the determination of larval nutritional condition and suggested that this has at least two unresolved limitations. One regards the low objectivity of some methods since the measures are mainly qualitative and rely on the experience of the observer. To date, quantitative data have been restricted to the measurement of cell heights of a few tissues, mainly gut and liver, and have proved useful for early larval stages of some species. However, some of these measurements are only obtainable from species with an elongated digestive duct, or have been restricted to particular larval stages. The second main problem with histological indices (extendable to any condition index) is the large dependence of condition on the experimental rearing parameters, with subsequent poor applicability to field studies. Until further evidence is supplied, there is a need to establish a relationship between survival and each condition measurement under laboratory conditions.

1.6.2 Digestive enzymes

Due to their essential role in metabolic reactions, enzymes can be good indicators for the condition of an organism. For fish larvae, the activity level of digestive enzymes is well suited as a biochemical indicator of the feeding activity. In addition, digestive enzymes are considered to be reliable indicators of the nutritional state of the individuals due to their species and age specificity, sensitivity, and short latency. Different digestive enzymes are used for this purpose, ranging from proteolytic pancreatic enzymes (Ueberschär and Clemmesen 1992; Lamarre et al. 2004; Cara et al. 2007) to intestinal brush border and cytosolic enzymes (Zambonino-Infante and Cahu 2007; Zambonino-Infante et al. 2008).

Pancreatic enzyme synthesis and secretion appear to be particularly sensitive to food deprivation and dietary composition in teleost larvae, and consequently, the pancreatic enzyme activity provides a reliable biochemical marker of larval fish development and condition (Zambonino-Infante and Cahu 2001). The pancreatic secretory process matures during the first 3 or 4 weeks after hatching in temperate marine fish larvae. This maturational process can be disrupted when larvae are fed diets that do not meet their specific needs (Cahu and Zambonino-Infante 1994): The earlier the feeding with such inadequate diets, the lower the pancreatic secretion level. On the other hand, some dietary components, like free amino acids (Zambonino-Infante and Cahu 1994a, 1994b) or some nonbiodegradable particles (Pedersen and Andersen 1992), can enhance pancreatic secretion, revealing the coexistence of chemical and neural mechanisms controlling secretion in larvae. Because protein is one of the major components of the fish larval diet, the activity levels of pancreatic proteolytic enzymes, for example, trypsin and chymotrypsin, are well suited as indicators of the nutritional condition of the organism. Secretion rate of pancreatic enzymes is related to feed intake, the stomach filling, and nutrient composition (Rønnestad and Morais 2007); thus, starvation, reduced feed intake, or an unbalanced diet in terms of free amino acids or protein content may result in a decrease in secretion and, consequently, activity of trypsin and chymotrypsin (Pedersen et al. 1987; Ueberschär 1995; Applebaum et al. 2001; Cara et al. 2007). In addition, some authors have suggested using the trypsin/chymotrypsin ratio as a better indicator of the larval nutritional condition since it might indicate to what extent chymotrypsin is activated by trypsin, and this in turn may indicate the growth potential of the fish (Cara et al. 2007). The higher the trypsin/chymotrypsin ratio, the higher the absorption rate of essential amino acids for protein synthesis and growth potential.

The morphoanatomical development and maturation of the intestine is characterized by a decrease in activity of the cytosolic enzyme activity of leucine-alanine peptidase, which is accompanied by an increase in activity of the brush border enzymes from the enterocytes. This maturation process is known to be nutrient sensitive; consequently, disparity between diet composition and larvae digestive features may delay or prevent the genetically programmed sequence of intestinal development (Zambonino-Infante and Cahu 2001). In this sense, intestinal maturation is often assessed by the alkaline phosphatase/leucine-alanine peptidase or aminopeptidase/leucine-alanine peptidase ratios (Zambonino-Infante and Cahu 1994a). These can be considered as nutritional condition indices for evaluating the switch from a primary or early to an adult mode of digestion. In any case, independent of the digestive enzyme activity considered, reference values for each species, developmental stage, and nutritional condition need to be standardized under laboratory-controlled conditions since the development of the digestive function varies among species, as do their basal levels of digestive enzyme activities, and it may turn out that some enzymes may be more informative than others.

1.6.3 Gene expression

Results of digestive enzyme gene expression analyses from recent studies on fish larvae (Darias 2005; Geurden et al. 2007; Sánchez-Amaya et al. 2009) suggest the possibility of including the molecular level as the fourth organization category (organism, tissue, cellular, and molecular) in the list of markers for nutritional conditions in fish. Knowledge of gene expression amount and pattern of digestive enzyme precursors constitutes a valuable tool that complements the information about the nutritional condition of an organism obtained through enzymatic indicators. This is particularly interesting in aquaculture, where nutritional requirements for fish larvae need to be optimized and the origin of the suboptimal larval growth and performance derived from food supply is often unknown. In this sense, the study of the molecular mechanisms underlying digestive system ontogeny and digestion would expand knowledge of larval physiology and facilitate finding solutions to nutritional problems by localizing the molecular pathways that have been disrupted. However, since gene expression does not always necessarily culminate in protein synthesis, both molecular and cellular indicators should be considered in order to obtain more comprehensive information about the physiological status of fish larvae.

The ontogeny of digestive enzyme gene expression is genetically programmed and their expression patterns are stage specific. Therefore, genes coding for digestive enzymes could be used as markers for fish larval development. For instance, the development of pepsinogen gene expression reveals the attainment of complete functionality of the gastric glands, hence constituting a suitable indicator of the transition from larval to juvenile stage (Segner et al. 1994; Darias et al. 2005). Besides, the nutritional condition of fish larvae could be reflected in the gene expression patterns of some digestive enzymes during ontogenesis. The simplest example is provided by differences in the amount of transcripts (i.e., amylase) found in starved larvae compared with fed ones as a result of triggered physiological mechanisms necessary to adapt the energetic balance to the different nutritional status (Darias 2005; Sánchez-Amaya et al. 2009) . Furthermore, digestive enzyme gene expression can be modulated depending on diet composition, at least during late larval stages. For instance, dietary protein amount and nature modulates trypsin mRNA transcription and translation in European sea bass larvae (Péres et al. 1996). Wang et al. (2006) also found that dietary protein level significantly affects trypsin mRNA level in yellow catfish (*Pelteobagrus fulvidraco*) larvae. Digestive enzyme gene expression can be modulated even during early larval development. Geurden et al. (2007) reported higher levels of α-amylase, maltase, and glucokinase gene expression during the yolk sac period of rainbow trout (*Oncorhynchus mykiss*) fed a hyperglucidic diet compared with a commercial diet. This indicates a very quick adaptation of this carnivorous species to the utilization of exogenous glucose and therefore could be suitable indicators of larval nutritional condition.

1.6.4 Indirect methods for assessing nutritional condition

The nutritional condition of a fish larva can also be indirectly determined. It is well known that nutrients can influence not only digestive system development, and hence survival and growth, but also skeletogenesis (Cahu et al. 2003; Lall and Lewis-McCrea 2007). Recent studies have demonstrated that the degree of fish larval ossification is influenced by diet (see Chapter 7) and is an adequate indicator of larval quality. The ossification status has been shown to be correlated with osteocalcin gene expression (Mazurais et al. 2008; Darias et al. 2010a). This gene is specifically localized in bone and constitutes the most specific marker for bone mineralization (Lian and Stein 1995). Moreover, its expression level can be correlated with dietary levels of several nutrients, thus providing a suitable molecular marker for larval nutritional condition (Darias et al. 2010b).

From nutritional studies using molecular approaches (Villeneuve et al. 2006; Mazurais et al. 2008, 2009; Darias et al. 2010b), other genes emerge as suitable markers for larval quality. For instance, transient receptor potential cation channel, subfamily V, member 6 (TRPV6) expression, which codes for the most important intestinal Ca^{2+} transporter, can be modulated by dietary vitamin D_3 levels, consequently affecting intestinal maturation and therefore larval development (Darias et al. 2010b). Low levels of vitamin mix have been shown to induce skeletal malformations correlated with the modulation of genes involved in osteoblast determination and differentiation such as bone morphogenetic protein 4 (BMP4), insulin growth factor 1 (IGF1), and peroxisome proliferator–activated receptor γ (PPARγ) (Mazurais et al. 2008). Similarly, inadequate dietary retinol levels alter morphogenesis through the modulation of homeobox protein Hox-D9 (Hoxd9) and retinoic acid receptor γ (RARγ) gene expression, provoking a variety of skeletal deformities (Villeneuve et al. 2006; Mazurais et al. 2009).

Genomic research technologies such as microarrays appear to be useful tools not only for studying mechanisms to explain phenotypes but also for exploratory interest, which is useful in the search for markers. With the application of the recent advances in genomics research, studies of larval fish nutrition will advance rapidly, improving our capabilities to assess the nutritional status of fish larvae under different nutritional and rearing conditions. Such resources will contribute to the ultimate goal of understanding digestive capabilities during ontogeny in fish larvae that can lead to successful weaning to microdiets.

Literature cited

Albrecht, M.P., Ferrera, M.F.N., and Caramaschi, E.P. 2001. Anatomical features and histology of the digestive tract of two related neotropical omnivorous fishes (Characiformes; Anostomidae). *Journal of Fish Biology* 58(3):419–430.

Allen, P.A., Cech, J.J. Jr., and Kültz, D. 2009. Mechanisms of seawater acclimation in a primitive, anadromous fish, the green sturgeon. *Journal of Comparative Physiology Part B* 179(7):903–920.

Alliot, E., Pastoureaud, A., and Trellu, J. 1980. Evolution des activities enzymatiques dans le tractus digestif au cour de la vie larvaire de la sole. Variations des proteinogrammes et des zymogrammes. *Biochemical Systematics and Ecology* 8:441–445.

Alvarez-González, A., Márquez-Couturier, G., Arias-Rodriguez, L., et al. 2008. Advances in the digestive physiology and nutrition of bay snook *Petenia splendida*. In: Cruz, E.L., Ricque, D., Tapia, M., et al. (eds.) *Avances Em Nutrición Acuícola IX*. Universidad Autónoma de Nuevo León, Monterrey, Mexico, pp. 135–235.

de Amorim, M.P., Campos Gomes, B.V., Martins, Y.S., et al. 2009. Early development of the silver catfish *Rhamdia quelen* (Quoy & Gaimard, 1824) (Pisces: Heptapteridae) from the São Francisco River Basin, Brazil. *Aquaculture Research* 40(2):172–180.

Applebaum, S.L., Perez, R., Lazo, J.P., et al. 2001. Characterization of chymotrypsin activity during early ontogeny of larval red drum (*Sciaenops ocellatus*). *Fish Physiology and Biochemistry* 25:291–300.

Arellano, J., Dinis, M.T., and Sarasquete, C. 1999. Histomorphological and histochemical characteristics of the intestine of the Senegal sole, *Solea senegalensis*. *European Journal of Histochemistry* 43:121–133.

Arellano, J., Storch, V., and Sarasquete, C. 2001. A histological and histochemical study of the oesophagus and oesogaster of the Senegal sole, *Solea senegalensis*. *European Journal of Histochemistry* 45(3):279–294.

Atencio García, V.J., Hernández-Muñoz, J., and Pardo-Carrasco, S.C. 2007. Alimentary tract of juvenile Rubio *Salminus affinis* (Pisces: Characidae) morphological description. *Acta Biológica Colombiana* 13(3):99–112.

Baglole, C.J., Murray, H.M., Goff, G.P., et al. 1997. Ontogeny of the digestive tract during larval development of yellowtail flounder: a light microscopic and mucous histochemical study. *Journal of Fish Biology* 51(1):120–134.

Baglole, C.L., Goff, G.P., and Wright, G.M. 1998. Distribution and ontogeny of digestive enzymes in larval yellowtail and winter flounder. *Journal of Fish Biology* 53:767–784.

Bal, H.S., and Ghoshal, N.G. 1992. Electron microscopy of the oxynticopeptic cells of the gastric glands and the intestinal glands of the caecum of the guinea pig. *Laboratory Animals* 26:7–52.

Balon, E.K. 1985. Early ontogeny of *Labeotropheus* Ahl, 1927 (Mbuna, Cichlidae, Lake Malawi), with a discussion on advanced protective styles in fish reproduction and development. In: Balon, E.K. (ed.) *The Early Life Histories of Fishes: New Developmental, Ecological and Evolutionary Perspectives*. Dr. W. Junk Publishers, Dordrecht, pp. 207–236.

Baragi, V., and Lovell, R. 1986. Digestive enzyme activities in striped bass from first feeding through larval development. *Transactions of the American Fisheries Society* 115:478–484.

Bardocz, S., Grant, G., Brown, D.S., et al. 1993. Polyamines in food—implications for growth and health. *The Journal of Nutritional Biochemistry* 4:66–71.

Beccaria, C., Diaz, J.P., Connes, R., et al. 1991. Organogenesis of the exocrine pancreas in the sea bass, *Dicentrarchus labrax* L., reared extensively and intensively. *Aquaculture* 99(3–4):339–354.

Bisbal, G.A., and Bengtson, D.A. 1995. Development of the digestive tract in larval summer flounder. *Journal of Fish Biology* 47:277–291.

Boulhic, M., and Gabaudan, J. 1992. Histological study of the organogenesis of the digestive system and swim bladder of the Dover sole, *Solea solea* (Linnaeus 1758). *Aquaculture* 102(4):373–396.

Buddington, R.K., and Diamond, J.M. 1987. Pyloric ceca of fish: a "new" absorptive organ. *American Journal of Physiology. Gastrointestinal and Liver Physiology* 252:65–76.

Buddington, R.K., and Diamond, J.M. 1989. Ontogenic development of intestinal nutrients transporters. *Annual Review of Physiology* 51:601–619.

Buddington, R.K., and Doroshov, S.I. 1986. Structural and functional relations of the white sturgeon alimentary canal *Acipenser transmontanus*. *Journal of Morphology* 190(2):201–213.

Caballero, M.J., López-Calero, G., Socorro, J., et al. 1999. Combined effect of lipid level and fish meal quality on liver histology of gilthead seabream (*Sparus aurata*). *Aquaculture* 179(1–4):277–290.

Cahu, C.L., and Zambonino-Infante, J.L. 1994. Early weaning of sea bass (*Dicentrarchus labrax*) larvae with a compound diet: effect on digestive enzymes. *Comparative Biochemistry and Physiology Part A* 109(2):213–222.

Cahu, C.L., and Zambonino-Infante, J.L. 1995. Effect of the molecular form of dietary nitrogen supply in sea bass larvae: response of pancreatic enzymes and intestinal peptidases. *Fish Physiological and Biochemical* 14:209–214.

Cahu, C.L., and Zambonino-Infante, J.L. 1997. Is the digestive capacity of marine fish larvae sufficient for compound diet feeding? *Aquaculture International* 5:151–160.

Cahu, C.L., and Zambonino-Infante, J.L. 2001. Substitution of live food by formulated diets in marine fish larvae. *Aquaculture* 200:161–180.

Cahu, C.L., Zambonino-Infante, J.L., and Barbosa, V. 2003. Effect of dietary phospholipid level and phospholipid/neutral lipid ratio on development of sea bass (*Dicentrarchus labrax*) fed compound diet. *British Journal of Nutrition* 90(1):21–28.

Cahu, C., Rønnestad, I., Grangier, V., et al. 2004. Expression and activities of pancreatic enzymes in developing sea bass larvae (*Dicentrarchus labrax*) in relation to intact and hydrolyzed dietary protein, involvement of cholecystokinin. *Aquaculture* 238:295–308.

Cara, J.B., Moyano, F.J., Cárdenas, S., et al. 2003. Assessment of digestive enzyme activities during larval development of white bream. *Journal of Fish Biology* 63:48–58.

Cara, B., Moyano, F.J., Zambonino-Infante, J.L., et al. 2007. Trypsin and chymotrypsin as indicators of nutritional status of post-weaned sea bass larvae. *Journal of Fish Biology* 70(6): 1798–1808.

Catalan, I.A. 2003. Condition indices and their relationship with environmental factors in fish larvae. PhD thesis, University of Barcelona.

Catalan, I.A., and Olivar, M.P. 2002. Quantification of muscle condition using digital image analysis in *Dicentrarchus labrax* larvae, and relationship with survival. *Journal of Marine Biological Association of the United Kingdom* 82(4): 649–654.

Chen, Q., Wternby, B., Åkesson, B., et al. 1990. Effects of human pancreatic lipasecolipase and carboxyl ester lipase on eicosapentanoic acid and arachidonic acid ester bounds of triacyglycerols rich in fish oil fatty acids. *Biochimica and Biophysica Acta* 1044:11–117.

Chey, W.Y. 1993. Hormonal control of pancreatic exocrine secretion. In: Go, V.L.W., Gardner, J.D., Brooks, F.P., et al. (eds.) *The Pancreas: Biology, Pathology and Disease*. Raven Press, New York, pp. 403–424.

Cousin, C.B., and Baudin-Laurencin, F. 1985. Morphogénèse de l'appareil digestif de la vessie gazeuse du turbot, *Scophthalmus maximus* L. *Aquaculture* 47(4):305–319.

Cousin, J.C.B., Baudin-Laurencin, F., and Gabaudan, J. 1987. Ontogeny of enzymatic enzyme activities in fed and fasting turbot, *Scophthalmus maximus* L. *Journal of Fish Biology* 30:15–33.

Dabrowski, K. 1979. The role of proteolitic enzymes in fish digestion. In: Styczunska-Jurewivcsk, E., Jaspers, T., and Persoone, E. (eds.) *Cultivation of Fish Fry and Its Live Food*, Vol. 4. European Mariculture Society, Belgium, pp. 107–126.

Dabrowski, K. 1984. The feeding of fish larvae: present "state of the art" and perspectives. *Reproduction Nutrition Development* 24(6): 807–833.

Darias, M.J. 2005. Balance energético y ontogenia del aparato digestivo durante el desarrollo larvario del pargo, *Pagrus pagrus* y del sargo,

Diplodus sargus, en cultivo. PhD thesis, University of Cádiz, Spain.

Darias, M.J., Murray, H.M., Gallant, J.W., et al. 2005. Gene expression of pepsinogen during the larval development of red porgy (*Pagrus pagrus*). *Aquaculture* 248:245–252.

Darias, M.J., Murray, H.M., Gallant, J.W., et al. 2006. Characterization of a partial α-amylase clone from red porgy (*Pagrus pagrus*) and its expression during the larval development. *Comparative Biochemistry and Physiology Part B* 143:209–218.

Darias, M.J., Ortiz-Delgado, J.B., Sarasquete, C., et al. 2007a. Larval organogenesis of *Pagrus pagrus* L., 1758 with special attention to the digestive system development. *Histology and Histopathology* 22:753–768.

Darias, M.J., Murray, H.M., Gallant, J.W., et al. 2007b. The spatiotemporal expression pattern of trypsinogen and bile salt–activated lipase during the larval development of red porgy (*Pagrus pagrus*, Pisces, Sparidae). *Marine Biology* 152:109–118.

Darias, M.J., Murray, H.M., Gallant, J.W., et al. 2007c. Ontogeny of pepsinogen and proton pump expression in red porgy (*Pagrus pagrus*): determination of stomach functionality. *Aquaculture* 270:369–378.

Darias, M.J., Lan Chow Wing, O., Mazurais, D., et al. 2010a. Alcian blue–alizarin red double staining technique for developing sea bass (*Dicentrachus labrax*) larvae. *Journal of Applied Ichthyology* 26:280–285.

Darias, M.J., Mazurais, D., Koumoundouros, G., et al. 2010b. Dietary vitamin D3 affects digestive system ontogenesis and ossification in European sea bass (*Dicentrachus labrax*, Linnaeus, 1758). *Aquaculture* 298:300–307.

Deplano, M., Connes, R., Díaz, J.P., et al. 1989. Intestinal steatosis in the farm-reared sea bass *Dicentrarchus labrax* L. *Diseases of Aquatic Organisms* 6:121–130.

Deplano, M., Diaz, J.P., Connes, R., et al. 1991. Appearance of lipid-absorption capacities of the sea bass *Dicentrarchus labrax* during transition to the exotrophic phase. *Marine Biology* 108(3):361–371.

Dettlaff, T.A., Ginsburg, A.S., and Schmalhausen, O.I. 1993. *Sturgeon Fishes. Developmental Biology and Aquaculture.* Springer-Verlag, Berlin.

Diaz, M., Moyano, F.J., Garcia-Carreno, F.L., et al. 1997a. Substrate-SDS-PAGE determination of protease activity through larval development in sea bream. *Aquaculture International* 5:461–471.

Diaz, J.P., Guyot, E., Vigier, S.M., et al. 1997b. First events in lipid absorption during post-embryonic development of the anterior intestine in gilt-head sea bream. *Journal of Fish Biology* 51(1):180–192.

Diaz, J.P., Mani-Ponset, L., Blasco, C., et al. 2002. Cytological detection of the main phases of lipid metabolism during early post-embryonic development in three teleost species: *Dicentrarchus labrax*, *Sparus aurata* and *Stizostedion lucioperca*. *Aquatic Living Resources* 15:169–178.

Domeneghini, C., Pannelli Straini, R., and Veggetti, A. 1998. Gut glycoconjugates in *Sparus aurata* L. (Pisces, Teleostei). A comparative histochemical study in larval and adult ages. *Histology and Histopathology* 13(2):359–372.

Dortch, Q. 1987. The biochemical composition of plankton in a subsurface chlorophyll maximum. *Deep-Sea Research* 34:705–712.

Douglas, S.E., and Gallant, J.W. 1998. Isolation of cDNAs for trypsinogen from the winter flounder, *Pleuronectes americanus*. *Journal of Marine Biotechnology* 6:214–219.

Douglas, S.E., Gawlika, A., Mandla, S., et al. 1999. Ontogeny of the stomach of winter flounder: characterization and expression of the pepsinogen and proton pump genes and determination of pepsin activity. *Journal of Fish Biology* 55:897–915.

Douglas, S.E., Mandla, S., and Gallant, J.W. 2000. Molecular analysis of the amylase gene and its expression during development in the winter flounder, *Pleuronectes americanus*. *Aquaculture* 190:247–260.

Drewe, K.E., Horn, M.H., Dickson, K.A., et al. 2004. Insectivore to frugivore: ontogenetic changes in gut morphology and digestive enzyme activity in the characid fish *Brycon guatemalensis* from Costa Rican rain forest streams. *Journal of Fish Biology* 64(4): 890–902.

Elbal, M.T., and Agulleiro, B. 1986. A histochemical and ultrastructural study of the gut of *Sparus aurata* (Teleostei). *Journal of Submicroscopy and Cytology* 18:335–347.

Elbal, M.T., García-Hernández, M.P., Lozano, M.T., et al. 2004. Development of the digestive tract of gilthead sea bream (*Sparus aurata* L.). Light and electron microscopic studies. *Aquaculture* 234:215–238.

Fange, R., and Grove, D. 1979. Digestion. In: Hoar, W.S., Randall, D.J., and Brett, J.R. (eds.) *Fish Physiology*, Vol. VIII. Academic Press, New York, pp. 353–405.

Faulk, C.K., Benninghoff, A.D., and Holt, G.J. 2007. Gut morphology and function in *Atherinops affinis* (Teleostei: Atherinopsidae), a stomachless omnivore feeding on macroalgae. *Journal of Fish Biology* 70(2):567–583.

Ferron, A., and Leggett, W.C. 1994. An appraisal of condition measures for marine fish larvae. *Advances in Marine Biology* 30:217–303.

Fioramonti, J., Fargeas, M.J., Bertrand, V., et al. 1994. Induction of postprandial intestinal motility and release of cholecystokinin by polyamines in rats. *American Journal of Physiology* 267:G960–G965.

Fishelson, L. 1995. Ontogenesis of cytological structures around the yolk sac during embryologic and early larval development of some cichlid fishes. *Journal of Fish Biology* 47(3):479–491.

Fyhn, H.J. 1993. Multiple functions of free amino acids during embryogenesis in marine fishes. In: Walther, B.T., and Jorgen-Fyhn, H. (eds.) *Physiological and Biochemical Aspects of Fish Development*. University of Bergen, Norway, pp. 299–308.

García-Hernández, M.P., Lozano, M.T., Elbal, M.T., et al. 2001. Development of the digestive tract of sea bass (*Dicentrarchus labrax* L). Light and electron microscopic studies. *Anatomy and Embryology* 204(1):39–57.

García-Jimenez, P., Rodrigo, M., and Robaina, R.R. 1998. Influence of plant growth regulators polyamines and glycerol interaction on growth and morphogenesis of carposporelings of *Grateloupia* cultured *in vitro*. *Journal of Applied Phycology* 10:95–100.

Gargiulo, A.M., Ceccarelli, C.P., Dall'aglio, C., et al. 1997. Ultrastructural study on the stomach of *Tilapia* spp. (Teleostei). *Anatomia Histologia Embryologia* 26(4):331–336.

Gawlicka, A., Teh, S.J., Hung, S.S.O., et al. 1995. Histological and histochemical changes in the digestive tract of white sturgeon larvae during ontogeny. *Fish Physiology and Biochemistry* 14(5):357–371.

Gawlicka, A., Leggiadro, C.T., Gallant, J.W., et al. 2001. Cellular expression of the pepsinogen and gastric proton pump genes in the stomach of winter flounder as determined by in situ hybridization. *Journal of Fish Biology* 58:529–536.

Geurden, I., Aramendi, M., Zambonino-Infante, J.L., et al. 2007. Early feeding of carnivorous rainbow trout (*Oncorhynchus mykiss*) with a hyperglucidic diet during a short period: effect on dietary glucose utilization in juveniles. *American Journal of Physiology Regulatory Integrative and Comparative Physiology* 292:R2275–R2283.

Gilloteaux, J., Oldham, C.K., and Biaginati-Risbourg, S. 1996. Ultrastructural diversity of the biliary tract and the gallbladder in fish. In: Datta Munshi, J.S., and Dutta, H.M. (eds.) *Fish Morphology: Horizon of New Research*. Science Publishers Inc., Lebanon, NH, pp. 95–110.

Gisbert, E., and Doroshov, S.I. 2003. Histology of the developing digestive system and the effect of food deprivation in larval green sturgeon (*Acipenser medirostris*). *Aquatic Living Resources* 16(2):77–89.

Gisbert, E., Rodríguez, A., Williot, P., et al. 1998. A histological study of the development of the digestive tract of Siberian sturgeon (*Acipenser baeri*) during early ontogeny. *Aquaculture* 167(3–4):195–209.

Gisbert, E., Sarasquete, M.C., Williot, P., et al. 1999. Histochemistry of the development of the digestive system of Siberian sturgeon (*Acipenser baeri*, Brandt) during early ontogeny. *Journal of Fish Biology* 55(3):596–616.

Gisbert, E., Piedrahita, R.H., and Conklin, D.E. 2004a. Ontogenetic development of the digestive system in California halibut (*Paralichthys californicus*) with notes on feeding practices. *Aquaculture* 232(1–4):455–470.

Gisbert, E., Piedrahita, R.H., and Conklin, D.E. 2004b. Effects of delayed first feeding on the nutritional condition and mortality of California halibut larvae. *Journal of Fish Biology* 64(1):116–132.

Gisbert, E., Villeneuve, L., Zambonino-Infante, J.L., et al. 2005. Dietary phospholipids are more efficient than neutral lipids for long chain polyunsaturated fatty acid supply in European sea bass *Dicentrarchus labrax* larval development. *Lipids* 40(6):609–618.

Gisbert, E., Ortiz-Delgado, J.B., Sarasquete, C. 2008. Nutritional cellular biomarkers in early life stages of fish. *Histology and Histopathology* 23:1525–1539.

Gjellesvik, D.R., Lombardo, D., and Walter, B.T. 1992. Pancreatic bile salt dependent lipase from cod (*Gadus morhua*): purification and properties. *Biochimica and Biophysica Acta* 1124: 123–134.

Govoni, J.J., Boehlert, G.W., and Watanabe, Y. 1986. The physiology of digestion in fish larvae. *Environmental Biology of Fishes* 16(1–3):59–77.

Grau, A., Crespo, S., Sarasquete, C., et al. 1992. The digestive tract of the amberjack *Seriola dumerili*, Risso: a light and scanning electron microscopy study. *Journal of Fish Biology* 41(2):287–303.

Green, B.S., and McCormick, M.I. 2001. Ontogeny of the digestive and feeding systems in the anemonefish *Amphiprion melanopus*. *Environmental Biology of Fishes* 61(1):73–83.

Guillaume, J., Métailler, R., Kaushik, S., et al. 2001. *Nutrition and Feeding of Fish and Crustaceans*. Praxis Publishing Ltd., Chichester, UK.

Hachero-Cruzado, I., Ortiz-Delgado, J.B., Borrega, B., et al. 2009. Larval organogenesis of flatfish brill *Scophthalmus rhombus* L: histological and histochemical aspects. *Aquaculture* 286(1–2):138–149.

Hamlin, H.J., Hunt Von Herbing, I., and King, L.J. 2000. Histological and morphological evaluations of the digestive tract and associated organs of haddock throughout post-hatching ontogeny. *Journal of Fish Biology* 57(3):716–732.

Helander, H.F. 1981. The cells of the gastric mucosa. *International Review of Cytology* 70:217–289.

Hellberg, H., and Bjerkås, I. 2005. Intestinal epithelium in *Anarhichas lupus* L., with emphasis on cell renewal. *Journal of Fish Biology* 66(5):1342–1356.

Hernández, D.R., Gianeselli, M.P., and Domitrovic, H.A. 2009. Morphology, histology and histochemistry of the digestive system of South American catfish (*Rhamdia quelen*). *International Journal of Morphology* 27(1): 105–111.

Hjelmeland, K., Pedersen, B.H., and Nilssen, E.M. 1988. Trypsin content in intestines of herring larvae, *Clupea harengus*, ingesting inert polystyrene spheres or live crustacean prey. *Marine Biology* 98:331–335.

Hoehne-Reitan, K., and Kjørsvik, E. 2004. Functional development of the liver and exocrine pancreas in teleost fish. In: Govoni, J.J. (ed.) *The Development of Form and Function in Fishes and the Question of Larval Adaptation*. American Fisheries Society, Symposium 40. American Fisheries Society, Bethseda, MD, pp. 9–36.

Hoehne-Reitan, K., Kjørsvik, E., and Gjellesvik, D.R. 2001a. Development of bile salt–dependent lipase in larval turbot. *Journal of Fish Biology* 58:737–745.

Hoehne-Reitan, K., Kjørsvik, E., and Reitan, K.I. 2001b. Bile salt–dependent lipase in larval turbot, as influenced by density and lipid content of fed prey. *Journal of Fish Biology* 58:746–754.

Holt, G.J. 1993. Feeding larval red drum on microparticulate diets in closed recirculating water system. *Journal of the World Aquaculture Society* 42:225–240.

Hossain, A.M., and Dutta, H.M. 1996. Assessment of structural and functional similarities and differences between caeca of the bluegill. *Journal of Fish Biology* 53(6):1317–1323.

Hrubec, T.C., and Caceci, T. 2001. Rodlet cells in the gall bladder of the black molly. In: Dezfuli, B.S., Manera, M., and Leiro, R. (eds.) *First International Rodlet Cell Workshop*. University of Ferrara, Ferrara, Italy, p. 9.

Huang, L.Y., Schreiber, A.M., Soffientino, B., et al. 1998. Metamorphosis of summer flounder (*Paralichthys dentatus*): thyroid status and the timing of gastric gland formation. *Journal of Experimental Zoology* 280(6):413–420.

Iijima, N., Tanaka, S., and Ota, Y. 1998. Purification and characterization of bile salt–activated lipase from the hepatopancreas of red sea bream, *Pagrus major*. *Fish Physiology and Biochemistry* 18:59–56.

Izquierdo, M., Socorro, J., Arantzamendi, L., et al. 2000. Recent advances in lipid nutrition in fish larvae. *Fish Physiology and Biochemistry* 22:97–107.

Kamisaka, Y., Fujii, Y., Yamamoto, S., et al. 2003. Distribution of cholecystokinin-immunoreactive cells in the digestive tract of the larval teleost, ayu, *Plecoglossus altivelis*. *General and Comparative Endocrinology* 134(2):116–121.

Kapoor, B.G., Smith, H., and Verighina, I.A. 1975. The alimentary channel and digestion in teleosts. *Advances in Marine Biology* 63: 301–308.

Kato, K., Ishimaru, K., Sawada, Y., et al. 2004. Ontogeny of digestive and immune system organs of larval and juvenile kelp grouper *Epinephelus bruneus* reared in the laboratory. *Fisheries Science* 70(6):1061–1069.

Kolkovski, S. 2001. Digestive enzymes in fish larvae and juveniles—implications and applications to formulated diets. *Aquaculture* 200: 181–201.

Kolkovski, S., Arieli, A., and Tandler, A. 1997. Visual and chemical cues stimulate microdiet ingestion in seabream larvae. *Aquaculture International* 5:527–536.

Koven, W., Rojas-García, C.R., Finn, R.N., et al. 2002. The stimulatory effect of ingested protein and/or free amino acids on the secretion of the gastro-endocrine hormone, cholecystokinin (CCK) and the protease, trypsin, in first feeding herring larvae, *Clupea harengus*. *Marine Biology* 140:1241–1247.

Kozarić, Z., Kužir, S., Petrinec, Z., et al. 2008. The development of the digestive tract in larval European catfish (*Silurus glanis* L.). *Anatomia Histologia Embryologia* 37(2):141–146.

Kramer, C.R., Kramer, A.J., and Konovalov, A. 2005. Rodlet cell distribution in the gall bladder epithelium of *Fundulus heteroclitus*. *Journal of Fish Biology* 67(2):555–560.

Kurokawa, T., and Suzuki, T. 1996. Formation of the diffuse pancreas and the development of digestive enzyme synthesis in larvae of the Japanese flounder *Paralichthys olivaceus*. *Aquaculture* 114:267–276.

Kurokawa, T., Shiraishi, M., and Suzuki, T. 1998. Quantification of exogenous protease derived from zooplankton in the intestine of Japanese sardine (*Sardinops melanotictus*) larvae. *Aquaculture* 161:491–499.

Kurokawa, T., Suzuki, T., and Andoh, T. 2000. Development of cholecystokinin and pancreatic polypeptide endocrine systems during the larval stage of Japanese flounder, *Paralichthys olivaceus*. *General Comparative Endocrinology* 120(1):8–16.

Kurokawa, T., Uji, S., and Suzuki, T. 2005. Identification of pepsinogen gene in the genome of stomachless fish (*Takifugu rubripes*). *Comparative Biochemistry and Physiology* 140B:133–140.

Lall, S.P., and Lewis-McCrea, L. 2007. Role of nutrients in skeletal metabolism and pathology in fish, an overview. *Aquaculture* 267:3–19.

Lamarre, S.G., Le François, N.R., Falk-Petersen, I., et al. 2004. Can digestive and metabolic enzyme activity levels predict growth rate and survival of newly hatched Atlantic wolffish (*Anarhichas lupus* Olafsen)? *Aquaculture Research* 35(6):608–613.

Lauf, M., and Hoffer, R. 1984. Proteolitic enzymes in fish development and the importance of dietary enzymes. *Aquaculture* 37:335–346.

Lazo, J.P., Holt, G.J., and Arnold, C.R. 2000a. Ontogeny of pancreatic enzymes in larval red drum (*Sciaenops ocellatus*). *Aquaculture Nutrition* 6:183–192.

Lazo, J.P., Dinis, M.T., Holt, G.J., et al. 2000b. Co-feeding microparticulate diets with algae: toward eliminating the need of zooplankton at first feeding in larvae red drum (*Sciaenops ocellatus*). *Aquaculture* 188:339–351.

Lazo, J.P., Holt, G.J., and Arnold, C.R. 2002. Towards the development of suitable microdiets for substitution of live prey in the rearing of red drum larvae: applications of studies on the digestive physiology. *Fisheries Science* 68(1):888–891.

Lazo, J.P., Mendoza, R., Holt, G.J., et al. 2007. Characterization of digestive enzymes during larval development of red drum (*Sciaenops ocellatus*). *Aquaculture* 265:194–205.

Lian, J.B., and Stein, G.S. 1995. Development of the osteoblast phenotype, molecular mechanisms mediating osteoblast growth and differentiation. *Iowa Orthopedic Journal* 15: 118–140.

Lingling, W., and Qianru, C. 1981. Observation of the embryonic and larval development of *Tilapia nilotica*. *Acta Zoologica Sinica* 27:327–336.

Loewe, H., and Eckmann, R. 1988. The ontogeny of the alimentary tract of coregonid larvae: normal development. *Journal of Fish Biology* 33(6):841–850.

Luizi, F.S., Gara, B., Shields, R.J., et al. 1999. Further description of the development of the digestive organs in Atlantic halibut (*Hippoglossus hippoglossus*) larvae, with notes on differential absorption of copepod and *Artemia* prey. *Aquaculture* 176:101–116.

Ma, P., Sivaloganathan, B., Reddy, K.P., et al. 2001. Ontogeny of alpha amylase gene expression in sea bass larvae (*Lates calcarifer*). *Marine Biotechnology* 3:463–469.

Ma, P., Liu, Y., Reddy, K.P., et al. 2004. Characterization of the seabass pancreatic α-amylase gene and promoter. *Genetic and Comparative Endocrinology* 137:78–88.

Mähr, K., Grabner, M., Hofer, R., et al. 1983. Histological and physiological development of the stomach in *Coregonus* sp. *Archiv für Hydrobiologie* 98(2):344–353.

Mai, K., Yu, H., Ma, H., et al. 2005. A histological study on the development of the digestive system of *Pseudosciaena crocea* larvae and juveniles. *Journal of Fish Biology* 67(4):1094–1106.

Manera, M., and Dezfuli, B.S. 2004. Rodlet cells in teleosts: a new insight into their nature and functions. *Journal of Fish Biology* 65(3):597–619.

Margulies, D. 1993. Assessment of the nutritional condition of larval and early juvenile tuna and Spanish mackerel (Pisces: Scombridae) in the Panamá Bight. *Marine Biology* 115(2):317–330.

Mathews, C.K., and van Holde, K.E. 1990. *Biochemistry*. The Benjamin/Cummins Publishing Company, Inc., New York.

Mazurais, D., Darias, M.J., Gouillou-Coustans, M.F., et al. 2008. Dietary vitamin mix levels influence the ossification process in European sea bass (*Dicentrarchus labrax*) larvae. *American Journal of Physiology. Regulatory Integrative and Comparative Physiology* 294:R520–R527.

Mazurais, D., Glynatsi, G., Darias, M.J., et al. 2009. Optimal levels of dietary vitamin A for reduced deformity incidence during development of European sea bass larvae (*Dicentrarchus labrax*) depend on malformation type. *Aquaculture* 294:262–270.

Meijide, F.J., and Guerrero, G.A. 2000. Embryonic and larval development of a substrate-brooding cichlid *Cichlasoma dimerus* (Heckel, 1940) under laboratory conditions. *Journal of Zoology* 252(4):481–493.

Micale, V., Garaffo, M., Genovese, L., et al. 2006. The ontogeny of the alimentary tract during larval development in common pandora *Pagellus erythrinus* L. *Aquaculture* 251(2–4):345–365.

Mobin, S.M.A., Kanai, K., and Yoshikoshi, K. 2000. Histopathological alterations in the digestive system of larval and juvenile Japanese flounder *Paralichthys olivaceus* reared on four feeding levels. *Journal of Aquatic Animal Health* 12(3):196–208.

Morais, S., Cahu, C., Zambonino-Infante, J.L., et al. 2004. Dietary TAG source and level affect performance and lipase expression in larval sea bass (*Dicentrarchus labrax*). *Lipids* 39:449–458.

Morrison, C.M. 1993. *Histology of the Atlantic Cod, Gadus morhua: An Atlas. Part Four. Eleutheroembryo and Larva.* Canadian Special Publication of Fisheries and Aquatic Sciences 119. National Research Council of Canada, Ottawa.

Morrison, C.M., Miyake, T., and Wright, J.R. Jr. 2001. Histological study of the development of the embryo and early larva of *Oreochromis niloticus* (Pisces: Cichlidae). *Journal of Morphology* 247(2):172–195.

Moyano, F.J., Díaz, M., Alarcón, F.J., et al. 1996. Characterization of digestive enzyme activity during larval development of gilthead seabream (*Sparus aurata*). *Fish Physiology and Biochemistry* 15:121–130.

Moyle, P., and Cech, J.J. Jr. 2000. *Fishes. An Introduction to Ichthyology*, 4th edition. Prentice Hall, Inc., Upper Saddle River, NJ.

Munilla-Moran, R., Stark, J.R., and Babour, A. 1990. The role of exogenous enzymes in digestion in cultured turbot larvae (*Scopththalmus maximus* L.). *Aquaculture* 88:337–350.

Murray, H.M., Wright, G.M., and Goff, G.P. 1994. A comparative histological and histochemical study of the stomach from three species of Pleuronectid, the Atlantic halibut *Hippoglossus hippoglossus*, the yellowtail flounder, *Pleuronectes ferruginea*, and the winter flounder, *Pleuronectes americanus*. *Canadian Journal of Zoology* 72(6):1199–1210.

Murray, H.M., Douglas, S.E., Gallant, J.W., et al. 2003. Ontogeny of lipase expression in winter flounder, *Pseuodopleuronectes americanus. Journal of Fish Biology* 62:816–833.

Murray, H.M., Pérez-Casanova, J.C., Gallant, J.W., et al. 2004. Trypsinogen expression during the development of the exocrine pancreas in winter flounder (*Pseudopleuronectes americanus*). *Comparative Biochemistry and Physiology. Part A* 138(1):53–59.

Nakagawa, H., Umino, T., Sekimoto, T., et al. 2002. Characterization of the digestive tract of wild ayu. *Fisheries Science* 68(2):341–346.

Olsen, R.E., Myklebust, R., Ringø, E., et al. 2000. The influences of dietary linseed oil and saturated fatty acids on caecal enterocytes in Arctic char (*Salvelinus alpinus* L.): a quantitative ultrastructural study. *Fish Physiology and Biochemistry* 22(2):207–216.

Önal, U., Langdon, C., and Celik, I. 2008. Ontogeny of the digestive tract of larval percula clownfish, *Amphiprion percula* (Lacepede 1802): a histological perspective. *Aquaculture Research* 39(11):1077–1086.

Ortiz-Delgado, J.B., Darias, M.J., Cañavate, J.P., et al. 2003. Organogenesis of the digestive tract in the white seabream, *Diplodus sargus*. Histological and histochemical approaches. *Histology and Histopathology* 18(4):1141–1154.

Osse, J.W.M., and van den Boogart, J.G.M. 2004. Allometric growth in fish larvae: timing and function. In: Govoni, J.J. (ed.) *The Development of Form and Function in Fishes and the Question of Larval Adaptation*. American Fisheries Society, Symposium 40. American Fisheries Society, Bethesda, MD, pp. 167–194.

Ostaszewska, T. 2005. Developmental changes of digestive system structures in pike-perch (*Sander lucioperca* L.). *Electronic Journal of Ichthyology* 2(2):65–78.

Patton, J.S., Warner, T.G., and Benson, A.A. 1977. Partial characterization of the bile-salt-dependent triacylglycerol lipase from the leopard shark pancreas. *Biochimica and Biophysica Acta* 486:322–330.

Pedersen, B.H., and Andersen, K.P. 1992. Induction of trypsinogen secretion in herring larvae (*Clupea harengus*). *Marine Biology* 112:559–565.

Pedersen, B.H., Nilssen, E.M., and Hjelmeland, K. 1987. Variations in the content of trypsin and trypsinogen in larval herring (*Clupea harengus*) digesting copepod nauplii. *Marine Biology* 94(2):171–181.

Peña, R., Dumas, S., Villalejo-Fuerte, M., et al. 2003. Ontogenetic development of the digestive tract in reared spotted sand bass *Paralabrax maculatofasciatus* larvae. *Aquaculture* 219(1-4):633–644.

Péres, A., Cahu, C., Zambonino-Infante, J.L., et al. 1996. Amylase and trypsin responses to intake of dietary carbohydrate and protein depend on the developmental stage in sea bass (*Dicentrarchus labrax*) larvae. *Fish Physiology and Biochemistry* 15(3):237–242.

Peres, A., Cahu, C.L., and Zambonino-Infante, J.L. 1997. Dietary spermine supplementation induces intestinal maturation in sea bass (*Dicentrarchus labrax*) larvae. *Fish Physiology and Biochemistry* 16:479–485.

Pérez-Casanova, J.C., Murray, H.M., Gallant, J.W., et al. 2004. Bile-salt activated lipase expression during larval development in the haddock (*Melanogrammus aeglefinus*). *Aquaculture* 235:601–617.

Petcoff, G.M., Díaz, A.O., Escalante, A.H., et al. 2006. Histology of the liver of *Oligosarcus jenynsii* (Ostariophysi, Characidae) from Los Padres Lake, Argentina. *Iheringia, Série Zoolologia* 96(2):205–208.

Ribeiro, L., Zambonino-Infante, J.L., Cahu, C.L., et al. 1999. Development of digestive enzymes in larvae of *Solea senegalensis*, Kaup 1858. *Aquaculture* 179(3–4):465–473.

Rønnestad, I., and Morais, S. 2007. Digestion. In: Fin, R.N., and Kapoor, B.G. (eds.) *Fish Larval Physiology*. Science Publishers, Enfield, NH, pp. 201–262.

Rønnestad, I., Kamisaka, Y., Conceição, L.E.C., et al. 2007. Digestive physiology of marine fish larvae: hormonal control and processing capacity for proteins, peptides and amino acids. *Aquaculture* 268:82–97.

Rust, M.B. 2002. Nutritional physiology. In: Halver, J.E., and Hardy, R.W. (eds.) *Fish Nutrition*. Academic Press, Amsterdam, pp. 367–452.

Sánchez-Amaya, M.I., Yúfera, M., and Martínez-Rodríguez, G. 2009. Expression of digestive enzyme precursors under different feeding conditions in *Sparus aurata* larvae. In: Hendry, C.I., Van Stappen, G., Wille, M. (eds.) *Larvi '09: Fish and Shellfish Larviculture Symposium*. European Aquaculture Society, Oostende, Belgium, p. 388, abstract 38.

Santamaría Rojas, C.A., Marín de Mateo, M., Traveset, R., et al. 2004. Organogenesis in larval common *Dentex dentex* L. (Sparidae): histological and histochemical aspects. *Aquaculture* 237:207–228.

Sarasquete, C., Polo, A., and Yúfera, M. 1995. Histology and histochemistry of the development of the digestive system of larval gilthead seabream, *Sparus aurata* L. *Aquaculture* 130:79–92.

Sarasquete, C., González de Canales, M.L., Arellano, J.M., et al. 1996. Histochemical aspects of the yolk-sac and digestive tract of larvae of the Senegal sole, *Solea senegalensis* (Kaup, 1858). *Histology and Histopathology* 11:881–888.

Sarasquete, C., Gisbert, E., Ribeiro, L., et al. 2001. Glycoconjugates in epidermal, branchial and digestive mucous cells and gastric glands of gilthead sea bream, *Sparus aurata*, Senegal sole, *Solea senegalensis* and Siberian sturgeon, *Acipenser baeri* development. *European Journal of Histochemistry* 45(3):267–278.

Sarieyyüpoglu, M., Girgin, A., and Köprücü, S. 2000. Histological study in the digestive tract on larval development of rainbow trout (*Oncorhynchus mykiss*, Walbaum, 1792). *Turkish Journal of Zoology* 24(2):199–205.

Scocco, P., Accili, D., Menghi, G., et al. 1998. Unusual glycoconjugates in the oesophagus of a tilapine polyhybrid. *Journal of Fish Biology* 53(1):39–48.

Segner, H., Rosch, R., Verreth, J., et al. 1993. Larval nutritional physiology: studies with *Clarias gariepinus*, *Coregonus lavaretus* and *Scophthalmus maximus*. *Journal of the World Aquaculture Society* 24(2):121–134.

Segner, H., Storch, V., Reinecke, M., et al. 1994. The development of functional digestive and metabolic organs in turbot, *Scophthalmus maximus*. *Marine Biology* 119(3):471–486.

Smallwood, W.M., and Smallwood, M.L. 1931. The development of the carp, *Cyprinus carpio*. I. The larval life of the carp, with special reference to the development of the intestinal canal. *Journal of Morphology* 52(1):217–231.

Srivastava, A.S., Kurokawa, T., and Suzuki, T. 2002. mRNA expression of pancreatic enzyme precursors and estimation of protein digestibility in first feeding larvae of the Japanese flounder, *Paralichthys olivaceus*. *Comparative Biochemistry and Physiology Part A* 132: 629–635.

Stevens, C.E., and Hume, I.D. 2005. *Comparative Physiology of the Vertebrate Digestive System.* Cambridge University Press, Cambridge.

Takashima, F., and Hibiya, T. 1995. *An Atlas of Fish Histology. Normal and Pathological Features*, 2nd edition. Gustav Fisher Verlag, Stuttgart.

Tanaka, M. 1973. Studies in the structure and function of the digestive system of teleost larvae. D.Agric. thesis, Kyoto University, Japan.

Tovar-Ramirez, D., Zambonino-Infante, J.L., Cahu, C.L., et al. 2002. Dietary incorporation level of live yeast influences European sea bass (*Dicentrarchus labrax*) development. *Aquaculture* 234:415–427.

Ueberschär, B. 1995. The use of tryptic enzyme activity measurement as a nutritional condition index: laboratory calibration data and field application. *ICES Marine Science Symposia* 201:119–129.

Ueberschär, B., and Clemmesen, C. 1992. A comparison of the nutritional condition of herring larvae as determined by two biochemical methods—tryptic enzyme activity and RNA/DNA ratio measurements. *ICES Journal of Marine Science* 49(2):245–249.

Veggetti, A., Rowlerson, A., Radaelli, G., et al. 1999. Post-hatching development of the gut and lateral muscle in the sole. *Journal of Fish Biology* 55(Suppl. A):44–65.

Verreth, J.A., Torrelle, E., Spazier, E., et al. 1992. The development of a functional digestive system in the African catfish, *Clarias garipinus*. *Journal of World Aquaculture Society* 23(4):286–298.

Villeneuve, L., Gisbert, E., Moriceau, J., et al. 2006. Intake of different levels of vitamin A and polyunsaturated fatty acids during different developmental periods modifies the expression of morphogenesis genes in European sea bass (*Dicentrarchus labrax*). *British Journal of Nutrition* 95:677–687.

Wallace, K.N., Akhter, S., Smith, E.N., et al. 2005. Intestinal growth and differentiation in zebrafish. *Mechanisms of Development* 122(2): 157–173.

Wang, C., Xie, S., Zhu, X., et al. 2006. Effects of age and dietary protein level on digestive enzyme activity and gene expression of

Pelteobagrus fulvidraco larvae. *Aquaculture* 254:554–562.

Yúfera, M., Fernández-Díaz, C., Vidaurreta, A., et al. 2004. Gastrointestinal pH and development of the acid digestion in larvae and early juveniles of *Sparus aurata* (Pisces: Teleostei). *Marine Biology* 144:863–869.

Zambonino-Infante, J.L., and Cahu, C. 1994a. Development and response to a diet change of some digestive enzymes in sea bass (*Dicentrarchus labrax*) larvae. *Fish Physiology and Biochemistry* 12(5):399–408.

Zambonino-Infante, J.L., and Cahu, C.L. 1994b. Influence of diet on pepsin and some pancreatic enzymes in sea bass (*Dicentrarchus labrax*) larvae. *Comparative Biochemistry and Physiology. Part A* 109(2):209–212.

Zambonino-Infante, J.L., and Cahu, C. 1999. High dietary lipid levels enhance digestive tract maturation and improve *Dicentrarchus labrax* larval development. *Journal of Nutrition* 129:1195–1200.

Zambonino-Infante, J.L., and Cahu, C.L. 2001. Ontogeny of the gastrointestinal tract of marine fish larvae. *Comparative Biochemistry and Physiology Part C* 130(4):477–487.

Zambonino-Infante, J.L., and Cahu, C.L. 2007. Dietary modulation of some digestive enzymes and metabolic processes in developing marine fish, applications to diet formulation. *Aquaculture* 268(1–4):98–105.

Zambonino-Infante, J.L., Cahu, C.L., Péres, A., et al. 1996. Sea bass (*Dicentrarchus labrax*) larvae fed different *Artemia* rations: growth, pancreas enzymatic response and development of digestive functions. *Aquaculture* 139:129–138.

Zambonino-Infante, J., Gisbert, E., Sarasquete, C., et al. 2008. Ontogeny and physiology of the digestive system of marine fish larvae. In: Cyrino, J.E.O., Bureau, D., and Kapoor, B.G. (eds.) *Feeding and Digestive Functions of Fish*. Science Publishers Inc, Enfield, NH, pp. 277–344.

Zimmer, G., Reuter, G., Schauer, R. 1992. Use of influenza c-virus for detection of acetylated sialic acids on immobilised conjugates by esterase activity. *European Journal of Biochemistry* 204:209–215.

Chapter 2

Lipids

Marisol Izquierdo and William Koven

2.1 Physiological role of essential fatty acids (EFAs)

In general, the main roles of fatty acids in fish larvae are consistent with those in juveniles and adults. Thus, they function as (1) a source of metabolic energy, (2) structural components in the phospholipids (PLs) of cellular membranes, and (3) precursors of bioactive molecules (Sargent et al. 1999a; Tocher 2003). In particular, long-chain polyunsaturated fatty acids (LCPUFAs) such as docosahexaenoic acid (DHA, 22:6n-3),[1] eicosapentaenoic acid (EPA, 20:5n-3), and arachidonic acid (ARA, 20:4n-6) have important physiological functions that can be broadly divided into two groups: (1) as critical structural components of the membrane PLs that facilitate key intramembrane reactions and processes, where DHA is particularly important; and (2) as precursors for eicosanoids, where ARA is generally recognized as one of the main substrates. Eicosanoids are physiologically active substances; they are local hormones (autacoids) with autocrine and paracrine functions targeting cells in the area where they are formed and involved in the regulation of a wide array of cellular pathways and cascades.

Another 20-carbon fatty acid, EPA is also a competitive precursor with ARA for eicosanoid synthesis, although its products are frequently considered less biologically active than ARA-derived eicosanoids (Tocher 2003). LCPUFAs can function in growing larvae as ligands, which are involved in gene transcription and expression. EPA, DHA, and ARA compete for acylation and incorporation into the membrane PLs of most cells as well as substrates for enzyme systems (Bell et al. 1991a, 1991b). Generally, DHA is preferentially conserved during food deprivation (Koven et al. 1989) and is selectively assimilated into neural tissues and brush border membranes, but it has been reported that EPA outcompetes DHA for the sn-2 position in the phosphoglycerides of marine fish larvae (Izquierdo et al. 2000; Villalta et al. 2008).

[1] n-3 (n-6, n-9, etc.): Fatty acids with the first double bond at the 3rd (6th, 9th, etc.) terminal carbon atom.

Larval Fish Nutrition, First Edition. Edited by G. Joan Holt.

Therefore, the overall impact of EFA on fish larvae physiology is directly related to the level and ratio of these compounds in the tissue PLs. This means that the ratio of EPA, DHA, and ARA in the diet largely determines the efficacy of many physiological processes during the rapid growth and dramatic development characterizing fish larval culture.

2.1.1 Regulation of membrane fluidity

A number of authors have shown that DHA promotes growth more effectively than EPA and ARA in marine fish larvae (Watanabe et al. 1989; Takeuchi et al. 1990; Toyota et al. 1991; Koven et al. 1993; Watanabe 1993; Wu et al. 2002). The contribution of DHA to weight gain lies in its structural function in the PL bilayer of the cellular membrane and its influence on membrane fluidity. There is an inverse relationship between the increasing number of double bonds in polyunsaturated fatty acids (PUFAs) and its decreasing gel–liquid transition temperature, or when the membrane transits from an ordered state to one that is more destabilized and fluid. Although increasing the carbon number in a fatty acid chain increases transition temperature, the six double bonds of the 22-carbon DHA results in the most potent LCPUFA resisting membrane packing and contributing to its fluidity. As membrane fluidity is thought to facilitate membrane-dependent functions such as receptor–ligand interaction, membrane transport, and membrane-bound adenylate cyclase activity (Hashimoto et al. 1999), having high DHA moiety in the tissue PLs would provide the correct milieu for these reactions to occur. However, concentration levels of DHA as well as other EFAs can vary among tissue phosphoglycerides or PLs. For example, larval DHA is highly represented in the PL phosphatidylethanolamine (PE) and correlates well with growth in larval gilthead sea bream (*Sparus aurata*) (Koven et al. 1993).

2.1.2 Sensorial and neural tissue development and behavior

PE is the largest PL class in neural tissue, particularly in the brain where DHA is selectively assimilated (Mourente and Tocher 1992), and this fact, coupled with the disproportionately large brain and eyes of developing larvae, suggests that DHA's role in growth may also be related to its physiological function in the membranes of main neural tissues. Larval eyes are particularly rich in DHA (Bell et al. 1995; Benítez-Santana et al. 2007) where di-22:6n-3 phosphoglycerides in the PE fish retina can be in excess of 70% (Bell et al. 1995), and this hints broadly at the role of DHA in vision. A number of authors have reported that increased dietary DHA or n-3 highly unsaturated fatty acid (HUFA) improve feeding behavior in herring larvae (Bell et al. 1995) and prey consumption in gilthead sea bream (Koven et al., unpublished data) when fed under low light intensity, whereas feeding DHA-deficient diets reduces visual capacity in yellowtail (*Seriola quinqueradiata*) (Masuda et al. 1999) and gilthead sea bream (Benítez-Santana et al. 2007). The substantial amount of di-22:6n-3 in the eye is predominantly found in the rod membranes of the retina. Although larvae at hatching generally have all-cone eyes (Blaxter and Jones 1967; Powers and Raymond 1990), the development and species-specific rate of appearance of rod cells is likely associated with progressively feeding in deeper water, which means hunting in the monochromatic conditions of lower light intensities. In fact, Shields et al. (1999) found that feeding DHA-rich copepods to halibut larvae resulted in significantly increased retinal rod/cone ratios compared with larvae fed *Artemia* nauplii containing less EFA—the industrial standard in commercial hatcheries. Moreover, a significant correlation between dietary n-3 HUFA and eye diameter, regardless of larval growth, together with a high density of photoreceptors has been found in gilthead sea bream (Roo et al. 1999; Izquierdo

2005). Nevertheless, other LCPUFAs may also play essential roles in neural tissues. For instance, EPA presence in the live prey enhances development of the brain (Furuita et al. 1998), whereas ARA functions as a retrograde signal to regulate the dynamic growth of retinal arbors.

Rhodopsin is the photopigment responsible for vision under low light intensities and is found in high concentrations in the rod membranes of fish larvae as well as other vertebrates. It consists of the protein opsin and a derivative of vitamin A, retinal. When rhodopsin absorbs a photon of light, 11-*cis*-retinal isomerizes to its *trans* form, which initiates a cascade of intramembrane events. These begin with the activation of the G protein transducin, which in turn activates phosphodiesterase, an enzyme that converts the nucleotide cyclic guanosine 3′-5′ monophosphate (cGMP) to 5′GMP. Under dark conditions, cGMP is a Na^+-gated channel protein, allowing an inward flow of Na^+ and depolarization of the rod cell. Once photoinduction begins, the resulting lower levels of cGMP close the gated ion channel, which subsequently blocks the influx of Na^+ into the cell. This leads to the hyperpolarization of the neural membrane, which ultimately causes the propagation of a signal down the central nervous system. It is generally argued that the DHA-modulated fluidity of the rod cell membranes facilitates the conformational change in rhodopsin when it is light stimulated, and therefore the processing of the visual stimulus (Brown 1994). Taken altogether, the correlation of DHA as a growth promoter during rapid larval development is due to, or at least heavily influenced by, its incorporation into the retinal PL of the rod membranes. This increases larval visual acuity, which improves hunting success and ultimately increases biomass.

On the other hand, the description of DHA's functionality in terms of its contribution to membrane fluidity may be somewhat oversimplified. Membrane fluidity depends not only on the level of unsaturation of its constituent PUFA but also on the level of cholesterol as well as on the cholesterol/PL ratio (Shinitzky and Henkart 1978). Nevertheless, it has been argued that enhanced membrane fluidity at colder temperatures in fish is a consequence of an increase in molecular species that include a monounsaturated fatty acid (MUFA), such as 18:1n-9, in position sn-1 being paired with DHA in sn-2 (Farkas et al. 2000; Jobling and Bendiksen 2003). Introducing a single double bond into a saturated fatty acid (SFA) markedly changes the character of membrane PLs (Hazel and Williams 1990; Farkas et al. 2001; Hochachka and Somero 2002). Moreover, the contribution of membrane DHA to facilitate conformational changes in rhodopsin may be due not only to membrane fluidity but also to direct DHA–rhodopsin interactions taking place at specific regions of the protein that interfere with membrane packing as well as contributing to photocycle kinetics (Farrens et al. 1996; Hubbell et al. 2003; Grosfield et al. 2006). In contrast, cholesterol and SFA chains are frequently tightly packed against rhodopsin (Stillwell and Wassall 2003), thereby reducing membrane fluidity and molecular conformational change.

2.1.3 Eicosanoid synthesis

The roles of eicosanoids in fish are quite diverse and include modulation of reproduction, hormone release, cardiovascular and neural function, osmoregulation (Mustafa and Srivastava 1989), and immune function (Kinsella et al. 1990). One of the main eicosanoid groups, the prostaglandins (PGs), regulate ion and electrolyte balance in the kidney (Brown and Bucknall 1986; Bell et al. 1993) and ion transport in the gills and opercular epithelium (Van Praag et al. 1987). ARA is thought to be the preferred precursor for cyclooxygenase (COX)-derived PGs, although EPA and dihomogammalinolenic acid (DHGLA, 20:3n-6) can modulate the biological efficacy of ARA-derived eicosanoids (Horrobin 1983; Bell et al. 1994; Ganga et al. 2005). Indeed, EPA can be an important source of eicosanoids in marine fish, particularly in an environment where ARA is not very abundant. For instance, PGE3 has been found to be the major PG produced in plasma of cultured gilthead sea bream (Ganga et al. 2005). Eicosanoids derived from EPA such as the leukotriene LTB5, which is involved in chemotaxis and have antibactericidal activity in neutrophils, can be present in haddock and halibut leukocytes in higher amounts than leukotrienes derived from ARA such as LTB4 (Izquierdo et al., unpublished data). However, in fish leukocytes, the affinity for ARA or EPA and sensitivity to dietary fatty acids depends on the type of lipoxygenase (LOX) considered. For instance, whereas 5-LOX (involved in chemotaxis in eosinophils and neutrophils) in both haddock and halibut frequently produce slightly higher levels of EPA derivatives, 15-LOX has a somewhat increased affinity for ARA over EPA, whereas 12-LOX has a strong affinity for ARA (Izquierdo et al., in preparation). These studies point out the importance of EPA as a precursor of eicosanoids in marine fish, at least for the correct functioning of blood cells, and it agrees well with the predominant role of this fatty acid in immune regulation in these species. In addition, the high content of DHA in cellular membranes may affect eicosanoid production (Nablone et al. 1990). This fatty acid is also recognized as a precursor of certain biologically active trioxylated derivatives (German et al. 1983; Hong et al. 2005).

2.1.4 Flatfish pigmentation

Dietary ARA fed during larval rearing has been shown to have a profound effect on juvenile pigmentation in flatfish (Estévez et al. 1997, 1999, 2001; McEvoy et al. 1998; Copeman et al. 2002; Willey et al. 2003; Bransden et al. 2005; Lund et al. 2008; Villalta et al. 2008). Elevated ARA fed during a crucial pre- (Seikai et al. 1987; Copeman et al. 2002) or early metamorphic "pigmentation window" (Næss and Lie 1998) produces malpigmentation. Conversely, dietary ARA fed during metamorphosis (35 days posthatch [dph]) and after the "window" improved pigmentation compared with fish feeding on high DHA rotifers and *Artemia* (Estévez et al. 2001). Lund et al. (2008) reported that sole larvae consuming high-ARA diets from 3–21 dph were almost all malpigmented, whereas fish fed the same diets from 11 dph exhibited moderate but significantly lower levels of hypomelanosis. However, other LCPUFAs seem to be also involved in the correct pigmentation of flatfish. For example, dietary EPA/ARA ratios of less than 4 fed during the "pigmentation window" hindered juvenile dorsal pigmentation in turbot, halibut, and Japanese flounder (McEvoy et al. 1998; Estévez et al. 1999, 2001). In addition, DHA has been associated with typical pigmentation in halibut (Hamre and Harboe 2008). Overall, an optimum DHA/EPA/ARA ratio seems to be necessary in early developing flatfish larvae in order to achieve normal metamorphosis and dorsal skin pigmentation (Sargent et al. 1999a). Thus, both absolute and relative dietary LCPUFA levels fed during specific periods of development appear to determine the success of pigmentation processes in flatfish.

However, the physiology underlying the developmental stage-dependent effect of LCPUFA on pigmentation could be closely related to PG production. ARA-derived prostanoids play an important role in melanocyte development and melanin production in mammals (Scott et al. 2005) and may modify the production of tyrosinase, a key enzyme involved in the L-tyrosine-to-melanin pathway (Bransden et al. 2005). Moreover, free ARA and/or leukotrienes and thromboxanes can modulate changes to pigmentation in mammalian cells (Tomita et al. 1992; Maeda and Naganuma 1997; Norris et al. 1998). Estévez et al. (1999) suggested that an imbalance in neural membrane composition and ARA-derived eicosanoids in the brain or pituitary during critical developmental periods prior to flatfish metamorphosis might have affected the neuroendocrine control of processes involved in pigmentation. However, these authors found no relationship between dietary ARA and levels of adenocorticotropin hormone (ACTH) or melanocyte-stimulating hormone (MSH) in the pituitary (Estévez et al. 2001). Hamre et al. (2007) hypothesized that high ARA incorporation in specific tissues during premetamorphosis may have lowered membrane concentrations of other LCPUFAs, which ligand with peroxisome proliferator–activated receptors (PPARs) and then dimerize with retinoic acid X receptors (RXRs). These dimers bind to DNA and modulate expression of genes involved in pigment cell development and differentiation during critical periods of larval development. Lower levels of these dimers would reduce the expression of key genes and consequently interfere with normal pigmentation.

2.1.5 Stress modulation

Fish kept under common production conditions are very sensitive to stressors, particularly during larval development and early fry stages, which can markedly contribute to the low survival rates found in commercial hatcheries. Unfortunately, studies on welfare and stress resistance in fish larvae have been limited in number and inconclusive due to the lack of appropriate and reliable indicators of larval welfare, which have primarily been restricted to the consequences of selected stressors on survival and growth. In marine fish, the few studies on the ontogeny of the stress response include gilthead sea bream (Szisch et al. 2005) and cod (King and Berlinsky 2006). A common thread in these papers is that they have been exclusively based on the dynamics of whole-body cortisol, a central hormone in stress regulation in teleosts.

Early studies on the value of LCPUFAs for fish larvae indicate the importance of these fatty acids for resistance to handling, temperature, salinity, or other types of acute stressors (Watanabe et al. 1984; Izquierdo et al. 1989; Liu et al. 2002). More recently, it has been shown that EPA, ARA, and DHA, through their COX and LOX derivatives, regulate cortisol production by modulating ACTH-stimulated interrenal cells in sea bream (Ganga et al., in press).

EPA markedly affects ACTH-induced cortisol production in sea bream interrenal cells (Ganga et al. 2006) in a dose-dependent manner, as well as fish stress resistance. Although elevation of dietary EPA increased red sea bream stress resistance to handling (Watanabe et al. 1989) and gilthead sea bream resistance to air exposure and temperature shock (Liu et al. 2002), excessive EPA, particularly in relation to ARA and DHA levels, reduced stress resistance to air exposure in Japanese flounder (Furuita et al. 1998). However, stress resistance after salinity shock was not improved by increasing dietary EPA in gilthead sea bream (Liu et al. 2002) or Japanese flounder (Furuita et al. 1999). This denotes a minor role of EPA in osmoregulation, or that this fatty acid was not mobilized in time to be effective. On the other hand, a higher content of ARA and lower EPA/DHA ratio were found in gills of the gilthead sea bream, the main osmoregulatory tissue for fish (Ganga et al., in press). ARA is a major component of phosphatidylinositol (PI) and, *in vitro*, was shown to be the preferred substrate for most COXs, the main enzymes for PG synthesis (Bell et al. 1995). On the other hand, EPA could also be an important substrate in marine fish *in vivo* (Ganga et al. 2005).

ARA has received increasing attention from several authors investigating its effect on whole-body cortisol or the response to various stressors (Koven et al. 2001a, 2003; Bransden et al. 2004; Van Anholt et al. 2004a, 2004b). Nevertheless, whole-body cortisol levels are not necessarily indicative of a stress response but may serve other physiological functions during specific developmental periods (residual cortisol in the egg, metamorphosis, development of melanophores, etc.)

The expression of certain genes involved with hormonal axes could be a useful tool to determine the welfare status of larvae. Steroid activity is primarily mediated through nuclear receptors and different enzymes involved in steroid synthesis. Recently, it has been found that dietary ARA increases expression of glucocorticoid receptors (GRs) and heat-shock protein 70 (HSP70) in European sea bass larvae, giving further evidence of the important role of this fatty acid in modulation of the stress response (Negrín et al., submitted). Feeding ARA prior to tank transfer or handling stress markedly improves survival in gilthead sea bream larvae (Koven et al. 2001a, 2003). Van Anholt et al. (2004a) found that increased ARA contents in *Artemia* markedly reduced poststress cortisol responses in 28- and 50-dph gilthead sea bream exposed to air for 90 seconds. Paradoxically, Koven et al. (2003) had found that an ARA dose-dependent increase in basal cortisol was correlated with reduced survival and growth in larvae exposed to daily salinity change. High basal cortisol can suppress the immune system (Barton and Iwama 1991; Wendelaar Bonga 1997), which may have increased mortality in the high-ARA fish. Van Anholt et al. (2004a) proposed that ARA supplementation had a differential effect on the synthesis of cortisol depending on the nature of the stressor.

The modulation of the hypothalamus–pituitary–interrenal (HPI) axis by ARA following exposure to a stressor has been generally attributed to an increased production of two-series PGs (Gupta et al. 1985; Harel et al. 2001; Van Anholt et al. 2003). PGs modulate the release of cortisol in ACTH-stimulated interrenal cells (Ganga et al. 2006) and, as shown in mammalian studies (Abou-Samra et al. 1986; Zacharieva et al. 1992; Nasushita et al. 1997), could also affect the hypothalamic corticotrophin-releasing hormone (CRH) and/or pituitary ACTH release, consequently changing the stress response (Nye et al. 1997; Bugajski et al. 2001; Di Luigi et al. 2001; Gadek-Michalska et al. 2002). Feeding gilthead sea bream with acetylsalicylic acid (ASA), an irreversible blocker of the COX pathway, partly reduced the peak cortisol response in acutely stressed fish fed low levels of ARA (Van Anholt et al. 2004c), suggesting that PGs are not the only metabolites mediating the ARA effect on cortisol production. Moreover, in high-ARA-fed fish, dietary ASA slightly increased the cortisol response, although at lower levels than fish fed with low-ARA diets. In agreement, Ganga et al. (2006) found that indometacin, another specific COX inhibitor, markedly reduces the cortisol release in ACTH-induced sea bream interrenal cells incubated with EPA and ARA. However, the regulation of cortisol by these fatty acids has also been found to be mediated by other eicosanoids derived from LOXs since the incubation with nordihydroguaiaretic acid, a potent LOX inhibitor, markedly reduces cortisol release (Ganga et al., in press).

The regulation of the stress response by fatty acids likely occurs at different levels of the HPI axis. Upon stimulation of the HPI axis, ACTH binds to membrane receptors on the interrenal cells of the head kidney, which activates the steroidogenic pathway leading to the release of cortisol (Wendelaar Bonga 1997; Mommsen et al. 1999; Hontela 2005). In European sea bass larvae, dietary ARA has been found to increase the expression of GRs (Negrín et al., submitted). Cortisol in interrenal cell mitochondria is regulated by the protein kinases (PKs) A and C, where activated PKA enhances cortisol synthesis while PKC inhibits steroidogenesis (Planas et al. 1997; Lacroix and Hontela 2001). PKA is activated by an increase of intracellular cyclic adenosine 3′-5′ monophosphate (cAMP), a second messenger, triggered by the binding of ACTH to membrane receptors. PKA then phosphorylates steroidogenic acute regulatory protein (StAR), which mediates the delivery of cholesterol, the substrate of cortisol synthesis. In fact, the transport of cholesterol from intracellular sites of storage to the inner mitochondrial membrane is the key rate-limiting step in steroid synthesis. The binding of ACTH also activates specific phospholipases (such as phospholipase A2 [PLA2]), which releases ARA as well as EPA from the membranes for eicosanoid synthesis.

PGs have been shown to enhance cAMP levels by affecting the G proteins that stimulate adenylate cyclase activity (Naor 1991; Cooke 1999) and thereby upregulate cortisol synthesis. Conversely, PKC is activated by increased levels of Ca^{2+}, which in turn are mediated by the second messengers' diacylglycerols (DAGs) and inositol triphosphate (IP3). These messengers are released by ACTH-induced activation of phospholipase C (PLC). Free ARA, acting as a first messenger, can stimulate PLC activity (Axelrod et al. 1988), while ARA and its eicosanoid derivatives are able to elevate cytosolic Ca^{2+} independently of IP3 formation. In addition, McPhail et al. (1984) found that low levels of free ARA enhanced the affinity of PKC for Ca^{2+}. All of these effects of ARA and its metabolites could ultimately downregulate cortisol synthesis.

Interestingly, high concentrations of free intracellular ARA, which may have been the case in chronically stressed sea bream fed a high-ARA diet, can have a reverse effect and inhibit PKC activity and therefore upregulate cortisol synthesis. Therefore, ARA in its free form as well as its various eicosanoid derivatives either up- or downregulates steroidogenesis plasma cortisol levels, probably depending on the activity of the different ARA metabolites on PKA and PKC. The type of stress (acute or chronic) and how it stimulates the HPI axis and/or the release of ARA and formation of its eicosanoid derivatives might also impact on regulating cortisol synthesis.

2.1.6 Gene regulation

In lipid homeostasis, a wide variety of genes control the balance between lipid uptake, storage, synthesis, and catabolism. LCPUFAs, in addition to their other physiological and nutritional roles, are able to modulate the transcription of genes involved in their metabolism in order to maintain lipid homeostasis. PUFA- or ARA-derived eicosanoids can directly activate nuclear transcriptional factors, such as the nuclear hormone receptors PPARs, which then form heterodimers with RXRs (Ross et al. 2000) and bind to DNA at sequence specific sites called peroxisomal proliferator response elements (PPREs). In mammals, Atlantic salmon, and plaice, there are three types of PPARs: PPARα, PPARβ, and PPARγ. Each of these receptors binds to a distinct but overlapping range of LCPUFA and has clear tissue distributions. PPARα is involved in the control of fatty acid transport and uptake by regulating the genes encoding for the fatty acid transport protein, fatty acid translocase, and the liver cytosolic fatty acid–binding protein (Motojima et al. 1998). In contrast, PPARγ is dominant in adipose tissue and regulates for adipocyte phenotype, a fatty acid–binding protein, and the glyceroneogenic enzyme phosphoenolpyruvate carboxykinase (Tontonoz et al. 1995). Shi et al. (2002) showed that PPARβ might compete with PPARα and PPARγ for binding at gene regulatory elements and then recruit proteins to depress their transcriptional activity. Taken altogether, in fish and mammals, PPARs appear to have critical roles in regulating genes involved in lipid homeostasis by acting as PUFA sensors and/or signal transducers.

Heterodimers formed from PUFA-activated PPARs and retinoids can also regulate genes for morphogenesis (Kliewer et al. 1997; Balmer and Blomhoff 2002). Villeneuve et al. (2005), studying the effect of PUFA in neutral and polar lipid classes on European sea bass larvae, concluded that high levels of EPA and DHA downregulate RXR and retinoic acid receptor (RAR) nuclear receptors, which adversely affected gene expression and ultimately induced severe skeletal malformations.

2.1.7 Immune system regulation

In general, the amount and type of dietary LCPUFA may affect the immune system by several mechanisms:

1. Altering an immune cell's fatty acid composition and hence membrane fluidity, which in turn interferes with intercellular interaction, receptor expression, nutrient transport, and signal transduction.
2. The eicosanoid type and concentration, since both ARA and EPA produce certain eicosanoids that can either stimulate or inhibit immune cells in a dose-dependent manner. Those derived from EPA are generally less potent mediators of inflammation, although PGI3 (derived from EPA) is equipotent to PGI2 (derived from ARA).
3. Through increase of oxidative stress. Increasing PUFA intake can increase oxidative stress if not compensated by antioxidant nutrients, which in turn can affect several genes involved in the inflammatory response, including those encoding for tumor necrosis factor (TNF)-α or interleukins (ILs) IL-1a, IL-1b, IL-6.
4. By modulation of synthesis of lipoproteins that in turn affect immune cell activity. An increase in n-3 LCPUFA and PLs has been found to promote very low-density lipoprotein (VLDL) synthesis in opposition to chylomicrons in marine fish juveniles (Caballero et al. 2002, 2006b) and larvae (Liu et al. 2002). Chylomicrons inhibit lymphocyte and monocyte functions through incorporating apolipoproteins B and E.
5. By activating nuclear transcription factors, which regulate a variety of anabolic and catabolic functions such as cell monocyte differentiation.

Thus, increased dietary n-3 LCPUFAs have been shown to enhance the immune system in fish (Montero et al. 1998; Lall 2000; Waagbø et al. 2003; Montero and Izquierdo 2010), enhancing head kidney macrophages (Sheldon and Blazer 1991) and alternative complement pathway activities (Montero et al. 1998) or increasing antibody production (Kiron et al. 1995). However, the effect of fatty acids on the immune system depends on the EFA requirements of each species and also on the levels of other nutrients associated with lipid metabolism, particularly antioxidants, as well as temperature or other stressors.

The effect of EFA on the larval immune system has scarcely been studied. At hatching, the fish immune system is very immature, and fully developed lymphoid organs will not appear until after metamorphosis (Zapata et al. 2006). Information regarding the immune system and the innate defense mechanisms during larval and juvenile fish development is not complete for cultured marine species, but the role that dietary EFA plays in immune defense during this period of the life cycle may shed some light on the marked effect of these fatty acids on larval survival.

2.2 Requirements for EFAs

2.2.1 LCPUFA requirements

The LCPUFAs—DHA, EPA, and ARA—play very important physiological roles in fish larvae, although fish, as all vertebrates, are incapable of their *de novo* synthesis (Sargent et al. 2002). Freshwater fish seem to possess sufficient Δ6- and Δ5-desaturase and elongase capability to produce ARA, EPA, and DHA from their shorter-chain precursors linoleic acid (LA; 18:2n-6) and linolenic acid (LNA; 18:3n-3) if they are present in the diet (Yu and Sinnhuber 1975). In contrast, marine species have very limited elongase and desaturase capability, requiring the ingestion of fully formed EFA for good growth and survival (Watanabe 1982; Sargent et al. 1989; Izquierdo 1996). The Δ6-desaturase-like genes isolated from zebrafish (*Danio rerio*) (Hastings et al. 2001) have been also found in marine fish species such as gilthead sea bream (Seiliez et al. 2003), although its expression is largely inhibited. Nevertheless, it has been shown that dietary lipids are able to regulate Δ6-desaturase expression in larvae of gilthead sea bream (Izquierdo et al. 2008), although the ability to synthesize DHA was insufficient to satisfy its requirement. Consequently, these LCPUFAs are considered essential (EFA) for both freshwater and, in particular, marine teleosts.

Therefore, suboptimal levels of these EFA in larval feeds may give rise to several behavioral and morphological alterations such as reduced feeding and swimming activity, poor growth and increasing mortality (particularly in the young stages), fatty livers, hydrops, deficient swim bladder inflation, abnormal pigmentation, disintegration of gill epithelia, immune deficiency, elevated basal cortisol levels, and skeleton deformities (Izquierdo 1996, 2005).

Studies on the specific dietary fatty acid requirements for fish larvae are difficult to conduct since larvae, in general, are less efficient at ingesting, digesting, and absorbing formulated diets from first feeding compared with live food. Although a degree of successful manipulation of EFA content is possible in live feed (*Brachionus* spp. and *Artemia* spp.) and is routinely carried out, it is far from precise. Moreover, the literature is inconsistent and at best can only estimate qualitative EFA requirements. In fact, the problem is further exacerbated by the inability to accurately control, in tandem with EFA enrichment, other nutrients such as carotenoids, vitamins, and antioxidants, whose levels may modulate EFA requirements. In addition, EFA requirements can vary throughout the series of profound morphological and physiological changes that occur from first feeding through metamorphosis. Therefore, determination of larval nutritional requirements is one of the most complicated aspects of fish nutrition. Nutritional reserves in fish larvae are very limited at first feeding, and hence their survival dramatically depends on exogenous high-quality feeds that contain all the necessary nutrients to match their requirements.

N-3 LCPUFA requirements during larval development of several fish species vary, from very low values in common sole (*Solea solea*) (Lund et al. 2007) and carp (*Cyprinus carpio*) (0.05% n-3 LCPUFA dry weight [DW] dietary content) (Radunz-Neto et al. 1993) to 3.9% in the fast-growing larvae of *Seriola quinqueradiata* (Izquierdo, unpublished data) and 4% in *Dentex dentex* (Mourente et al. 1999) (Table 2.1). On average, the requirements for n-3 LCPUFAs in larvae of marine fish are about 3% DW diet or prey, tending to be lower in freshwater or flatfish larvae and higher in fast-growing species such as yellowtails and jacks. Nevertheless, the precise determination of the requirements may be affected by several factors. For instance, during larval development, several authors have shown a requirement of EFA for gilthead sea bream that is very close to 1.5% n-3 HUFA DW diet when larvae were fed either live prey (Rodríguez et al. 1998) or microdiets (Salhi et al. 1999), independent of dietary lipid levels (Salhi et al. 1994). However, much higher requirements are estimated in the literature for EPA, which can reach two to three times higher than those of DHA (Rodríguez et al. 1994, 1997. This is due to the very high incorporation of EPA into and the displacement of DHA from certain polar lipids

Table 2.1 Best dietary essential fatty acid contents assayed during larval feeding of several fish species (% DW in rotifers, *Artemia*, or microdiets).

Species	ARA	EPA	DHA	HUFA	Reference
Pagrus major				3.5	Izquierdo et al. (1989); Watanabe et al. (1989)
Longirostris delicatissimus				>3	Izquierdo et al. (1989)
Paralichthys olivaceus				1.8	Izquierdo et al. (1992)
Cyprinus carpio				0.05	Radunz-Neto et al. (1993)
Psetta maxima		0.7	0.8	1.5	Reitan et al. (1994)
Gadus morhua	0.5				Zheng et al. (1996)
Sparus aurata		0.7	0.8	1.5	Rodríguez et al. (1997, 1998); Salhi et al. (1999)
	0.5–1				Bessonart et al. (1999)
	0.6				Fountoulaki et al. (2003)
Sciaenops ocellatus				1.5–6	Brinkmeyer and Holt (1998)
Acanthochromys poliacanthus		1	0.5		Southgate and Kavanagh (1999)
Dentex dentex		1.6	2.4	4	Mourente et al. (1999)
Pagrus pagrus			1.5	3.4	Hernández-Cruz et al. (1999)
Centropomus parallelus		1.3	>0.6		Seiffert et al. (2001)
Limanda ferruginea	<1.3		About 2.5		Copeman et al. (2002)
Latris lineata					Bransden et al. (2003)
Paralichthys dentatus	0.1				Willey et al. (2003)
Dicentrarchus labrax				2.3	Gisbert et al. (2005)
	1.2				Atalah et al. (2010a)
Poecilia reticulata	2.5				Khozin-Goldberg et al. (2006)
Solea solea			Low	Low	Lund et al. (2007)
Hippoglossus hippoglossus			2.5		Hamre and Harboe (2008)
Thunus orientalis				2.1	Seoka et al. (2007)
Oplegniatus fasciatus				3	Izquierdo (unpublished data)
Seriola quinqueradiata				>3.9	Izquierdo (unpublished data)

(Izquierdo et al. 2000). However, provided other nutrients such as antioxidants are balanced, the increase of dietary n-3 HUFA up to 8% DW diet while keeping a DHA/EPA ratio of 1.7 further improves larval growth and survival (Liu et al. 2002). Therefore, the high n-3 HUFA requirements estimated for species such as red porgy (3.39% at 1.35 DHA/EPA; Hernández-Cruz et al. 1999) or *Dentex dentex* (Mourente et al. 1999) could be related to other nutrient imbalances. For instance, in the *Dentex dentex* study, the high EPA content in *Artemia* may have caused an overestimation of the requirements as was demonstrated in gilthead sea bream (Rodríguez et al. 1997).

Since environmental factors such as temperature (Farkas et al. 1980; Olsen et al. 1999), salinity (Borlongan and Benítez 1992), and light (Ota and Yamada 1971) affect lipid composition of fish tissue, EFA requirements could also be affected by these abiotic conditions. Reductions in water temperature have been associated with an increase in PUFA content in carp tissues (Kayama et al. 1986) or with an increase in DHA content in salmon (Olsen and Skjervold 1995). This is likely related to the positive correlation between the degree of unsaturation of fatty acids and membrane fluidity, which would be particularly important at low culture temperatures (Bowden et al. 1996). Reduction in water temperature affects both lipolytic and lipogenic enzymes, increasing the oxidative capacity in rainbow trout (Guederley et al. 1997) or the specific activity of Δ9-desaturase in carp (Kayama et al. 1986). Higher contents of tissue PUFA, particularly DHA and ARA, have also been found in fish cultured with increased salinity such as guppy (*Poecilia reticulata*) (Daikoku et al. 1982), milkfish (*Chanos chanos*) (Borlongan and Benítez 1992), and puye (*Galaxias maculatus*) (Dantagnan et al. 2007), denoting the important role of some of these fatty acids in osmotic regulation (Sampekalo et al. 1992). Therefore, dietary requirements for EFAs for a given fish species may be dependent on the salinity at which that species is reared, as was demonstrated in *Galaxias maculatus* where larvae reared at 0‰ showed higher requirements for EPA, whereas increased salinity augmented the dietary requirement for DHA (Dantagnan and Izquierdo 2007; Dantagnan et al. 2007, in press).

EFA requirements seem to be higher during larval development and early postmetamorphosis than for the broodstock or juvenile stages. This may be related not only to the higher metabolic activity and growth rate of very young fish but also to the rapid development of specific EFA-dense tissues such as brain, sensory organs, and gills.

2.2.2 DHA requirements

The relevance of DHA in diets for marine fish larvae has been well documented (Watanabe et al. 1989; Izquierdo 1996; Sargent et al. 1999b) and its positive effect on survival has been related to its important role in stress control (Watanabe 1993; Izquierdo 2005; Ganga et al. 2006), immune system development (Montero et al. 2003), and improvement of health and bacterial resistance in fish larvae (Bransden et al. 2003). However, apart from DHA's contribution to improved survival and stress resistance during early larval development, DHA, more than other EFAs, appears to be the most effective larval growth promoter (Watanabe et al. 1989; Watanabe and Kiron 1994) and necessary for normal behavior (Masuda et al. 1999). Indeed, DHA incorporation into larval tissues affects intercellular interaction, receptor expression, nutrient transport, and signal transduction, all of which affect cell growth. DHA is specifically retained in starved or low-EFA-fed fish (Izquierdo et al. 1989; Izquierdo 1996) due to its lower β-oxidation rate in comparison with other LCPUFAs (Madsen et al. 1999). Being one of the most abundant fatty acids in marine fish eggs, regardless of species, and an important component of biomembranes, requirements for DHA are quantitatively higher than those for the other LCPUFAs in most of the species studied (Izquierdo 1996) (Table 2.1). Therefore DHA is considered to be more essential as an EFA than EPA in marine fish larvae (Watanabe et al. 1989; Watanabe 1993).

In species such as gilthead sea bream (*Sparus aurata*) (Izquierdo et al. 2005), red sea bream (*Pagrus major*) (Izquierdo et al. 1989), striped trumpeter (*Latris lineata*) (Bransden et al. 2005), or common dentex (*Dentex dentex*) (Mourente et al. 1999), the minimum DHA requirement for optimum growth has been reported to be 0.8, 1.2, 2.3, and 2.0% DW DHA, respectively. Optimum DHA levels in larval feeds to promote growth and survival range from 0.5% for *Acanthochromys poliacanthus* (Southgate and Kavanagh 1999) to 2.5% for Atlantic halibut (Hamre and Harboe 2008). On the other hand, the DHA requirement for species such as common carp is probably lower, although this has to be more precisely determined. In first feeding, marine fish larvae fed a diet deficient in DHA causes very high mortalities and reduced growth rates in only 10–15 days (Izquierdo et al. 1989; Sargent et al. 1997). However, supplementation of DHA improves growth and survival in a dose-dependent manner. As the requirement for this EFA is high in marine fish larvae and DHA sources (primarily marine oils) are rapidly diminishing as well as becoming increasingly expensive, excessive dietary levels of this fatty acid are rarely achieved and their possible deleterious consequences have not been clearly determined. For instance, high levels of dietary DHA (5% DW, provided mainly as a triglyceride form) in microdiets for gilthead sea bream resulted in high growth and larval survival and did not cause any pathology (Liu et al. 2002). Similarly, in red porgy (*Pagrus pagrus*), the single elevation of DHA in rotifers from 2% up to 4.5% DW not only improved larval survival but also reduced the incidence of cranial deformities and vertebral fusions (Roo et al. 2009).

In general, doubling the DHA content of rotifers repeatedly reduced by 50% the occurrence of deformed red porgy, underscoring the important role of this EFA during early larval feeding for bone development (Izquierdo et al. 2010). A reduced incidence of opercular deformities in *Chanos chanos* has also been correlated to higher DHA content in live prey (Gapasin and Duray 2001). Despite vertebral fusion and other related deformities that develop later on in development (Witten et al. 2005), many skeletal anomalies originate at the time of chondrogenesis and osteogenesis during the early larval stages (Faustino 2002) where DHA may be a determinant factor. In general, LCPUFAs likely play an important role in bone formation. In other vertebrates, fatty acid composition of bone and cartilage is markedly affected by dietary lipids (Kokkinos et al. 1993; Watkins et al. 1991, 1997; Liu et al. 2004) where bone formation and resorption can be regulated by locally produced eicosanoids (Meghji et al. 1988; Raisz et al. 1993).

Although, both ARA and EPA are the main eicosanoid precursors in fish, DHA directly or indirectly affects eicosanoid production (Nablone et al. 1990), being a substrate for LOXs and forming active trioxylated derivatives (Hong et al. 2005) that regulate cell response in fish (Ganga et al., in press). Finally, the fish expression of osteogenic markers such as osteocalcin (OC) and matrix carboxyglutamic acid protein (MGP) are affected by LCPUFA (Karsenty and Wagner 2002; Pombinho et al. 2004), which also may interfere in osteoblast differentiation through regulation of retinoid X receptor alpha (RXRα) and bone morphogenetic protein 4 (BMP4) (Villeneuve et al. 2005, 2006).

However, up- or downregulation of these genes and the subsequent positive or negative effect of these fatty acids seem to be dependent on the species studied, the period of feeding, and the LCPUFA dietary levels and type of lipid source. Thus, in opposition to the reduction in bone anomalies found in *Pagrus pagrus* (Roo et al. 2009) and *Chanos chanos* (Gapasin and Duray 2001) by feeding DHA-rich live prey, in European sea bass, elevation of both DHA and EPA over 2% in the neutral lipid (NL) fraction of compound diets led to skeletal abnormalities and larval mortality. These negative effects were associated with an upregulation of RXRα expression 16 dph followed by downregulation at 23 dph (Villeneuve et al. 2005). Deleterious effects of excessive DHA content in microdiets for sea bass have also been found by Betancor et al. (2010). In this study, elevation of dietary DHA up to 5% in microdiets for sea bass larvae caused a high deposition of DHA, which reached almost 30% of total lipids in fish tissues, and accompanied high mortality (Betancor et al. 2010). High dietary DHA content markedly increased the risk of peroxidation and the subsequent proliferation of free radicals and toxic oxidized compounds, which may have also been a causal agent of mortality. Oxidation of PUFAs produces compounds such as fatty acid hydroxyperoxides, fatty acid hydroxides, aldehydes, and hydrocarbons. Several of these metabolites are toxic as they bind to proteins, amino groups, nucleic acids, and PL bases, ultimately damaging membrane lipids, proteins, and DNA (Frankel 1998). Indeed, increased DHA amplified by a factor of two, four, or eight the oxidation potential of the diet compared with ARA, LNA, or LA, respectively. In fact, high dietary DHA content markedly increased the incidence of muscular lesions in sea bass, including hyaline degeneration and fragmentation of myofibrils, which denotes severe damage in both membrane and structural proteins, necrosis, and mononuclear infiltrates. These pathologies together with the presence of ceroid pigment within hepatocytes are indicative of an imbalance between anti- and prooxidants that were prevented when dietary vitamin E was added. This suggests that optimum DHA levels may also be dependent on dietary levels of other nutrients (Betancor et al. 2010).

As discussed previously, DHA selectively accumulates in neural tissues of fish (Mourente and Tocher 1993; Benítez-Santana et al. 2007) where it is involved in functions such as neurocyte myelination and synapse construction. DHA has been found in large amounts in the telencephalon, optic tectum, and cerebellum of fish species such as rainbow trout (*Oncorhynchus mykiss*), carp (*Cyprinus carpio*), or skipjack tuna (*Katsuwonus pelamis*) (Ushio et al. 1996). Consequently, DHA-deficient larval feeds can negatively affect brain development as well as fish behavior. In larval Japanese flounder (*Paralichthys olivaceus*), dietary DHA content has been found to increase brain and, particularly, cerebellum volume (Furuita et al. 1998), whereas in striped jack (*Longirostris delicatissimus*), this EFA promoted the development of white and gray zones on the optic tectum (Masuda 1995). In gilthead sea bream, DHA accumulated in the brain and eyes during the first 20 dph in larvae fed rotifers with high DHA content (Benítez-Santana et al. 2007), whereas it is also retained in other tissues when larvae are fed low-LCPUFA rotifers. Moreover, feeding gilthead sea bream larvae with rotifers deficient in DHA significantly affected fish behavior, reducing cruise swimming speed, inhibiting larvae response to a visual stimulus, and postponing the appearance of behavior patterns (Benítez-Santana et al. 2007), suggesting a delay in neural development in DHA-deficient larvae.

2.2.3 ARA requirements

Despite its important roles in fish metabolism, dietary ARA requirements for fish larvae have been less studied. In fact, early on, researchers found that ARA was preferentially retained in the tissues together with DHA during starvation, suggesting a metabolic priority for this fatty acid (Ostrowski and Divakaran 1990; Rainuzzo et al. 1994; Izquierdo 1996). Quantitatively, these fatty acids are much less represented in the tissues of marine fish where lipids can typically have an n-3/n-6 ratio of 10–15:1.

Elevation of dietary ARA increased larval survival and stress resistance (Bessonart et al. 1999; Harel et al. 2001; Koven et al. 2001a), modulated cortisol response (Koven et al. 2003; Van Anholt et al. 2004a, 2004b, 2004c), as well as elevated GR and HSP70 genes expression (Negrín et al., submitted) and pigmentation (Estévez et al. 1997, 1999; McEvoy et al. 1998; Copeman et al. 2002; Willey et al. 2003). ARA seems to have a lower direct impact on larval growth compared with DHA and EPA, which may be tied to its reduced presence in the natural zooplankton diet, limiting its impact on membrane fluidity. Thus, increasing dietary ARA from 0.2 to 0.5% or 1.8% did not affect growth, respectively, in Senegalese sole (*Solea senegalensis*, Villalta et al. 2005; Lund et al. 2008), gilthead sea bream (Koven et al. 2001b), or summer flounder (*Paralichthys dentatus*, Willey et al. 2003). In contrast, ARA elevation up to 1 and 1.2%, respectively, improved growth in gilthead sea bream (Bessonart et al. 1999) and European sea bass (Atalah et al. 2010b), as well as in larval Japanese flounder (Estévez et al. 1997), probably by indirectly modulating growth-related factors. In gilthead sea bream larvae, increasing ARA up to 1% DW diet enhanced survival and growth when DHA and EPA dietary contents were 1.3 and 0.7% DW, respectively (Izquierdo 1996; Bessonart et al. 1999). However, elevation of ARA contents up to 1.2% did not improve growth if dietary EPA was low (Atalah et al. 2010a). Similarly, in European sea bass (Atalah et al. 2010b), elevation of dietary ARA up to 1.2% showed a positive correlation with larval survival and a significant improvement in growth as well as increased expression of GR and HSP70 (Negrín et al., submitted). Dietary ARA requirements for marine larval fish range from 0.5 to 1.2% DW (Table 2.1), whereas for freshwater fish it can be as high as 2.5% (*Poecilia reticulata*, Khozin-Goldberg et al. 2006).

However, an excess of ARA can also exert a negative impact on marine fish. For instance, increasing ARA up to 0.6% in diets for Japanese flounder broodstock (Furuita et al. 2003) improved larval survival, but higher levels of 1.2% markedly reduced it. Similarly, growth and survival of cod were reduced when ARA levels increased from 0.5 to 3.7% and 7.6% DW in *Artemia* (Zheng et al. 1996). In *Limanda ferruginea* larvae and Senegalese sole, increasing dietary ARA levels significantly increased the incidence of malpigmented fish (Copeman et al. 2002; Villalta et al. 2004). This negative effect may be related to an eicosanoid imbalance due to the competition of ARA and EPA for the LOX and COX enzymes (Izquierdo et al. 2000; Furuita et al. 2003; Izquierdo 2005). Interestingly, although an increase in *Artemia* ARA similarly reduced growth in white bass (*Morone chrysops*) larvae, this effect was reversed if the DHA level was increased (Harel et al. 2000). Clearly, when determining the efficacy of ARA, the presence of other EFA must be taken into account, such as EPA (McEvoy et al. 1998; Bessonart et al. 1999; Estévez et al. 1999; Sargent et al. 1999a; Copeman et al. 2002).

2.2.4 EPA requirements

EPA is a good substrate for both COXs and LOXs, and in marine fish, it can be an important source of certain eicosanoids. In fish-oil-fed gilthead sea bream, the plasma PGE3 is present in higher amounts than PGE2 (Ganga et al. 2005). Although some eicosanoids derived from EPA can have a lower biological activity than those derived from ARA, others are equally active. In mammals, EPA has been shown to have a neuroprotective role, suggesting its importance for brain development, aging, and behavior as has been demonstrated for DHA. Apart from being very abundant in the natural prey of marine larvae, EPA is a preferred substrate for digestive lipolytic enzymes (Iijima et al. 1990; Izquierdo et al. 2000). Moreover, it may constitute a potential important energy reserve in developing larvae, particularly during starvation, although to a lesser degree than SFA and MUFA, since EPA is a preferential substrate over DHA and ARA for mitochondrial β-oxidation (Froyland et al. 1997; Madsen et al. 1999). In agreement, food deprivation in marine fish larvae causes a preferential conservation of DHA over EPA (Koven et al. 1989; Tandler et al. 1989; Ako et al. 1991; Van der Meeren et al. 1991; Rainuzzo et al. 1994; Rodríguez 1994; Takeuchi 1997), and EPA has been found to be catabolized over ARA in certain species (Rainuzzo et al. 1994). Moreover, EPA has been found to be necessary for complete eye migration in *Limanda ferruginea* (Copeman et al. 2002). Finally, although Kanazawa et al. (1982) showed that labeled exogenous EPA is incorporated into a variety of tissues, including the gallbladder, swim bladder, liver, and pyloric cecum, its physiological function in these tissues has not yet been elucidated.

The benefits from feeding EPA to marine fish larvae are similar to those obtained from DHA and ARA. Thus, an increase in EPA as the sole LCPUFA in the larval food for red sea bream (Watanabe et al. 1989), flatfish (Furuita et al. 1998), or yellowtail (Furuita et al. 1996) improves larval performance in terms of survival, stress resistance, and growth, although EPA was less effective than DHA. When feeding combinations of LCPUFAs, it has been found that best growth, survival, and resistance to stress were obtained with diets containing 0.7–1.6% EPA (Table 2.1).

To a large extent, the metabolic impact of EPA is markedly dependent on DHA and ARA dietary levels as excessive dietary EPA in relation to those EFA can result in an imbalance of tissue eicosanoids, a displacement of DHA from membrane lipids, or an increased oxidation risk. An increase in dietary EPA up to 2.9% DW while DHA/EPA level is high (1.72) and ARA content is relatively low at 0.05% significantly improved growth, survival, and resistance to shock temperature stress in gilthead sea bream (Liu et al. 2002). Under these conditions, the increase in dietary EPA level did not affect incorporation of dietary DHA or ARA into larval lipids. On the contrary, an increase in dietary EPA of up to 1.8% reduced growth when ARA is at a similar level and the DHA/EPA ratio is about 1.3, suggesting that EPA is dependent on the levels of the other two EFA (Izquierdo 2005). An excess of EPA (30% of total fatty acids in *Artemia* lipid or 6% of DW) has also been found to be detrimental for survival of Senegalese sole (*Solea senegalensis*) (Villalta et al. 2008).

2.2.5 Interactions among fatty acids

2.2.5.1 EPA/ARA

Dietary ARA is more efficiently incorporated into larval tissues than EPA (Figures 2.1 and 2.2); therefore, EPA/ARA ratios become lower in larval tissues than in diets. This efficient incorporation of ARA into larval tissues could reflect the higher affinity for this fatty acid by enzymes from the glycerol-3-phosphate pathway that synthesize triacylglycerols (TAGs) and PLs and are stimulated by n-6 fatty acids (Caballero et al. 2006a). Inhibition of EPA incorporation by dietary ARA also

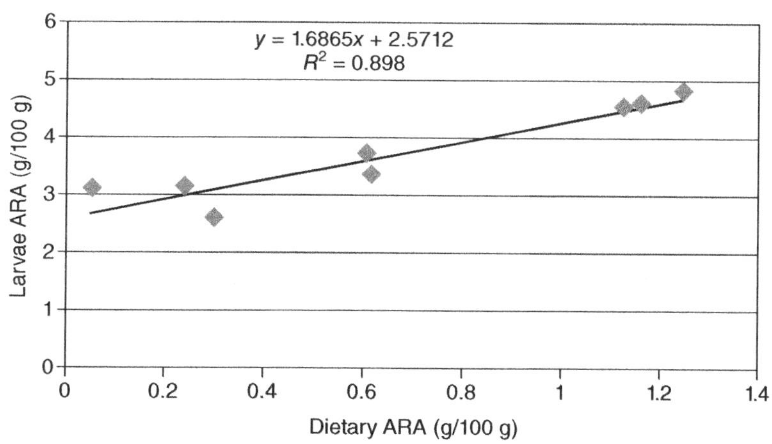

Figure 2.1 Incorporation of dietary ARA into larval gilthead sea bream tissues.

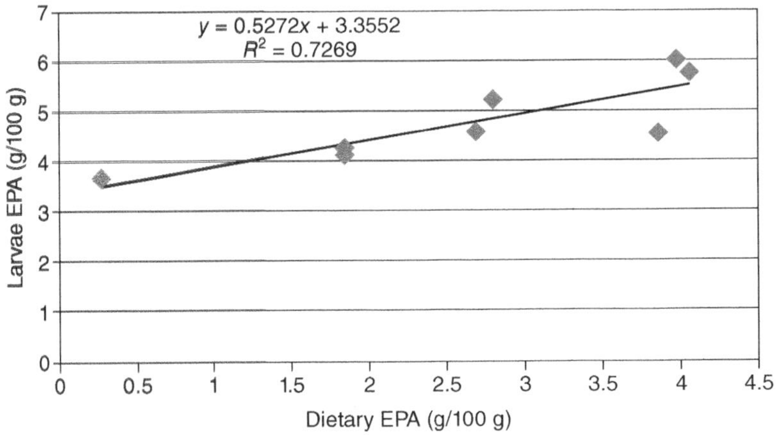

Figure 2.2 Incorporation of dietary EPA into larval gilthead sea bream tissues.

Table 2.2 Best dietary ratios between essential fatty acids assayed during larval feeding of several fish species.

Species	n-3/n-6	EPA/ARA	DHA/EPA	Reference
Scophthalmus maximus			2	Reitan et al. (1994)
		4		McEvoy et al. (1998)
		>1		Estévez et al. (1999)
Sparus aurata			1.2	Rodríguez et al. (1997, 1998); Salhi et al. (1999)
			1.7	Liu et al. (2002)
		3.3–4		Atalah et al. (2010a)
Sciaenops ocellatus			>2.5	Brinkmeyer and Holt (1998)
Dentex dentex			1.5	Mourente et al. (1999)
Pagrus pagrus			1.4	Hernandez-Cruz et al. (1999)
Latris lineata		5	2	Bransden et al. (2005)
Hippoglossus hippoglossus		3.5		Hamre and Harboe (2008)
Scardinius erythrophthalmus	1:05			Kamler et al. (2008)

has been found in liver and brain PI (Bell et al. 1995). Moreover, competition for incorporation into the different lipid classes will also differ among tissues since fatty acid composition of each lipid class markedly differs among cellular types (Lie et al. 1992). Nevertheless, the higher mitochondrial β-oxidation of EPA (Froyland et al. 1997) could be also responsible for the more efficient incorporation and retention of dietary ARA since EPA is a better substrate for mitochondrial carnitine acyltransferase I.

The optimum EPA/ARA ratios described in the literature range from 3.5 to 5 (Table 2.2), although generally they are closer to 4. For example, in turbot, the optimum EPA/ARA ratio approximates 4 (McEvoy et al. 1998), while 3.5 is the optimum for halibut pigmentation (Hamre and Harboe 2008). In agreement, Estévez et al. (1999) found that PI

EPA/ARA of turbot larvae exhibit ratios ≥1, which are related to normal skin pigmentation. Atalah et al. (2010b) tested a wide range of EPA and ARA levels as well as their ratios. They found that, when DHA requirements are satisfied, there is a significant improvement in growth and survival only when increased EPA and ARA maintain a ratio ranging from 3.3 to 4.0.

2.2.5.2 *DHA/EPA*

An increase in total n-3 LCPUFA in *Artemia* from 2.5 to 6% DW, with a DHA/EPA ratio of 0.6, had no significant effect on growth and survival in turbot larvae (Reitan et al. 1994). This suggests that the EFA requirements for this species were satisfied by the lowest n-3 LCPUFA level, although it caused a marked increase in malpigmented fish. On the other hand, a further increase in n-3 LCPUFA of up to 8.8% DW, with a DHA/EPA ratio of 1.82 (DHA increased from 2.18 to 5.4% DW and EPA decreased from 3.6 to 2.8% DW), significantly improved the percentage of correctly pigmented fish (Reitan et al. 1994), suggesting the need for a suitable DHA/EPA balance for normal pigmentation of developing flatfish. Nevertheless, the improved pigmentation in turbot could also be related to the increase in DHA, which has been shown to enhance pigmentation in halibut when increased up to 2.5% (Hamre and Harboe 2008).

Therefore, since increased dietary EPA has been found to reduce dietary DHA incorporation into larval PLs, and increased dietary DHA markedly reduced incorporation of EPA into the second position of PE (Izquierdo et al. 2000), suitable DHA/EPA ratios are required for optimizing the dietary utilization of both fatty acids. In general, the DHA/EPA ratio of marine fish eggs is close to 2 in several species studied and this has been suggested as a general value for first-feeding marine larvae (Sargent et al. 1997; Table 2.2). On the other hand, optimum dietary DHA/EPA ratios ranged from around 1.2 for gilthead sea bream (Rodríguez et al. 1997, 1998) and 2 for red drum (Brinkmeyer and Holt 1998), to 8 for yellowtail flounder, *Limanda ferruginea* (Copeman et al. 2002).

Optimal dietary DHA/EPA ratios also can be dependent on salinity. Dantagnan and Izquierdo (2007) found that reducing the DHA/EPA ratio from 1.4 to 0.5 markedly increased larval growth in *Galaxias maculatus* when fish were reared at 0 or 10 ppm, whereas 15 ppm significantly reduced weight gain. In agreement, Brinkmeyer and Holt (1998) found that increasing the dietary DHA/EPA ratio over 2.5 improved larval survival after exposure to a salinity stress test, whereas raising dietary EPA did not have the same effect (Liu et al. 2002).

2.3 Utilization of dietary lipids

2.3.1 TAGs and free fatty acids (FFAs)

EFA-rich PLs are highly represented in lipids of copepods, the major prey species for fish larvae in nature (Sargent et al. 1997). However, rotifers and *Artemia*, which are the zooplankton fed to marine fish larvae in commercial hatcheries worldwide, have low and unfavorable ratios of DHA and EPA. These zooplankters must be enriched in EFA in order to ensure reasonable larval growth and survival. This is largely accomplished with oil-based emulsions or enrichment preparations rich in TAGs containing these EFA. These enrichment protocols result in prey NL levels that are very high or even excessive (Morais et al. 2007).

In copepods, the PL fraction generally serves as the dominant reservoir of LCPUFA, which include the EFA, and phosphoglycerides designated for membrane synthesis. On the other hand, NL mainly provides nonessential fatty acids as an energy source. Elevated NL fractions in enriched rotifers and

Artemia will not only contain inappropriately high levels and ratios of DHA and EPA but will also provide significant levels of SFAs and MUFAs for β-oxidation. These zooplankter NL and fatty acid imbalances may exact an extra metabolic price from the larvae in terms of digestion and absorption efficiency.

From the onset of exogenous feeding, the larval digestive tract is essentially fully capable of the mechanical breakdown and enzymatic activity required for the digestion of dietary lipids (Cousin et al. 1987; Izquierdo et al. 2000; Pérez-Casanova et al. 2004; Morais et al. 2005a; Murray et al. 2006; see also Chapter 1). In mammals, the hydrolysis of TAG in the digestive tract is carried out by 1,3-specific pancreatic lipase. This enzyme cleaves the outer or α positions of TAG, yielding a mixture of α- and β-DAGs, β-monoacylglycerols (MAGs), and FFAs that are eventually absorbed in the intestinal wall of the jejunum. However, in studies on a number of fish species, the main pancreatic lipase was found to be a nonspecific bile salt–dependent pancreatic lipase (BSDL) (Gjellesvik et al. 1992; Koven et al. 1994a, 1994b; Izquierdo and Henderson 1998; Iijima et al. 1998; Olsen et al. 1998) that is capable of hydrolyzing all α- and β-TAG positions, yielding mostly FFA and glycerol, which would be easily and efficiently absorbed. This would be particularly advantageous in fish larvae as the digestive tract is short and not fully developed, requiring the absorption of lipolytic breakdown products to be as efficient as possible. Moreover, PUFAs, which are highly represented in marine lipids, are poor substrates for the enzymatic breakdown by 1,3-specific lipase when esterified to TAG, irrespective of position (Bottino et al. 1967; Chen et al. 1990).

The accumulating evidence suggests, contrary to mammals, that the major digestive lipase in teleosts and presumably their early developmental stages, appears to be nonspecific and bile salt dependent, while the 1,3-specific pancreatic lipase exists at lower concentration or is nonexistent (Lie et al. 1987; Gjellesvik et al. 1992; Koven et al. 1994a, 1994b).

It has been reported that the efficiency of absorbing digestion products in the fish digestive tract increases with decreasing ration size, which would extend the residence time of the food bolus (Ryer and Boehlert 1983; Boehlert and Yoklavich 1984; Hoehne-Reitan et al. 2001). This was supported by work on herring (*Clupea harengus*) larvae (Morais et al. 2005a), which found that reducing the volume of ^{14}C-TAG injected into the digestive tract increased the absorption of the label and decreased any evacuated loss of ^{14}C. On the other hand, Hoehne-Reitan et al. (2001b) showed that the BSDL content of turbot larvae was a function of ingestion rate. These authors argued that diet-activated intestinal receptors determine enzyme secretion, suggesting that stimulation would increase with the length of the intestine exposed to ingested food. Morais et al. (2004) concluded that the source of dietary TAG and not the amount affects lipase activity in sea bass larvae. This is in agreement with the higher BSDL activity found by Izquierdo et al. (2000) when larvae were fed TAG-rich rotifers containing higher EFA levels. This means that the chain length and degree of unsaturation of the constituent fatty acids of TAG can affect digestion and, ultimately, absorption (Austreng et al. 1979; Koven et al. 1994b; Olsen et al. 1998). The specificity of fish lipases hydrolyzing TAG in the lumen and absorption of the resulting FFA decrease with increasing chain length and increase with unsaturation (Austreng et al. 1979; Olsen et al. 1998; Morais et al. 2006).

In other words, the order of preferred fish lipolytic activity can be described as PUFA > MUFA > SFA. Olsen et al. (1998), working on Arctic char, and Morais et al. (2004), working on European sea bass larvae, found the shorter acyl chain length of fatty acids in coconut oil diets induced high lipase activity that can be expressed as 12:0 > 14:0 and 16:0 > 18:0. In fact, TAG esterified to short-chain fatty acids (12:0 and 14:0) can be absorbed into the enterocyte even in the absence of pancreatic lipase and bile salts (Bach and Babayan 1982). Moreover, despite the high PUFA moiety in fish oil, lipolytic activity may vary among different marine sources due to the inclusion of long-chain MUFA such as 22:1n-11, which characterizes northern-latitude marine oils (capelin, sand eels, herring, sprat, and mackerel).

2.3.2 PLs

In fish, as in other vertebrates, phosphoglycerides are critical components in the structure of cellular biomembranes and lipoproteins (Tocher 1995). Moreover, phosphoglycerides can act as emulsifiers in the formation of mixed micelles in the digestive tract (Olsen and Ringø 1997). Although TAGs are commonly considered as the main source of fatty acids metabolized for energy, phosphoglycerides can also serve as an energy source during egg development and endogenous feeding in the prelarval stages of marine fish. Larvae from various species such as Atlantic herring (*Clupea harengus*), cod (*Gadus morhua*), halibut (*Hippoglossus hippoglossus*), and plaice (*Pleuronectes platessa*) (Tocher et al. 1985; Fraser et al. 1988; Rainuzzo et al. 1992; Finn et al. 1995; Rønnestad et al. 1995) that hatch from PL-rich eggs are able to metabolize yolk phosphatidylcholine (PC) for energy production. However, EFAs such as DHA that are bound to decreasing yolk PC are not necessarily catabolized and may be mobilized for synthesis of NL and/or PE.

The beneficial effects of dietary PLs on growth and survival have been demonstrated in the larval and juvenile stages of various marine species such as red sea bream, knife jaw (Kanazawa et al. 1983b), Japanese flounder (Kanazawa 1993), and Atlantic salmon (Poston 1990b, 1991) as well as freshwater species that include ayu (Kanazawa et al. 1981, 1983a; Kanazawa 1985), rainbow trout (Poston 1990a), and common carp (Geurden et al. 1995, 1998; Fontagné et al. 2000). However, the growth-promoting effects of dietary PL diminish with age (Kanazawa et al. 1981; Kanazawa 1993; Radünz-Neto et al. 1994; Coutteau et al. 1997) and it appears to have little or no effect in adult fish (Kanazawa 1985; Hung and Lutes 1988; Poston 1991). Moreover, PC is considered to have the greatest influence on growth and larval fish performance compared with the other PL classes (Takeuchi et al. 1992; Kanazawa 1993; Hadas 1998).

Koven et al. (1998) found in gilthead sea bream that dietary supplementation of PC and not PE (Hadas 1998) temporarily stimulated feeding activity in larvae up to 26 dph but did not elicit a significant ($P > 0.05$) response in 30-dph fish. These authors reasoned that the trimethyl group of the choline base of PC stimulated a gustatory response by binding to receptor cells in the mouth and/or on the fish body surface.

A number of hypotheses have been put forth to explain the age-dependent effect of dietary polar lipid. One possibility is that dietary PL is enhancing the absorption of ingested fats by acting as temporary emulsifiers (Kanazawa 1993; Koven et al. 1993). PL is digested by pancreatic PLA2, which results in 1-acyl-lysophospholipids, which have the ability to emulsify and could conceivably contribute to the digestion of other lipids during larval development. However, this has been deemed unlikely in research on common carp (Geurden et al. 1995; Fontagné et al. 1998), and later in gilthead sea bream (Koven et al. 1998; Hadas et al. 2003).

Recent studies have argued more compellingly that dietary PC contributes to the assimilation of dietary lipid through increased lipoprotein synthesis (Geurden et al. 1997; Coutteau et al. 1997; Fontagné et al. 1998, 2000; Liu et al. 2002; Hadas et al. 2003). Moreover, dietary PL not only increases synthesis and release of lipoproteins in the lamina propria by enterocytes but also significantly reduces lipoprotein size by promoting VLDL synthesis (higher in PL) rather than chylomicron production (higher in NL) (Liu et al. 2002; Caballero et al. 2003, 2006b). Insufficient levels of dietary PL increase lipid droplets in the enterocytes of marine larvae fed a PL-deficient diet (Fontagné et al. 1998; Salhi et al. 1999; Morais et al. 2005b). If these droplets are found in a supranuclear location in the enterocyte, they can be interpreted as accumulations of NL, whereas accumulation in the basal part of the enterocyte or intercellular space is related to accumulation of chylomicrons rather than VLDL (Caballero et al. 2003, 2006a), which are more efficiently transported in fish. Both cases seem to depend on the type and amount of dietary PL and its relation to the dietary NL content, and indicate an insufficient amount of PL for lipoprotein synthesis.

Although PL can be effectively synthesized by the enterocyte mainly by the glycerol-3-phosphate pathway, as well as the MAG pathway, the efficiency of these pathways can be markedly impaired by dietary lipids (Caballero et al. 2006a). Moreover, the poor development of cell organelles where both reacylation pathways occur in the enterocyte of developing larvae (Deplano et al. 1991) can also reduce their PL synthesis potential. Nevertheless, an age-dependent deficiency in PL synthesis in marine fish larvae becomes a liability only during larviculture where rotifers and *Artemia* nauplii are ubiquitously used as live food. These zooplankters are not representative of their natural diet and are PL poor compared with copepods (Van der Meeren et al. 2008).

Among the different PLs, PC seems to be particularly important, being the main product of PL synthesis in fish enterocyte (Caballero et al. 2006a) and comprising up to 95% of the PL found in VLDL (Lie et al. 1993).

Interestingly, Field et al. (1994) showed, *in vitro*, that cellular concentrations of PC above a specific level were critical to elicit the synthesis of apolipoprotein B, an integral part of low-density lipoprotein particles, whereas the presence of other phosphoglycerides did not trigger the same response (Field and Mathur 1995). On the other hand, Kanazawa et al. (1981) demonstrated that the dietary incorporation of total PL also reduced malformation, especially twist of jaw and scoliosis, in ayu larvae, *Plecoglossus altivelis*. Geurden et al. (1997, 1998), working on carp, concluded that PI promoted growth less well than PC but reduced skeletal deformity more effectively. These findings were reinforced by Cahu et al. (2003), who reported that a diet having a PC/PI ratio of 2.18 (1.6% of DW diet) fed to European sea bass from first feeding prevented deformities during development. It is becoming increasingly apparent that other phosphoglycerides may influence fish performance albeit through different biochemical pathways.

Sandel et al. (2010) found a very high correlation between developmental age of gilthead sea bream fed high PI and gene expression for OC, a protein used as a marker for bone and cartilage growth. PI, as the second messenger IP3, mobilizes calcium by regulating the entry of this ion into the cell from the endoplasmic reticulum (Cahu et al. 2003; Tocher et al. 2008). OC or bone Gla protein (BGP) is a vitamin K–requiring protein in bone hydroxyapatite that has a strong affinity for calcium and has been correlated with the mineralization of bone (Szulc et al. 1994; Simes et al. 2008). These authors argued that dietary PI may be increasing the availability of calcium for bone mineralization, which stimulates OC production and normal bone development. In support of this, Nishimoto et al. (2003) found in carp that bone hydroxyapatite binding of BGP is enhanced in the presence of calcium ions.

2.4 Summary

In summary, EFAs have a variety of vital functions for fish larvae, and since they cannot be synthesized at adequate rates, they must be provided by the diet. Suboptimal levels of EFAs in larval feeds may give rise to reduced feeding and swimming activity, poor growth and increasing mortality, fatty livers, hydrops, deficient swim bladder inflation, abnormal pigmentation, disintegration of gill epithelia, immune deficiency, elevated basal cortisol levels, skeleton deformities, and so on. EFA requirements for fish larvae are specific and may depend on the ratios among these fatty acids, the presence of other nutrients, such as antioxidants, or several environmental factors. Described n-3 LCPUFA requirements during larval development varied from 0.05 to 3.9%. On average, the requirements for n-3 LCPUFAs in larvae of marine fish are about 3% DW diet or live prey.

Optimum DHA levels in larval feeds range from 0.5 to 2.5%. DHA plays also an important physiological function in the membranes of the major neural tissues, affecting brain and sense organs and promoting growth. Deleterious effects of excess amounts of DHA on upregulation of certain genes or severe damage of cellular structures have been only scarcely described and have been frequently associated with an imbalance between anti- and pro-oxidant substances.

Reported requirements for EPA range from 0.7 to 1.6% EPA. The metabolic impact of EPA is dependent on DHA and ARA levels as excessive dietary EPA in relation to those EFA can result in an imbalance of tissue eicosanoids, a displacement of DHA from membrane lipids, or an increased oxidation risk.

Dietary ARA requirements for marine larval fish range from 0.5 to 1.2%, whereas for freshwater fish it can be as high as 2.5%. Dietary ARA increases expression of certain genes involved in stress modulation such as GRs and heat-shock proteins. Both ARA and EPA produce certain eicosanoids that can either stimulate or inhibit immune cells in a dose-dependent manner. However, an excess of ARA can also exert important negative effects on marine fish.

The optimum EPA/ARA ratios described in the literature range from 3.5 to 5, whereas optimum dietary DHA/EPA ratios range from 1.2 to 8. Dietary ARA and DHA, as well as EPA/ARA ratios, have a profound effect on juvenile pigmentation in flatfish. Besides, LCPUFAs are important for resistance to handling, temperature, salinity, or other types of acute stressors. Even though EFA-rich PLs are highly represented in lipids of the prey species for fish larvae in nature, enrichment protocols result in very high NL levels in rotifers or *Artemia*. The beneficial effects of dietary PLs on growth and survival have been demonstrated in larval stages of various species.

Literature cited

Abou-Samra, A.B., Catt, K.J., and Auilera, G. 1986. Role of arachidonic acid in the regulation of adreno-corticotropin release from rat anterior pituitary cell cultures. *Endocrinology* 119:1427–1431.

Ako, H., Kraul, S., and Tamaru, C. 1991. Pattern of fatty acid loss in several warmwater fish species during early development. In: Lavens, P., Sorgeloos, P., Jaspers, E., and Ollevier, F. (eds.) *Larvi '91: Fish and Crustacean Larviculture Symposium*. Special Publication No. 15. European Aquaculture Society, Ghent, Belgium, pp. 23–25.

Atalah, E., Hernández-Cruz, C.M., Roo, J., Fernández-Palacios, H., and Izquierdo, M. 2010a. Importance of relative levels of dietary ARA and EPA for culture performance of gilthead seabream (*Sparus aurata*) larvae. *Aquaculture Research* doi:10.1111/j.1365-2109.2010.02716.x.

Atalah, E., Hernández-Cruz, C.M., Ganuza, E., Benítez-Santana, T., Ganga, R., Roo, J., Montero, D., and Izquierdo, M.S. 2010b. Importance of dietary arachidonic acid for survival, growth and stress resistance of larval European sea bass (*Dicentrarchus labrax*) fed high dietary docosahexaenoic and eicosapentaenoic acids. *Aquaculture Research* doi:10.1111/j.1365-2109.2010.02714.x.

Austreng, E., Skrede, A., and Eldegard, Å. 1979. Effect of dietary fat source on the digestibility of fat and fatty acids in rainbow trout and mink. *Acta Agriculturae Scandinavica* 29:119–126.

Axelrod, J., Burch, A., and Jelsema, C.L. 1988. Receptor-mediated activation of phospholipase A2 via GTP-binding proteins: arachidonic acid and its metabolites as second messengers. *Trends in Neurosciences* 11:117–123.

Bach, A.C., and Babayan, V.K. 1982. Medium-chain triglycerides: an update. *The American Journal of Clinical Nutrition* 36:950–962.

Balmer, J.E., and Blomhoff, R. 2002. Gene expression regulation by retinoic acid. *Journal of Lipid Research* 43:1773–1808.

Barton, B.A., and Iwama, G.K. 1991. Physiological changes in fish from stress in aquaculture with emphasis on the stress response and effects of corticosteroids. *Annual Reviews of Fish Diseases* 1:3–36.

Bell, J.G., Raynard, R.S., and Sargent, J.R. 1991a. The effect of dietary linoleic acid on the fatty acid composition of individual phospholipids and lipoxygenase products from gills and leucocytes of Atlantic salmon (*Salmo salar*). *Lipids* 26:445–450.

Bell, J.G., Mc Vicar, A.H., Park, M.T., and Sargent, J.R. 1991b. High dietary linoleic acid affects the fatty acid compositions of individual phospholipids from tissues of Atlantic salmon (*Salmo salar*): association with stress susceptibility and cardiac lesions. *The Journal of Nutrition* 121:1163–1172.

Bell, J.G., Dick, J.R., McVicar, A.H., Sargent, J.R., and Thompson, K.D. 1993. Dietary sunflower, linseed and fish oils affect phospholipid fatty acid composition, development of cardiac lesions, phospholipase activity and eicosanoid production in Atlantic salmon (*Salmo salar*). *Prostaglandins, Leucotrienes and Eicosanoid Fatty Acids* 49:665–673.

Bell, J.G., Tocher, D.R., and Sargent, J.R. 1994. Effect of supplementation with 20:3(n-6), 20:4(n-6) and 20:5(n-3) on the production of prostaglandins E and F of the 1-, 2- and 3-series in turbot (*Scophthalmus maximus*) brain astroglial cells in primary culture. *Biochimica Biophysica Acta* 1211:335–342.

Bell, M.V., Batty, R.S., Dick, J.R., Fretwell, K., Navarro, J.C., and Sargent, J.R. 1995. Dietary deficiency of docosahexaenoic acid impairs vision at low light intensities in juvenile herring (*Clupea harengus* L.). *Lipids* 30:443–449.

Benítez-Santana, T., Masuda, R., Juárez Carrillo, E., Ganuza, E., Valencia, A., Hernández-Cruz, C.M., and Izquierdo, M.S. 2007. Dietary n-3 HUFA deficiency induces a reduced visual response in gilthead seabream *Sparus aurata* larvae. *Aquaculture* 264:408–417.

Bessonart, M., Izquierdo, M.S., Salhi, M., Hernandez-Cruz, C.M., Gonzalez, M.M., and Fernandez-Palacios, H. 1999. Effect of dietary arachidonic acid levels on growth and fatty acid composition of gilthead sea bream (*Sparus aurata* L.) larvae. *Aquaculture* 179:265–275.

Betancor, M.B., Atalah, E., Caballero, M.J., Benitez-Santana, T., Roo, J., Montero, D., and Izquierdo, M.S. 2010. Dietary α-tocopherol in weaning diets for European sea bass (*Dicentrarchus labrax*) improves survival and tissue damage caused by excess dietary DHA contents. *Aquaculture Nutrition* doi:10.1111/j.1365-2095.2009.00741x.

Blaxter, J.H.S., and Jones, M.P. 1967. The development of the retina and retinomotor responses in the herring. *Journal of the Marine Biological Association of the United Kingdom* 47: 677–697.

Boehlert, G.W., and Yoklavich, M.M. 1984. Carbon assimilation as a function of ingestion rate in larval pacific herring, *Clupea harengus pallasi* Valenciennes. *Journal of Experimental Marine Biology and Ecology* 79:251–262.

Borlongan, I.G., and Benitez, L.V. 1992. Lipid and fatty acid composition of milkfish (*Chanos chanos* Forksskal) grown in freshwater and seawater. *Aquaculture* 104:79–89.

Bottino, N.R., Vandenburg, G.A., and Reiser, R. 1967. Resistance of certain long chain polyunsaturated fatty acids of marine oils to pancreatic lipase hydrolysis. *Lipids* 2:489–493.

Bowden, L.A., Restall, C.J., and Rowley, A.F. 1996. The influence of environmental temperature on membrane fluidity, fatty acid composition and lipoxygenase product generation in head kidney leucocytes of the rainbow trout, *Oncorhynchus mykiss*. *Comparative Biochemistry and Physiology* 115B:375–382.

Bransden, M.P., Carter, C.G., and Nichols, P.D. 2003. Replacement of fish oil with sunflower oil in feeds for Atlantic salmon (*Salmo salar* L.): effect on growth performance, tissue fatty acid composition and disease resistance. *Comparative Biochemistry and Physiology, Part B* 135:611–625.

Bransden, M.P., Dunstan, G.A., Battaglene, S.C., Cobcroft, J.M., Morehead, D.T., Kolkovski, S., and Nichols, P.D. 2004. Influence of dietary n-3 long chain PUFA on body concentrations of 20:5n-3, 22:5n-3 and 22:6n-3 in larvae of a marine teleost fish from Australian waters, the striped trumpeter (*Latris lineata*). *Lipid* 39: 215–222.

Bransden, M.P., Battaglene, S.C., Morehead, D.T., Dunstan, G.A., and Nichols, P.D. 2005. Effect of dietary 22:6n-3 on growth, survival and tissue fatty acid profile of striped trumpeter (*Latris lineata*) larvae fed enriched *Artemia*. *Aquaculture* 243:331–344.

Brinkmeyer, R., and Holt, G.J. 1998. Highly unsaturated fatty acids in diets for red drum (*Sciaenops ocellatus*) larvae. *Aquaculture* 161:253–268.

Brown, J.A., and Bucknall, R.M. 1986. Antidiuretic and cardiovascular actions of prostaglandin E2 in the rainbow trout *Salmo gairdneri*. *General and Comparative Endocrinology* 61:330–337.

Brown, M.F. 1994. Modulation of rhodopsin function by properties of the membrane bilayer. *Chemistry and Physics of Lipids* 73:159–180.

Bugajski, J., Glod, R., Gadek-Michalska, A., and Bugajski, A.J. 2001. Involvement of constitutive (COX-1) and inducible cyclooxygenase (COX-2) in the adrenergic-induced ACTH and corticosterone secretion. *Journal of Physiology and Pharmacology* 52:795–809.

Caballero, M.J., Obach, A., Rosenlund, G., Montero, D., Gisvold, M., and Izquierdo, M.S. 2002. Impact of different dietary lipid sources on growth, lipid digestibility, tissue fatty acid composition and histology of rainbow trout, *Oncorhynchus mykiss*. *Aquaculture* 214:253–271.

Caballero, M.J., Izquierdo, M.S., Kjorsvik, E., Montero, D., Socorro, J., Fernández, A., and Rosenlund, G. 2003. Morphological aspects of the intestinal cells from gilthead seabream (*Sparus aurata*) fed diets containing different lipid sources. *Aquaculture* 225:325–340.

Caballero, M.J., Gallardo, G., Robaina, L., Montero, D., Fernández, A., and Izquierdo, M.S. 2006a. Vegetable lipid sources affect *in vitro* biosynthesis of triacylglycerols and phospholipids in the intestine of sea bream (*Sparus aurata*). *British Journal of Nutrition* 95:448–454.

Caballero, M.J., Torstensen, B., Robaina, L., Montero, D., and Izquierdo, M.S. 2006b. Vegetable oils affect composition of lipoproteins in sea bream (*Sparus aurata*). *British Journal of Nutrition* 96:830–839.

Cahu, C., Zambonino Infante, J.L., and Barbosa, V. 2003. Effect of dietary phospholipid level and phospholipid: neutral lipid value on the development of sea bass (*Dicentrarchus labrax*) larvae fed a compound diet. *British Journal of Nutrition* 90:21–28.

Chen, Q., Sternby, B., Åkesson, B., and Nilsson, Å. 1990. Effects of human pancreatic lipase-colipase and carboxyl ester lipase on eicosapentaenoic and arachidonic acid ester bonds of triacylglycerols rich in fish oil fatty acids. *Biochimica Biophysica Acta* 1044:111–117.

Cooke, B.A. 1999. Signal transduction involving cyclic AMP–dependent and cyclic AMP–independent mechanisms in the control of steroidogenesis. *Molecular and Cellular Endocrinology* 151:25–35.

Copeman, L.A., Parrsh, C.C., Brown, J.A., and Harel, M. 2002. Effects of docosahexaenoic, eicosapentaenoic, and arachidonic acids on the early growth, survival, lipid composition and pigmentation of yellowtail flounder (*Limada ferruginea*): a live food enrichment experiment. *Aquaculture* 210:285–304.

Cousin, J.C.B., Baudin-Laurencin, F., and Gabaudan, J. 1987. Ontogeny of enzymatic activities in fed and fasting turbot, *Scophthalmus maximus* L. *Journal of Fish Biology* 30: 15–33.

Coutteau, P., Geurden, I., Camara, M.R., Bergot, P., and Sorgeloos, P. 1997. Review on the dietary effects of phospholipids in fish and crustacean larviculture. *Aquaculture* 155:149–164.

Daikoku, T., Yano, I., and Masui, M. 1982. Lipid and fatty acid composition and their changes in the different organs and tissue of guppy, *Poecillia reticulata*, on sea water adaptation. *Comparative Biochemistry and Physiology* 73a:167–174.

Dantagnan, P., and Izquierdo, M.S. 2007. Requerimientos de ácidos grasos en larvas de peces: efecto de factores ambientales. In: Dantagnan, P., Bórquez, A., Valdebenito, I., and Hernández, A. (eds.) *Producción De Larvas De Peces. Innovación Y Avances En La Nutrición Para Contribuir Al Mejoramiento Y*

Escalamiento De Los Cultivos. UC Temuco, Temuco, Chile, pp. 149–161.

Dantagnan, H., Bórquez, A.S., Valdebenito, I.N., Salgado, I.A., Serrano, E.A., and Izquierdo, M.S. 2007. Lipid and fatty acid composition during embryo and larval development of puye *Galaxias maculatus* Jenyns, 1842, obtained from estuarine, freshwater and cultured populations. *Journal of Fish Biology* 70:770–781.

Dantagnan, P., Bórquez, A., Hernández, A., and Izquierdo, M. In press. Effect of EPA/DHA ratios on growth and survival of *Galaxias maculatus* (Jenyns, 1842) larvae reared under different salinity regimes. *Aquaculture Research*.

Deplano, M., Díaz, J.P., Connes, R., Kentouri-Divanach, M., and Cavalier, F. 1991. Appearance of lipid-absorption capacities in larvae of the sea bass *Dicentrarchus labrax* during transition to the exotrophic phase. *Marine Biology* 108:361–371.

Di Luigi, L., Guidetti, L., Romanelli, F., Baldari, C., and Conte, D. 2001. Acetylsalicylic acid inhibits the pituitary response to exercise-related stress in humans. *Medicine and Science in Sports and Exercise* 33:2029–2035.

Estévez, A., Ishikawa, M., and Kanazawa, A. 1997. Effects of arachidonic acid on pigmentation and fatty acid composition of Japanese flounder, *Paralichthys olivaceus* (Temminck and Schlegel). *Aquaculture Research* 28: 279–289.

Estévez, A., McEvoy, L.A., Bell, J.G., and Sargent, J.R. 1999. Growth, survival, lipid composition and pigmentation of turbot larvae fed live-prey enriched in arachidonic (ARA) and eicosapentaenoic (EPA) acids. *Aquaculture* 180: 321–343.

Estévez, A., Kaneko, T., Seikai, T., Tagawa, M., and Tanaka, M. 2001. ACTH and MSH production in Japanese flounder (*Paralichthys olivaceus*) larvae fed arachidonic acid–enriched prey. *Aquaculture* 192:309–319.

Farkas, T., Csegri, I., Majors, F., and Olah, J. 1980. Metabolism of fatty acids in fish. III. Combined effect of environmental temperature and diet on formation and deposition of fatty acids in the carp, *Cyprinus carpio* Linnaeus, 1758. *Aquaculture* 20:29–40.

Farkas, T., Kitajka, K., Fodor, E., Csengeri, I., Lahdes, E., Yeo, Y.K., Krasznai, Z., and Halver, J.E. 2000. Docosahexaenoic acid–containing phospholipid molecular species in brains of vertebrates. *Proceedings of the National Academy of Sciences of the United States of America* 97(12):6362–6366.

Farkas, T., Fodor, E., Kitajka, K., and Halver, J.E. 2001. Response of fish membranes to environmental temperature. *Aquaculture Research* 32:645–655.

Farrens, D.L., Altenbach, C., Yang, K., Hubbell, W.L., and Khorana, H.G. 1996. Requirement of rigid-body motion of transmembrane helices for light activation of rhodopsin. *Science* 274:768–770.

Faustino, M. 2002. Developmental osteology of the gilthead seabream (*Sparus aurata* L.). Thesis, Faro University, Portugal.

Field, F.G., and Mathur, S.N. 1995. Intestinal lipoprotein synthesis and secretion. *Progress in Lipid Research* 34:185–198.

Field, F.J., Born, E., Chen, H., Murthy, S., and Mathur, S.N. 1994. Regulation of apolipoprotein B secretion by biliary lipids in CaCo-2 cells. *Journal of Lipid Research* 35:749–762.

Finn, R.N., Henderson, R.J., and Fyhn, H.J. 1995. Physiological energetics of developing embryos and yolk-sac larvae of Atlantic cod (*Gadus morhua*). II. Lipid metabolism and enthalpy balance. *Marine Biology* 124:371–379.

Fontagné, S., Geurden, I., Escaffre, A.M., and Bergot, P. 1998. Histological changes induced by dietary phospholipids in intestine and liver of common carp (*Cyprinus carpio* L.) larvae. *Aquaculture* 161:213–223.

Fontagné, S., Robin, J., Corraze, G., and Bergot, P. 2000. Growth and survival of European sea bass (*Dicentrarchus labrax*) larvae fed from first feeding on compound diets containing medium-chain triacylglycerols. *Aquaculture* 190:261–271.

Fountoulaki, E., Alexis, M.N., Nengas, I., and Venon, B. 2003. Effects of dietary arachidonic acid (20:4n-6), on growth, body composition, and tissue fatty acid profile of gilthead bream fingerlings (*Sparus aurata* L.). *Aquaculture* 225:309–323.

Frankel, S. 1998. The more distantly related of the actin-related proteins. In: Vale, R., and Kreis, T. (eds.) *Guidebook to the Cytoskeletal and Motor Proteins*. Oxford University Press, Oxford, pp. 49–51.

Fraser, A.J., Gamble, J.C., and Sargent, J.R. 1988. Changes in lipid content, lipid class composition and fatty acid composition of developing eggs and unfed larvae of cod (*Gadus morhua*). *Marine Biology* 99:307–313.

Froyland, L., Madsen, L., Vaagenes, H., Totland, G.K., Auwerx, J., Kryviv, H., Staels, B., and Berge, R.K. 1997. Mitochondrion is the principal target for nutritional and pharmacological control of triglyceride metabolism. *Journal of Lipid Research* 38(9):1851–1858.

Furuita, H., Takeuchi, T., Watanabe, T., Fujimoto, H., Sekiya, S., and Imaizumi, K. 1996. Requirements of larval yellowtail for eicosapentaenoic acid, docosahexaenoic acid and n-3 highly unsaturated fatty acids. *Fisheries Science* 62:372–379.

Furuita, H., Takeuchi, T., and Uematsu, K. 1998. Effect of eicosapentaenoic and docosahexaenoic acids on growth, survival and brain development of larval Japanese flounder *Paralichthys olivaceus*. *Aquaculture* 161:269–279.

Furuita, H., Konishi, K., and Takeuchi, T. 1999. Effect of different levels of eicosapentaenoic acid and docosahexaenoic acid in *Artemia* nauplii on growth, survival and salinity tolerance of larvae of the Japanese flounder, *Paralichthys olivaceus*. *Aquaculture* 170: 59–69.

Furuita, H., Yamamoto, T., Shima, T., Suzuki, N., and Takeuchi, T. 2003. Effect of arachidonic acid levels in broodstock diet on larval and egg quality of Japanese flounder *Paralichthys olivaceus*. *Aquaculture* 220:725–735.

Gadek-Michalska, A., Bugajski, J., Bugajski, A.J., and Glod, R. 2002. Effect of adrenergic antagonists and cyclooxygenase inhibitors on the nicotine-induced hypothalamic–pituitary–adrenocortical activity. *Journal of Physiology and Pharmacology* 53:275–287.

Ganga, R., Bell, J.G., Montero, D., Robaina, L., Caballero, M.J., and Izquierdo, M.S. 2005. Effect of feeding gilthead seabream (*Sparus aurata*) with vegetable lipid sources on two potential immunomodulator products: prostanoids and leptins. *Comparative Biochemistry and Physiology* 142:410–418.

Ganga, R., Tort, L., Acerete, L., Montero, D., and Izquierdo, M.S. 2006. Modulation of ACTH-induced cortisol release by polyunsaturated fatty acids in interrenal cells from gilthead seabream, *Sparus aurata*. *Journal of Endocrinology* 190:39–45.

Ganga, R., Bell, J.G., Montero, D., Vraskou, Y., Ganuza, E., Tort, L., Vaquero, A., and Izquierdo, M.S. In press. ACTH-stimulated cortisol-release by the head-kidney-interrenal tissue from sea bream (*Sparus aurata*) fed with linseed and soybean oil. *British Journal of Nutrition*.

Gapasin, R.S.J., and Duray, M.N. 2001. Effects of DHA-enriched live food on growth, survival and incidence of opercular deformities in milkfish (*Chanos chanos*). *Aquaculture* 193:49–63.

German, B., Bruckner, G., and Kinsella, J. 1983. Evidence against a PGF4 prostaglandin structure in trout tissue: a correction. *Prostaglandins* 6:207–210.

Geurden, I., Coutteaux, P., and Sorgeloos, P. 1995. Dietary phospholipids for European sea bass (*Dicentrarchus labrax* L.) during first ongrowing. In: Lavens, P., Jaspers, E., and Roelants, I. (eds.) *Larvi '95: Fish and Shellfish Larviculture Symposium*, EAS Special Publication 24. European Aquaculture Society, Ghent, Belgium, pp. 175–178.

Geurden, I., Charlon, N., Marion, D., and Bergot, P. 1997. Influence of purified soybean phospholipids on early development of common carp. *Aquaculture International* 5:137–149.

Geurden, I., Marion, D., Charlon, N., Coutteau, P., and Bergot, P. 1998. Comparison of different soybean phospholipidic fraction as dietary supplements for common carp *Cyprinus carpio* larvae. *Aquaculture* 161:213–223.

Gisbert, E., Villeneuve, L., Zambonino Infante, J.L., Quazuguel, P., and Cahu, C.L. 2005. Dietary phospholipids are more efficient than neutral lipids for long chain polyunsaturated fatty acid supply in European sea bass *Dicentrarchus labrax* larval development. *Lipids* 40:609–618.

Gjellesvik, D.R., Lombard, D., and Walther, B.T. 1992. Pancreatic bile salt dependent lipase from cod (*Gadus morhua*): purification and properties. *Biochimica et Biophysica Acta* 1124: 123–134.

Grosfield, A., Feller, S.E., and Pitman, M.C. 2006. A role for direct interactions in the modulation of rhodopsin by n-3 polyunsaturated lipids. *Proceedings of the National Academy of Sciences of the United States of America* 103(13):4888–4893.

Guederley, H., St. Pierre, J., Couture, P., and Hulbert, A.J. 1997. Plasticity of the properties of mitochondria from rainbow trout red muscle with seasonal acclimatization. *Fish Physiology and Biochemistry* 16:531–541.

Gupta, O.P., Lahlou, B., Botella, J., and Porthé-Nibelle, J. 1985. *In vivo* and *in vitro* studies on the release of cortisol from interrenal tissue in trout. I. Effects of ACTH and prostaglandins. *Experimental Biology* 43:201–212.

Hadas, E. 1998. The influence of dietary phospholipids on feeding rate and absorption of fatty acids in the larvae of the gilthead seabream (*Sparus aurata*). MSc thesis, Hebrew University of Jerusalem, Rehovot, Israel.

Hadas, E., Koven, W., Sklan, D., and Tandler, A. 2003. The effect of dietary phosphatidylcholine on the assimilation and distribution of ingested free oleic acid (18:1n-9) in gilthead sea bream (*Sparus aurata*) larvae. *Aquaculture* 217:577–588.

Hamre, K., and Harboe, T. 2008. Critical levels of essential fatty acids for normal pigmentation in Atlantic halibut (*Hippoglossus hippoglossus* L.) larvae. *Aquaculture* 227:101–108.

Hamre, K., Holen, E., and Moren, M. 2007. Pigmentation and eye migration in Atlantic halibut (*Hippoglossus hippoglossus* L.) larvae: new findings and hypotheses. *Aquaculture Nutrition* 13:65–80.

Harel, M., Lund, E., Gavasso, S., Herbert, R., and Place, A.R. 2000. Modulation of arachidonate and docosahexaenoate in *Morone chrysops* larval tissues and the effect on growth and survival. *Lipids* 35:1269–1280.

Harel, M., Gavasso, S., Leshin, J., Gubernatis, A., and Place, A.R. 2001. The effect of tissue docosahexaenoic and arachidonic acids levels on hypersaline tolerance and leucocyte composition in striped bass (*Morone saxatilis*) larvae. *Fish Physiology and Biochemistry* 24:113–124.

Hashimoto, M., Hossain, M.S., Yamasaki, H., Yazawa, K., and Masumura, S. 1999. Effects of eicosapentaenoic acid and docosahexaenoic acid on plasma membrane fluidity of aortic endothelial cells. *Lipids* 34:1297–1304.

Hastings, N., Agaba, M., Tocher, D.R., Leaver, M.J., Sargent, J.R., and Teale, A.J. 2001. A vertebrate fatty acid desaturase with delta5 and delta6 activity. *Proceedings of the National Academy of Sciences of the United States of America* 98:14304–14309.

Hazel, J.R., and Williams, E.E. 1990. The role of alterations in membrane lipid composition in enabling physiological adaptation of organisms to their physical environment. *Progress in Lipid Research* 29:167–227.

Hernández-Cruz, C.M., Salhi, M., Bessonart, M., Izquierdo, M.S., Gonzalez, M.M., and Fernandez-Palacios, H. 1999. Rearing techniques for red porgy (*Pagrus pagrus*) during larval development. *Aquaculture* 179:489–497.

Hochachka, P.W., and Somero, G.N. 2002. *Biochemical Adaptation*. Oxford University Press, Oxford, UK.

Hoehne-Reitan, K., Kjørsvik, E., and Reitan, K.I. 2001. Bile salt dependent lipase in larval turbot, as influenced by density and lipid content of fed prey. *Journal of Fish Biology* 58:746–754.

Hong, S., Tjonahen, E., Morgan, E.L., Lu, Y., Serhan, C.N., and Rowley, A.F. 2005. Rainbow trout (*Oncorhynchus mykiss*) brain cells biosynthesize novel docosahexaenoic acid–derived resolvins and protectins-mediator lipidomic analysis. *Prostaglandins & Other Lipid Mediators* 78:107–116.

Hontela, A. 2005. Adrenal toxicology: Environmental pollutants and the HPI axis. *Biochemistry and Molecular Biology of Fishes* 6:331–363.

Horrobin, D.F. 1983. The regulation of prostaglandin biosynthesis by the manipulation of essential fatty acid metabolism. *Journal of Review of Pure and Applied Sciences* 4:339–383.

Hubbell, W.L., Altenbach, C., Hubbell, C.M., and Khorana, H.G. 2003. Rhodopsin structure, dynamics, and activation: a perspective from crystallography, site-directed spin labeling, sulfhydryl reactivity, and disulfide cross-linking. *Advances in Protein Chemistry* 63:243–290.

Hung, S.S.O., and Lutes, P.B. 1988. A preliminary study on the nonessentiality of lecithin for hatchery produced juvenile white sturgeon (*Acipenser transmontanus*). *Aquaculture* 68: 353–360.

Iijima, N., Nakamura, M., Uematsu, K., and Kayama, M. 1990. Partial purification and characterization of phospholipase A2 from the hepatopancreas of red sea bream, *Pagrus major*. *Nippon Suisan Gakkaishi* 56:1331–1339.

Iijima, N., Tanaka, S., and Ota, Y. 1998. Purification and characterization of bile salt–activated lipase from the hepatopancreas of red sea bream, *Pagrus major*. *Fish Physiology and Biochemistry* 18:59–69.

Izquierdo, M.S. 1996. Review article: essential fatty acid requirements of cultured marine fish larvae. *Aquaculture Nutrition* 2:183–191.

Izquierdo, M.S. 2005. Essential fatty acid requirements in Mediterranean fish species. *Cahiers Options Mediterraneennes* 63:91–102.

Izquierdo, M.S., and Henderson, R.J. 1998. The determination of lipase and phospholipase activities in gut contents of turbot (*Scophthalmus maximus*) by fluorescence-based assays. *Fish Physiology and Biochemistry* 19:153–162.

Izquierdo, M.S., Watanabe, T., Takeuchi, T., Arakawa, T., and Kitajima, C. 1989. Requirement of larval red seabream *Pagrus major* for essential fatty acids. *Bulletin of the Japanese Society of Scientific Fisheries* 55(5):859–867.

Izquierdo, M.S., Arakawa, T., Takeuchi, T., Haroun, R., and Watanabe, T. 1992. Effect of n-3 HUFA levels in *Artemia* on growth of larval Japanese flounder (*Paralichthys olivaceus*). *Aquaculture* 105:73–82.

Izquierdo, M.S., Socorro, J., Arantzamendi, L., and Hernández-Cruz, C.M. 2000. Recent advances in lipid nutrition in fish larvae. *Fish Physiology and Biochemistry* 22:97–107.

Izquierdo, M.S., Montero, D., Robaina, L., Caballero, M.J., Rosenlund, G., and Ginés, R. 2005. Alterations in fillet fatty acid profile and flesh quality in gilthead seabream (*Sparus aurata*) fed vegetable oils for a long term period. Recovery of fatty acid profiles by fish oil feeding. *Aquaculture* 250:431–444.

Izquierdo, M.S., Robaina, L., Juárez, E., Oliva, R., Hernández-Cruz, C.M., and Afonso, J.M. 2008. Regulation of growth, fatty acid composition and delta 6 desaturase expression by dietary lipids in gilthead seabream larvae (*Sparus aurata*). *Fish Physiology and Biochemistry* 34:117–127.

Izquierdo, M., Socorro, J., and Roo, J. 2010. Studies on the appearance of skeleton anomalies in red porgy: effect of culture intensiveness, feeding habits and nutritional quality of live preys. *Journal of Applied Ichthyology* 26(2):320–326.

Jobling, M., and Bendiksen, E.A. 2003. Dietary lipids and temperature interact to influence tissue fatty acid compositions of Atlantic salmon, *Salmo salar* L., parr. *Aquaculture Research* 34:1423–1441.

Kanazawa, A. 1985. Essential fatty acid and lipid requirement of fish. In: Cowey, C.B., Mackie, A.M., and Bell, J.G. (eds.) *Nutrition and Feeding in Fish*. Academic Press, London, pp. 281–298.

Kanazawa, A. 1993. Nutritional mechanisms involved in the occurrence of abnormal pigmentation in hatchery-reared flatfish. *Journal of the World Aquaculture Society* 24:162–166.

Kanazawa, A., Teshima, S., Inamori, S., Iwashita, T., and Nagao, A. 1981. Effects of phospholipids on growth, survival rate and incidence of malformation in the larval ayu. *Memoirs of the Faculty of Fisheries of Kagoshima University* 30:301–309.

Kanazawa, A., Teshima, S.-I., Imatanaka, M., Imada, O., and Inouw, A. 1982. Tissue uptake of radioactive eicosapentaenoic acid in the red sea bream. *Bulletin of the Japanese Society of Scientific Fisheries* 48(10):1441–1444.

Kanazawa, A., Teshima, S., Kobayashi, T., Takae, M., Iwashita, T., and Uehara, R. 1983a. Necessity of dietary phospholipids for growth of the larval ayu. *Memoirs of the Faculty of Fisheries of Kagoshima University* 32:115–120.

Kanazawa, A., Teshima, S., Inamori, S., and Matsubara, H. 1983b. Effects of dietary phospholipids on growth of the larval red sea bream and knife jaw. *Memoirs of the Faculty of Fisheries of Kagoshima University* 32:109–114.

Kamler, E., Wolnicki, J., Kamiński, R., and Sikorska, J. 2008. Fatty acid composition, growth and morphological deformities in juvenile cyprinid, *Scardinius erythrophthalmus* fed formulated diet supplemented with natural food. *Aquaculture* 278:69–76.

Karsenty, G., and Wagner, E.F. 2002. Reaching a genetic and molecular understanding of skeletal development. *Developmental Cell* 2:389–406.

Kayama, M., Hirata, M., and Hisai, T. 1986. Effect of water temperature on the desaturation of fatty acids in carp. *Bulletin of the Japanese Society of Scientific Fisheries (Nissuishi)* 52(5):853–857.

Khozin-Goldberg, I., Cohen, Z., Pimenta-Leibowitz, M., Nechev, J., and Zilberg, D. 2006. Feeding with arachidonic acid–rich triacylglycerols from the microalga *Parietochloris incisa* improved recovery of guppies from infection with *Tetrahymena* sp. *Aquaculture* 255:142–150.

King, W., and Berlinsky, D. 2006. Whole-body corticosteroid and plasma cortisol concentrations in larval and juvenile Atlantic cod *Gadus morhua* L., following acute stress. *Aquaculture Research* 37:1282–1289.

Kinsella, J.E., Lokesh, B., Broughton, S., and Whelan, J. 1990. Dietary polyunsaturated fatty acids and eicosanoids: potential effects on the modulation of inflammatory and immune cells. An overview. *Nutrition* 6:24–44.

Kiron, V., Fukuda, H., Takeuchi, T., and Watanabe, T. 1995. Essential fatty acid nutrition and defence mechanisms in rainbow trout *Oncorhynchus mykiss*. *Comparative Biochemistry and Physiology* 111A:361–367.

Kliewer, S.A., Sundseth, S.S., and Jones, S.A. 1997. Fatty acids and eicosanoids regulate gene expression through direct interactions with peroxisome proliferator–activated receptors alpha and gamma. *Proceedings of the National Academy of Sciences of the United States of America* 94:4318–4323.

Kokkinos, P.P., Shaye, R., and Alam, B.S. 1993. Dietary lipids, prostaglandin E2 levels, and tooth movement in alveolar bone of rats. *Calcified Tissue International* 53:333–337.

Koven, W.M., Kissil, G.W., and Tandler, A. 1989. Lipid and n-3 requirement of *Sparus aurata* larvae during starvation and feeding. *Aquaculture* 79:185–191.

Koven, W.M., Tandler, A., Sklan, D., and Kissil, G.W. 1993. The association of eicosapentaenoic and docosahexaenoic acids in the main phospholipids of different-age *Sparus aurata* larvae with growth. *Aquaculture* 116:71–82.

Koven, W.M., Henderson, R.J., and Sargent, J.R. 1994a. Lipid digestion in turbot (*Scophthalmus maximus*). I: lipid class and fatty acid composition of digesta from different segments of the digestive tract. *Fish Physiology and Biochemistry* 13:69–79.

Koven, W.M., Henderson, R.J., and Sargent, J.R. 1994b. Lipid digestion in turbot (*Scophthalmus maximus*). II: lipolysis *in vitro* of 14C-labeled triacylglycerol, cholesterol ester and phosphatidylcholine by digesta from different segments of the digestive tract. *Fish Physiology and Biochemistry* 13:275–283.

Koven, W.M., Parra, G., Kolkovski, S., and Tandler, A. 1998. The effect of dietary phosphatidylcholine and its constituent fatty acids on microdiet ingestion and fatty acid absorption rate in gilthead sea bream, *Sparus auratus*, larvae. *Aquaculture Nutrition* 4:39–45.

Koven, W.M., Barr, Y., Lutzky, S., Ben Atia, I., Weiss, R., Harel, M., Behrens, P., and Tandler, A. 2001a. The effect of dietary arachidonic acid (20:4n-6) on growth, survival and resistance to handling stress in gilthead sea bream (*Sparus aurata*) larvae. *Aquaculture* 193:107–122.

Koven, W., Van Anholt, R., Lutzky, S., Ben-Atia, I., Gamsiz, K., Weiss, R., and Tandler, R. 2001b. The importance of arachidonic acid, as a modulator of stress resistance through the hypothalamus–pituitary–interrenal axis, in different aged gilthead seabream larvae. In: Hendry, C.I., Van Stappen, G., Wille, M., and Sorgeloos, P. (eds.) *Larvi '01: Fish and Shellfish Larviculture Symposium*, Special Publication 30. European Aquaculture Society, Oostende, Belgium, pp. 292–293.

Koven, W., Van Anholt, R., Lutzky, S., Ben Atia, I., Nixon, O., Ron, B., and Tandler, A. 2003. The effect of dietary arachidonic acid on growth, survival, and cortisol levels in different-age gilthead sea bream larvae (*Sparus aurata*) exposed to handling or daily salinity change. *Aquaculture* 228:307–320.

Lacroix, M., and Hontela, A. 2001. Regulation of acute cortisol synthesis by cAMP-dependent protein kinase A and protein kinase C in a teleost species, the rainbow trout (*Oncorhynchus mykiss*). *Journal of Endocrinology* 169:71–78.

Lall, S.P. 2000. Nutrition and health of fish. In: Cruz-Suárez, L.E., Ricque-Marie, D., Tapia-Salazar, M., Olvera-Novoa, M.A., and Civera-Cerecedo, R. (eds.) *Advances En Nutrición Acuícola V. Memorias Del V Simposium Internacional De Nutrición Acuícola*, November 19–22, 2000, Mérida, Yucatán, Mexico, pp. 18–33.

Lie, J., De Crom, R., Van Gent, T., Van Haperen, R., Scheek, L., Lankhuizen, I., and Van Tol, A. 1993. Elevation of plasma phospholipid transfer protein in transgenic mice increases VLDL secretion. *Journal of Lipid Research* 43:1875–1880.

Lie, Ø., Lied, E., and Lambertsen, G. 1987. Lipid digestion in cod (*Gadus morhua*). *Comparative Biochemistry and Physiology* 88B:697–700.

Lie, Ø., Hemre, G.I., and Lambertsen, G. 1992. Influence of dietary fatty acids on the glycerophospholipid composition in organs of cod (*Gadus morhua*). *Lipid* 27:770–775.

Liu, D., Veit, H.P., and Denbow, D.M. 2004. Effects of long-term dietary lipids on mature bone mineral content, collagen, crosslinks, and prostaglandins E2 production in Japanese quail. *Poultry Science* 83:18767–11883.

Liu, J., Caballero, M.J., El-Sayed Ali, T., Izquierdo, M.S., Hernández Cruz, C.M., Valencia, A., and Fernández-Palacios, H. 2002. Necessity of dietary lecithin and eicosapentaenoic acid for growth, survival, stress resistance and lipoprotein formation in gilthead sea bream (*Sparus aurata*). *Fisheries Science* 68:1165–1172.

Lund, I., Steenfeld, S.J., and Hansen, B.W. 2007. Effect of dietary arachidonic acid, eicosapentaenoic acid and docosahexaenoic acid on survival, growth and pigmentation in larvae of common sole (*Solea solea* L). *Aquaculture* 273:532–544.

Lund, I., Steenfeldt, S.J., Banta, G., and Winding Hansen, B. 2008. The influence of dietary concentrations of arachidonic acid and eicosapentaenoic acid at various stages of larval ontogeny on eye migration, pigmentation and prostaglandin content of common sole larvae (*Solea solea* L). *Aquaculture* 276:143–153.

Madsen, L., Rustan-Arild, C., Vaagenes, H., Berge, K., Dyroy, E., and Berge, R.K. 1999. Eicosapentaenoic and docosahexaenoic acid affect mitochondrial and peroxisomal fatty acid oxidation in relation to substrate preference. *Lipid* 34(9):951–963.

Maeda, K., and Naganuma, M. 1997. Melanocyte-stimulating properties of secretory phospholipase A2. *Photochemistry and Photobiology* 65:145–149.

Masuda, R. 1995. The ontogeny of schooling behaviour in the striped jack *Pseudocaranx dentex*. D.Phil. thesis, University of Tokyo, Tokyo, Japan. (in Japanese)

Masuda, R., Takeuchi, T., Tsukamoto, K., Sato, H., Shimizu, K., and Imaizumi, K. 1999. Incorporation of dietary docosahexaenoic acid into the central nervous system of the yellowtail *Seriola quinqueradiata*. *Brain, Behavior and Evolution* 53:173–179.

McEvoy, L.A., Estevez, A., Bell, J.G., Shields, R.J., Gara, B., and Sargent, J.R. 1998. The influence of dietary levels of eicosapentaenoic and arachidonic acids on the pigmentation success of turbot (*Scophthalmus maximus* L.) and halibut (*Hippoglossus hippoglossus* L.). *Bulletin of the Aquaculture Association of Canada* 4:17–20.

McPhail, L., Clayton, C.C., and Snyderman, R. 1984. A potential second messenger role for unsaturated fatty acids: activation of Ca^{2+}-dependent protein kinase. *Science* 224: 622–625.

Meghji, J.A., Sandy, J.R., Scatt, A.M., Harvey, W., and Harris, M. 1988. Stimulation of bone resorption by lipoxygenase metabolites of arachidonic acid. *Prostaglandins* 36:131–149.

Mommsen, T.P., Vijayan, M.M., and Moon, T.W. 1999. Cortisol in teleosts: dynamics, mechanisms of action, and metabolic regulation. *Review in Fish Biology and Fisheries* 9:211–268.

Montero, D., and Izquierdo, Y.M. 2010. Welfare and health of fish fed vegetable oils as alternative lipid sources to fish oil. In: Turchini, G., Ng, W., and Tocher, D. (eds.) *Fish Oil Replacement and Alternative Lipid Sources in Aquaculture Feeds*. CRC Press, Cambridge, UK, pp. 439–486.

Montero, D., Tort, L., Izquierdo, M.S., Robaina, L., and Vergara, J.M. 1998. Depletion of serum alternative complement pathway activity in gilthead seabream caused by α-tocopherol and n-3 HUFA dietary deficiencies. *Fish Physiology and Biochemistry* 18:399–407.

Montero, D., Kalinowski, T., Obach, A., Robaina, L., Tort, L., Caballero, M.J., and Izquierdo, M.S. 2003. Vegetable lipid sources for gilthead seabream (*Sparus aurata*): effects on fish health. *Aquaculture* 225:353–370.

Morais, S., Cahu, C., Zambonino-Infante, J.L., Robin, J., Rønnestad, I., Dinis, M.T., and Conceição, L.E.C. 2004. Dietary triacylglycerol source and level affects performance and lipase expression in larval seabass (*Dicentrarchus labrax*). *Lipids* 39:449–458.

Morais, S., Rojas-Garcia, C.R., Conceição, L.E.C., and Rønnestad, I. 2005a. Digestion and absorption of a pure triacylglycerol and a free fatty acid by *Clupea harengus* L. larvae. *Journal of Fish Biology* 67:223–238.

Morais, S., Koven, W., Ronnestad, I., Dinis, M.T., and Conceicao, L.C.C. 2005b. Dietary protein : lipid ratio and lipid nature affects fatty acid absorption and metabolism in a teleost larva. *British Journal of Nutrition* 93: 813–820.

Morais, S., Caballero, M.J., Conceicao, L.E.C., Izquierdo, M.S.Y., and Dinis, M.T. 2006. Dietary neutral lipid level and source in Senegalese sole (*Solea senegalensis*) larvae: effect on growth, lipid metabolism and digestive capacity. *Comparative Biochemistry and Physiology* 144B:57–69.

Morais, S., Conceicao, L.E.C., Ronnestad, I., Koven, W., Cahu, C., Zambonino-Infante, J.L., and Dinis, M.T. 2007. Dietary neutral lipid level and source in marine fish larvae: effects on digestive physiology and food intake. *Aquaculture* 268:106–122.

Motojima, K., Passilly, P., Peters, J.M., Gonzales, F.J., and Latruffe, N. 1998. Expression of putative fatty acid transporter genes are regulated by peroxisome proliferator activated receptors alpha and gamma activators in a tissue- and inducer-specific manner. *The Journal of Biological Chemistry* 273:16710–16714.

Mourente, G., and Tocher, D.R. 1992. Effects of weaning onto a pelleted diet on docosahexaenoic acid (22:6n-3) levels in brain of developing turbot (*Scophthalmus maximus* L.). *Aquaculture* 105:363–377.

Mourente, G., and Tocher, D.R. 1993. The effects of weaning on to a dry pellet diet on brain lipid and fatty acid compositions in post-larval gilthead sea bream *Sparus aurata* L. *Comparative Biochemistry and Physiology* 104A:605–611.

Mourente, G., Tocher, D.R., Diaz-Salvago, E., Grau, A., and Pastor, E. 1999. Study of the high n-3 highly unsaturated fatty acids requirement and antioxidant status of *Dentex dentex* larvae at the *Artemia* feeding state. *Aquaculture* 179:291–307.

Murray, H.M., Gallant, J.W., Johnson, S.C., and Douglas, S.E. 2006. Cloning and expression analysis of three digestive enzymes from Atlantic halibut (*Hippoglossus hippoglossus*) during early development: predicting gastrointestinal functionality. *Aquaculture* 252:394–408.

Mustafa, T., and Srivastava, K.C. 1989. Prostaglandins (eicosanoids) and their role in ectothermic organisms. In: Gilles, R. (ed.) *Advances in Comparative and Environmental Physiology*, Vol. 5, Springer, Berlin, pp. 157–207.

Nablone, G., Grynberg, A., Chevalier, A., Leonardi, J., Termine, E., and Lafont, H. 1990. Phospholipase A activity of cultured rat ventricular myocytes is affected by the nature of cellular polyunsaturated fatty acids. *Lipids* 25:301–306.

Næss, T., and Lie, Ø. 1998. A sensitive period for the determination of pigmentation pattern in Atlantic halibut, *Hippoglossus hippoglossus* L. juveniles: the role of diet. *Aquaculture Research* 29:925–934.

Naor, Z. 1991. Is arachidonic acid a second messenger in signal transduction? *Molecular and Cellular Endocrinology* 80:C181–C186.

Nasushita, R., Watanobe, H., and Takebe, K. 1997. A comparative study of adrenocorticotropin-releasing activity of prostaglandins E1, E2, F2a and D2 in the rat. *Prostaglandins, Leukotrienes, and Essential Fatty Acids* 56:165–168.

Negrín, D., Montero, D., Izquierdo, M., Atalah, E., and Afonso, M. Submitted. Dietary arachidonic acid modulates expression of stress response–related genes in European sea bass *Dicentrarchus labrax* larvae. *Aquaculture.*

Nishimoto, N.K., Waite, J.H., and Kriwacki, R.W. 2003. Structure, activity, and distribution of fish osteocalcin. *Journal of Biology and Chemistry* 278:11843–11848.

Norris, D.A., Morelli, J.G., and Fujita, M. 1998. Melanocyte interactions in skin. In: Nordlund, J.J., Biossy, R.E., Hearing, V.J., King, R.A., and Ortinne, J.-P. (eds.) *The Pigmentary System.* Oxford University Press, New York, pp. 123–133.

Nye, E.J., Hockings, G.I., Grice, J.E., Torpy, D.J., Walters, M.M., Crosbie, G.V., Wagenaar, M., Cooper, M., and Jakson, R.V. 1997. Aspirin inhibits vasopressin-induced hypothalamic-pituitary-adrenal activity in normal humans. *Journal of Clinical Endocrinology and Metabolism* 82:812–817.

Olsen, R.E., and Ringø, E. 1997. Lipid digestibility in fish: a review. *Recent Research Development in Lipid Research* 1:199–265.

Olsen, Y., and Skjervold, H. 1995. Variation in content of omega 3 fatty acids in farmed

Atlantic salmon, with special emphasis on effect of non dietary factor. *Aquaculture International* 3:2–35.

Olsen, R.E., Henderson, R.J., and Ringø, E. 1998. The digestion and selective absorption of dietary fatty acids in Arctic charr, *Salvelinus alpinus*. *Aquaculture Nutrition* 4:13–21.

Olsen, R.E., Lovaas, E., and Lie, O. 1999. The influence of temperature, dietary polyunsaturated fatty acids, alpha-tocopherol and spermine on fatty acid composition and indices of oxidative stress in juvenile Artic char, *Salvelinus alpinus* (L.). *Fish Physiology and Biochemistry* 20:13–29.

Ostrowski, A.C., and Divakaran, S. 1990. Survival and bioconversion of n-3 fatty acids during early development of dolphin (*Coryphaena hippurus*) larvae fed oil-enriched rotifers. *Aquaculture* 89:273–285.

Ota, T., and Yamada, M. 1971. Lipids of masu salmon *Oncorhynchus masou*. I. Variations of the lipid content and fatty acid composition of juvenile masu salmon during the period of smolt transformation, and the influence of light upon those variations. *Bulletin of the Faculty of Fisheries, Hokkaido University* 22:151–158.

Pérez-Casanova, J.C., Murray, H.M., Gallant, J.W., Ross, N.W., Douglas, S.E., and Johnson, S.C. 2004. Bile-salt activated lipase expression during larval development in the haddock (*Melanogrammus aeglefinus*). *Aquaculture* 235:601–617.

Planas, J.V., Goetz, F.W., and Swanson, P. 1997. Stimulation of brook trout ovarian steroidogenesis by gonadotropins I and II by the cyclic adenosine 3′-5′-monophosphate/protein kinase A pathway. *Biology of Reproduction* 57:647–654.

Pombinho, A.R., Laize, V., Molha, D.M., Marques, S.M.P., and Cancela, M.L. 2004. Development of two bone-derived cell lines from the marine teleost *Sparus aurata*: evidence for extracellular matrix mineralization and cell-type-specific expression of matrix Gla protein and osteocalcin. *Cell and Tissue Research* 315:393–406.

Poston, H.A. 1990a. Performance of rainbow trout fry fed supplemental soy lecithin and choline. *Progress in Fish Culture* 52:218–225.

Poston, H.A. 1990b. Effect of body size on growth, survival and chemical composition of Atlantic salmon fed soy lecithin and choline. *Progress in Fish Culture* 52:226–230.

Poston, H.A. 1991. Response of Atlantic salmon fry to feed-grade lecithin and choline. *Progress in Fish Culture* 53:224–228.

Powers, M.K., and Raymond, P.A. 1990. Development of the visual system. In: Douglas, R.H., and Djamgoz, M.B.A. (eds.) *The Visual System of Fish*. Chapman and Hall, London, pp. 419–442.

Radunz-Neto, J., Corraze, G., Charlon, N., and Bergot, P. 1993. Essential n-3 fatty acid requirements of carp (*Cyprinus carpio*) larvae. In: Carrillo, M., Dahle, L., Morales, J., Sorgeloos, P., Svennevig, N., and Wyban, J. (eds.) *From Discovery to Commercialization*. Special Publication No. 19. European Aquaculture Society, Ghent, Belgium, p. 187.

Radünz-Neto, J., Corraze, G., Charlon, N., and Bergot, P. 1994. Lipid supplementation of casein-based purified diets for carp (*Cyprinus carpio* L.) larvae. *Aquaculture* 128:153–161.

Rainuzzo, J.R., Reitan, K.I., and Jørgensen, L. 1992. Comparative study on the fatty acid and lipid composition of four marine fish larvae. *Comparative Biochemistry and Physiology* 103B:21–26.

Rainuzzo, J.R., Reitan, K.I., Jorgensen, L., and Olsen, Y. 1994. Lipid composition in turbot larvae fed live feed cultured by emulsions of different lipid classes. *Comparative Biochemistry and Physiology* 107:699–710.

Raisz, L.G., Pilbeam, C.C., and Fall, P.M. 1993. Prostaglandins mechanisms of action and regulation of production in bone. *Osteoporosis International* 3(Suppl.1):S316–S140.

Reitan, K.I., Rainuzzo, J.R., and Olsen, Y. 1994. Influence of lipid composition of live feed on growth, survival and pigmentation of turbot larvae. *Aquaculture International* 2: 33–48.

Rodríguez, C. 1994. Estudio de los requerimientos de acidos grasos esenciales de la dorada europea *Sparus aurata* L. durante las dos primeras semanas de alimentacion. PhD thesis, University of La Laguna, Spain.

Rodríguez, C., Pérez, J.A., Lorenzo, A., Izquierdo, M.S., and Cejas, J.R. 1994. N-3 HUFA requirement of larval gilthead seabream *Sparus aurata* when using high levels of eicosapentaenoic acid. *Comparative Biochemistry and Physiology* 107A:693–698.

Rodríguez, C., Pérez, J.A., Díaz, M., Izquierdo, M.S., Fernandez-Palacios, H., and Lorenzo, A. 1997. Influence of the EPA/DHA ratio in rotifers on gilthead seabream *Sparus aurata* larval development. *Aquaculture* 150:77–89.

Rodríguez, C., Pérez, J.A., Badía, P., Izquierdo, M.S., Fernández-Palacios, H., and Lorenzo Hernández, A. 1998. The n-3 highly unsaturated fatty acids requirements of gilthead sea bream (*Sparus aurata* L.) larvae when using an appropriate DHA/EPA ratio in the diet. *Aquaculture* 169:9–23.

Rønnestad, I., Finn, R.N., Lein, I., and Lie, Ø. 1995. Compartmental changes in the contents of total lipid, lipid classes and their associated fatty acids in developing yolk-sac larvae of Atlantic halibut, *Hippoglossus hippoglossus* (L.). *Aquaculture Nutrition* 1:119–130.

Roo, F.J., Socorro, J., Izquierdo, M.S., Caballero, M.J., Hernandez-Cruz, C.M., Fernandez, A., and Fernandez-Palacios, H. 1999. Development of red porgy *Pagrus pagrus* visual system in relation with changes in the digestive tract and larval feeding habits. *Aquaculture* 179: 499–512.

Roo, J., Hernández-Cruz, C.M., Socorro, J., Fernández-Palacios, H., Montero, D., and Izquierdo, M.S. 2009. Effect of DHA content in rotifers on the occurrence of skeletal deformities in red porgy *Pagrus pagrus* (Linnaeus, 1758). *Aquaculture* 287:84–93.

Ross, S.A., McCaffery, P.J., Drager, U.C., and De Luca, L.M. 2000. Retinoids in embryonal development. *Physiological Review* 80: 1021–1054.

Ryer, C.H., and Boehlert, G.W. 1983. Feeding chronology, daily ration, and the effects of temperature upon gastric evacuation in the pipefish, *Syngnathus fuscus*. *Environmental Biology of Fishes* 9:301–306.

Salhi, M., Izquierdo, M.S., Hernández-Cruz, C.M., Gonzalez, M., and Fernández-Palacios, H. 1994. Effect of lipid and n-3 HUFA levels in microdiets on growth survival and fatty acid composition of larval gilthead seabream *Sparus aurata*. *Aquaculture* 124:275–282.

Salhi, M., Izquierdo, M.S., Hernández-Cruz, C.M., Bessonart, M., and Fernández-Palacios, H. 1999. Effect of different dietary polar lipid levels and different n-3 HUFA content in polar lipids on the gut and liver histological structure of seabream *Sparus aurata* larvae. *Aquaculture* 179:253–264.

Sampekalo, J., Takeuchi, T., and Watanabe, T. 1992. Comparison of gill lipids between freshwater fish. *Journal of the Tokyo University of Fisheries* 79:71–76.

Sandel, E., Nixon, O., Lutzky, S., Ginsberg, B., Geva, I., Tandler, A., Uni, Z., and Koven, W. 2010. The effect of dietary phosphatidylcholine/ phosphatidylinositol ratio on malformation in larvae and juvenile gilthead sea bream (*Sparus aurata*). *Aquaculture Nutrition* 304:42–48.

Sargent, J., Henderson, R.J., and Tocher, D.R. 1989. The lipids. In: Halver, J.E. (ed.) *Fish Nutrition*. Academic Press, London, UK, pp. 154–218.

Sargent, J.R., McEvoy, L.A., and Bell, J.G. 1997. Requirements, presentation and sources of polyunsaturated fatty acids in marine fish larval feeds. *Aquaculture* 155:117–127.

Sargent, J.R., Bell, J.G., McEvoy, L.A., Tocher, D.R., and Estevez, A. 1999a. Recent developments in the essential fatty acid nutrition of fish. *Aquaculture* 177:191–199.

Sargent, J.R., McEvoy, L.A., Estevez, A., Bell, J.G., Bell, M.V., Henderson, R.J., and Tocher, D.R. 1999b. Lipid nutrition of marine fish during early development: current status and future directions. *Aquaculture* 179:217–229.

Sargent, J.R., Tocher, D.R., and Bell, J.G. 2002. The lipids. In: Halver, H.E., and Hardy, R.W. (eds.) *Fish Nutrition*, 3rd edition. Elsevier, New York, pp. 181–257.

Scott, G., Jacobs, S., Leopardi, S., Anthony, F.A., Learn, D., Malaviya, R., and Pentland, A. 2005. Effects of PGF2a on human melanocytes and regulation of the FP receptor by ultraviolet radiation. *Experimental Cell Research* 304: 407–416.

Seiffert, M.E.B., Cerqueira, V.R., and Madureira, L.A.S. 2001. Effect of dietary (n-3) highly unsaturated fatty acids on growth and survival of fat snook (*Centropomus parallelus*, Pisces: Centropomidae) larvae during first feeding. *Brazilian Journal of Medical and Biological Research* 34:645–651.

Seikai, T., Shimozaki, M., and Watanabe, T. 1987. Estimation of larval stage determining the appearance of albinism in hatchery-reared juvenile flounder *Paralichthys olivaceus*. *Nippon Suisan Gakkaishi* 53:1107–1114.

Seiliez, I., Panserat, S., Corraze, G., Kaushik, S., and Bergot, P. 2003. Cloning and nutritional regulation of a Δ6-desaturase-like enzyme in marine teleost gilthead seabream (*Sparus aurata*). *Comparative Biochemistry and Physiology. Part B, Biochemistry and Molecular Biology* 135:449–460.

Seoka, M., Kurata, M., and Kumai, H. 2007. Effect of docosahexaenoic acid enrichment in *Artemia* on growth of Pacific bluefin tuna *Thunnus orientalis* larvae. *Aquaculture* 270:193–199.

Sheldon, W.M.J., and Blazer, V.S. 1991. Influence of dietary lipid and temperature on bactericidal activity of channel catfish macrophages. *Journal of Aquatic Animal Health* 3:87–93.

Shi, Y., Hon, M., and Evans, R.M. 2002. The peroxisome proliferator–activated receptor δ, an integrator of transcriptional repression and nuclear receptor signaling. *Proceedings of the National Academy of Science of the United States of America* 99:2613–2618.

Shields, R.J., Bell, J.G., Luizi, F.S., Gara, B., Bromage, N.R., and Sargent, J.R. 1999. Natural copepods are superior to enriched *Artemia* nauplii as feed for halibut larvae (*Hippoglossus hippoglossus*) in terms of survival, pigmentation and retinal morphology: relation to dietary essential fatty acids. *Journal of Nutrition* 129:1186–1194.

Shinitzky, M., and Henkart, P. 1979. Fluidity of cell membranes: current concepts and trends. *International Review of Cytology* 60:121–147.

Simes, D.C., Viegas, C.B., Williamson, M.K., Price, P.A., and Cancela, L. 2008. Purification of matrix Gla protein and osteocalcin from the Adriatic sturgeon (*Acipenser naccarrii*), and ancient bony fish with a cartilaginous endoskeleton. *Bone* 42:265.

Southgate, P., and Kavanagh, K.D. 1999. Effects of n-3 fatty-acid composition of enhanced *Artemia* diet on growth and mortality of *Acanthochromis polyacanthus*. *Aquatic Living Resources* 12:31–36.

Stillwell, W., and Wassall, S.R. 2003. Docosahexaenoic acid: membrane properties of a unique fatty acid. *Chemistry and Physics of Lipids* 126:1–27.

Szisch, V., Papandroulakis, N., Fanouraki, E., and Pavlidis, M. 2005. Ontogeny of the thyroid hormones and cortisol in the gilthead seabream, *Sparus aurata*. *General and Comparative Endocrinology* 142:186–192.

Szulc, P., Arlot, M., Chapuy, M.-C., Duboeuf, F., Meunier, P.J., and Delmas, P.D. 1994. Serum undercarboxylated osteocalcin correlates with hip bone mineral density in elderly women. *Journal of Bone and Mineral Research* 9:1591–1595.

Takeuchi, T. 1997. Essential fatty acid requirements of aquatic animals with emphasis on fish larvae and fingerlings. *Review in Fisheries Science* 5:1–25.

Takeuchi, T., Toyota, M., Satoh, S., and Watanabe, T. 1990. Requirement of juvenile red sea bream *Pagrus major* for eicosapentaenoic and docosahexaenoic acids. *Nippon Suisan Gakkaishi* 56:1263–1269.

Takeuchi, T., Arakawa, T., Satoh, S., and Watanabe, T. 1992. Supplemental effect of phospholipids and requirement of eicosapentaenoic acid and docosahexaenoic acid for juvenile striped jack. *Nippon Suisan Gakkaishi* 58:707–713.

Tandler, A., Watanabe, T., Satoh, S., and Fukusho, K. 1989. The effect of food deprivation on the fatty acid and lipid profile of red seabream (*Pagrus major*) larvae. *British Journal of Nutrition* 62:349–361.

Tocher, D.R. 1995. Glycerophospholipid metabolism. In: Hochachka, P.W., and Mommsen, T.P. (eds.) *Biochemistry and Molecular Biology of Fishes, Vol. 4. Metabolic and Adaptational Biochemistry*. Elsevier Press, Amsterdam, pp. 119–157.

Tocher, D.R. 2003. Metabolism and functions of lipids and fatty acids in teleost fish. *Review in Fisheries Science* 11:107–184.

Tocher, D.R., Fraser, A.J., Sargent, J.R., and Gamble, J.C. 1985. Fatty acid composition of phospholipids and neutral lipids during embryonic and early development in Atlantic herring (*Clupea harengus* L.). *Lipids* 20:69–74.

Tocher, D.R., Bendiksen, E.A., Campbell, P.J., and Bell, J.G. 2008. The role of phospholipids in nutrition and metabolism of teleost fish. *Aquaculture* 280:21–34.

Tomita, Y., Maeda, K., and Tagami, H. 1992. Melanocyte-stimulating properties of arachidonic acid metabolites: possible role in postinflammatory pigmentation. *Pigment Cell Research* 5:357–361.

Tontonoz, P., Hu, E., and Spiegelman, B.M. 1995. Regulation of adipocyte gene expression and differentiation by peroxisome proliferator activated receptor gamma.

Toyota, M., Takeuchi, T., and Watanabe, T. 1991. Dietary value to larval yellowtail of *Artemia* nauplii enriched with EPA and DHA. Abstracts of the Annual Meeting of the Japanese Society of Scientific Fisheries, April, Tokyo, Japan.

Ushio, H., Ohshima, T., and Koizumi, C. 1996. Fatty acid compositions in glycerophospholipids from brain lobes of rainbow trout, carp and skipjack tuna. *Fisheries Science* 62:126–133.

Van Anholt, R.D., Spanings, T., Koven, W.M., and Wendelaar Bonga, S.E. 2003. Effects of acetylsalicylic acid treatment on thyroid hormones, prolactins, and the stress response of tilapia (*Oreochromis mossambicus*). *American Journal of Physiology* 285:R1098–R1106.

Van Anholt, R.D., Koven, W.M., Lutzky, S., Tandler, A., and Wendelaar Bonga, S.E. 2004a. Dietary supplementation with arachidonic acid alters the stress response of gilthead seabream (*Sparus aurata*) larvae. *Aquaculture* 238: 369–383.

Van Anholt, R.D., Spanings, F.A.T., Koven, W.M., Nixon, O., and Wendelaar Bonga, S.E. 2004b. Arachidonic acid reduces the stress response of gilthead sea bream (*Sparus auratus* L.). *Journal of Experimental Biology* 207:3419–3430.

Van Anholt, R.D., Spanings, T., Koven, W.M., and Wendelaar Bonga, S.E. 2004c. Dietary supplementation with arachidonic acid in tilapia (*Oreochromis mossambicus*) reveals physiological effects not mediated by prostaglandins. *General and Comparative Endocrinology* 139:215–226.

Van der Meeren, T., Klungsoyr, J., Wilhelmsen, S., and Kvensenth, P.G. 1991. Fatty acid composition of unfed and growing cod larvae, *Gadus morhua* L., feeding on natural plankton in large enclosures. In: Walther, B.T., and Fyhn, H.J. (eds.) *Physiological and Biochemical Aspects of Fish Development*. University of Bergen, Bergen, Norway, pp. 34–35.

Van Der Meeren, T., Olsen, R.E., Hamre, K., and Fyhn, H.J. 2008. Biochemical composition of copepods for evaluation of feed quality in production of juvenile marine fish. *Aquaculture* 274:375–397.

Van Praag, D., Farber, D.V., Minkin, S.J., and Primor, N. 1987. Production of eicosanoids by the killifish gills and opercular epithelia and their effect on active transport of ions. *General and Comparative Endocrinology* 67:50–57.

Villalta, C.M., Estévez, G.A., and Bransden, M. 2004. Arachidonic acid enriched live prey induce albinism in Senegal sole (*Solea senegalensis*). 11th International Symposium on Nutrition and Feeding in Fish, Phuket, Thailand, May 2–7, 2004.

Villalta, M., Estévez, A., and Bransden, M.P. 2005. Arachidonic acid enriched live prey induces albinism in Senegal sole (*Solea senegalensis*) larvae. *Aquaculture* 245:193–209.

Villalta, M., Estévez, A., Bransden, M.P., and Bell, J.G. 2008. Effects of dietary eicosapentaenoic acid on growth, survival, pigmentation, and fatty acid composition in Senegal sole (*Solea senegalensis*) larvae during the *Artemia* feeding period. *Aquaculture Nutrition* 14:232–241.

Villeneuve, L., Gisbert, E., Zambonino-Infante, J.L., Quazuguel, P., and Cahu, C.L. 2005. Effect of nature of dietary lipids on European sea bass morphogenesis: implication of retinoid receptors. *British Journal of Nutrition* 94: 877–884.

Villeneuve, L., Gisbert, E., Moriceau, J., Cahu, C., and Zambonino Infante, J.L. 2006. Intake of different levels of vitamin A and polyunsaturated fatty acids during different developmental periods modifies the expression of morphogenesis genes in European sea bass (*Dicentrarchus labrax*). *British Journal of Nutrition* 95: 677–687.

Waagbø, R., Bjerkâs, E., Hamre, K., Berge, R., Wathne, E., Lie, Ø., and Torstensen, B. 2003. Cataract formation in Atlantic salmon, *Salmo salar* L., smolts relative to dietary pro- and antioxidant and lipid level. *Journal of Fish Diseases* 26:213–229.

Watanabe, T. 1982. Lipid nutrition in fish. *Comparative Biochemistry and Physiology* 73:3–15.

Watanabe, T. 1993. Importance of docosahexaenoic acid in marine larval fish. *Journal of the World Aquaculture Society* 24:152–161.

Watanabe, T., and Kiron, V. 1994. Prospects in larval fish dietetics. *Aquaculture* 124: 223–251.

Watanabe, T., Arakawa, T., Kitajima, C., and Fujita, S. 1984. Effect of nutritional quality of broodstock diets on reproduction of red sea bream. *Nippon Suisan Gakkaishi* 50:495–501.

Watanabe, T., Izquierdo, M.S., Takeuchi, T., Satoh, S., and Kitajima, C. 1989. Comparison between eicosapentaenoic and docosahexaenoic acids in terms of essential fatty acid efficacy in larval red sea bream. *Nippon Suisan Gakkaishi* 55: 1635–1640.

Watkins, B.A., Whitehead, C.C., and Duff, S.R.I. 1991. Hydrogenated oil decreases tissue concentrations of n-6 polyunsaturated fatty acids and may contribute to dyschondroplasia in broilers. *British Poultry Science* 32:1109–1119.

Watkins, B.A., Shen, C.L., McMurtry, J.P., Xu, H., Bain, S.D., Allen, K.G.D., and Seifert, M.F. 1997. Dietary lipids modulate bone prostaglandin E2 production, insulin-like growth factor-I concentration and formation rate in chicks. *Journal of Nutrition* 127:1084–1091.

Wendelaar Bonga, S.E. 1997. The stress response in fish. *Physiological Reviews* 77:591–625.

Willey, S., Bengtson, D.A., and Harel, M. 2003. Arachidonic acid requirements in larval summer flounder, *Paralichthys dentatus*. *Aquaculture International* 11:131–149.

Witten, P.E., Gil-Martens, L., Hall, B.K., Huysseune, A., and Obach, A. 2005. Compressed vertebrae in Atlantic salmon *Salmo salar*: evidence for metaplastic chondrogenesis as a skeletogenic response late in ontogeny. *Diseases of Aquatic Organisms* 64:237–246.

Wu, F.-C., Ting, Y.-Y., and Chen, H.Y. 2002. Docosahexaenoic acid is superior to eicosapentaenoic acid as the essential fatty acid for growth of grouper, *Epinephelus malabaricus*. *Journal of Nutrition* 132:72–79.

Yu, T.C., and Sinnhuber, R.O. 1975. Effect of dietary linolenic and linoleic acids upon growth and lipid metabolism of rainbow trout (*Salmo gairdneri*). *Lipids* 10:63–66.

Zacharieva, S., Borissova, A.-M., Andonova, K., Stoeva, I., and Matrozov, P. 1992. Role of prostaglandin E2 (PGE2) on the corticotropin-releasing hormone (CRH)-induced ACTH release in healthy men. *Hormone and Metabolic Research* 24:336–338.

Zapata, A., Diez, B., Cejalvo, T., Gutierrez-de Frias, C., and Cortes, A. 2006. Ontogeny of the immune system of fish. *Fish & Shellfish Immunology* 20:126–136.

Zheng, F., Takeuchi, T., Yoseda, K., Kobayashi, M., Hirokawa, J., and Watanabe, T. 1996. Requirement of larval cod for arachidonic acid, eicosapentaenoic acid, and docosahexaenoic acid using by their enriched *Artemia* nauplii. *Nippon Suisan Gakkaishi* 62:669–676.

Chapter 3

Proteins

Luis Conceição, Cláudia Aragão, and Ivar Rønnestad

3.1 Introduction

Fish larvae have tremendous growth potential, with relative growth rates much higher than juvenile and adult fish (Kamler 1992; Otterlei et al. 1999; Conceição et al. 2003a). However, to fully express such growth potential, protein of the right quality must be provided in sufficient quantity. In fact, fish growth is primarily deposition of muscle protein (Houlihan et al. 1995a; Carter and Houlihan 2001). In addition, it is well established that amino acids (AA) are a major energy source during the larval stage of most marine teleost species (Conceição et al. 1993; Rønnestad and Fyhn 1993; Parra et al. 1999; Rønnestad et al. 1999, 2003; Finn et al. 2002). Still, the AA requirements of fish larvae are poorly understood, and the available knowledge is more of a qualitative nature than precise requirements (Conceição et al. 2003a). In addition, the fast growth of fish larvae is paradoxical when one considers its poorly developed gut (Rønnestad and Conceição 2005; Zambonino-Infante et al. 2008).

Dietary AA are mostly absorbed as free AA (FAA) or as small peptides (Rønnestad and Morais 2008). However, these are quickly polymerized into proteins, or used otherwise by larval metabolism. In fact, tissue concentrations of FAA are kept within narrow limits (Houlihan et al. 1995a). The yolk sac contains the large FAA pool present in marine pelagic fish eggs, which may account for up to 60% of the total larval AA (Rønnestad and Fyhn 1993; Finn 1994; Rønnestad et al. 1999). This FAA pool acts both as an osmolyte and as an important energy fuel, in addition to the supply of AA for protein synthesis, before larvae initiate first feeding (Fyhn 1993; Wright and Fyhn 2001).

Absorbed dietary AA that are not polymerized into proteins can be catabolized for energy production, may be transaminated into other AA, used in gluconeogenesis or lipogenesis, or used in the synthesis of other nitrogen-containing molecules such as purines, pyrimidines, or hormones. Furthermore, protein is in continuous turnover, in a very dynamic transfer of AA between the FAA and the protein pools.

Larval Fish Nutrition, First Edition. Edited by G. Joan Holt.

In order to meet the high protein requirements of fish larvae, the nutrient flux involved is notorious. For instance, during the first 3 weeks of feeding of Senegalese sole (*Solea senegalensis*), conservative estimates of *Artemia* protein intake range between 18 and 60% per day of body protein mass (Engrola et al. 2009). From this protein intake, 70–85% is digestible and 11–30% of the absorbed AA is used in catabolism (Morais et al. 2004a; Engrola et al. 2009). These large fluxes of AA demonstrate the importance of understanding AA metabolism in order to meet the requirements of fish larvae so that their high growth potential can be fully used.

This chapter reviews the current knowledge of protein and AA requirements of fish larvae, taking into account the available understanding of protein digestion and AA absorption, as well as metabolism of larval AA and protein pools.

3.2 Protein digestion

The larval-type digestive tract has a processing capacity (capability to digest and absorb dietary nutrients) that can support very high growth rates, given that suitable feed is available. Digestion of dietary proteins involves coordination of a range of basic processes in the digestive tract, including enzymatic and fluid secretions (B), digestion (C), absorption (D, E), and motility (including evacuation, F). The efficiency of the overall digestive pro cesses is a key factor that determines the delivery of AA to the rapidly growing larval tissues, particularly during the first critical stages when the larvae start exogenous feeding. The following chapter will mainly deal with protein in teleosts that do not possess a stomach at the onset of first feeding but acquire a functional stomach during metamorphosis (altricial species). The particular morphological and physiological characteristics of the larval digestive system (see Chapters 1 and 9), as well as the changes occurring during larval ontogeny, are believed to be critical factors that determine the capacity for digesting and absorbing dietary proteins, peptides, and AA.

3.2.1 Enzymatic secretion and digestion

From the onset of exogenous feeding, digestion of dietary proteins in altricial fish larvae starts in the midgut when the ingested feeds (Figure 3.1-A) are mixed with pancreatic secretions and bile from the gallbladder (Figure 3.1-B). Bile has not been shown to interfere with protein digestion, but the pancreatic secretions contain a variety of proteolytic enzymes in an alkaline solution. In qualitative terms, the pancreatic proteases in fish are similar to those of higher vertebrates (Ash 1985; Sire and Vernier 1992; Suzuki et al. 2002), with trypsin and chymotrypsin as the major components (Dabrowski 1983; Gildberg 2004). Other proteolytic enzymes, such as elastase and carboxypeptidases A and B, are also found in pancreatic secretions. In the species investigated, these enzymes in most cases have several paralogues. For instance, in Japanese flounder (*Paralichthys olivaceus*) and winter flounder (*Pleuronectes americanus*), there are three variants of trypsinogen (Suzuki et al. 2002; Murray et al. 2006), while in Senegalese sole, six trypsin paralogues have been described (Manchado et al. 2008). As in all vertebrates, the pancreatic proteolytic enzymes in fish are secreted as inactive proenzymes and are only activated in the midgut lumen. Trypsin is a key factor in the activation of the pancreatic enzymes, and trypsin is itself activated when it gets in contact with brush border–bound enteropeptidase. Both trypsin and chymotrypsin are endopeptidases, and while trypsin is a serine protease that cleaves peptide bonds on the carboxyl side of the basic AA, chymotrypsin primarily hydrolyzes peptide bonds next to the aromatic AA (Jobling and Hjelmeland

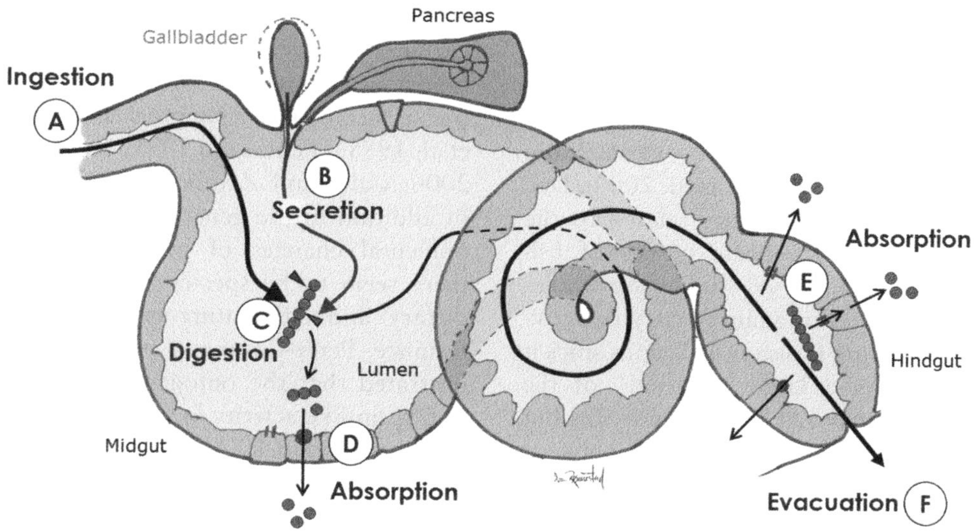

Figure 3.1 Digestion and absorption of dietary proteins are involved in coordination of a range of basic processes in the digestive tract. In agastric fish larvae, these includes enzymatic secretions (B), digestion (C), absorption (D, E), and motility (including evacuation, F). The gut transit time is a function of rates of ingestion and evacuation, taking the gut volume into account.

1992). The combined effects of endopeptidase activity yields protein fragments and small peptides in the luminal contents (Figure 3.1-C), and further hydrolysis is performed by exopeptidases, secreted by both the pancreas and the apical membrane of the enterocytes (Buddington et al. 1997). Peptidases in the enterocyte membranes contribute to the end result of digestion, which are small peptides and FAA. Cytoplasmatic peptidases will perform additional breakdown of very small peptides toward FAA, but it is unknown if smaller peptides are transferred into the systemic circulation.

During metamorphosis, the stomach gradually becomes functional and the digestion of proteins is initiated in this compartment. The gastric degradation of proteins includes acid denaturation and enzymatic cleavage by pepsin. Pepsin is produced and secreted in its inactive form (pepsinogen) by oxynticopeptic cells in the mucosa (Morrison 1987; Sire and Vernier 1992), and in fish these cells also secrete hydrochloric acid (HCl), which makes the necessary acid environment for pepsinogen activation. Pepsin is an endopeptidase and it hydrolyzes proteins at bonds between aromatic and acidic AA (Jobling and Hjelmeland 1992). Pepsin digestion yields polypeptides of very diverse sizes but few FAA (Lied and Solbakken 1984; Jobling and Hjelmeland 1992; Espeland 2006). The combined effects of HCl and pepsin make the stomach a highly efficient organ for degrading complex proteins, such as collagen-rich connective tissues (Gildberg 2004). Thereby, the acquisition of a functional stomach enables the inclusion of a wider range of feed ingredients in fish diets. In addition, the stomach possesses strong peristaltic and contractile activity that contributes to a mechanical degradation of the ingested feed and to an enhanced mixing of the luminal gastric contents, allowing easier exposure of dietary proteins to acid proteolytic conditions. Peristalsis is also responsible for transporting the degraded proteins and peptides in the chyme onward to the anterior intestine for further hydrolysis and processing by a new set of proteases.

There is a large body of literature quantifying ontogenetic patterns and dietary effects on the proteolytic enzyme activity (e.g., Cousin et al. 1987; Pedersen et al. 1990; Ribeiro et al. 1999; Gawlicka et al. 2000; Lazo et al. 2000; Cahu and Zambonino-Infante 2001; Alvarez-González et al. 2006; Bolasina et al. 2006; Gisbert et al. 2009). Due to methodological difficulties related to the extremely tiny and delicate nature of fish larvae, many of the larval studies use whole-body homogenates to quantify enzyme activity. This procedure is disadvantageous since it not only causes a large "dilution" of the enzyme being analyzed by other proteins but mostly it also does not distinguish between enzymatic activities of different origins (intracellular enzymes or those present in tissues other than in the digestive tract, and enzymes from ingested prey or the gut microflora). In recent years, the development and more widespread use of sensitive molecular tools has enabled the identification, localization, and quantification of gene transcripts for enzymes, in conjunction with biochemical assays (Douglas et al. 1999; Murray et al. 2004, 2006; Perez-Casanova et al. 2004, 2006; Marza et al. 2005).

Studies using highly sensitive polymerase chain reaction (PCR) techniques have shown that mRNA transcripts of several proteolytic enzymes are already present from hatch, which is for most species several days before the onset of exogenous feeding (Srivastava et al. 2002; Lo and Weng 2006; Perez-Casanova et al. 2006; Darias et al. 2007). There appear to be temporal differences between expressions of the different enzymatic paralogues. For instance, in winter flounder, trypsin 2 expression was detected from 5 days posthatch (dph), while trypsin 1 levels of expression only became significant in late larval stages and during metamorphosis and trypsin 3 showed expression only after 20 dph (Murray et al. 2004). Whether these ontogenetic differences in mRNA expression for the different paralogues have functional implications for the digestion of proteins is not known.

In general, the activity of proteolytic enzymes in the midgut is low when marine fish larvae commence exogenous feeding, but it rises as metamorphosis approaches (Hjelmeland et al. 1988; Hjelmeland 1995; Gawlicka et al. 2000; Cahu and Zambonino-Infante 2001). In addition to the general patterns of developmental changes of proteolytic enzymes, there seem to be species-specific as well as dietary- and temperature-induced effects. For instance, Perez-Casanova et al. (2006) demonstrated that the ontogenetic development of trypsin-like activity in whole-body homogenates of haddock (*Melanogrammus aeglefinus*) and Atlantic cod (*Gadus morhua*) reared on the same protocol was very different. In haddock, the activity was high at mouth opening and then declined and remained at a low level until 530 degree-days. In Atlantic cod, trypsin-like activity was low at mouth opening, but it increased after a switch in prey from nonenriched to enriched rotifers, and declined again when *Artemia* was gradually introduced (Perez-Casanova et al. 2006).

The secretion of pancreatic enzymes is believed to be under both hormonal and neural control, although there is little experimental evidence to substantiate the mechanisms controlling the secretion, particularly in larval stages (see Chapter 9). In Atlantic herring (*Clupea harengus*) larvae, the amount of pancreatic trypsin and trypsinogen secreted into the gut increases as the number of ingested prey increases, until a plateau is reached at high levels of feed intake (Pedersen et al. 1987; Pedersen and Andersen 1992). There are also data that indicate a link between the digestive hormone cholecystokinin and trypsin activity in herring (Koven et al. 2002).

Proteolytic enzymes in the enterocytes include peptidases in the brush border membrane as well as intracellular forms. The activities of brush border membrane enzymes of the enterocytes in Atlantic cod and Atlantic halibut (*Hippoglossus hipoglossus*) are low at first feeding but increase during larval development (Kvåle et al. 2007a). The activity of cytosolic (intracellular) peptidases, proposed to participate in the degradation of protein after pinocytoses, such as leucine-alanine peptidase, are high in some species around first feeding but tend to decrease as larvae develop, concurrent with rising levels of alkaline phosphatase (Cahu and Zambonino-Infante 2001; Kolkovski 2001). This picture is less clear in cod than in halibut (Kvåle et al. 2007a). The ontogenetic patterns of trypsin, membranous, and cytosolic peptidase activity during larval development have been proposed to reflect changes in the mode of larval digestion, which is believed to become increasingly dependent on luminal digestion and less so on intracellular digestion (Cahu and Zambonino-Infante 2001).

Several authors have discussed whether the enzymatic content of the ingested prey (exogenous source) contributes to the digestive process in fish larvae (e.g., Dabrowski and Glogowski 1977; Munilla-Moran et al. 1990; Walford and Lam 1993). However, considerable controversy exists around this notion as several other studies have quantified the contribution of live prey proteolytic enzymes to the overall digestive process and have concluded that this contribution is negligible (Pedersen et al. 1987; Cahu et al. 1995; Cahu and Zambonino-Infante 1997; Kurokawa et al. 1998; Perez-Casanova et al. 2006).

3.2.2 Protein digestibility

Understanding the digestibility of specific feed ingredients is essential for the formulation of optimized diets to be used in larval fish culture systems. Digestibility estimates of dietary proteins need to consider the *in vivo* conditions found in the larval midgut lumen (Figure 3.1-C). Surprisingly, little is known about the digestibility of various protein sources commonly used in the formulation of inert microdiets for marine fish larvae. Estimates of the digestibility for such ingredients derived from studies with juvenile fish and other animal models are of some but limited relevance, and they need to be applied with caution since there exist major differences in the digestive physiology of larval and juvenile/adult stages of fish.

True larval digestibility estimates for protein-rich ingredients must be assessed *in vivo*, and there are recent methods, using fluorescent microspheres (Hansen et al. 2009) and rare earth oxides (Johnson et al. 2009), that enable the evaluation of protein quality in larvae. However, *in vitro* methods to evaluate the digestibility of feeds and feedstuffs have often been used for rapid, easy, and reproducible screening (Hsu et al. 1977; Satterlee et al. 1979; Lazo et al. 1998; Tonheim et al. 2007). A comparison of three different methods for assessing dry matter and protein digestibility in discus (*Symphysodon aequifasciata*) showed that *in vitro* digestibility is a good screening approach (Chong et al. 2002). In their study, *in vitro* protocols (Hsu et al. 1977; Satterlee et al. 1979; Lazo et al. 1998) were compared with *in vivo* digestibility assessed in feeding trials with fish. Chong et al. (2002) found that relative digestibility estimates obtained by simple methods involving only a few proteases in a single reaction step correlated well with digestibility measured *in vivo*. *In vitro* studies may be based on intestinal extract from a species of particular interest (e.g., Jany 1976; Nankervis and Southgate 2006) or on commercially available crystalline proteases (e.g., Hsu et al. 1977; Satterlee et al. 1979). Therefore, *in vitro* digestibility can be a very useful approach to selecting promising candidate protein-rich ingredients for subsequent *in vivo* larval studies.

Protein solubility has been suggested as an important determinant of digestibility in fish larvae (Carvalho et al. 2004). Larvae of common carp (*Cyprinus carpio*) grew better and had higher survival when insoluble casein was replaced by soluble Na^{+}-caseinate in a starter diet (Carvalho et al. 2004). The protein sources used in that study were very similar, except for water solubility. In an *in vitro* digestibility trial simulating larval midgut conditions, Tonheim et al. (2007) demonstrated that a soluble reference protein was digested faster than the similar but insoluble reference protein, although their final digestibility was the same.

Protein solubility may, at least partly, explain the difficulties in using inert microdiets in marine fish larvae since commonly used live feeds, unlike formulated feeds, contain a high proportion of water-soluble protein (Fyhn et al. 1993; Hamre et al. 2002; Helland et al. 2003; van der Meeren et al. 2008; Conceição et al. 2010).

3.2.3 Absorption

The major site of absorption of dietary proteins, peptides, and FAA in larval fish has been proposed to be the midgut (Figure 3.1-D; Rønnestad and Conceição 2005), as in other vertebrates (Stevens and Hume 1996). However, the current understanding of the mechanisms and ontogenetic changes of protein, peptide, and FAA absorption in fish is limited. AA absorption involves a complex array of transport processes, some of them with overlapping functions. Numerous AA transporters (AATs) are expressed in the brush border membrane of adult enterocytes (Ray et al. 2002). The transport rates for FAA observed *in vivo* (Rønnestad et al. 2000a, 2000b; Applebaum and Rønnestad 2004) depend on several factors, including the luminal concentration of AA, the transport affinity and capacity for each AAT, and the amount of each AAT present in the epithelium. While several of the genes coding for AATs have been recently cloned in mammals, this work has only been started in fish (Narawane et al. 2009). AATs normally comprise a diverse group of membrane proteins sharing several attributes (Ray et al. 2002; Wipf et al. 2002). They typically exhibit broad substrate specificity (several AA may be transported by one type of transport system), allowing them to recognize, bind, and transfer AA across cell membranes. Existing data suggest that there are many genetic and splice variants within a group of what were once thought to be distinct AATs. Recent advances include the *in vivo* characterization of FAA absorption rates in fish larvae (Rønnestad et al. 2000a; Applebaum and Rønnestad 2004; Rønnestad and Rojas-García 2006) and also the molecular characterization of AATs' molecular and functional properties (Narawane et al. 2009). A saturable component for FAA absorption has not been demonstrated yet using available *in vivo* data (Applebaum and Rønnestad 2004), and further studies using higher FAA concentrations are needed in order to better understand the absorption kinetics of AA.

Protein digestion is not complete prior to absorption, and recent studies from a variety of vertebrate species suggest that entry of peptides over the apical membrane of the enterocytes is an important route of absorption of digested proteins (Daniel 2004). Peptides that are transported over the brush border membrane are normally hydrolyzed by cytoplasmatic peptidases, while FAA are mainly transported out of the basal membrane of the enterocytes and into the portal circulation. However, small peptides that are resistant to hydrolysis may enter the systemic circulation. Peptide transport may therefore be an important route for absorption of both dietary proteins and peptides in teleost fishes, also in the young stages.

Vertebrate enterocytes express a variety of transporters responsible for AA transport, but so far only one intestinal transporter has been described for peptides; the oligopeptide transporter 1 (PepT1; Nielsen and Brodin 2003; Daniel 2004). The molecular characterization of peptide transport has just begun in fish (Verri et al. 2010). The complete mRNA coding sequence for PepT1 has recently been reported in teleosts, including zebrafish

(*Danio rerio*; Verri et al. 2003), Atlantic cod (Rønnestad et al. 2006), European sea bass (*Dicentrarchus labrax*; Sangaletti et al. 2009), and Atlantic salmon (*Salmo salar*; Rønnestad et al. 2008).

In teleosts, the presently available data suggests that the PepT1 transporter is expressed at hatching, both in zebrafish (Verri et al. 2003) and in Atlantic cod (Rønnestad et al. 2006; Amberg et al. 2008). The study with zebrafish demonstrated that PepT1 is highly expressed in the proximal intestine. In Atlantic cod, PepT1 mRNA was expressed prior to the onset of exogenous feeding and PepT1 was expressed throughout the digestive system, except the esophagus and sphincter regions (Figure 3.2; Amberg et al. 2008). *In situ* hybridization data from Atlantic cod confirmed that the PepT1 transporter is expressed in the enterocytes (Amberg et al. 2008). Functional analysis by *in vitro* expression systems categorized the zebrafish and salmon PepT1 as classical low-affinity/high-capacity systems (Verri et al. 2003; Rønnestad et al. 2010). The PepT1 characterized to date show that teleost PepT1 is a H^+-dependent peptide cotransporter, but in contrast to higher vertebrates where the maximal transport activity is independent of extracellular pH, the maximal transport rates increase at alkaline extracellular pH (Verri et al. 2003; Rønnestad et al. 2010).

Fish larvae have been reported to absorb large molecules, including proteins, by endocytosis, followed by intracellular digestion in the hindgut mucosa (Figure 3.1-E; Iwai and

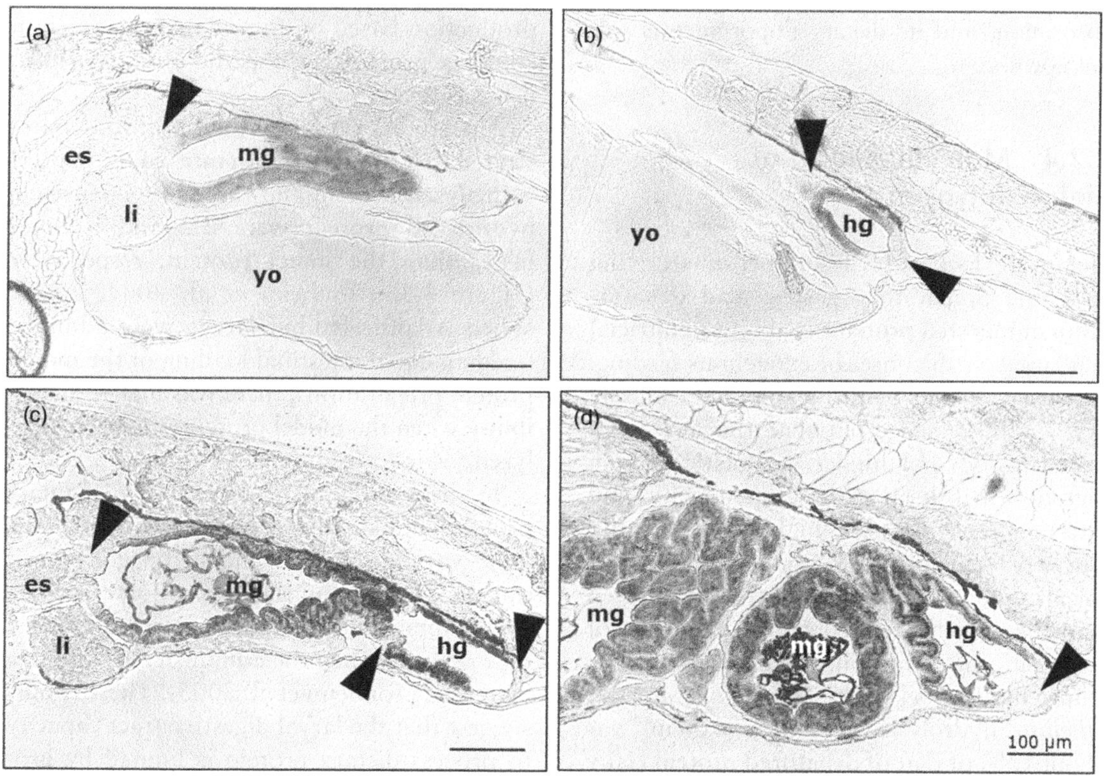

Figure 3.2 *In situ* expression pattern of the peptide transporter PepT1 mRNA in the digestive tract of Atlantic cod larvae. (a, b) Larva at 0 day posthatch (dph); (c) larva fed zooplankton at 4 dph; (d) larva fed zooplankton at 22 dph. Arrowheads point to the sphincter regions with no PepT1 expression. Scale bars indicate 100 µm. es = esophagus, li = liver, mg = midgut, hg = hindgut, yo = yolk. (Adapted from Amberg et al. 2008.)

Tanaka 1968; Watanabe 1984; Kurokawa et al. 1996; Luizi et al. 1999). Based on these findings, it has been suggested that low intestinal proteolytic capacity in the larval stages may be compensated by hindgut protein endocytosis. Watanabe (1984) showed that a protein absorbed by hindgut pinocytosis, took from 10 to 24 hours to be degraded intracellularly in larvae of pond smelt (*Hypomesus olidus*). Rønnestad et al. (2001a, 2007) argued that such processing rates of dietary proteins are much too low to satisfy the high metabolic and anabolic demands for AA of rapidly growing fish larvae. McLean et al. (1999) suggested that a primary role of macromolecule absorption in the fish hindgut may be antigen sampling, as found in mammals. A similar conclusion was drawn for zebrafish (Wallace et al. 2005). Experimentally, it has not been possible to quantify this route of AA absorption, and its dietary importance is still unknown.

3.2.4 Molecular form of dietary nitrogen

Due to the lack of HCl- and pepsin-secreting cells, the preparatory gastric acid denaturation of ingested proteins is absent in altricial fish larvae at the onset of exogenous feeding. However, Cahu and Zambonino-Infante (1997, 2001) suggested that the lack of a stomach does not hinder enzymatic protein digestion in fish larvae since pancreatic and intestinal enzymes are highly active. Still, native proteins present a smaller surface area for alkaline enzymatic attack than denatured proteins, which may reduce the efficiency of pancreatic and intestinal hydrolysis. In the stomachless gibel carp (*Carassius auratus gibelio*), hydrolysis of intact protein was about 12% of that of denatured protein (Jany 1976). In addition, studies on Atlantic halibut (Tonheim et al. 2004, 2005), striped bass (*Morone saxatilis*), and walleye (*Sander vitreus*; Rust 1995) suggest that the lack of a functional stomach in larvae affects the larval ability to utilize dietary proteins. Specific experiments using Atlantic halibut (Tonheim et al. 2004, 2005; Rønnestad et al. 2008) demonstrated that administration of an equal intestinal loading of preparations, with various degrees of hydrolyzation, resulted in a higher absorption efficiency when the model protein was prehydrolyzed prior to feeding (Figure 3.3a). The absorption efficiency of prehydrolyzed protein preparations was relatively high and independent of the amount of protein administered (Tonheim et al. 2005). On the other hand, the absorption efficiency of the intact model protein decreased dramatically when the administered dose size increased. These results support the view that in some species the larval capacity to digest and absorb dietary proteins is limited and may constitute a bottleneck to intensive larval production based on formulated diets rich in complex proteins such as the ones present in fish meal.

The studies on larval halibut also demonstrated that the absorption rates of the pepsin-hydrolyzed protein and the more extensively hydrolyzed protein were 2.2 and 3 times faster than the intact protein, respectively (Figure 3.3b; Tonheim et al. 2004, 2005). When Atlantic halibut larvae were administered an equal intestinal loading of the model protein preparations, there was higher digestibility when the model protein was prehydrolyzed prior to feeding (Tonheim et al. 2005).

In these studies, the digestibility of intact protein, unlike prehydrolyzed protein, was quite variable and dependent on gut transit time (see later), and the absorption efficiency of the intact/native model protein decreased dramatically when the administered dose size increased (Tonheim et al. 2005). These results suggest that the larval digestive tract capacity to process dietary protein is limited by proteolytic rather than by its absorptive capacity, and this agrees with previous findings that indicated a rapid and efficient absorption of FAA, peptides, and hydrolyzed proteins in

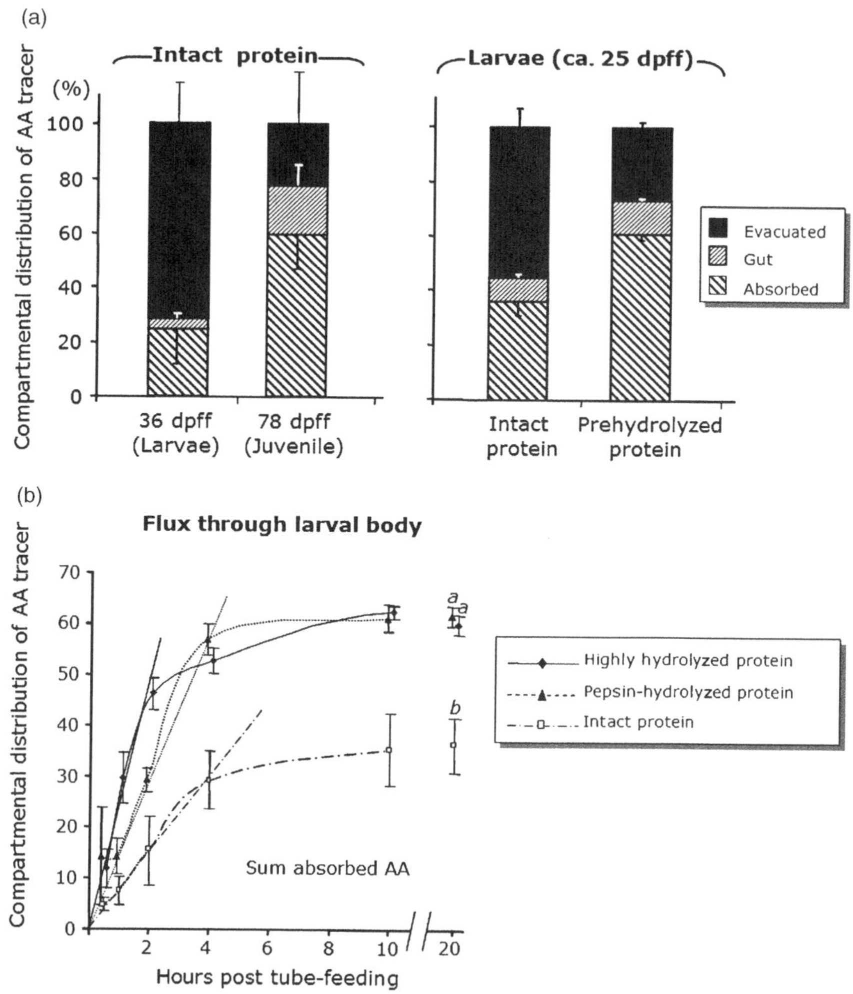

Figure 3.3 (a) Compartmental distribution of a radioactive tracer (^{14}C) in larvae and juvenile Atlantic halibut, *Hippoglossus hippoglossus*, 20 hours after administration of ^{14}C-labeled model protein (salmon serum protein [SSP]) by tube feeding. The left panel (Intact protein) shows distribution in *larvae* and *juvenile* administered the protein in intact form in equal dose size relative to body mass wet weight. The right panel (Larvae ca. 25 dpff) shows distribution of ^{14}C in larvae 20 hours after being administered *intact* ^{14}C-SSP or *prehydrolyzed* ^{14}C-SSP. (Based on Tonheim et al. 2004, 2005.) dpff = days post first feeding. (b) Absorption of ^{14}C tracer in Atlantic halibut larvae administered a bolus of ^{14}C-SSP that was either intact or hydrolyzed to two different levels (partly hydrolyzed and highly hydrolyzed). (Adapted from Tonheim et al. 2005.)

larvae (Rust et al. 1993; Rust 1995; Rønnestad et al. 1999, 2000a, 2001a; Rojas-García and Rønnestad 2003; Applebaum and Rønnestad 2004).

It is generally believed that the digestive capacity of fish larvae and their ability to digest intact protein may increase gradually throughout development (Rust 1995; Tonheim et al. 2004, 2005; Rønnestad et al. 2008). Many of the experimental studies behind these data depend on model studies where larvae are removed from their rearing environments and one nutrient studied at a time, sometimes using tube feeding. However, *Artemia* nauplii and metanauplii have also been used to examine ontogenetic changes in the digestive processing capacity. Studies with Senegalese sole larvae (Morais et al. 2004a; Engrola et al. 2009) established that even young stages (8 days after hatching [dah]), have a high capacity for digesting live *Artemia* (at least when given a single meal). The digestibility of *Artemia* was high in all tested ages, with label absorption varying between 70 and 85% at 24 hours after feeding. Rapid digestion and absorption of *Artemia* AA were noted, with most of the absorption into the body occurring during the first 3 hours after feeding. It can be hypothesized that the relatively higher absorption of AA-label in studies using radio-labeled *Artemia* as the source of protein instead of tube-feeding a liquid tracer could be partly related to the physical nature of the diet, and thus the experimental conditions. It is likely that *Artemia* are retained in the larval gut for a longer period of time than a liquid test solution, thereby allowing more time for protein digestion and absorption. Similarly, very high ^{14}C absorption (67–99%) has been recorded in larval spot (*Leiostomus xanthurus*) using radiolabeled rotifers containing 66% of the ^{14}C-label in the protein fraction (Govoni et al. 1982).

Recent studies have demonstrated that rainbow trout (*Oncorhynchus mykiss*) may be reared from the onset of exogenous feeding on diets using synthetic dipeptides as the source of AA (Dabrowski et al. 2003, 2005). Another recent study has also demonstrated that the supply of peptides to balance the dietary AA profile increases AA retention in fish (Aragão et al. 2004a). Several reports suggest that the dietary inclusion of hydrolyzed protein may eventually compensate for deficiencies in the larval capacity for processing dietary intact proteins (Dabrowski 1983; Carvalho et al. 1997; Zambonino-Infante and Cahu 2001). Still, only moderate, but not high, levels of protein hydrolysates have been shown to improve growth and increase survival in young stages of different fish species (Zambonino-Infante et al. 1997; Carvalho et al. 1997; Day et al. 1997; Cahu et al. 1999; Hamre et al. 2001).

3.2.5 The role of gut transit time on dietary protein utilization

Gut transit time appears to be an important aspect in the fish larvae's ability to utilize the ingested dietary proteins. The gut transit time is a function of rates of ingestion (Figure 3.1-A) and evacuation (Figure 3.1-F), also taking the gut volume into account. It is not clear how, and to what extent, the intestinal transit time is regulated in fish larvae. At least in some species the early larvae appear to continue to ingest feeds, with no apparent anorectic feedback signal from gut fullness (e.g., Atlantic halibut). Under these conditions, ingestion rates appear to equal evacuation rates. In general, short and variable passage times have been reported in fish larvae (Govoni et al. 1986), and a reduced gut transit time has been found to have reduced the efficiency of absorption and utilization of the ingested feeds (Werner and Blaxter 1980; Govoni et al. 1986).

One of the main problems of a fast gut passage is the short time available for efficient digestion and absorption of proteins from the ingested feeds. Proteins with slow digestibility may therefore be lost in the feces in larvae with a short gut retention time, while in larvae with a longer transit time these proteins will be absorbed. As discussed by Rønnestad et al. (2007), the problems associated with rapid gut transit in a continuously feeding (and evacuating) larva may potentially also include lowering luminal proteolytic activity due to overloading of the enzyme producing capacity in the exocrine pancreas. Although this remains to be demonstrated, it is evident that gut transit that is too fast may limit any recovery of the enzymes (as well as bile) from the chyme.

In the study of Tonheim et al. (2005), there was a significantly higher fraction of the intact model protein absorbed by a group of larvae at 31 days post first feeding [dpff]) than by larvae at 25 dpff. A more detailed analysis of the fecal evacuation in these larvae indicated that the poor protein utilization in the younger group was due to rapid fecal evacuation, which in some larvae was more than 50% evacuated by 4 hours post diet administration, at a time when the processes of digestion and absorption were far from complete (Tonheim et al. 2005). This indicates that fecal evacuation is a critical factor in the utilization of slowly digested and absorbed dietary proteins by fish larvae, while it will have a lower effect on the utilization efficiency of rapidly absorbed feed constituents such as FAA, peptides, and prehydrolyzed protein.

A recent study on Atlantic halibut larvae demonstrated that control of light condition strongly affected the feeding regime and also enabled successful eye migration in Atlantic halibut juveniles (Harboe et al. 2009). In this species, continuous light results in continuous feeding and high feed ingestion rates, while combinations of light and dark periods result in defined meals. It was speculated that a fast gut transit in continuously feeding larvae may also cause slowly digestible and essential compounds to be lost in the feces, while in larvae which received a meal based feeding regime, a longer transit time may enable nutrients critical for development to be absorbed. Further work is necessary to determine the effects of gut transit time on feeding regimes combined with prey availability in the rearing systems. These studies also need to focus on the bioavailability of different nutrients.

3.3 Protein metabolism

3.3.1 AA metabolism

The intestinal absorption of individual AA and small peptides depends on different transport systems (Jürss and Bastrop 1995; Rønnestad and Morais 2008), and it has been proposed that absorption of individual AA proceeds at different rates (Dabrowski 1983). In addition, the absorption efficiency of various AA is different, and may also change with species and developmental stage (Rønnestad et al. 2001b; Conceição et al. 2002; Saavedra et al. 2008a, 2008b). Methionine and arginine are more efficiently absorbed than tryptophan, tyrosine, and especially lysine in white sea bream (*Diplodus sargus*) larvae (Saavedra et al. 2008a, 2008b). Different rates of absorption between AA may lead to transitory AA imbalances in the cellular FAA pool where the protein synthesis is carried out, leading to increased AA catabolism.

Indispensable (or essential) AA (IAA) must be provided through the diet (Figure 3.4). The other AA can be synthesized *de novo* from α-keto acids or through transamination and other reactions from other AA and thus are termed dispensable (or nonessential) AA (DAA). Glutamate has a pivotal role, being involved in the synthesis of most other DAA, either as a precursor or as an amino group donor (see Bender 1985; Stryer 1995). Synthesis of DAA from glucose has been demonstrated in adult fish (Cowey and Walton 1989); however, it is unknown whether this *de novo* synthesis of AA is of quantitative significance in adults, and whether fish larvae also have that capability.

Transaminases and other AA-converting enzymes are present in juvenile and adult fish (for a review, see Cowey and Walton 1989), but their importance in AA flux is largely unknown. Transaminase activities have been shown to change both with development and with type of diet in African catfish (*Clarias gariepinus*) larvae (Segner and Verreth 1995). However, in early life stages of two marine teleost species fed diets with different AA compositions, no differences in transaminase activity could be detected (Aragão et al. 2003; Saavedra et al. 2009b). Transamination may

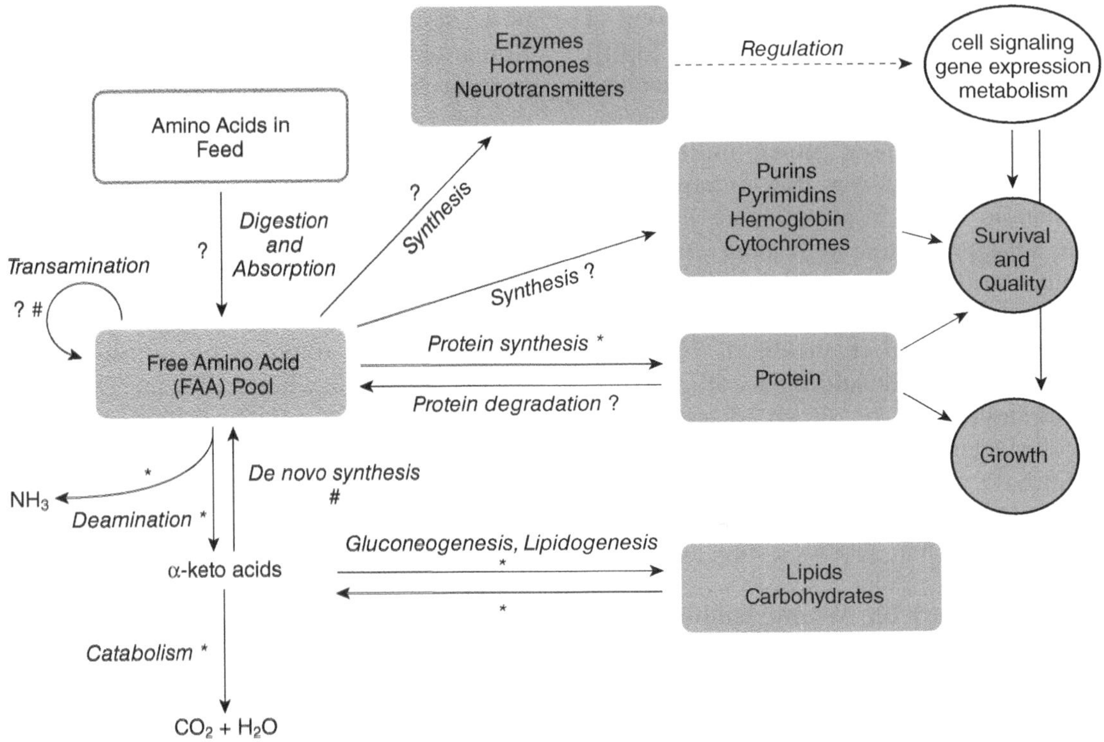

Figure 3.4 Main routes of amino acid (AA) metabolism in fish larvae, and how they may affect larval fish performance. Current knowledge of effects of dietary AA in these routes is given as follows: * = affected by dietary AA composition, ? = unknown if affected by dietary AA composition, # = unknown if relevant in meeting dispensable AA requirements.

be important in improving protein utilization as it may compensate for DAA imbalances in dietary protein. However, it remains to be established whether in fast-growing fish larvae transamination is sufficient to make such compensation.

The first step in AA catabolism is deamination, the removal of the amino group (Figure 3.4). After deamination, resulting α-keto acids may be oxidized to carbon dioxide and water via the tricarboxylic acid (TCA) cycle, or be used in lipid or carbohydrate synthesis. The pathways through which the carbon backbones of individual AA enter the TCA cycle are complex and sometimes multiple (for details, see Cowey and Walton 1989; Stryer 1995). Even if AA are a major energy source during the larval stage of most marine teleost species (Rønnestad 1992; Finn 1994), it is unclear whether larval fish have a higher AA catabolism compared with older fish. It has been proposed that any higher AA usage for energy purposes in fish larvae is caused by a reduced catabolic adaptability associated with the strict carnivorous nature of fish larvae (Dabrowski 1986). However, Morais et al. (2004a) showed that 12-dah Senegalese sole larvae had significantly lower AA catabolism compared with older fish (22 and 35 dah). Still, studies determining AA oxidation (using a $^{14}CO_2$-metabolic trap) in fish larvae show that in general larvae have a high AA catabolism—for example, about 20% of AA intake oxidized 10 hours after ingestion in larval spot (Govoni et al. 1982), 17% at 13 hours after tube feeding in Atlantic halibut postlarvae (Rønnestad et al. 2001a), 39% at 24 hours after feeding in Atlantic herring larvae (Morais et al. 2004b), and 11–30% at 24 hours after feeding in Senegalese sole larvae (Morais et al. 2004a, 2005; Engrola et al. 2009).

Few studies have quantified the fraction of total energy expenditure derived from AA catabolism in fed larvae. AA have been estimated to supply two-thirds of the total energy required in first-feeding Atlantic halibut larvae (Rønnestad and Naas 1993), and were also estimated to account for 70–95% of the total energy expenditure in Atlantic cod larvae at 3–4 weeks of age. After that, lipids gain importance as fuel, but even in 40–60-mm standard-length juveniles, AA still represent more than 50% of the energy substrates (Finn et al. 2002).

It has been demonstrated that larvae have a good capacity to discriminate between individual AA (Conceição et al. 2003a), through studies using tube-feeding of single ^{14}C-labeled AA. Senegalese sole postlarvae (Rønnestad et al. 2001b), Atlantic herring larvae (Conceição et al. 2002), and early metamorphosing Atlantic halibut larvae (Applebaum and Rønnestad 2004) all use DAA preferentially to IAA as energy substrates, and catabolism rates vary between individual IAA. Therefore, fish are able to spare IAA from very early stages of development and have a tight control over AA catabolism. These results show that fish larvae and postlarvae have control of AA catabolism comparable with that of juvenile fish (Cowey and Sargent 1979; Kim et al. 1992) and other animals (Tanaka et al. 1995; Heger et al. 1998; Roth et al. 1999), in contrast to what was previously believed (Conceição et al. 1997a, 1998). Various enzymes are involved in catabolism and transamination of AA (Cowey and Walton 1989; Jürss and Bastrop 1995), allowing regulation for the differential use of individual AA. In fact, catabolism was higher for tyrosine, intermediate for methionine, tryptophan, and lysine, and lower for arginine in white sea bream larvae (Saavedra et al. 2008a, 2008b).

AA are also precursors for the synthesis of several nonproteinic N-containing molecules (Figure 3.4). However, little quantitative significance has been attributed to AA usage through these pathways in vertebrates (Simon 1989). AA are required for the synthesis of many important biomolecules, such as purines and pyrimidines of the nucleic acids, the porphyrine nucleus of hemoglobin and cytochromes, phosphocreatine, as well as peptide- or AA-derived hormones (for details, see Bender 1985; Stryer 1995). AA have also been proposed to be the best precursors for lipid and carbohydrate synthesis in fish (Nagai and Ikeda 1972, 1973); AA deamination results in the production of α-keto acids that can be used for gluconeogenesis and/or lipogenesis. Depending on the carbon backbone, an AA may be termed glucogenic (precursor for carbohydrates) and/or ketogenic (precursor of lipids; for details, see Stryer 1995). Still, gluconeogenesis is believed to be a minor pathway of AA metabolism in fish (van Waarde 1988).

3.3.2 Protein turnover

Protein turnover is the renewal and cycling of tissue proteins and is the dynamic balance between protein synthesis and protein degradation (Figure 3.4). Any protein deposition (or growth) is the net result of this balance. Protein growth may result from an increase in the rate of protein synthesis and/or from a decrease in the rate of protein degradation. Protein turnover equals protein degradation in a growing organism, is the same as both protein synthesis and degradation at maintenance, and is equivalent to protein synthesis during starvation (Wiesner and Zak 1991).

Protein turnover can be divided into an obligatory component associated with the maintenance of cell functions, and a variable component related to growth (Reeds 1989). The main functions of protein turnover have been suggested to be (Hawkins 1991) (1) the ability to dynamically change the amounts (and activities) of specific enzymes; (2) the mobilization of protein during fasting, either to provide AA as precursors for protein synthesis or as substrates for energy; (3) the removal of abnormal proteins; and (4) the precondition for restructuring cells during cell development combined with changes in cell functions.

Protein synthesis and protein degradation are complex processes, and the mechanisms are described in standard textbooks (e.g., Bender 1985; Stryer 1995). The different methodologies to determine the rates of protein synthesis in fish larvae are described and discussed in Conceição et al. (2007). Protein synthesis in fish larvae seems to follow the general trends observed in adult fish and mammals (Houlihan et al. 1995b). Protein synthesis increases with growth rate (Fauconneau et al. 1986a, 1986b), dietary protein level (Fauconneau et al. 1986b), temperature (Fauconneau et al. 1986a), and ration size (Fauconneau et al. 1986a, 1986b; Houlihan et al. 1992). However, while protein degradation increases concurrently with protein synthesis in adult fish and mammals, in fish larvae this does not seem to happen (Houlihan et al. 1992). The result is that at higher growth rates, fish larvae may be more efficient in depositing protein compared with slower-growing larvae. It has been proposed that fish larvae may decrease the rate of protein turnover to reduce the costs of protein synthesis, in order to respond to strong selective pressure to increase efficiency of protein deposition (Kiørboe et al. 1987; Kiørboe 1989; Conceição et al. 1997b).

Fractional rates of protein synthesis (% per day), protein turnover, and protein degradation tend to decrease with body size and with age in mammals (Simon 1989), juvenile and adult fish (Houlihan et al. 1986, 1995a, 1995c), and fish larvae (Fauconneau et al. 1986a; Houlihan et al. 1995b). This decrease in protein synthesis with age may be explained by the increase in the ratio of white muscle tissue to tissues with higher metabolic activity, such as the liver and the digestive tract (Dabrowski 1986). Rates of protein synthesis rates are known to vary considerably between tissues. In adult fish (Fauconneau and Arnal 1985; Houlihan and Laurent 1987; Houlihan et al. 1988), as in mammals (Simon 1989), protein synthesis rates are about 10-fold higher in the liver, gill, digestive tract, kidney, and spleen compared with the heart and muscle. Moreover, the turnover rate of a particular protein may differ among tissues (Houlihan et al. 1995a). How these different turnover rates are regulated is still largely unknown (Simon 1989).

Protein turnover may affect larval viability and survival. A high turnover has been demonstrated to increase metabolic plasticity and response to environmental stress (Hawkins et al. 1987). Turbot (*Scophthalmus maximus*) larvae fed an immunostimulant showed increased rates of protein synthesis and turnover (Conceição et al. 2001), which may result in higher survival in conditions of environmental/disease stress. However, as protein turnover is energetically costly, an increase in protein turnover may cause a trade-off between fast growth and stress resistance/survival in fish larvae (Conceição et al. 2001).

The efficiency of utilization of absorbed AA will depend on the rates of protein synthesis and protein turnover. On the one hand, higher protein synthesis rates tend to reduce AA catabolism, as absorbed AA will leave the free pool faster, which would bring higher AA utilization efficiency. On the other hand, the synthesis of proteins is one of most energy-demanding cellular processes; this may result in increased concurrent use of AA for energy production. Increased protein turnover usually decreases AA utilization efficiency, as a higher amount of AA will be available for catabolism. However, protein turnover may also have an important short-term role, in compensating dietary AA imbalances during postprandrial metabolism through the endogenous release of AA. In fish larvae, the daily turnover rate of proteins can be more than 50% (Houlihan et al. 1993; Conceição et al. 1997b, 2001). Nevertheless, protein turnover is just the recycling of AA (Houlihan et al. 1995a), and therefore AA released by protein turnover will have no net contribution in terms of meeting AA requirements. Protein growth depends exclusively on dietary AA.

3.4 AA requirements

3.4.1 IAA and DAA

In 1957, an AA test diet was successfully developed for fingerling salmonids and allowed nutritionists to identify the essentiality of each AA for fish (Halver 1957). Consequently, it was concluded that 10 out of the 20 AA that constitute protein are indispensable for salmonids: arginine, histidine, isoleucine, leucine, lysine, methionine, phenylalanine, threonine, tryptophan, and valine (Halver et al. 1957; Halver and Shanks 1960; Table 3.1). Further studies using a radioactively labeled carbon source ([U-^{14}C]-glucose) have confirmed that plaice (*Pleuronectes platessa*) and Dover sole (*Solea solea*) were unable to biosynthesize the same AA (Cowey et al. 1970). A dietary source of such IAA is needed to meet the fish requirements for normal growth.

Table 3.1 Indispensable and dispensable amino acids for fish according to Halver et al. (1957).

Indispensable	Conditionally indispensable	Dispensable
Arginine	Cysteine	Alanine
Histidine	Tyrosine	Asparagine
Isoleucine		Aspartic acid
Leucine		Glutamic acid
Lysine		Glutamine
Methionine		Glycine
Phenylalanine		Proline
Threonine		Serine
Tryptophan		
Valine		

Difficulties in working with fish larvae (i.e., small-sized organisms, very fragile to manipulation, constraints in developing well-accepted microdiets with a well-defined nutritional composition at the time of ingestion, limitations in manipulating the AA composition of live feeds) have hampered the use of the traditional dose–response methodology applied to juvenile studies. In addition, some of the same difficulties have thus far prevented the use of the AA oxidation method (Walton et al. 1984). Thus, AA considered indispensable for juvenile and adult fish have been considered indispensable for the larvae. This may not be absolutely valid, as the incipient degree of development and the high growth rates of early fish larvae may limit the biosynthesis of some AA considered dispensable.

Recently, a dipeptide-based diet has been successfully used with rainbow trout alevins (Dabrowski et al. 2003, 2005; Terjesen et al. 2006), opening new perspectives for the identification of the essentiality of AA for fish larvae. This dipeptide-based diet was used to confirm that arginine is an IAA in first-feeding rainbow trout alevins (Terjesen et al. 2006). During the first stages of development, the ornithine–urea cycle enzymes are expressed in trout alevins and in larvae of several fish species (e.g., Wright et al. 1995; Chadwick and Wright 1999; Terjesen et al. 2000, 2001), which theoretically could enable arginine biosynthesis. However, this pathway apparently is involved in ammonia detoxification during the early fish stages (Wright et al. 1995; Terjesen 2008), and even if some arginine biosynthesis may occur, a dietary source is needed to cover larval requirements.

Cysteine and tyrosine are considered conditionally IAA for fish since their sole precursors are IAA—methionine and phenylalanine, respectively. For the purpose of estimating the AA requirements of fish, a sulfur and an aromatic AA requirement is usually considered more pertinent, representing the sum of methionine with cysteine and phenylalanine with tyrosine, respectively (e.g., Mambrini and Kaushik 1995). Lately, it has been suggested that proline may be conditionally indispensable for larval fish (Dabrowski et al. 2005; Terjesen et al. 2006), but this hypothesis needs to be further tested in controlled experiments.

Contrary to what was suggested in the past, it is now known that fish larvae are able to distinguish between IAA and DAA. Hence, IAA are preferentially spared for growth purposes, while DAA are preferentially used as an energy substrate (Rønnestad and Conceição 2005). Therefore, besides their high growth rates and their high energy requirements, fish larvae have a tight regulation of their AA metabolism. Hence, the dietary IAA:DAA ratios may have an impact on larval growth and nitrogen metabolism, as has been shown in fish juveniles (e.g., Green et al. 2002; Peres and Oliva-Teles 2006), but studies with fish larvae are still needed.

3.4.2 AA profiles as indicators of requirements

AA requirements of fish larvae are not easily determined due to difficulties in the use of formulated diets and in the manipulation of the live feed protein profiles (see Chapter 11 for discussions). Thus, traditional methods used in nutritional studies with juvenile and adult fish, such as dose–response studies, are difficult to apply to larvae. López-Alvarado and Kanazawa (1994) reported the quantitative dietary arginine requirement for red sea bream (*Pagrus major*) larvae (20 days old) using dose–response trials, but most studies with fish larvae used AA profiles as an indicator of their requirements. Fiogbé and Kestemont (1995) assessed the protein and AA requirements in goldfish (*Carassius auratus*) larvae based on dose–response methods and concluded that, despite larvae being fed diets differing in their protein levels, the larval AA composition remained stable and independent from the dietary protein content. Thus, AA profiles are a valuable tool as a starting point for the determination of larval AA requirements.

In fish juveniles, several studies have determined the tissue AA profile that should be used as a guideline for studying AA requirements. The AA profiles of fish eggs have been used as an indicator of the larval AA requirements (e.g., Dendrinos and Thorpe 1987; Shcherbina et al. 1988); however, several authors have concluded that whole-body IAA profiles better reflect the fish AA requirements, rather than the profiles of the egg stages (Ogata et al. 1983; Wilson and Poe 1985; Mambrini and Kaushik 1995; Alam et al. 2005). Marine pelagic fish eggs have a high FAA content, and in most cases, these are only partly used for protein synthesis (e.g., Rønnestad et al. 1992, 1994; Finn et al. 1995, 1996; Parra et al. 1999; Kamler 2008). Total AA content of fish eggs may then lead to erroneous estimations of the larval AA requirements since, in exogenous feeding larvae, the FAA pool represents only a small fraction of the total AA pool (Conceição et al. 1997a; Brown et al. 2005). Furthermore, studies aiming to estimate the larval AA requirements should consider that larval growth is alometric, which suggests that different proteins may be synthesized at different periods during ontogenesis (Osse and van den Boogart 1995). Physiological and morphological changes occurring throughout larval life may also affect considerably the AA profile. For instance, in species with a marked metamorphosis, such as Senegalese sole, changes in the AA profile during ontogeny are more pronounced than in species that have a smoother metamorphosis, such as gilthead sea bream (*Sparus aurata*; Aragão et al. 2004b). For a better estimation of the AA requirements of fish larvae, an analysis of the whole-body AA profile at several points of development seems worthwhile.

Based on the ontogenetical changes found for larval AA profile (Table 3.2), the AA

Table 3.2 Ontogenetic changes in larval amino acid (AA) profiles of several teleost species.

Species	Reference	Nr points[a]	Arg	Lys	His	Ile	Leu	Thr	Val	Met	Phe	Trp	Tyr	Cys
African catfish (*Clarias gariepinus*)	Conceição et al. (1998)	3	⇧	⇧	⇩							*	⇩	
Turbot (*Scophthalmus maximus*)	Conceição et al. (1997a)	3		⇩			⇩			⇧	⇩	*		*
Senegalese sole (*Solea senegalensis*)	Aragão et al. (2004b)	8	⇳	⇧		⇧	⇳	⇩	⇳	⇩	⇩	*	⇩	⇳
Gilthead sea bream (*Sparus aurata*)	Aragão et al. (2004b)	6	⇳	⇳						⇩		*	⇳	⇧
Dentex (*Dentex dentex*)	Tulli and Tibaldi (1997)	3		⇧	⇧	⇳	⇧		⇳	⇧[b]	⇧	*		⇧[b]
White sea bream (*Diplodus sargus*)	Saavedra et al. (2006)	8										*		
Sharpsnout sea bream (*Diplodus puntazzo*)	Saavedra et al. (2007)	6										*		

Note: ⇧ = contribution to AA profile increases during ontogeny; ⇳ = contribution to AA profile fluctuates during ontogeny; ⇩ = contribution to AA profile decreases during ontogeny; * = not determined. Absence of symbol means that no significant differences in contribution of the AA to the AA profile were detected during larval ontogeny.

[a] Number of developmental points sampled;
[b] Met and Cys data are given together in source.

requirements have been suggested to change during ontogenesis of some fish species, such as dentex (*Dentex dentex*; Tulli and Tibaldi 1997), African catfish (Conceição et al. 1998), Asian sea bass (*Lates calcarifer*; Syama Dayal et al. 2003), turbot (Conceição et al. 1997a), gilthead sea bream, and Senegalese sole (Aragão et al. 2004b). However, for other species, such as the white sea bream (*Diplodus sargus*; Saavedra et al. 2006) and sharpsnout sea bream (*Diplodus puntazzo*; Saavedra et al. 2007), the larval AA profile is rather stable after the onset of exogenous feeding, suggesting no major changes in AA requirements during ontogeny. It should be noted that even what may seem like small changes in the IAA profile may have important implications in terms of IAA requirements. For instance, a 0.5% decrease in the contribution of methionine to the IAA profile of African catfish was estimated to potentially result in an increase of up to 21% of the methionine requirement (Conceição et al. 1998).

As the whole-larval IAA profile is a rough estimation of the IAA requirements of larvae, several authors have compared this with the IAA of larval feed in order to detect possible dietary imbalances. Most common live prey used in aquaculture—rotifers and *Artemia*—do not meet the dietary requirements of larvae from several fish species, such as African catfish (Conceição et al. 1998), turbot (Conceição et al. 2003a), white sea bream (Saavedra et al. 2006), gilthead sea bream, and Senegalese sole (Aragão et al. 2004b; Figure 3.5). Nevertheless, this appears to not hold true for all species since those prey seem to meet the AA requirements of larvae from species such as dentex (Tulli and Tibaldi 1997) and striped trumpeter (*Latris lineata*; Brown et al. 2005).

3.4.3 AA bioavailabilities and requirements

Although AA profiles are a good starting point to define the IAA requirements of larval fish, these are only a rough estimation of their requirements. This approach only takes into consideration the AA requirements for protein synthesis, and not the requirements for routine metabolic demands or for other metabolic or physiological purposes. This method

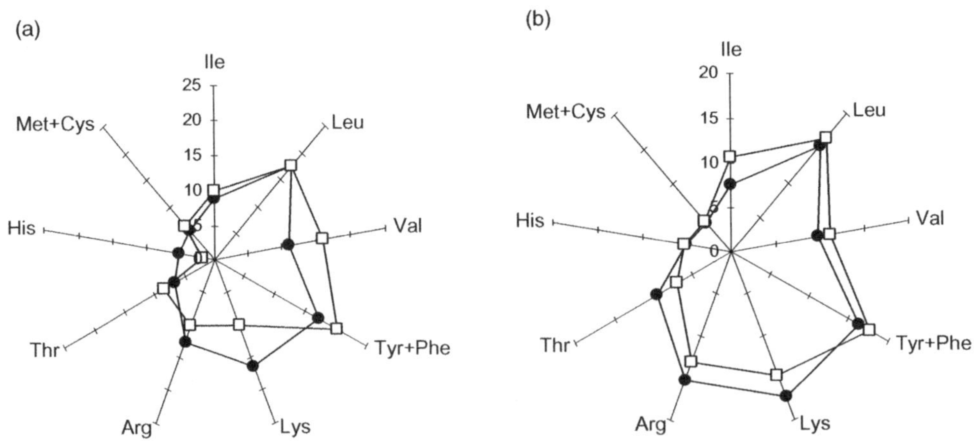

Figure 3.5 Comparison of the indispensable amino acid profiles of fish larvae (filled circles) and live prey (open squares) for an estimation of dietary amino acid imbalances. (a) *Sparus aurata* 7 days after hatching and rotifers (adapted from Aragão et al. 2004b); (b) *Diplodus sargus* 17 days after hatching and *Artemia* nauplii. (Adapted from Saavedra et al. 2006.)

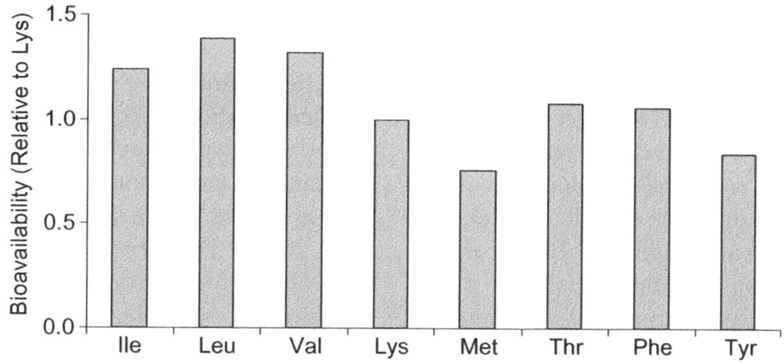

Figure 3.6 Relative bioavailability of several amino acids in sharpsnout sea bream (*Diplodus puntazzo*) larvae determined using ^{15}N-enriched rotifers in combination with GC-IRMS. Bioavailability of lysine is set to 1, and other amino acids are expressed in relation to lysine. (Based on data from Saavedra et al. 2007.)

is more robust if differences in the bioavailability of individual AA, due to selective absorption and/or catabolism, are taken into account (Conceição et al. 2003a, 2003b, 2007). The gut absorption of individual AA depends on different transport systems (Jürss and Bastrop 1995), and thereby individual AA may have different absorption rates and efficiencies (Dabrowski 1983). Absorption and/or catabolism of individual AA has been shown to vary in larvae of several fish species (Rønnestad et al. 2001b; Conceição et al. 2002; Saavedra et al. 2008a, 2008b).

A method combining high-resolution ^{13}C-nuclear magnetic resonance (NMR) spectroscopy and the use of ^{13}C-labeled live feed has been used to study simultaneously the relative bioavailability of several individual AA in gilthead sea bream larvae (Conceição et al. 2003b). Relative bioavailability is a combined measure of absorption efficiency and rate of catabolism for each AA when compared with other AA. Threonine and lysine were shown to have relative bioavailabilities of 0.64 and 1.49, respectively, meaning that threonine is retained less efficiently by larvae when compared with other IAA, while lysine is retained more efficiently (Conceição et al. 2003b). A similar study using ^{15}N-enriched rotifers in combination with gas chromatography–isotope ratio mass spectrometry (GC-IRMS) in sharpsnout sea bream larvae showed higher relative bioavailabilities for isoleucine, leucine, and valine compared with lysine, while methionine and tyrosine had lower relative bioavailabilities (Saavedra et al. 2007; Figure 3.6).

It is important to realize that AA relative bioavailabilities, just as larval IAA profiles and thereby AA requirements, may change between species and also during development for a given species (Conceição et al. 2003a). The capacity and type of intestinal transporters change during fish ontogeny (Buddington 1992), and catabolism of lysine was shown to change during ontogeny of Atlantic herring larvae (Conceição et al. 2002). In any case, the combination of larval protein AA profiles with bioavailability data is probably the most promising tool to study the IAA requirements of larval fish. These two methodologies can be used to define the ideal dietary IAA profile for a given species, that is, the qualitative AA requirements for growth (Conceição et al. 2003a). The ideal dietary IAA profile can be calculated by dividing the contribution of each IAA to the larval IAA profile by the relative bioavailability of that same IAA (Conceição et al. 2003b). Comparison of the IAA profile alone of larval gilthead sea bream with the IAA profile of rotifers suggests that, due to dietary imbalances, larvae would lose by catabolism 45% of the total AA absorbed (Conceição et al. 2003a). However, if the IAA profile had been corrected for relative bioavailabilities, the estimated AA losses would be only 15% (Conceição et al. 2003a). It should be noted that using larval IAA profiles without correcting for bioavailability may lead to either overestimation of IAA requirements (as in the example above) or underestimation, depending on the bioavailabilities of the more imbalanced IAA in the diet.

The combination of larval protein AA profiles with bioavailability data can also be instrumental in reducing the efforts needed to study AA requirements in fish larvae. Providing the requirement for a single AA is established for a given species by dose–response or oxidation studies, IAA profiles corrected for bioavailability may be used to quickly estimate requirements of all other IAA.

3.4.4 Requirements for uses other than growth

Few studies have analyzed the AA requirements for growth in fish larvae, when compared with studies focusing on juvenile or adult fish. However, even fewer have analyzed the AA requirements for purposes other than growth. Some recent literature suggests that the importance of meeting larval AA requirements goes further than maximizing growth performance. For instance, a balanced AA profile increased survival in gilthead sea bream larvae (Aragão et al. 2007) and decreased skeletal deformities in white sea bream larvae (Saavedra et al. 2009b).

Beneficial effects of AA other than larval growth enhancement have also been demonstrated in studies using dietary AA supplementation. White sea bream larvae fed balanced AA diets supplemented with aromatic AA had fewer skeletal deformities and increased resistance to a stress test when compared with a nonsupplemented balanced diet (Saavedra 2008). Dietary tryptophan supplementation increased brain serotonin (5-hydroxytryptamine) concentration, contributing to mitigation of cannibalism in 38-dah groupers (*Epinephelus coioides*; Hseu et al. 2003). Therefore, it has been suggested that dietary formulations based on the larval AA profiles may be insufficient to cover requirements other than growth.

There is also some evidence that during certain periods of development, larvae may have higher requirements for specific AA. For instance, in flatfish species, metamorphosis is a highly complex and physiologically demanding process occurring during larval development. Hence, Senegalese sole larvae retain a higher proportion of dietary tyrosine during pre- and metamorphic stages than after metamorphosis (Pinto et al. 2009a), which suggests higher tyrosine requirements during this period. Interestingly, short-term tyrosine supplementation affected aromatic AA (phenylalanine and tyrosine) metabolism only during the metamorphic stage of Senegalese sole larvae. When these larvae were tube-fed tyrosine supplements at pre- or postmetamorphic stages, no significant effects were observed (Pinto et al. 2009b).

Studies on the AA requirements of fish had focused mostly on the AA that structure proteins. However, many AA are only found in the free form and often are ignored by nutritionists. Several FAA, such as taurine, ornithine, or γ-amino-*n*-butyric acid (GABA), have important functions in fish.

Among the AA only present in the free form, taurine is probably the one most studied in terms of requirements for fish. Nevertheless, its role and importance in fish are still not fully understood, especially in the larval stages. Taurine is one of the most abundant low-molecular-weight organic constituents in fish. Lacking a carboxyl group, taurine cannot be incorporated into protein and, in animals, it cannot be oxidized for energetic purposes.

However, taurine participates in several physiological functions in vertebrates, such as bile salt synthesis, membrane stabilization, modulation of neurotransmitters, antioxidation, and detoxification, and it has effects on early development of the brain, eyes, and heart (Huxtable 1992). A positive correlation between taurine levels in live feed and larval growth rates has been suggested (Conceição et al. 1997a), and some results indicate that taurine might be an essential nutrient during early fish developmental stages (Takeuchi et al. 2001). Further studies testing the effects of dietary taurine supplementation on fish larvae have shown that rotifers enriched with taurine increased growth in red sea bream, Pacific cod (*Gadus macrocephalus*), and Japanese flounder larvae (Chen et al. 2004, 2005; Matsunari et al. 2005). Besides this positive effect on larval growth, dietary taurine supplements also increased resistance to starvation in red sea bream larvae (Chen et al. 2004). Furthermore, for flatfish species going through a marked metamorphosis during their ontogenesis, taurine supplements enhanced larval metamorphosis, although this may, at least in part, be linked to the increased growth observed in these larvae (Chen et al. 2005; Pinto et al. 2008). The capacity of fish to biosynthesize taurine remains unclear, and it has been suggested that this may be species specific and even stage specific (Yokoyama et al. 2001; Kim et al. 2008).

Many more AA exist in the free form and their functions are largely or completely unknown. Furthermore, emerging evidence shows that physiological functions of AA go further than being key precursors of hormones and several low-molecular-weight nitrogenous substances; AA are also cell signaling molecules and regulators of gene expression (Li et al. 2009; Wu 2009). Fish studies on these subjects are very scarce and almost nonexistent for fish larvae, so the requirements for purposes other than growth are still far for being understood. This almost certainly means there is still much room for improving the performance and quality of larval fish.

3.4.5 Dietary supply of AA

Larvae of most marine fish species, and even many freshwater species, rely for periods ranging from 1 week to 1 or more months on zooplankton for feed (see review by Conceição et al. 2010). Despite considerable progress in microdiet technology in recent years, the feeding regime for larvae of most marine fish species starts with live rotifers and/or *Artemia*. The problem is that the AA profile of rotifers and *Artemia* probably do not meet the dietary AA requirements of larvae of most fish species studied so far (see Section 3.4.2). Despite some studies reporting the use of microdiets to study larval requirements for AA (López-Alvarado et al. 1994; López-Alvarado and Kanazawa 1995; Saavedra et al. 2009a, 2009b) or other nutrients (Yúfera et al. 2000; Koven et al. 2001, 2002; Seiliez et al. 2006) with reasonable survival rates, growth rates are typically low compared with when live feeds are used. However, the use of inert microdiets in advanced larval stages is a reality for most commercially reared species. The problems that hamper further progress in the use of microdiets in early larval stages of fish include low attractiveness of microdiets and consequent lower ingestion rates; poor digestibility; high leaching losses of soluble molecules such as FAA, peptides, vitamins, and minerals; and difficulties formulating complete and well-balanced diets due to lack of knowledge of nutritional requirements.

The problem of microdiet attractiveness has been partly overcome by the inclusion of protein hydrolysates or FAA, which are known to have a role as attractants (Kolkovski et al. 1997; Cahu and Zambonino-Infante 2001; Koven et al. 2001). The FAA alanine, glycine, and arginine, as well as betaine, have been identified to stimulate feed intake in gilthead sea bream larvae (Kolkovski et al. 1997).

Koven et al. (2001) have also demonstrated that supplementation of microdiets with phospholipids, in particular phosphatidylcholine, stimulates feed intake. The main challenge in protein nutrition of fish larvae is to develop a highly digestible microdiet while controlling leaching losses (see Chapter 13). Fish larvae have difficulties digesting complex proteins (see Section 3.2.4), while soluble proteins, peptides, and FAA may be lost to a large extent due to leaching (Yúfera et al. 2002; Kvåle et al. 2006, 2007b). Encapsulation techniques that prevent leaching tend to make microdiets very difficult to digest. Several microdiet types have been developed and tested with only limited success, in attempts to deal with these challenges (see review by Langdon 2003 and Chapter 12).

3.5 Conclusions

Fish larvae have very high growth potential. Despite the relatively simple digestive system of marine fish larvae, the larval gut has the capacity to degrade and absorb dietary nutrients that can support very high growth rates, provided that a suitable feed is available. Therefore, in order to meet the high protein requirements of fish larvae and optimize growth, it is important to understand the way proteins, peptides, and FAA are digested, absorbed, and metabolized by fish larvae. Despite considerable progress in recent years, many questions remain largely unanswered, and several research avenues remain open.

It is well established that the molecular form in which dietary nitrogen is given to fish larvae is a major determinant of its utilization, but the ideal molecular composition that should be included in practical diets is still to be determined. It is clear that there are ontogenetic changes in protein, peptide, and FAA absorption in fish, but the mechanisms that explain these ontogenetic changes are still poorly understood. The factors that are limiting and regulating the absorption kinetics of AA and peptides are largely unknown. Peptide transport is probably an important route for absorption of both dietary proteins and peptides, but its quantitative importance remains to be established. Gut transit time appears to be an important aspect in the use of ingested dietary proteins in feeds. However, it is unclear how and to what extent intestinal transit time is regulated in fish larvae, and how the bioavailability of dietary proteins changes with time and degree of digestion. The *in vivo* digestibility for the dietary protein source will also be an important factor that determines their utilization by fish larvae.

Fish larvae have tight control over AA metabolism, and actively spare IAA from very early stages of development, rather using DAA for energy production. Still, it is unknown whether fish larvae have transamination capacity to meet their requirements for all AA considered dispensable for older fish. Some of these may actually be IAA for fish larvae. Protein turnover probably has important roles in the endogenous mobilization of AA to compensate for dietary AA imbalances during postprandial metabolism, or for the synthesis of proteins required during stress or disease. However, the magnitude and relevance of these roles, and their energetic implications, remain to be established.

Optimization of dietary AA utilization will depend on defining the ideal dietary IAA profile. Determination of the larval AA requirements is not an easy task, and traditional methods used in nutrition are difficult to apply. The combination of protein AA profiles of larvae with bioavailability data can significantly reduce the effort to study AA requirements in fish larvae. However, it is largely unknown to what extent bioavailabilities of individual AA change with species, ontogeny, and even dietary composition. Furthermore, AA requirements for functions other than growth and energy production are very poorly understood in fish larvae but should not be overlooked. Beneficial effects of AA supplementation on larval development and resistance to stress have been documented. Moreover, AA that are only found in the free form and not in proteins, such as taurine, have important functions in fish and may be required in the diet for larval stages. AA requirements for functions other than growth and energy production, including precursors of hormones, cell signaling molecules, and regulators of gene expression, may prove crucial in solving performance and quality problems of larval fish.

Technological problems still exist in supplying dietary protein and AA to fish larvae. A major challenge is to develop a highly digestible microdiet with controlled leaching losses of soluble dietary nitrogen and that overrides the difficulties that fish larvae have in digesting complex proteins. Microencapsulation techniques may have an important role in this respect.

Literature cited

Alam, M.S., Teshima, S.-I., Yaniharto, D., Sumule, O., Ishikawa, M., and Koshio, S. 2005. Assessment of reference dietary amino acid pattern for juvenile red sea bream, *Pagrus major*. *Aquaculture International* 13:369–379.

Alvarez-González, C.A., Cervantes-Trujano, M., Tovar-Ramírez, D., Conklin, D.E., Nolasco, H., Gisbert, E., and Piedrahita, R. 2006. Development of digestive enzymes in California halibut (*Paralichthys californicus*) larvae. *Fish Physiology and Biochemistry* 31:83–93.

Amberg, J.J., Myr, C., Kamisaka, Y., Jordal, A.E.O., Rust, M.B., Hardy, R.W., Koedijk, R., and Rønnestad, I. 2008. Expression of the oligopeptide transporter, PepT1, in larval Atlantic cod (*Gadus morhua*). *Comparative Biochemistry and Physiology. Part B* 150:177–182.

Applebaum, S.L., and Rønnestad, I. 2004. Absorption, assimilation and catabolism of individual free amino acids by late larval Atlantic halibut (*Hippoglossus hippoglossus*). *Aquaculture* 230:313–322.

Aragão, C., Conceição, L.E.C., Dias, J., Marques, A.C., Gomes, E., and Dinis, M.T. 2003. Soy protein concentrate as a protein source for Senegalese sole (*Solea senegalensis* Kaup 1858) diets: effects on growth and amino acid metabolism of postlarvae. *Aquaculture Research* 34: 1443–1452.

Aragão, C., Conceição, L.E.C., Martins, D., Rønnestad, I., Gomes, E., and Dinis, M.T. 2004a. A balanced dietary amino acid profile improves amino acid retention in post-larval Senegalese sole (*Solea senegalensis*). *Aquaculture* 233:293–304.

Aragão, C., Conceição, L.E.C., Fyhn, H.J., and Dinis, M.T. 2004b. Estimated amino acid requirements during early ontogeny in fish with different life styles: gilthead seabream (*Sparus aurata*) and Senegalese sole (*Solea senegalensis*). *Aquaculture* 242:589–605.

Aragão, C., Conceição, L.E.C., Lacuisse, M., Yúfera, M., and Dinis, M.T. 2007. Do dietary amino acid profiles affect performance of larval gilthead seabream? *Aquatic Living Resources* 20:155–161.

Ash, R. 1985. Protein digestion and absorption. In: Cowey, C.B., Mackie, A., and Bell, J.B. (eds.) *Nutrition and Feeding in Fish*. Academic Press, London, pp. 69–91.

Bender, D.A. 1985. *Amino Acid Metabolism*. John Wiley & Sons, Inc., Great Britain.

Bolasina, S., Pérez, A., and Yamashita, Y. 2006. Digestive enzymes activity during ontogenetic development and effect of starvation in Japanese flounder, *Paralichthys olivaceus*. *Aquaculture* 252:503–515.

Brown, M.R., Battaglene, S.C., Morehead, D.T., and Brock, M. 2005. Ontogenetic changes in amino acid and vitamins during early larval stages of striped trumpeter (*Latris lineata*). *Aquaculture* 248:263–274.

Buddington, R.K. 1992. Intestinal nutrient transport during ontogeny of vertebrates. *American Journal of Physiology* 262:R503–R509.

Buddington, R.K., Krogdahl, Å., and Bakke-McKellep, A.M. 1997. The intestines of carnivorous fish: structure and functions and the relations with diet. *Acta Physiologica Scandinavica* 161:67–80.

Cahu, C.L., and Zambonino-Infante, J.L. 1997. Is the digestive capacity of marine fish larvae sufficient for compound diet feeding? *Aquaculture International* 5:151–160.

Cahu, C., and Zambonino-Infante, J. 2001. Substitution of live food by formulated diets in marine fish larvae. *Aquaculture* 200: 161–180.

Cahu, C.L., Zambonino-Infante, J.L., Le Gall, M.M., and Quazuguel, P. 1995. Early weaning of seabass: are digestive enzymes limiting? In: Lavens, P., Jaspers, E., and Roelants, I. (eds.) *Larvi '95: Fish and Shellfish Larviculture Symposium*, Special Publication No. 24. European Aquaculture Society, Ghent, Belgium, pp. 268–271.

Cahu, C.L., Zambonino-Infante, J.L., Quazuguel, P., and Le Gall, M.M. 1999. Protein hydrolysate vs. fish meal in compound diets for 10-day old sea bass *Dicentrarchus labrax* larvae. *Aquaculture* 171:109–119.

Carter, C.G., and Houlihan, D.F. 2001. Protein synthesis. In: Wright, P.A., and Anderson, P.M. (eds.) *Nitrogen Excretion*. Academic Press, San Diego, CA, pp. 31–76.

Carvalho, A.P., Escaffre, A.M., Teles, A.O., and Bergot, P. 1997. First feeding of common carp larvae on diets with high levels of protein hydrolysates. *Aquaculture International* 5:361–367.

Carvalho, A.P., Sá, R., Oliva-Teles, A., and Bergot, P. 2004. Solubility and peptide profile affect the utilization of dietary protein by common carp (*Cyprinus carpio*) during early larval stages. *Aquaculture* 234:319–333.

Chadwick, T.D., and Wright, P.A. 1999. Nitrogen excretion and expression of urea cycle enzymes in the Atlantic cod (*Gadus morhua* L.): a comparison of early life stages with adults. *Journal of Experimental Biology* 202:2653–2662.

Chen, J.-N., Takeuchi, T., Takahashi, T., Tomoda, T., Koiso, M., and Kuwada, H. 2004. Effect of rotifers enriched with taurine on growth and survival activity of red sea bream *Pagrus major* larvae. *Nippon Suisan Gakkaishi* 70:542–547.

Chen, J.-N., Takeuchi, T., Takahashi, T., Tomoda, T., Koiso, M., and Kuwada, H. 2005. Effect of rotifers enriched with taurine on growth in larvae of Japanese flounder *Paralichthys olivaceus*. *Nippon Suisan Gakkaishi* 71:342–347.

Chong, A.S.C., Hashim, R., and Ali, A.B. 2002. Assessment of dry matter and protein digestibility of selected raw ingredients by discus fish (*Symphysodon aequifasciata*) using *in vivo* and *in vitro* methods. *Aquaculture Nutrition* 8:229–238.

Conceição, L., Verreth, J., Scheltema, T., and Machiels, M. 1993. A simulation model for the metabolism of yolk sac larvae of the African catfish, *Clarias gariepinus* (Burchell). *Aquaculture Fisheries Management* 24:431–433.

Conceição, L.E.C., van der Meeren, T., Verreth, J.A.J., Evjen, M.S., Houlihan, D.F., and Fyhn, H.J. 1997a. Amino acid metabolism and protein turnover in larval turbot (*Scophthalmus maximus*) fed natural zooplankton or *Artemia*. *Marine Biology* 129:255–265.

Conceição, L.E.C., Houlihan, D.F., and Verreth, J.A.J. 1997b. Fast growth, protein turnover and costs of protein metabolism in yolk-sac larvae of the African catfish (*Clarias gariepinus*). *Fish Physiology and Biochemistry* 16:291–302.

Conceição, L.E.C., Ozório, R.O.A., Suurd, E.A., and Verreth, J.A.J. 1998. Amino acid profiles and amino acid utilization in larval African catfish (*Clarias gariepinus*): effects of ontogeny and temperature. *Fish Physiology and Biochemistry* 19:43–57.

Conceição, L.E.C., Skjermo, J., Skjak-Braek, G., and Verreth, J.A.J. 2001. Effect of an immunostimulating alginate on protein turnover of turbot (*Scophthalmus maximus* L.) larvae. *Fish Physiology and Biochemistry* 24:207–212.

Conceição, L.E.C., Rønnestad, I., and Tonheim, S.K. 2002. Metabolic budgets for lysine and glutamate in unfed herring (*Clupea harengus*) larvae. *Aquaculture* 206:305–312.

Conceição, L.E.C., Grasdalen, H., and Rønnestad, I. 2003a. Amino acid requirements of fish larvae and post-larvae: new tools and recent findings. *Aquaculture* 227:221–232.

Conceição, L.E.C., Grasdalen, H., and Dinis, M.T. 2003b. A new method to estimate the relative bioavailability of individual amino acids in fish larvae using ^{13}C-NMR spectroscopy. *Comparative Biochemistry and Physiology. Part B* 134:103–109.

Conceição, L.E.C., Morais, S., and Rønnestad, I. 2007. Tracers in fish larvae nutrition: a review of methods and applications. *Aquaculture* 267:62–75.

Conceição, L.E.C., Yúfera, M., Makridis, P., Morais, S., and Dinis, M.T. 2010. Live feeds for early stages of fish rearing. *Aquaculture Research* 41:613–640. doi:10.1111/j.1365-2109.2009.02242.x.

Cousin, J.C.B., Baudin-Laurencin, F., and Gabaudan, J. 1987. Ontogeny of enzymatic activities in fed and fasting turbot, *Scophthalmus maximus* L. *Journal of Fish Biology* 30:15–33.

Cowey, C., and Sargent, J.R. 1979. Nutrition. In: Hoar, W.S., Randall, D.J., and Brett, J.R. (eds.) *Fish Physiology*. Academic Press, New York, pp. 1–70.

Cowey, C.B., and Walton, M.J. 1989. Intermediary metabolism. In: Halver, J.E. (ed.) *Fish Nutrition*. Academic Press, New York, pp. 259–329.

Cowey, C.B., Adron, J., and Blair, A. 1970. Studies on the nutrition of marine flatfish. The essential amino acid requirements of plaice and sole. *Journal of the Marine Biological Association of the United Kingdom* 50:87–95.

Dabrowski, K. 1983. Digestion of protein and amino acid absorption in stomachless fish, common carp (*Cyprinus carpio* L). *Comparative Biochemistry and Physiology. Part A* 74: 409–415.

Dabrowski, K. 1986. Protein digestion and amino acid absorption along the intestine of the common carp (*Cyprinus carpio* L), a stomachless fish—an *in vivo* study. *Reproduction, Nutrition, Development* 26:755–766.

Dabrowski, K., and Glogowski, J. 1977. Studies on the role of exogenous proteolytic enzymes in digestion processes in fish. *Hydrobiologia* 54:129–134.

Dabrowski, K., Lee, K.J., and Rinchard, J. 2003. The smallest vertebrate, teleost fish, can utilize synthetic dipeptide-based diets. *Journal of Nutrition* 133:4225–4229.

Dabrowski, K., Terjesen, B.F., Zhang, Y., Phang, J.M., and Lee, K.J. 2005. A concept of dietary dipeptides: a step to resolve the problem of amino acid availability in the early life of vertebrates. *Journal of Experimental Biology* 208:2885–2894.

Daniel, H. 2004. Molecular and integrative physiology of intestinal peptide transport. *Annual Review of Physiology* 66:361–384.

Darias, M.J., Murray, H.M., Gallant, J.W., Douglas, S.E., Yúfera, M., and Martinez-Rodriguez, G. 2007. The spatiotemporal expression pattern of trypsinogen and bile salt–activated lipase during the larval development of red porgy (*Pagrus pagrus*, Pisces, Sparidae). *Marine Biology* 152:109–118.

Day, O.J., Howell, B.R., and Jones, D.A. 1997. The effect of dietary hydrolyzed fish protein concentrate on the survival and growth of juvenile Dover sole, *Solea solea* (L.), during and after weaning. *Aquaculture Research* 28:911–921.

Dendrinos, P., and Thorpe, J.P. 1987. Experiments on the artificial regulation of the amino acid and fatty acid contents of food organisms to meet the assessed nutritional requirements of larval, post-larval and juvenile Dover sole [*Solea solea* (L.)]. *Aquaculture* 61:121–154.

Douglas, S.E., Gawlicka, A., Mandla, S., and Gallant, J.W. 1999. Ontogeny of the stomach in winter flounder: characterization and expression of the pepsinogen and proton pump genes and determination of pepsin activity. *Journal of Fish Biology* 55:897–915.

Engrola, S., Mai, M., Dinis, M.T., and Conceição, L.E.C. 2009. Co-feeding of inert diet from mouth opening does not impair protein utilization by Senegalese sole (*Solea senegalensis*) larvae. *Aquaculture* 287:185–190.

Espeland, S. 2006. Postprandial changes in *intestinal*, plasma, urine, and bile free amino acids of Atlantic cod (*Gadus morhua*) fed a pelleted commercial feed. MSc thesis, University of Bergen, Bergen, Norway.

Fauconneau, B., and Arnal, M. 1985. *In vivo* protein synthesis in different tissues and the whole-body of rainbow trout (*Salmo gairdnerii* R.)—influence of environmental temperature. *Comparative Biochemistry and Physiology. Part A* 82:179–187.

Fauconneau, B., Aguirre, P., and Bergot, P. 1986a. Protein synthesis in early life of coregonids: influence of temperature and feeding. *Archiv für Hydrobiologie—Beiheft Ergebnisse der Limnologie* 22:171–188.

Fauconneau, B., Aguirre, P., Dabrowski, K., and Kaushik, S.J. 1986b. Rearing of sturgeon (*Acipenser baeri* Brandt) larvae. 2. Protein metabolism—influence of fasting and diet quality. *Aquaculture* 51:117–131.

Finn, R.N. 1994. Physiological energetics of developing marine fish embryos and larvae. Dr. Scient. thesis, University of Bergen, Bergen, Norway.

Finn, R.N., Fyhn, H.J., and Evjen, M.S. 1995. Physiological energetics of developing embryos and yolk-sac larvae of Atlantic cod (*Gadus *morhua*). I. *Respiration* and nitrogen metabolism. *Marine Biology* 124:355–369.

Finn, R.N., Fyhn, H.J., Henderson, R.J., and Evjen, M.S. 1996. The sequence of catabolic substrate oxidation and enthalpy balance of developing embryos and yolk-sac larvae of turbot (*Scophthalmus maximus* L.). *Comparative Biochemistry and Physiology. Part A* 115:133–151.

Finn, R.N., Rønnestad, I., Meeren, T., and Fyhn, H.J. 2002. Fuel and metabolic scaling during the early life stages of Atlantic cod *Gadus morhua*. *Marine Ecology Progress Series* 243:217–234.

Fiogbé, E.D., and Kestemont, P. 1995. An assessment of the protein and amino acid requirements in goldfish (*Carassius auratus*) larvae. *Journal of Applied Ichthyology* 11:282–289.

Fyhn, H.J. 1993. Multiple functions of free amino acids during embryogenesis in marine fishes. In: Walther, B.T., and Fyhn, H.J. (eds.) *Physiological and Biochemical Aspects of Fish Development*. University of Bergen, Bergen, Norway, pp. 299–308.

Fyhn, H.J., Finn, R.N., Helland, S., Rønnestad, I., and Lømsland, E. 1993. Nutritional value of phyto- and zooplankton as live food for marine fish larvae. In: Reinertsen, H., Dahle, L.A., Jørgensen, L., and Tvinnerheim, K. (eds.) *Fish Farming Technology*. Balkema, Rotterdam, pp. 121–126.

Gawlicka, A., Parent, B., Horn, M.H., Ross, N., Opstad, I., and Torrissen, O.J. 2000. Activity of digestive enzymes in yolk-sac larvae of Atlantic halibut (*Hippoglossus hippoglossus*): Indication of readiness for first feeding. *Aquaculture* 184:303–314.

Gildberg, A. 2004. Digestive enzyme activities in starved pre-slaughter farmed and wild-captured, Atlantic cod (*Gadus morhua*). *Aquaculture* 238:343–353.

Gisbert, E., Gimenez, G., Fernandez, I., Kotzamanis, Y., and Estevez, A. 2009. Development of digestive enzymes in common dentex *Dentex dentex* during early ontogeny. *Aquaculture* 287:381–387.

Govoni, J.J., Peters, D.S., and Merriner, J.V. 1982. Carbon assimilation during larval development of the marine teleost *Leiostomus xanthurus* Lacépède. *Journal of Experimental Marine Biology and Ecology* 64:287–299.

Govoni, J.J., Boehlert, G.W., and Watanabe, Y. 1986. The physiology of digestion in fish larvae. *Environmental Biology of Fishes* 16:59–77.

Green, J.A., Hardy, R.W., and Brannon, E.L. 2002. The optimum dietary essential: nonessential amino acid ratio for rainbow trout (*Oncorhynchus mykiss*), which maximizes nitrogen retention and minimizes nitrogen excretion. *Fish Physiology and Biochemistry* 27:109–115.

Halver, J.E. 1957. Nutrition of salmonoid fishes: IV. An amino acid test diet for chinook salmon. *Journal of Nutrition* 62:245–254.

Halver, J.E., and Shanks, W.E. 1960. Nutrition of salmonoid fishes: VIII. Indispensable amino acids for sockeye salmon. *Journal of Nutrition* 72:340–346.

Halver, J.E., DeLong, D.C., and Mertz, E.T. 1957. Nutrition of salmonoid fishes: V. Classification of essential amino acids for chinook salmon. *Journal of Nutrition* 63:95–105.

Hamre, K., Naess, T., Espe, M., Holm, J.C., and Lie, Ø. 2001. A formulated diet for Atlantic halibut (*Hippoglossus hippoglossus*, L.) larvae. *Aquaculture Nutrition* 7:123–132.

Hamre, K., Opstad, I., Espe, M., Solbakken, J., Hemre, G.I., and Pittman, K. 2002. Nutrient composition and metamorphosis success of Atlantic halibut (*Hippoglossus hippoglossus*, L.) larvae fed natural zooplankton or *Artemia*. *Aquaculture Nutrition* 8:139–148.

Hansen, J.M., Lazo, J.P., and Kling, L.J. 2009. A method to determine protein digestibility of microdiets for larval and early juvenile fish. *Aquaculture Nutrition* 15:615–626.

Harboe, T., Mangor-Jensen, A., Moren, M., Hamre, K., and Rønnestad, I. 2009. Control of light condition affects the feeding regime and enables successful eye migration in Atlantic halibut juveniles. *Aquaculture* 290:250–255.

Hawkins, A.J.S. 1991. Protein turnover—a functional appraisal. *Funct. Ecol.* 5:222–233.

Hawkins, A.J.S., Wilson, I.A., and Bayne, B.L. 1987. Thermal responses reflect protein turnover in *Mytilus edulis* L. *Functional Ecology* 1:339–351.

Heger, J., Mengesha, S., and Vodehnal, D. 1998. Effect of essential:total nitrogen ratio on protein utilization in the growing pig. *British Journal of Nutrition* 80:537–544.

Helland, S., Terjesen, B.F., and Berg, L. 2003. Free amino acid and protein content in the planktonic copepod *Temora longicornis* compared to *Artemia franciscana*. *Aquaculture* 215: 213–228.

Hjelmeland, K. 1995. Trypsin in fish—studies of the enzyme and its inhibitors in the digestive system and epidermis of fish. PhD thesis, The Norwegian College of Fishery Science, Tromsø, Norway.

Hjelmeland, K., Pedersen, B.H., and Nilssen, E.M. 1988. Trypsin content in intestines of herring larvae, *Clupea harengus*, ingesting inert polystyrene spheres or live crustacea prey. *Marine Biology* 98:331–335.

Houlihan, D.F., and Laurent, P. 1987. Effects of exercise training on the performance, growth, and protein-turnover of rainbow-trout (*Salmo gairdneri*). *Canadian Journal of Fisheries and Aquatic Sciences* 44:1614–1621.

Houlihan, D.F., McMillan, D.N., and Laurent, P. 1986. Growth rates, protein synthesis, and protein degradation rates in rainbow trout—effects of body size. *Physiological Zoology* 59:482–493.

Houlihan, D.F., Hall, S.J., Gray, C., and Noble, B.S. 1988. Growth rates and protein turnover in Atlantic cod, *Gadus morhua*. *Canadian Journal of Fisheries and Aquatic Sciences* 45:951–964.

Houlihan, D.F., Wieser, W., Foster, A., and Brechin, J. 1992. *In vivo* protein synthesis rates in larval nase (*Chondrostoma nasus* L). *Canadian Journal of Zoology* 70:2436–2440.

Houlihan, D.F., Pannevis, M., and Heba, H. 1993. Protein synthesis in juvenile tilapia *Oreochromis mossambicus*. *Journal of the World Aquaculture Society* 24:145–151.

Houlihan, D.F., Carter, C., and McCarthy, I. 1995a. Protein synthesis in fish. In: Hochachka, P., and Mommsen, T. (eds.) *Fish Molecular Biology and Biochemistry*. Elsevier Press, Amsterdam, pp. 191–220.

Houlihan, D.F., McCarthy, I.D., Carter, C.G., and Marttin, F. 1995b. Protein turnover and amino acid flux in fish larvae. *ICES Marine Science Symposia* 201:87–99.

Houlihan, D.F., Pedersen, B.H., Steffensen, J.F., and Brechin, J. 1995c. Protein synthesis, growth and energetics in larval herring (*Clupea harengus*) at different feeding regimes. *Fish Physiology and Biochemistry* 14:195–208.

Hseu, J.R., Lu, F.I., Su, H.M., Wang, L.S., Tsai, C.L., and Hwang, P.P. 2003. Effect of exogenous tryptophan on cannibalism, survival and growth in juvenile grouper, *Epinephelus coioides*. *Aquaculture* 218:251–263.

Hsu, H.W., Vavak, D.L., Satterlee, L.D., and Miller, G.A. 1977. A multi-enzyme technique for estimating protein digestibility. *Journal of Food Science* 42:1269–1273.

Huxtable, R.J. 1992. Physiological actions of taurine. *Physiological Reviews* 72:101–144.

Iwai, T., and Tanaka, M. 1968. The comparative study of digestive tract of teleost larvae. III. Epithelial cells in the posterior gut of halfbeak larvae. *Bulletin of the Japanese Society of Scientific Fisheries* 34:44–48.

Jany, K.D. 1976. Studies on the digestive enzymes of the stomachless bonefish *Carassius auratus gibelo* (Bloch). Endopeptidases. *Comparative Biochemistry and Physiology. Part B* 53: 31–38.

Jobling, M., and Hjelmeland, K. 1992. Ernæring og metabolisme. In: Døving, K., and Reimers, E. (eds.) *Fiskens Fysiologi*. John Grieg Forlag, Bergen, Norway, pp. 234–276.

Johnson, R.B., Cook, M.A., Nicklason, P.M., and Rust, M.B. 2009. Determination of apparent protein digestibility of live *Artemia* and a microparticulate diet in 8-week-old Atlantic cod *Gadus morhua* larvae. *Aquaculture* 288: 290–298.

Jürss, K., and Bastrop, R. 1995. Amino acid metabolism in fish. In: Hochachka, P., and Mommsen, T. (eds.) *Fish Molecular Biology and Biochemistry*. Elsevier Press, Amsterdam, pp. 159–189.

Kamler, E. 1992. *Early Life History of Fish: An Energetic Approach*. Chapman & Hall, London.

Kamler, E. 2008. Resource allocation in yolk-feeding fish. *Reviews in Fish Biology and Fisheries* 18:143–200.

Kim, K.I., Grimshaw, T.W., Kayes, T.B., and Amundson, C.H. 1992. Effect of fasting or feeding diets containing different levels of protein or amino acids on the activities of the liver amino acid–degrading enzymes and amino acid oxidation in rainbow trout (*Oncorhynchus mykiss*). *Aquaculture* 107:89–105.

Kim, S.-K., Matsunari, H., Takeuchi, T., Yokoyama, M., Furuita, H., Murata, Y., and Goto, T. 2008. Comparison of taurine biosynthesis ability between juveniles of Japanese flounder and common carp. *Amino Acids* 35:161–168.

Kiørboe, T. 1989. Growth in fish larvae: are they particularly efficient? *Rapports et Procès-Verbaux des Réunions du Conseil International pour l' Exploration de la Mer* 191:383–389.

Kiørboe, T., Munk, P., and Richardson, K. 1987. Respiration and growth of larval herring *Clupea harengus*—relation between specific dynamic action and growth efficiency. *Marine Ecology Progress Series* 40:1–10.

Kolkovski, S. 2001. Digestive enzymes in fish larvae and juveniles—implications and applications to formulated diets. *Aquaculture* 200:181–201.

Kolkovski, S., Koven, W., and Tandler, A. 1997. The mode of action of *Artemia* in enhancing utilization of microdiet by gilthead seabream *Sparus aurata* larvae. *Aquaculture* 155: 193–205.

Koven, W., Kolkovski, S., Hadas, E., Gamsiz, K., and Tandler, A. 2001. Advances in the development of microdiets for gilthead seabream, *Sparus aurata*: a review. *Aquaculture* 194: 107–121.

Koven, W., Rojas-Garcia, C.R., Finn, R.N., Tandler, A., and Rønnestad, I. 2002. Stimulatory effect of ingested protein and/or free amino acids on the secretion of the gastro-endocrine hormone cholecystokinin and on tryptic activity, in early-feeding herring larvae, *Clupea harengus*. *Marine Biology* 140:1241–1247.

Kurokawa, T., Tanaka, H., Kagawa, H., and Otha, H. 1996. Absorption of protein molecules by the rectal cells in eel larvae *Anguilla japonica*. *Fisheries Science* 62:832–833.

Kurokawa, T., Shiraishi, M., and Suzuki, T. 1998. Quantification of exogenous protease derived from zooplankton in the intestine of Japanese sardine (*Sardinops melanotictus*) larvae. *Aquaculture* 161:491–499.

Kvåle, A., Yúfera, M., Nygard, E., Aursland, K., Harboe, T., and Hamre, K. 2006. Leaching properties of three different microparticulate diets and preference of the diets in cod (*Gadus morhua* L.) larvae. *Aquaculture* 251:402–415.

Kvåle, A., Mangor-Jensen, A., Moren, M., Espe, M., and Hamre, K. 2007a. Development and characterisation of some intestinal enzymes in Atlantic cod (*Gadus morhua* L.) and Atlantic halibut (*Hippoglossus hippoglossus* L.) larvae. *Aquaculture* 264:457–468.

Kvåle, A., Nordgreen, A., Tonheim, S.K., and Hamre, K. 2007b. The problem of meeting dietary protein requirements in intensive aquaculture of marine fish larvae, with emphasis on Atlantic halibut (*Hippoglossus hippoglossus* L.). *Aquaculture Nutrition* 13:170–185.

Langdon, C. 2003. Microparticle types for delivering nutrients to marine fish larvae. *Aquaculture* 227:259–275.

Lazo, J.P., Romaire, R.P., and Reigh, R.C. 1998. Evaluation of three *in vitro* enzyme assays for estimating protein digestibility in the Pacific white shrimp *Penaeus vannamei*. *Journal of the World Aquaculture Society* 29:441–450.

Lazo, J.P., Holt, G.J., and Arnold, C.R. 2000. Ontogeny of pancreatic enzymes in larval red drum *Sciaenops ocellatus*. *Aquaculture Nutrition* 6:183–192.

Li, P., Mai, K., Trushenski, J., and Wu, G. 2009. New developments in fish amino acid nutrition: towards functional and environmentally oriented aquafeeds. *Amino Acids* 37:43–53.

Lied, E., and Solbakken, R. 1984. The course of protein digestion in Atlantic cod (*Gadus morhua*). *Comparative Biochemistry and Physiology. Part A* 77:503–506.

Lo, M.-J., and Weng, C.-F. 2006. Developmental regulation of gastric pepsin and pancreatic serine protease in larvae of the euryhaline teleost, *Oreochromis mossambicus*. *Aquaculture* 261:1403–1412.

López-Alvarado, J., and Kanazawa, A. 1994. Effect of dietary arginine levels on growth of red sea bream larvae fed diets supplemented with crystalline amino acids. *Fisheries Science* 60:435–439.

López-Alvarado, J., and Kanazawa, A. 1995. Optimum levels of crystalline amino acids in diets for larval red sea bream (*Pagrus major*). *ICES Marine Science Symposia* 201:1–5.

López-Alvarado, J., Langdon, C.J., Teshima, S., and Kanazawa, A. 1994. Effects of coating and encapsulation of crystalline amino acids on leaching in larval feeds. *Aquaculture* 122:335–346.

Luizi, F.S., Gara, B., Shields, R.J., and Bromage, N.R. 1999. Further description of the development of the digestive organs in Atlantic halibut (*Hippoglossus hippoglossus*) larvae, with notes on differential absorption of copepod and *Artemia* prey. *Aquaculture* 176:101–116.

Mambrini, M., and Kaushik, S.J. 1995. Indispensable amino acid requirements of fish: correspondence between quantitative data and amino acid profiles of tissue proteins. *Journal of Applied Ichthyology* 11:240–247.

Manchado, M., Infante, C., Asensio, E., Crespo, A., Zuasti, E., and Canavate, J.P. 2008. Molecular characterization and gene expression of six trypsinogens in the flatfish Senegalese sole (*Solea senegalensis* Kaup) during larval development and in tissues. *Comparative Biochemistry and Physiology. Part B* 149:334–344.

Marza, E., Barthe, C., André, M., Villeneuve, L., Hélou, C., and Babin, P.J. 2005. Developmental expression and nutritional regulation of a zebrafish gene homologous to mammalian microsomal triglyceride transfer protein large subunit. *Developmental Dynamics* 232:506–518.

Matsunari, H., Arai, D., Koiso, M., Kuwada, H., Takashi, T., and Takeuchi, T. 2005. Effect of feeding rotifers enriched with taurine on growth performance and body composition of Pacific cod larvae *Gadus macrocephalus*. *Aquaculture Science* 53:297–304.

McLean, E., Rønsholdt, B., Sten, C., and Najamuddin, J.F. 1999. Gastrointestinal delivery of peptide and protein drugs to aquacultured teleosts. *Aquaculture* 177:231–247.

van der Meeren, T., Olsen, R.E., Hamre, K., and Fyhn, H.J. 2008. Biochemical composition of copepods for evaluation of feed quality in production of juvenile marine fish. *Aquaculture* 274:375–397.

Morais, S., Lacuisse, M., Conceição, L.E.C., Dinis, M.T., and Rønnestad, I. 2004a. Ontogeny of the digestive capacity of Senegalese sole (*Solea senegalensis* Kaup 1858), with respect to digestion, absorption and metabolism of amino acids from *Artemia*. *Marine Biology* 145:243–250.

Morais, S., Conceição, L.E.C., Dinis, M.T., and Rønnestad, I. 2004b. A method for radiolabeling *Artemia* with applications in studies of food intake, digestibility, protein and amino acid metabolism in larval fish. *Aquaculture* 231:469–487.

Morais, S., Koven, W., Rønnestad, I., Dinis, M.T., and Conceição, L.E.C. 2005. Dietary protein/lipid ratio affects growth and amino acid and fatty acid absorption and metabolism in Senegalese sole (*Solea senegalensis* Kaup 1858) larvae. *Aquaculture* 246:347–357.

Morrison, C.M. 1987. Histology of the Atlantic cod, *Gadus morhua*: an atlas. Part one. Digestive tract and associated organs. *Canadian Special Publication of Fisheries and Aquatic Sciences* 98:219.

Munilla-Moran, R., Stark, J.R., and Barbour, A. 1990. The role of exogenous enzymes in digestion in cultured turbot larvae (*Scophthalmus maximus* L.). *Aquaculture* 88:337–350.

Murray, H.M., Perez-Casanova, J.C., Gallant, J.W., Johnson, S.C., and Douglas, S.E. 2004. Trypsinogen expression during the development of the exocrine pancreas in winter flounder (*Pleuronectes americanus*). *Comparative Biochemistry Physiology. Part A* 138:53–59.

Murray, H.M., Gallant, J.W., Johnson, S.C., and Douglas, S.E. 2006. Cloning and expression analysis of three digestive enzymes from Atlantic halibut (*Hippoglossus hippoglossus*) during early development: predicting gastrointestinal functionality. *Aquaculture* 252:394–408.

Nagai, M., and Ikeda, S. 1972. Carbohydrate metabolism in fish. 3. Effect of dietary composition on metabolism of glucose-U-C-14 and glutamate-U-C-14 in carp. *Bulletin of the Japanese Society of Scientific Fisheries* 38:137–143.

Nagai, M., and Ikeda, S. 1973. Carbohydrate metabolism in fish. 4. Effect of dietary composition on metabolism of acetate-U-^{14}C and L-alanine-U-^{14}C in carp. *Bulletin of the Japanese Society of Scientific Fisheries* 39:633–643.

Nankervis, L., and Southgate, P.C. 2006. An integrated assessment of gross marine protein sources used in formulated microbound diets for barramundi (*Lates calcarifer*) larvae. *Aquaculture* 257:453–464.

Narawane, S., Rønnestad, I., Fjose, A., and Ellingsen, S. 2009. Spatiotemporal distribution and developmental role of cationic amino acid transporter 1 (CAT1) in zebrafish (*Danio rerio*). 6th European Zebrafish Genetics and Development Meeting, July 15–19. Rome, Italy.

Nielsen, C.U., and Brodin, B. 2003. Di/tri-peptide transporters as drug delivery targets: regulation of transport under physiological and patho-physiological conditions. *Current Drug Targets* 4:373–388.

Ogata, H., Arai, S., and Nose, T. 1983. Growth responses of cherry salmon *Oncorhynchus masou* and amago salmon *O. rhodurus* fry fed purified casein diets supplemented with amino acids. *Bulletin of the Japanese Society of Scientific Fisheries* 49:1381–1385.

Osse, J.W.M., and van den Boogart, J.G.M. 1995. Fish larvae, development, allometric growth, and the aquatic environment. *ICES Marine Science Symposia* 201:21–34.

Otterlei, E., Nyhammer, G., Folkvord, A., and Stefansson, S.O. 1999. Temperature- and size-dependent growth of larval and early juvenile Atlantic cod (*Gadus morhua*): a comparative study of Norwegian coastal cod and northeast Arctic cod. *Canadian Journal of Fisheries and Aquatic Sciences* 56:2099–2111.

Parra, G., Rønnestad, I., and Yúfera, M. 1999. Energy metabolism in eggs and larvae of the Senegal sole. *Journal of Fish Biology* 55A:205–214.

Pedersen, B.H., and Andersen, K.P. 1992. Induction of trypsinogen secretion in herring larvae (*Clupea harengus*). *Marine Biology* 112: 559–565.

Pedersen, B.H., Nilssen, E.M., and Hjelmeland, K. 1987. Variations in the content of trypsin and trypsinogen in larval herring (*Clupea harengus*) digesting copepod nauplii. *Marine Biology* 94:171–181.

Pedersen, B.H., Ugelstad, I., and Hjelmeland, K. 1990. Effects of a transitory, low food supply in the early life of larval herring (*Clupea harengus*) on mortality, growth and digestive capacity. *Marine Biology* 107:61–66.

Peres, H., and Oliva-Teles, A. 2006. Effect of the dietary essential to non-essential amino acid ratio on growth, feed utilization and nitrogen metabolism of European sea bass (*Dicentrarchus labrax*). *Aquaculture* 256:395–402.

Perez-Casanova, J.C., Murray, H.M., Gallant, J.W., Ross, N.W., Douglas, S.E., and Johnson, S.C. 2004. Bile salt–activated lipase expression during larval development in the haddock (*Melanogrammus aeglefinus*). *Aquaculture* 235:601–617.

Perez-Casanova, J.C., Murray, H.M., Gallant, J.W., Ross, N.W., Douglas, S.E., and Johnson, S.C. 2006. Development of the digestive capacity in larvae of haddock (*Melanogrammus aeglefinus*) and Atlantic cod (*Gadus morhua*). *Aquaculture* 251:377–401.

Pinto, W., Figueira, L., Ribeiro, L., Yúfera, M., Dinis, M.T., and Aragão, C. 2008. Dietary taurine supplementation enhances metamorphosis and growth potential in *Solea senegalensis* larvae. XIII International Symposium on Fish Nutrition and Feeding, Florianópolis, Brazil, pp. 107.

Pinto, W., Figueira, L., Dinis, M.T., and Aragão, C. 2009a. How does fish metamorphosis affect aromatic amino acid metabolism? *Amino Acids* 36:177–183.

Pinto, W., Figueira, L., Rodrigues, V., Dinis, M.T., and Aragão, C. 2009b. The importance of aromatic amino acids during fish ontogeny. In: Hendry, C.I., Van Stappen, G., Wille, M., and Sorgeloos, P. (eds.) *Larvi '09: Fish and Shellfish Larviculture Symposium*, Special Publication No. 38. European Aquaculture Society, Ghent, Belgium, pp. 329–332.

Ray, E.C., Avissar, N.E., and Sax, H.C. 2002. Growth factor regulation of enterocyte nutrient transport during intestinal adaptation. *American Journal of Surgery* 183:361–371.

Reeds, P.J. 1989. Regulation of protein turnover. In: Campion, D.R., Hausman, G.J., and Martin, R.J. (eds.) *Animal Growth and Regulation*. Plenum Press, New York, pp. 183–210.

Ribeiro, L., Zambonino-Infante, J.L., Cahu, C., and Dinis, M.T. 1999. Development of digestive enzymes in larvae of *Solea senegalensis*, Kaup 1858. *Aquaculture* 179:465–473.

Rojas-García, C.R., and Rønnestad, I. 2003. Assimilation of dietary free amino acids, peptides and protein in post-larval Atlantic halibut (*Hippoglossus hippoglossus*). *Marine Biology* 142:801–808.

Rønnestad, I. 1992. Utilization of free amino acids in marine fish eggs and larvae. Dr.Scient. thesis, University of Bergen, Bergen, Norway.

Rønnestad, I., and Conceição, L.E.C. 2005. Aspects of protein and amino acids digestion and utilization by marine fish larvae. In: Starck, J.M., and Wang, T. (eds.) *Physiological and Ecological Adaptations to Feeding in Vertebrates*. Science Publishers, Enfield, NH, pp. 389–416.

Rønnestad, I., and Fyhn, H.J. 1993. Metabolic aspects of free amino acids in developing marine fish eggs and larvae. *Reviews in Fisheries Science* 1:239–259.

Rønnestad, I., and Morais, S. 2008. Digestion. In: Finn, R.N., and Kapoor, B.G. (eds.) *Fish Larval Physiology*. Science Publishers, Enfield, NH, pp. 201–262.

Rønnestad, I., and Naas, K.E. 1993. Oxygen consumption and ammonia excretion in larval Atlantic halibut (*Hippoglossus hippoglossus* L.) at first feeding: a first step towards an energetic model. In: Walther, B.T., and Fyhn, H.J. (eds.) *Physiology and Biochemical Aspects of Fish Development*. University of Bergen, Bergen, Norway pp. 279–284.

Rønnestad, I., and Rojas-García, C.R. 2006. Absorption kinetics of free amino acids in larval Atlantic herring (*Clupea harengus*). Abstracts from the AQUA 2006 Annual Meeting of the World Aquaculture Society, May 9–13, Fortezza da Basso Convention Centre, Firenze (Florence), Italy.

Rønnestad, I., Finn, R.N., Groot, E.P., and Fyhn, H.J. 1992. Utilization of free amino acids related to energy metabolism of developing eggs and larvae of lemon sole *Microstomus kitt* reared in the laboratory. *Marine Ecology Progress Series* 88:195–205.

Rønnestad, I., Koven, W.M., Harel, M., and Fyhn, H.J. 1994. Energy metabolism during development of eggs and larvae of gilthead sea bream (*Sparus aurata*). *Marine Biology* 120: 187–196.

Rønnestad, I., Thorsen, A., and Finn, R.N. 1999. Fish larval nutrition: a review of recent advances in the roles of amino acids. *Aquaculture* 177: 201–216.

Rønnestad, I., Conceição, L.E.C., Aragão, C., and Dinis, M.T. 2000a. Free amino acids are absorbed faster and assimilated more efficiently than protein in postlarval Senegal sole (*Solea senegalensis*). *Journal of Nutrition* 130: 2809–2812.

Rønnestad, I., Peréz Dominguez, R., and Tanaka, M. 2000b. Ontogeny of digestive tract functionality in Japanese flounder (*Paralichthys olivaceus*) studied by *in vivo* microinjection: pH and assimilation of free amino acids. *Fish Physiology and Biochemistry* 22:225–235.

Rønnestad, I., Rojas-Garcia, C.R., Tonheim, S.K., and Conceição, L.E.C. 2001a. *In vivo* studies of digestion and nutrient assimilation in marine fish larvae. *Aquaculture* 201:161–175.

Rønnestad, I., Conceição, L.E.C., Aragão, C., and Dinis, M.T. 2001b. Assimilation and catabolism of dispensable and indispensable free amino acids in post-larval Senegal sole (*Solea senegalensis*). *Comparative Biochemistry and Physiology. Part C* 130:461–466.

Rønnestad, I., Tonheim, S.K., Fyhn, H.J., Rojas-Garcia, C.R., Kamisaka, Y., Koven, W., Finn, R.N., Terjesen, B.F., Barr, Y., and Conceição, L.E.C. 2003. The supply of amino acids during early feeding stages of marine fish larvae: a review of recent findings. *Aquaculture* 227: 147–164.

Rønnestad, I., Gavaia, P., Viegas, C., and Cancela, L.M. 2006. PepT1 in a coldwater marine teleost larvae—Atlantic cod: cloning and preliminary studies of expression and phylogeny. *Acta BioMedica* 77(Suppl. 3):82.

Rønnestad, I., Kamisaka, Y., Conceição, L.E.C., Morais, S., and Tonheim, S.K. 2007. Digestive physiology of marine fish larvae: hormonal control and processing capacity for proteins, peptides and amino acids. *Aquaculture* 268: 82–97.

Rønnestad, I., Murashita, K., and Verri, T. 2008. Oligopeptide transporter PepT1 in Atlantic salmon (*Salmo salar*): cloning, tissue expression, comparative aspects and effects of starvation. Transporters 2008, August 27–30. Centre Loewenberg, Murten, Switzerland.

Rønnestad, I., Murashita, K., Kottra, G., Jordal, A.E., Narawane, S., Jolly, C., Daniel, H., and Verri, T. 2010. Molecular cloning and functional expression of Atlantic salmon peptide transporter 1 in *Xenopus* oocytes reveals efficient intestinal uptake of lysine-containing and other bioactive di- and tripeptides in teleost . *Journal of Nutrition* 140:893–900.

Roth, F.X., Gotterbarm, G.G., Windisch, W., and Kirchgessner, M. 1999. Influence of dietary level of dispensable amino acids on nitrogen balance and whole-body protein turnover in growing pigs. *Journal of Animal Physiology and Animal Nutrition* 81:232–238.

Rust, M.B. 1995. Quantitative aspects of nutrient assimilation in six species of fish larvae. PhD thesis, School of Fisheries, University of Washington, Seattle, WA.

Rust, M.B., Hardy, R.W., and Stickney, R.R. 1993. A new method for force-feeding larval fish. *Aquaculture* 116:341–352.

Saavedra, M. 2008. Amino acid requirements of white seabream (*Diplodus sargus*) larvae: effects on growth and performance. PhD thesis, Universidade do Algarve, Faro, Portugal.

Saavedra, M., Conceição, L.E.C., Pousão-Ferreira, P., and Dinis, M.T. 2006. Amino acid profiles of *Diplodus sargus* (L., 1758) larvae: implications for feed formulation. *Aquaculture* 261:587–593.

Saavedra, M., Beltran, M., Pousao-Ferreira, P., Dinis, M.T., Blasco, J., and Conceição, L.E.C. 2007. Evaluation of bioavailability of individual amino acids in *Diplodus puntazzo* larvae: towards the ideal dietary amino acid profile. *Aquaculture* 263:192–198.

Saavedra, M., Conceição, L.E.C., Helland, S., Pousao-Ferreira, R., and Dinis, M.T. 2008a. Effect of lysine and tyrosine supplementation in the amino acid metabolism of *Diplodus sargus* larvae fed rotifers. *Aquaculture* 284:180–184.

Saavedra, M., Conceição, L.E.C., Pousao-Ferreira, P., and Dinis, M.T. 2008b. Metabolism of tryptophan, methionine and arginine in Diplodus sargus larvae fed rotifers: effect of amino acid supplementation. *Amino Acids* 35:59–64.

Saavedra, M., Barr, Y., Pousão-Ferreira, P., Helland, S., Yúfera, M., Dinis, M.T., and Conceição, L.E.C. 2009a. Supplementation of tryptophan and lysine in *Diplodus sargus* larval diet: effects on growth and skeletal deformities. *Aquaculture Research* 40:1191–1201.

Saavedra, M., Pousão-Ferreira, P., Yúfera, M., Dinis, M.T., and Conceição, L.E.C. 2009b. A balanced amino acid diet improves *Diplodus sargus* larval quality and reduces nitrogen excretion. *Aquaculture Nutrition* 15:517–524. doi:10.1111/j.1365-2095.2008.00618.x.

Sangaletti, R., Terova, G., Peres, A., Bossi, E., Corà, S., and Saroglia, M. 2009. Functional expression of the oligopeptide transporter PepT1 from the sea bass (*Dicentrarchus labrax*). *Pflügers Archiv. European Journal of Physiology* 459:47–54.

Satterlee, L.D., Marshall, H.F., and Tennyson, J.M. 1979. Measuring protein quality. *Journal of the American Oil Chemists Society* 56:103–109.

Segner, H., and Verreth, J. 1995. Metabolic enzyme activities in larvae of the African catfish, *Clarias gariepinus*: changes in relation to age and nutrition. *Fish Physiology and Biochemistry* 14:385–398.

Seiliez, I., Bruant, J.S., Infante, J.L.Z., Kaushik, S., and Bergot, P. 2006. Effect of dietary phospholipid level on the development of gilthead sea bream (*Sparus aurata*) larvae fed a compound diet. *Aquaculture Nutrition* 12:372–378.

Shcherbina, M.A., Burlachenko, I.V., and Sergeeva, N.T. 1988. Chemical composition of eggs and amino acid requirements of two species of Black Sea mullets, *Mugil cephalus* and *Liza aurata*. *Journal of Ichthyology* 28:63–69.

Simon, O. 1989. Metabolism of protein and amino acid. In: Bock, H.D., Eggum, B.O., Low, A.G., Simon, O., and Zebrowska, T. (eds.) *Protein Metabolism in Farm Animals*. Oxford University Press, Oxford, pp. 273–366.

Sire, M.F., and Vernier, J.M. 1992. Intestinal absorption of protein in teleost fish. *Comparatvie Biochemistry and Physiology. Part A* 103: 771–781.

Srivastava, A.S., Kurokawa, T., and Suzuki, T. 2002. mRNA expression of pancreatic enzyme precursors and estimation of protein digestibility in first feeding larvae of the Japanese flounder, *Paralichthys olivaceus*. *Comparative Biochemistry and Physiology. Part A* 132:629–635.

Stevens, C.E., and Hume, I.D. 1996. *Comparative Physiology of the Vertebrate Digestive System*. Cambridge University Press, Cambridge.

Stryer, L. 1995. *Biochemistry*. Freeman, New York.

Suzuki, T., Srivastava, A.S., and Kurokawa, T. 2002. A homologue of human placental protein, PP11, and mouse T cell–specific protein, Tcl-30, in exocrine pancreas of a teleost (*Paralichthys olivaceus*). *Comparative Biochemistry and Physiology. Part B* 133:325–329.

Syama Dayal, J., Ahamad Ali, S., Thirunavukkarasu, A.R., Kailasam, M., and Subburaj, R. 2003. Nutrient and amino acid profiles of egg and larvae of Asian seabass, *Lates calcarifer* (Bloch). *Fish Physiology and Biochemistry* 29:141–147.

Takeuchi, T., Park, G.S., and Yokoyama, M. 2001. Taurine content in Japanese flounder *Paralichthys olivaceus* T. & S. and red sea bream *Pagrus major* T. & S. during the period of seed production. *Aquaculture Research* 32(Suppl.1):244–248.

Tanaka, H., Shibata, K., Mori, M., and Ogura, M. 1995. Metabolism of essential amino-acids in growing rats at graded levels of soybean protein osolate. *Journal of Nutrition Science and Vitaminology* 41:433–443.

Terjesen, B.F. 2008. Nitrogen excretion. In: Finn, R.N., and Kapoor, B.G. (eds.) *Fish Larval Physiology*. Science Publishers, Enfield, NH, pp. 263–302.

Terjesen, B.F., Rønnestad, I., Norberg, B., and Anderson, P.M. 2000. Detection and basic properties of carbamoyl phosphate synthetase III during teleost ontogeny: a case study in the Atlantic halibut (*Hippoglossus hippoglossus* L.). *Comparative Biochemistry and Physiology. Part B* 126:521–535.

Terjesen, B.F., Chadwick, T.D., Verreth, J.A.J., Ronnestad, I., and Wright, P.A. 2001. Pathways for urea production during early life of an air-breathing teleost, the African catfish *Clarias gariepinus* Burchell. *Journal of Experimental Biology* 204:2155–2165.

Terjesen, B.F., Lee, K.J., Zhang, Y., Failla, M., and Dabrowski, K. 2006. Optimization of dipeptide–protein mixtures in experimental diet formulations for rainbow trout (*Oncorhynchus mykiss*) alevins. *Aquaculture* 254:517–525.

Tonheim, S.K., Espe, M., Raae, A.J., Darias, M.J., and Rønnestad, I. 2004. *In vivo* incorporation of [U]C-^{14}-amino acids: an alternative protein labelling procedure for use in examining larval digestive physiology. *Aquaculture* 235:553–567.

Tonheim, S.K., Espe, M., Hamre, K., and Rønnestad, I. 2005. Pre-hydrolysis improves utilisation of dietary protein in the larval teleost Atlantic halibut (*Hippoglossus hippoglossus* L.). *Journal of Experimental Marine Biology and Ecology* 321:19–34.

Tonheim, S.K., Nordgreen, A., Høgøy, I., Hamre, K., and Rønnestad, I. 2007. *In vitro* digestibility of water-soluble and water-insoluble protein fractions of some common fish larval feeds and feed ingredients. *Aquaculture* 262:426–435.

Tulli, F., and Tibaldi, E. 1997. Changes in amino acids and essential fatty acids during early larval rearing of dentex. *Aquaculture International* 5:229–236.

Verri, T., Kottra, G., Romano, A., Tiso, N., Peric, M., Maffia, M., Boll, M., Argenton, F., Daniel, H., and Storelli, C. 2003. Molecular and functional characterisation of the zebrafish (*Danio rerio*) Pept1-type peptide transporter. *FEBS Letters* 549:115–122.

Verri, T., Romano, A., Barca, A., Kottra, G., Daniel, H., and Storelli, C. 2010. Transport of di- and tripeptides in teleost fish intestine. *Aquaculture Research* 41:641–653. doi:10.1111/j.1365-2109.2009.02270.x.

Walford, J., and Lam, T.J. 1993. Development of digestive tract and proteolytic enzyme activity in seabass (*Lates calcarifer*) larvae and juveniles. *Aquaculture* 109:187–205.

Wallace, K.N., Akhter, S., Smith, E.M., Lorent, K., and Pack, M. 2005. Intestinal growth and differentiation in zebrafish. *Mechanisms Development* 122:157–173.

Walton, M.J., Cowey, C.B., and Adron, J.W. 1984. The effect of dietary lysine levels on growth and metabolism of rainbow-trout (*Salmo gairdneri*). *British Journal of Nutrition* 52:115–122.

van Waarde, A. 1988. Biochemistry of non-protein nitrogenous compounds in fish including the use of amino acids for anaerobic energy production. *Comparative Biochemistry and Physiology. Part B* 91:207–228.

Watanabe, Y. 1984. Morphological and functional changes in rectal epithelium cells of pond smelt during post embryonic development. *Bulletin of the Japanese Society of Scientific Fisheries* 50: 805–814.

Werner, R.G., and Blaxter, J.H.S. 1980. Growth and survival of larval herring (*Clupea harengus*) in relation to prey density. *Canadian Journal of Fisheries and Aquatic Sciences* 37:1063–1069.

Wiesner, R.J., and Zak, R. 1991. Quantitative approaches for studying gene-expression. *American Journal of Physiology* 260: L179–L188.

Wilson, R.P., and Poe, W.E. 1985. Relationship of whole body and egg essential amino acid patterns to amino acid requirement patterns in channel catfish, *Ictalurus punctatus*. *Comparative Biochemistry and Physiology. Part B* 80:385–388.

Wipf, D., Ludewig, U., Tegeder, M., Rentsch, D., Koch, W., and Frommer, W.B. 2002. Conservation of amino acid transporters in fungi, plants and animals. *Trends in Biochemical Sciences* 27:139–147.

Wright, P.A., and Fyhn, H.J. 2001. Ontogeny of nitrogen metabolism and excretion. In: Wright, P.A., and Anderson, P.M. (eds.) *Nitrogen*

Excretion. Academic Press, San Diego, CA, pp. 149–200.

Wright, P., Felskie, A., and Anderson, P. 1995. Induction of ornithine-urea cycle enzymes and nitrogen metabolism and excretion in rainbow trout (*Oncorhynchus mykiss*) during early life stages. *Journal of Experimental Biology* 198:127–135.

Wu, G. 2009. Amino acids: metabolism, functions, and nutrition. *Amino Acids* 37:1–17.

Yokoyama, M., Takeuchi, T., Park, G.S., and Nakazoe, J. 2001. Hepatic cysteinesulphinate decarboxylase activity in fish. *Aquaculture Research* 32:216–220.

Yúfera, M., Fernandez-Diaz, C., Pascual, E., Sarasquete, M.C., Moyano, F.J., Diaz, M., Alarcon, F.J., Garcia-Gallego, M., and Parra, G. 2000. Towards an inert diet for first-feeding gilthead seabream *Sparus aurata* L. larvae. *Aquaculture Nutrition* 6:143–152.

Yúfera, M., Kolkovski, S., Fernandez-Diaz, C., and Dabrowski, K. 2002. Free amino acid leaching from a protein-walled microencapsulated diet for fish larvae. *Aquaculture* 214:273–287.

Zambonino-Infante, J.L., and Cahu, C.L. 2001. Ontogeny of the gastrointestinal tract of marine fish larvae. *Comparative Biochemistry and Physiology. Part C* 130:477–487.

Zambonino-Infante, J.L., Cahu, C.L., and Peres, A. 1997. Partial substitution of di- and tripeptides for native proteins in sea bass diet improves *Dicentrarchus labrax* larval development. *Journal of Nutrition* 127:608–614.

Zambonino-Infante, J.L., Gisbert, E., Sarasquete, C., Navarro, I., Gutiérrez, J., and Cahu, C.L. 2008. Ontogeny and physiology of the digestive system of marine fish larvae. In: Cyrino, J.E.P., Bureau, D., and Kapoor, B.G. (eds.) *Feeding and Digestive Functions in Fishes*. Science Publishers, Enfield, NH, pp. 281–348.

Chapter 4

Micronutrients

Mari Moren, Rune Waagbø, and Kristin Hamre

4.1 General introduction to fish micronutrient history

The history of micronutrient requirement research in fish often demonstrates initiatives initialized by crisis and the need for rapid solutions in a developing fish farming industry. For vitamins, most focus has been on vitamins C and E due to their unstable chemical nature as antioxidants, and thereby their loss in feed ingredients and formulated feeds (Waagbø 2009). Other vitamins, such as thiamine, have been related to heavy mortalities in wild salmon yolk sac fry. For most marine fish species, aquaculture is still in its infancy and much information on the requirements relies on knowledge from other species. In order to produce efficient diets for fish larvae, there is a need for nutritional knowledge on which nutrient, which type, how much, and when. The small size of fish larvae and the need for a start-feeding period with live feed present practical challenges for micronutrient supply. Natural live feed such as zooplankton has been regarded as the gold standard for marine fish larvae (van der Meeren et al. 2008) based on acceptable growth, survival, and lack of developmental malformations. Rotifers, *Artemia*, and other live feed organisms have been regarded as suitable alternatives, but they need to be enriched with some micronutrients to be comparable with copepods for larvae development and performance. For formulated diets in intensive larva nutrition, one may normally afford well-balanced high-quality feed ingredients, where inherent micronutrients are expected to cover the requirements for growth and survival. However, new feed concepts and feeding regimes that focus on optimizing macronutrient supply introduce nutritional and technological challenges. In many cases, this means that the producer cannot count on fulfilling the requirements for many micronutrients. Since fish larvae micronutrient research on fish larvae has often been initiated in relation to shortcomings and disease problems in commercial larvae production (Kanazawa 1995), this has to a great extent influenced research approaches. The experimental conditions and approaches in the literature therefore reflect the need for a rapid solution in suboptimal

Larval Fish Nutrition, First Edition. Edited by G. Joan Holt.

production units, rather than a well-designed dedicated micronutrient requirement experiment. This chapter reviews the micronutrient knowledge for marine fish larvae, where dominating literature covers vitamins A, E, and C, and selenium and iodine.

4.2 Micronutrients in larval feeds

4.2.1 Live feed

Marine fish larvae feed mainly on copepods in the wild, while in the hatchery they are fed live feed at first feeding. Early fish larvae have an immature digestive tract and the use of formulated diets for these stages gives low performance due to low bioavailability of nutrients and high leakage rates of water-soluble nutrients from the small feed particles. The feed organisms used for early marine fish larvae in captivity are mainly rotifers (*Brachionus* sp.) and *Artemia* (see more on live feeds in Chapter 11).

There are numerous differences in nutrient composition between copepods, rotifers, and *Artemia*, and some of the differences in performance between larvae from pond systems, where they feed on copepods, and intensively reared larvae are probably caused by differences in micronutrient nutrition. The concentrations of vitamins (Table 4.1) in rotifers vary according to the concentrations of vitamins in the culture and enrichment diets. At insufficient enrichments, rotifers may contain very low to undetectable levels of vitamins A, C, E, and thiamine (vitamin B_1). The other B vitamins are present in high concentrations in yeast, and yeast-cultured rotifers will also contain high levels of these vitamins. However, the concentrations in rotifers cultured on algae-based diets are not known. Rotifers generally have low concentrations of trace elements (Table 4.2; Hamre et al. 2008a), but it is possible to increase the concentrations by adding the elements, either to the culture or to the enrichment diet.

Artemia have low concentrations of thiamine, iodine, and zinc compared with copepods (Tables 4.1 and 4.2; Hamre et al. 2007). Their concentration of vitamin C is also lower than that in copepods, but it is still high at 4–500 mg/kg, and probably not deficient. It is possible to enrich *Artemia* with iodine (Moren

Table 4.1 Vitamin levels (mg/kg dry weight) in different live feeds, compared with the requirements in cold-water fish given by the NRC (1993).

	Rotifers[a]	*Artemia*[b]	Copepods[a]	NRC (1993)
Vitamin C	117–576	400–500	600–1,000	50
Riboflavin	22–44	30–60	14–27	4–7
Thiamine (B_1)	2–125	6–12	13–23	1.00
Folic acid	4.0–5.7	6–10	3–5	1
Pyridoxine (B_6)	20–53	2–13	2–6	3–6
Biotin	1.5–1.8	4.0	0.6–0.9	0.15–1
Cobalamin (B_{12})	23–61	3.9	1–2	0.01
Niacin	191–267	202	100–150	10–28
Vitamin E	85–889	100–500	50–200	50
Carotenoids	4–15	654–752	630–750	–
Vitamin A	0–9.5	0	0	2.4[b]

[a] From Hamre et al. (2008a).
[b] From Hamre et al. (2007).

Table 4.2 Minerals in different live feeds, compared with requirements in cold-water fish given by the NRC (1993) (mg/kg dry weight) and in water (μg/L).

	Rotifers[a]	*Artemia*[b]	Copepods[a]	NRC (1993)	Freshwater[c]	Saltwater[c]
Iodine	3.0–7.9	0.5–2.0[*]	50–350	0.6–1.1	0.2–10	2.5–60[‡]
Manganese	3.9–5.1	4.5–27	8–25	13	12	2
Copper	2.7–8.1	8.5–34	12–38	3–5	10	3
Zinc	62–64	160–188[*]	340–570	20–30	10	10
Selenium	0.08–0.09	2.2[†]	3–5	0.25–0.3	<20	0.09
Iron	57–114	88–127	85–371	30–150	670	10
Phosphorus	9,400	12,050–17,000	12,400–15,010	4,500–6,000	5	70
Calcium	1,880	1,535–2,030	1,070–2,370	Not determined	15,000	400,000
Magnesium	4,840	1,975–3,560	2,350–3,140	400–600	4,100	1,350,000
Sodium					6,300	10,500,000
Potassium					2,300	380
Chlorine					7,800	19,000,000

[a] Hamre et al. (2008a).
[b] Rønnestad (unpublished data).
[c] Lorentzen et al. (2001), except:
[*] Hamre et al. (2007), and
[†] Solbakken et al. (2002).
[‡] Iodide (I^-, the bioavailable form).

et al. 2006) and thiamine (Hamre et al., unpublished data). On the other hand, enrichment with zinc led to high mortality of *Artemia* (Hamre and Harboe, unpublished data).

4.2.2 Formulated feed

Vitamins and minerals are added to formulated diets as premixes, probably in the same form in larval diets as in feed for larger fish. This means that minerals and trace elements are added in inorganic forms in the dry feeds, while in live prey, they are most often present in organic forms. This will have consequences for the bioavailability and perhaps the metabolism of the minerals in the larval body. Vitamins are also often supplemented in formulated feed in forms other than those naturally present in live feed and we do not know how this affects bioavailability.

Another factor that affects micronutrient nutrition in marine fish larvae is the high leaching rates of water-soluble nutrients from formulated diets. Many species are weaned onto formulated diets while the larvae are still very small, and they require feed particle sizes down to 0.1 mm. The large surface-to-volume ratio of the particles facilitates leaching; as an example, more than 90% of free amino acids may leach from formulated diets after less than 2 minutes of immersion in water (Lopez-Alvarado et al. 1994; Önal and Langdon 2000; Langdon 2003; Hamre 2006; Hamre and Mangor-Jensen 2006). Similar leaching rates have been determined for water-soluble vitamins and minerals (Nordgreen et al. 2008), while lipid-soluble vitamins (Nordgreen et al. 2008) and unsoluble compounds such as L-ascorbate-2-polyphosphate (stabilized form of vitamin C) (unpublished data) do not seem to be subject to leaching. Clearly, new-generation microfeeds needs to face these technical challenges.

4.3 Requirements versus recommendations

With few exceptions, we do not know the nutritional requirements of marine fish larvae. This is especially true for micronutrients, where only very few direct measurements of requirements have been performed. Most of our knowledge in larval nutrition stems from indirect measurements of requirements. There are several approaches, for example, analyses of nutrient composition of eggs and ovaries (Brækkan 1958). Second, by recording the mass transfer of vitamins from the yolk to the larvae body compartment in high-quality larvae, it is possible to estimate the rate of transfer, retention, and loss of nutrients during ontogeny (Rønnestad et al. 1997; Mæland et al. 2003). Since the endogenous feeding period can be regarded as a closed system with respect to vitamin input, one can describe the transfer of the vitamins and estimate the requirements from the loss and respective growth in the period between hatching and start-feeding. A third approach is to analyze pond-cultured copepods for micronutrients. Analyses of copepods are the most complete source for requirement estimates (Hamre et al. 2008a; van der Meeren et al. 2008). The last approach is extrapolation of dose–response studies in very young juveniles, as was the purpose of the study by Moren et al. (2004b).

It is often assumed that fish larvae have higher requirements for micronutrients than juvenile and adult fish since larvae have very high growth rates and go through the demanding process of metamorphosis (Otterlei et al. 1999; Finn et al. 2002; Power et al. 2008). Furthermore, the natural prey of fish larvae, mainly copepods, generally contain very high levels of micronutrients compared with the requirements of fish given by the National Research Council (NRC 1993; Tables 4.1 and 4.2). It is possible that the larvae have adapted to the very high levels of micronutrients found in copepods. However, the concentrations may vary among copepod species, between locations, and over time, and fish larvae should be able to cope with micronutrient levels that are much lower than those in copepods cultured in lagoons, which probably have a higher nutrient supply compared with open waters. The indirect measurements of micronutrient requirements in fish larvae can therefore only be taken as an indication of the true requirements, which probably are somewhere in the range between established fish requirements (NRC 1993) and the micronutrient levels found in pond-cultured copepods.

4.4 Vitamins

4.4.1 Lipid-soluble vitamins

4.4.1.1 Vitamin A

The term vitamin A is applied to a group of compounds that possess the same biological activity as retinol. Retinoid is a wider generic term that, in addition to vitamin A, includes compounds that structurally resemble retinol but have no biological function. There are two aspects of vitamin A function: vision and gene expression regulation. 11-*cis*-Retinal is the light-absorbing component in vision. Light entering the eye excites the retinal molecules bound to the protein opsin. The excited retinal changes to all-*trans*-retinal and dissociates from opsin, and this eventually changes the membrane potential and thereby alters the nerve signals (reviewed by Rando 1994). The other aspect of vitamin A is its roles in gene regulation. Two groups of nuclear receptors exist: the retinoic X receptors (RXRs α, β, and γ) and retinoic acid receptors (RARs α, β, and γ). Both groups are activated by all-*trans*-retinoic acid. 9-*cis*-Retinoic acid appears to activate only the RXRs, but this is uncertain (Heyman et al. 1992; Kurlandsky et al. 1994). The receptors recognize retinoic acid response elements (RAREs) in promoter regions of genes responsive to retinoids, and

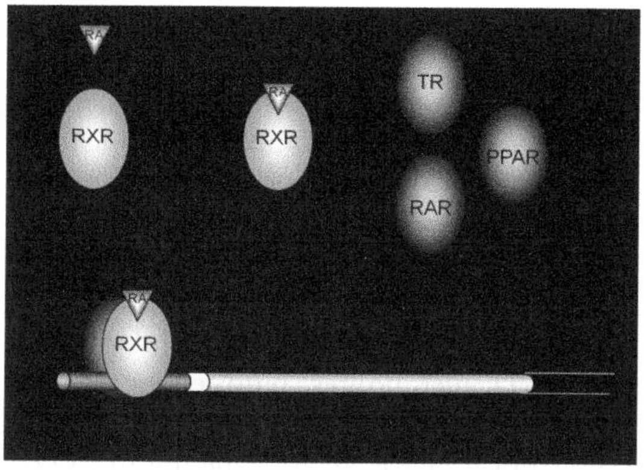

Figure 4.1 Vitamin A in gene regulation. Interactions between vitamin A, as retinoic acid, fatty acids, and thyroid hormones through interaction between their nuclear receptors: (1) retinoic acid (RA) binds to the retinoic X receptor (RXR), which activates the receptor; (2) the activated RXR can bind to several other receptors, for example, peroxisomal proliferator–activated receptor (PPAR) activated by fatty acid derivatives and thyroid hormone receptor (TR) activated by T_3, forming a dimer; (3) the dimer will bind to specific response elements on DNA, exemplified here with the retinoic acid response element (RARE). This leads to transcription of genes downstream of RARE.

gene expression is altered when activated receptors form dimers and bind to RAREs (reviewed by Balmer and Blomhoff 2002). RXR does not only form dimers with itself and with RAR, it also does with other nuclear receptors activated by vitamin D (vitamin D receptor [VDR]), thyroid hormones (THs; thyroid hormone receptor), and fatty acid derivatives (peroxisome proliferator–activated receptors [PPARs]). Through this interaction, vitamin A coregulates a wide range of hormone- and nutrient-responsive genes (Figure 4.1). Villeneuve et al. (2004) sequenced and analyzed the expression of RARα, RARγ, and RXRα in developing European sea bass larvae. They found that these receptors are differently regulated during development. RARα and RARγ were upregulated, while RXRα was downregulated in the 7–37-days posthatch (dph) larvae. It was concluded that these receptors play different roles at different stages of development. Further, the transcription of the retinoic receptors are altered both by high vitamin A levels and by high docosahexaenoic acid (DHA) and eicosapentaenoic acid (EPA) content of the phospholipid fraction of the diet (Villeneuve et al. 2005a, 2005b). One of the most investigated effects of excess vitamin A in fish larvae is the alterations seen in the developing skeleton and in pigmentation (Haga et al. 2002; Martinez et al. 2007; see Chapter 7 for more details). Multiple forms of deformities and a variety of pigmentation patterns may appear.

Fish larvae obtain vitamin A from the diet, either as different forms of vitamin A (retinol, retinal, and retinyl esters) or as carotenoids, which are abundant precursors of vitamin A in invertebrate prey organisms. The main enzyme cleaving carotenoids in two, producing two molecules of retinal, is the β-carotene 15,15′-monooxygenase. This enzyme is most abundant in the intestine and liver but has been detected in many other organs, such as the kidneys, testes, and ovaries (Levi et al. 2008). Recent studies have shown that it has a spatial and temporal expression pattern in zebrafish embryo, indicating a pattern formation function (Lampert et al. 2003). The enzyme can be downregulated by retinoic acid (Takitani et al. 2006), indicating that retinol homeostasis regulation occurs at the level of carotenoid cleavage. The cleavage product, retinal, is either metabolized into the active form retinoic acid or converted to the alcohol form of vitamin A, retinol. Retinol can then be esterified and stored in specific vitamin A–storing cells in the liver called stellate cells for later use (reviewed by Blomhoff et al. 1990).

Copepods contain very little or no vitamin A (Moren et al. 2005). The same is true for rotifers not enriched with vitamin A (Hamre et al. 2008a). Unenriched *Artemia* is reported to have low levels of vitamin A, if any (Estevez and Kanazawa 1995; Takeuchi et al. 1995, 1998; Rønnestad et al. 1998; Moren et al. 2004a). Larvae fed copepods or *Artemia*, where carotenoids are found in high abundance (Table 4.1), are likely to obtain sufficient vitamin A levels from these carotenoids (van der Meeren et al. 2008). Moren et al. (2004a) showed that halibut larvae fed either copepods or *Artemia* had an increase in retinyl esters, the storage form of vitamin A, and that the levels of retinol and retinal did not differ between dietary groups. This suggests that both live prey supplied sufficient amounts of carotenoids to cover the vitamin A requirement. Commercial enrichments for both rotifers and *Artemia* contain vitamin A. While vitamin A enrichment of rotifers is required due to low carotenoid and vitamin A levels (van der Meeren et al. 2008; Hamre et al. 2008a), it might not be necessary to enrich *Artemia*. Hamre et al. (2008a) tested different rotifer enrichments containing different levels of vitamin A (0.13, 1.16, 15.4, and 205 mg/kg dry weight [DW]), and only the highest level gave rotifers containing detectable levels of vitamin A (highest level: 9.5 mg vitamin A/kg DW). Fernandez et al. (2008) enriched rotifers with emulsions containing from 1,700 to 17,000 mg of vitamin A/kg (DW), leading to a range of 75–723 mg/kg rotifers (DW); the lowest level is 100 times higher than the NRC (1993) recommendation (see Table 4.1). All the larval groups had deformities, and the amount correlated with the vitamin A level in the diet. Moren et al. (2004b) estimated the requirement for Atlantic halibut juveniles to be between 0.75 and 2.5 mg of vitamin A/kg and Hernandez et al. (2005) found that juveniles of Japanese flounder should have a diet with 2.7 mg vitamin A/kg.

4.4.1.2 Vitamin D

The physiological function of active vitamin D metabolites (25-OH-vitamin D_3 and 1,25-$(OH)_2$ vitamin D_3) in Ca and P homeostasis in fishes continues to be debated (O'Connell and Gatlin 1994; Rao and Raghuramulu 1999). Recently, Lock et al. (2010) reviewed the role of vitamin D and active metabolites in fish. He discussed actions mediated by the vitamin D endocrine system in fish on, among others, plasma Ca and P regulation, intestinal Ca uptake, Ca transport via the gills, renal phosphate reabsorption, and bone tissue mineralization and resorption. The action of vitamin D involves the vitamin D–binding protein (DBP), which is important for binding, solubilization, and transport of vitamin D and its metabolites. On the cellular level, vitamin D metabolites bind either to the VDR, which is a ligand-activated nuclear transcription factor, or to a membrane VDR ($_m$VDR) with a more rapid activation of nongenomic signal transduction pathways in target cells (Lock et al. 2010). The vitamin D endocrine system acts in concert with a series of other calcium- and phosphorus-regulating hormones, which make it difficult to sort out its significance in the Ca and P homeostasis of fish. In feeding experiments with Atlantic salmon smolt in seawater, suboptimal P nutrition seems to trigger the vitamin D endocrine system, seen as elevated plasma vitamin D metabolites (Fjelldal et al. 2009).

Vitamin D deficiency was described in amago salmon (*Oncorhynchus rhodurus*, Jordan & McGregor) fed a vitamin D–free diet, as compared with a diet containing 0.5 mg/kg vitamin D (Taveekijakarn et al. 1996a). The minimum dietary requirement for vitamin D in salmonid feeds is estimated to be 0.06 mg/kg (2,400 IU/kg) (NRC 1993). The marine food web produces and accumulates vitamin D, and therefore marine feed ingredients normally contain high and sufficient levels of vitamin D to cover the requirement. However, vitamin D recorded in fish meals and fish oils varied 50- and 32-fold, respectively (Horvli and Lie 1994; Opstvedt et al. 1997), indicating that fish diets and enrichment cultures based on traditional marine ingredients may contain high levels of vitamin D. According to present knowledge, megadoses of vitamin D_3 supplementation are not harmful for fish. Feeding experiments with vitamin D doses of 57.5 mg/kg did not affect growth and health in Atlantic salmon

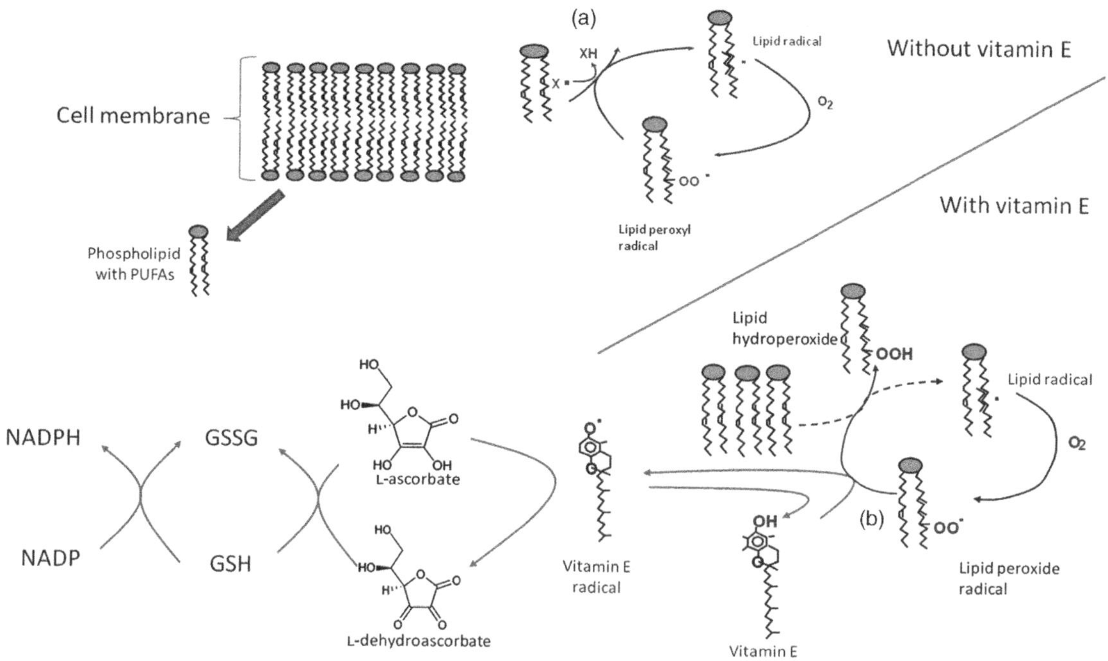

Figure 4.2 Function of vitamin E and vitamin C. (a) Without vitamin E: Auto-oxidation of lipids will occur when a free radical (x•: OH•, $O2^-$•, or others) abstracts a hydrogen atom from a PUFA; the PUFA radical formed reacts with oxygen to form a peroxide radical. Then the peroxide radical abstracts a hydrogen from a new PUFA, and enters a new turn in the reaction cycle. (b) With vitamin E: Vitamin E will interrupt the oxidation cycle by a reaction with the lipid peroxide radical, leaving a lipid hydroperoxide. The vitamin E is now a radical but may be "recycled" when vitamin C is present. Vitamin C in its reduced form (L-ascorbate) reacts with the vitamin E radical and donates a hydrogen to vitamin E. The oxidized form of vitamin C (L-dehydroascorbate) will in turn be "recycled" by glutathione (GSH; reduced form, GSSG; oxidized form), which finally will be reduced by NADP.

(*Salmo salar*) (Graff et al. 2002a). According to the Norwegian feed legislation (LD/FD 2002), vitamin D_3 supplementation should be below 0.075 mg/kg (3,000 IU/kg), while inherent vitamin D is not included in the upper limit unless a supplementation is performed. In natural live feed for fish larvae, van der Meeren et al. (2008) reported undetectable vitamin D_3 levels in copepods, while rotifers and *Artemia* contained 0.9–1.8 (μg/g vitamin D_3 DW) due to the fish oil used in the enrichment emulsion. The authors suggested that indoor rearing of larval fish in the absence of UV light might require dietary vitamin D_3.

4.4.1.3 Vitamin E

Vitamin E is a generic term for tocopherols and tocotrienols, where α-tocopherol (α-TOH) has the highest biological activity. Vitamin E protects lipids against peroxidation, breaking the chain of lipid auto-oxidation by donating a hydrogen molecule to the lipid peroxide radical formed in the chain (Figure 4.2; Frankel 1998). This reaction gives a tocopheryl radical, which can probably be regenerated to tocopherol by ascorbic acid (AA) (Tappel 1962; Packer et al. 1979). The ascorbate radical formed may be regenerated by glutathione, a reaction catalyzed by a group of selenium-dependent enzymes, the glutathione peroxidases (GPxs). Oxidized glutathione can be reduced by NADPH formed in energy metabolism (Mårtensson and Meister 1991). At very high concentrations and under conditions where vitamins C and E radicals are allowed to accumulate, these antioxidants may act as pro-oxidants (Ingold et al. 1987; Bowry et al. 1992). Vitamin E is thus part of the larval defense against lipid oxidation as is the case in other vertebrates. Deficiency symptoms of vitamin E in fish are the accumulation of lipid oxidation products in the tissues, increased cell membrane fragility, degeneration of muscle cells, anemia, and reduced growth and survival (Hamre et al. 1994, 1997; Betancor et al. 2008). Vitamin E also participates in modulation of eicosanoid synthesis (Cornwell and Panganamala 1993), inhibits proliferation of certain cell types, apparently through inhibition of protein kinase C (Azzi et al. 1993), and is thought to affect disease resistance and health through modulation of immune responses (Waagbø et al. 1994).

Live feed production and enrichment is performed under highly pro-oxidative conditions, with high levels of n-3 polyunsaturated fatty acids (PUFA), air or oxygen addition to the culture water, high temperature, and bright light. Formulated diets for marine fish larvae also contain high levels of PUFA and pro-oxidants, for example, in the form of minerals. The high surface-to-volume ratio of the feed particles also favors oxidation. It is therefore important to supplement marine fish larval diets with vitamin E, but vitamin E at high levels, in the absence of sufficient amount of vitamin C, has been shown to increase mortality and tissue lipid oxidation in Atlantic salmon, Atlantic halibut juveniles, and in sea bream larvae (Hamre et al. 1997, unpublished data). The concentration ratio of the two vitamins in copepods (110 mg/kg vitamin E and 500 mg/kg vitamin C; Hamre et al. 2008a) may give a guideline for supplementation of larval diets.

There is an interaction between vitamin E and the dietary level of highly unsaturated fatty acids (HUFA) in marine fish larvae as in other vertebrates. Betancor et al. (2008) fed diets with different ratios of DHA (22:6n-3) to vitamin E to sea bass larvae and found that increasing the level of DHA increased the incidence of muscular degeneration, while adding extra vitamin E at high DHA levels reduced the muscular pathology. Furthermore, Atalah et al. (2008) found similar variation in mortality, both under normal culture conditions and in response to stress, in sea bass larvae. This suggests that high DHA, which is the fatty acid most susceptible to oxidation due to the high number of unsaturated bindings, will cause oxidative damage to cellular membranes, which in turn will be protected by vitamin E.

Vitamin E is most often given as α-tocopheryl acetate in formulated diets. In enrichment diets, it may be given as acetate or in the free form; in the latter case it would also function as an antioxidant in the enrichment diet. It is not known to what extent fish larvae digest the acetate form of tocopherol.

The requirement of vitamin E in marine fish larvae is not known, but Atalah et al. (2008) suggested an optimal level of 3 g/kg dry diet because these high levels reduced mortality after stress but not the mortality during standard rearing conditions. This is in line with the general opinion that vitamins C and E in larval diets should be at the g/kg level, while the requirements given by the NRC (1993) for fish are 30 and 50 mg/kg, for vitamins C and E, respectively. It is possible that the oxidative stress that the intensively reared larvae must face justifies the high supplementation levels. However, further studies are needed to establish the vitamin E requirement in marine fish larvae and better understand how lipid oxidation and the interaction of vitamin E with other nutrients affect this requirement.

4.4.1.4 Vitamin K

Vitamin K plays vital roles in blood coagulation and bone mineralization (Udagawa 2000; Krossøy 2009). The main role of vitamin K is to act as a cofactor in the post-translational enzymatic carboxylation of glutamic acid residues of Gla- (gamma-carboxyglutamate residues) proteins. This facilitates calcium binding and thereby protein activation. Classical Gla-proteins belong to the coagulation cascade and bone tissue (bone Gla protein or osteocalcin, matrix Gla protein), and vitamin K deficiency therefore affects blood clotting and bone mineralization in fish. The minimum requirement is suggested to be 2.5 mg/kg using menadione sodium bisulfite (MSB; vitamin K_3) as the vitamin K source. Amago salmon (*Oncorhynchus mason*) fed a vitamin K–deficient, purified diet developed deficiency symptoms such as mortality, anemia, increased blood clotting time, and histopathological changes in liver and gills; the fish recovered from the symptoms after they had been fed a diet containing vitamin K (Taveekijakarn et al. 1996b). Feeding 10 mg/kg of MSB increased the active form menakinone (MK-4; vitamin K_2) in Atlantic salmon liver compared with no supplementation (0.05 mg phyllokinones/kg; vitamin K_1). However, the unsupplemented diet did not result in vitamin K deficiency (Graff et al. 2002b). MSB is unstable during feed processing and storage (Marchetti et al. 1999), and the dietary content may reach critical low levels under unfavorable production and storage conditions. Menadione nicotinamide bisulfite (MNB) has been suggested as a more stable and less toxic alternative for vitamin K supplementation (Marchetti et al. 1995). Recent research questions the availability of this form of vitamin K in feed for salmonids (Krossoy et al. 2009). The minimum requirement is probably <0.2 mg/kg, while the recommended level is 2.5 mg/kg (Grahl-Madsen and Lie 1997). However, there are analytical difficulties in correctly assessing inherent (phyllokinones) and supplemented forms (MSB or MNB) of vitamin K. No focus has been given to vitamin K deficiency as a risk factor for deformities that are triggered early in development but expressed at a later stage (Waagbø et al. 2005).

4.4.2 Water-soluble vitamins

The water-soluble vitamins cover eight B vitamins (thiamine, riboflavin, niacin, pantothenic acid, vitamin B_6, biotin, folate, and vitamin B_{12}) and vitamin C (AA). According to the name, this rather diverse group of vitamins represents essential water-soluble compounds. Another common characteristic among the B vitamins lies in their main action in the intermediate cellular metabolism as coenzymes. Water-soluble vitamins are generally poorly stored in organisms, which means that excess intake leads to tissue saturation and excretion (Albrektsen 1994; Schaeffer et al. 1995; Mæland et al. 1998). On the other hand, fish larvae have less organ storage capacity than in later developmental stages and will therefore be more vulnerable to deficiencies in water-soluble vitamins.

Since there is a lack of information on individual water-soluble vitamins in fish, "B vitamins" is commonly used as a general term. The diet for any organism should, however, be well balanced and cover the requirements for all the individual water-soluble vitamins. The large variation in requirements reported in Table 4.1 illustrates uncertainties in the estimates and calls for more exact reassessments. All vitamins are essential, meaning that the organism itself cannot synthesize them and that there is a need for dietary supply. It has been speculated that intestinal microorganisms can supply the fish with some of the water-soluble vitamins, such as biotin and vitamin B_{12} (Sugita et al. 1991, 1992).

B vitamins function as coenzymes in the main energy-producing metabolic pathways, such as glycolysis, citric acid cycle, and the respiratory chain, and in the degradation and synthesis of nucleic acids, proteins, lipids, and carbohydrates (Figure 4.3). High concentrations of the vitamins are therefore found in metabolically active organs. Obviously, deficiency states of individual vitamins lead to severe metabolic dysfunctions and mortalities.

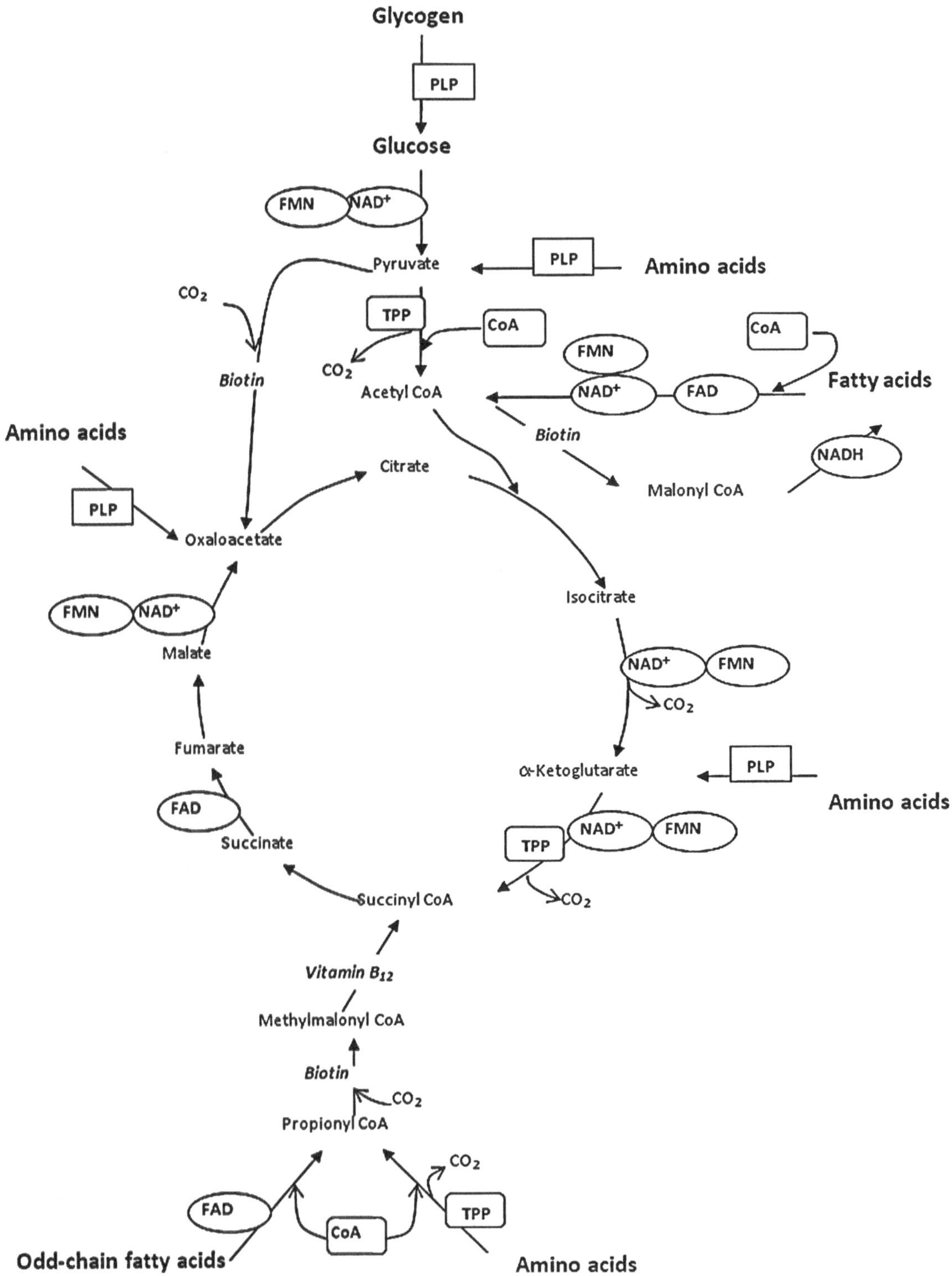

Figure 4.3 Krebs cycle. Examples of main metabolic pathways where B vitamins participate as coenzymes. Abbreviations for active coenzyme forms: thiamine (TPP); riboflavin (FAD and FMN); niacin (NAD^+/NADH); pantothenic acid (CoA); vitamin B_6 (PLP). (Modified from Mæland 2000.)

The most striking historical example is probably thiamine deficiency and the heavy mortalities observed in wild salmon yolk sac fry suffering from the early mortality (EMS) and M74 syndromes (Amcoff et al. 2002).

General unspecific deficiency symptoms of B vitamins in fish include loss of appetite, reduced growth, and mortalities (NRC 1993). More specific deficiency symptoms, such as activity of specific enzymes, need to be included to verify a suggested vitamin deficiency. In general, reduced enzyme activities in target organs are observed, while *in vitro* addition of the active vitamin coenzyme in question restores the activity. High % activation (activation coefficient) of specific enzymes during analysis has been regarded as specific indicators of deficiency of individual B vitamins (Albrektsen et al. 1993; Mæland et al. 1998; Shiau and Chin 1999; Truesdale et al. 2000).

4.4.2.1 Thiamine

Thiamine (vitamin B_1) is a coenzyme vitamin (active form is thiamine pyrophosphate [TPP]) participating in several biochemical pathways in the metabolism of carbohydrates and lipids (Figure 4.3). Thiamine deficiency is described in several salmonid and carp species, with heavy mortality following initial behavioral and homeostatic disturbances (Morito et al. 1986; NRC 1993; Woodward 1994; Halver 2002). Juveniles are more susceptible to deficiency due to rapid metabolism and lack of storage capacity, while larger fish show more general and obscure symptoms. Development of thiamine deficiency is accelerated at elevated temperatures relative to increased metabolism, and by increased dietary carbohydrates. Thiamine deficiency may be explored by studying the activity of key thiamine-dependent enzymes, such as α-ketoglutarate dehydrogenase and transketolase, in selected tissues. The M74 syndrome, a reproduction disorder of Baltic salmon manifested as death of developing yolk sac fry, seems to be related to thiamine deficiency (Amcoff et al. 2002). A similar disorder, "Cayuga syndrome," has been described in Atlantic salmon in the United States (Fisher et al. 1995). Low salmon egg thiamine concentration is probably mediated through imbalanced broodstock nutrition or increased degradation of thiamine by thiaminase from the wild prey in the broodfish intestine. Consequently, broodfish have been injected with thiamine to improve egg thiamine status (Fitzsimons et al. 2005), while M74-affected fry have successfully recovered after a thiamine bath treatment (Bylund and Lerche 1995).

The thiamine contribution from marine ingredients is variable, probably depending on thiaminase activity in the minced fish raw material during meal production prior to heating. The dietary thiamine requirement in fish is uncertain but is suggested to be considerably lower than 10 mg/kg, earlier suggested by the NRC (1993). Woodward suggested 1 mg/kg, while Morris and Davies estimated a thiamine requirement in excess of 5 mg/kg for gilthead sea bream (*Sparus aurata* L.) fed a diet with moderately elevated lipid content (Woodward 1994; Morris and Davies 1995b).

Thiamine concentrations in copepods are higher than in rotifers and *Artemia* (Table 4.1). However, copepod levels are relatively high compared with suggested requirements. Thiamine-enriched *Artemia* increased growth in Atlantic halibut larvae (Hamre, unpublished data). Rotifer content of thiamine may fall below the thiamine requirement given for fish (Table 4.1), and enrichment is therefore important.

4.4.2.2 Riboflavin

Riboflavin functions as a coenzyme in several important metabolic pathways. Figure 4.3 shows where riboflavin functions as a coenzyme (flavin adenine dinucleotide [FAD] and flavin mononucleotide [FMN]) in energy metabolism (glycolysis and Krebs cycle) and the final oxidative phosphorylation. Lack of riboflavin causes poor growth, reduced feed intake, with subsequently increased mortality in farmed fish species. Corneal and lenticular abnormalities due to riboflavin deficiency have been reported in a number of fish species, as well as nervousness, abnormal swimming behavior, fin erosion, and abnormal skin coloration (Tacon 1992; NRC 1993). However, cataracts or corneal damage is not always observed in riboflavin deficiency (Woodward 1984). Several enzymatic activities have been used as biochemical markers of riboflavin status in rainbow trout, for example, erythrocyte glutathione reductase (Hughes et al. 1981) and hepatic D-amino acid oxidase (Amezaga and Knox 1990). The coenzyme function in oxidase systems makes riboflavin particularly important in respiration within poorly vascularized tissues such as the eye cornea and eye lens.

The dietary riboflavin requirement, as measured by saturation of the riboflavin storage in the liver, enzyme data, weight gain, and absence of deficiency symptoms, has been estimated to be between 3 and 6 mg/kg for most fish species, including rainbow trout (Takeuchi et al. 1980; Hughes et al. 1981; Woodward 1985; Amezaga and Knox 1990), carp (Aoe et al. 1957), blue tilapia, (*Oreochromis aureus*; Soliman and Wilson 1992), red hybrid tilapia (Lim et al. 1993), channel catfish (*Ictalurus punctatus*; Serrini et al. 1996), and Atlantic salmon (Brønstad et al. 2002). Elevated water temperature or genetic differences in growth rate does not affect the requirement for riboflavin, and this has been used as support for the fairly equal requirement among species (Woodward 1985; NRC 1993). According to historical data, the ovaries show the highest riboflavin concentration (5–19 mg/kg wet weight) among selected organs of several wild-caught marine fish species (Brækkan 1959), indicating its importance in embryogenesis and for the developing larvae.

Riboflavin concentrations in rotifers, *Artemia*, and copepods (Table 4.1) seem to be in several-fold excess of known requirements for fish species, leaving no problems with riboflavin after start-feeding (Brown et al. 1997; van der Meeren et al. 2008; Hamre et al. 2008a).

4.4.2.3 Vitamin B_6

Vitamin B_6, or pyridoxine, has a major role in amino acid and protein metabolism, for example, as a coenzyme (as pyridoxal 5′-phosphate) in tissue transaminases (Albrektsen et al. 1994). A suggested uniform requirement among species, 15 nmol vitamin B_6 per gram wet weight gain (Woodward 1994), has been directly related to growth in young fish. In line with this, the estimated requirement for vitamin B_6 in Atlantic halibut larvae, measured as B_6 utilization from yolk versus body growth, was approximately the same, and equivalent to 3 mg/kg dry feed (Rønnestad et al. 1997). Delayed kinetics of vitamin B_6 transfer, compared with main yolk dry matter in their study, suggests a differential need for the vitamin for transaminase activity and utilization of protein and amino acids later in development (Sato et al. 1987; Albrektsen et al. 1994; Rønnestad et al. 1997). These studies may serve not only as a basis for first-feed formulations for halibut larvae but also for feeds for broodstocks.

4.4.2.4 Niacin

Niacin (vitamin B_3) functions as coenzyme in several energy-related metabolic pathways. Figure 4.3 illustrates some cellular biochemical pathways where niacin functions as the coenzymes nicotinamide adenine dinucleotide (NAD) and nicotinamide adenine dinucleotide phosphate (NADP). Niacin deficiency symptoms in fish include reduced growth, appetite, and feed utilization, dark pigmentation, skin lesions, muscular weakness, behavioral changes, edemas, and mortality (Chuang

1991; Halver 2002). Since niacin participates in plentiful enzymatic reactions, deficiency may develop relatively rapidly. Requirements in salmonids vary between 10 and 175 mg/kg, reflecting the degree of uncertainty in the estimates. A requirement level of 63–83 mg/kg was calculated for gilthead sea bream using a semipurified diet (Morris and Davies 1995a), while channel catfish required 7.4 mg/kg (Ng et al. 1997) and rainbow trout 10 mg/kg (Poston and Wolfe 1985). The wide range in requirements among species suggests that the niacin requirement may vary according to dietary composition and activity of the fish species. In contrast to land-living animals, tryptophan has been shown to be inefficient as a niacin precursor in channel catfish (Ng et al. 1997). Hashimoto (1953) reported on interactions or sparing actions among B vitamins. Young carp fed diets containing a vitamin mixture based on the established minimum requirements of the B vitamins showed deficiency symptoms, including reduced growth. This was prevented by the addition of niacin, thiamine, and riboflavin to the diets. Similarly, niacin seems to be influenced by the concentration of related vitamins.

The high content of niacin (>100 µg/g) in live feed (Table 4.1) most probably covers the requirement for the larvae during ontogeny, even though one may expect a higher requirement at later developmental stages.

4.4.2.5 Biotin

Biotin functions as a coenzyme for carboxylases in the metabolism of carbohydrates, lipids, and some amino acids (Dakshinamurti and Cauhan 1989). Biotin deficiency symptoms in fish are reduced growth rate and increased mortality, as well as more specific symptoms such as abnormalities in skin, intestine, and gill tissue (including "blue slime patch disease" in trout) and reduced activity of biotin-dependent enzymes in metabolically active tissues (Phillips et al. 1950; Lovell and Buston 1984; Woodward and Frigg 1989; Koppe 1993). One of the biotin-dependent enzymes, pyruvate carboxylase (PC), converts pyruvic acid to oxaloacetic acid and has been shown to be a sensitive indicator of biotin status in fish (Woodward and Frigg 1989). Avidin, a protein in raw egg white, binds strongly to biotin and makes the vitamin inaccessible for absorption in the intestine (Mock 1999). Biotin deficiency due to intake of raw egg white is a part of the history of the discovery of biotin as a vitamin (reviewed by Mock 1999). Supplementation of raw egg white or avidin also induces biotin deficiency symptoms in several fish species (Poston 1976; Casteldine et al. 1978; Lovell and Buston 1984; Mæland et al. 1998). The minimum dietary requirement for biotin for optimal growth in salmonids, such as rainbow trout and lake trout (*Salvelinus namaycush*) has been estimated to be 0.14 and 0.10 mg/kg, respectively (Poston 1976; Woodward and Frigg 1989). Practical feeds based on marine raw materials, containing 0.11 mg biotin/kg diet, covered the requirement for salmonid juveniles (Koppe 1993; Mæland et al. 1998). However, higher levels were needed for optimal lysozyme levels in serum and mucus (0.15 mg biotin/kg diet) (Koppe 1993). Elevated biotin levels, up to 1 mg/kg, may be favorable during salmon smoltification and seawater transfer (Waagbø et al. 1994). The bioavailability of inherent biotin in feedstuffs varies greatly, between 0 and 100% in animal and plant ingredients (Frigg 1976, 1984).

The levels of biotin in live rotifers and *Artemia* are similar to the levels found in copepods and are above the anticipated requirements (NRC 1993). However, variable bioavailability of biotin from live and formulated diets may be an issue in nutrition of marine fish larvae.

4.4.2.6 Pantothenic acid

Pantothenic acid is a part of coenzyme A (CoA), essential in the transfer of acetyl and acyl groups as substrates in energy production (Figure 4.3), the synthesis of the neurotransmitter acetylcholine, acetylation of aromatic amines, and synthesis of cholesterol. Mitochondria-rich tissues such as the kidney and gills are especially sensitive to pantothenic acid deficiency (NRC 1993; Halver and Hardy 2002). The NRC requirement is estimated to be in the range of 10–50 mg/kg (NRC 1993), while 10–15 mg/kg was suggested by Smith and Song (1996) in their review on the comparative requirement of pantothenic acid. Fish fed a diet devoid of the vitamin show severe anorexia, growth depression, high mortality rates, clubbed gills, anemia, and damaged skin and fins (Takeshi and Andrews 1977). In blue tilapia (*Tilapia aurea*), the requirement was estimated to be 10 mg/kg based on growth and lack of pathology (Roem et al. 1991). Natural feed ingredients such as fish meal contain between 8 and 30 mg pantothenic acid/kg, while plant materials contain somewhat less. The vitamin is added in the form of a stable calcium salt. Pantothenic acid has a role in lipid and energy metabolism, and was suggested as the first vitamin to cause visible deficiency signs in the gills of rainbow trout fed a vitamin premix-free diet (Barrows et al. 2008). In wild tunny (*Thunnus thynnus*) caught off the coast of Norway, Brækkan (1955) reported as high as 245 µg pantothenic acid/g in the ovaries, and suggested an important role for this vitamin in reproduction. Care should therefore be taken to fulfill the requirements in the rapidly growing fish larvae, both through broodstock (Sandnes et al. 1998) and larval nutrition. Even with large variations in concentration, live feed used for marine fish larvae seems to fulfill the requirement for pantothenic acid.

4.4.2.7 Folate

The major role of folate is related to one-carbon transfer reactions in nucleic acid and amino acid metabolism. Consequently, folate requirement is related to growth and cellular proliferation, essential for fish larvae ontogeny. Folate is essential to teleosts, and deficiency symptoms such as impaired hematopoiesis and reduced growth and survival are observed in several fish species, including Atlantic salmon (*Salmo salar*; Mæland and Sandnes, unpublished data), coho salmon (*Oncorhynchus kisutch*; Smith 1968), rainbow trout, (*Oncorhynchus mykiss*) and channel catfish (*Ictalurus punctatus*; Duncan et al. 1993). Hematological effects of folate deficiency in teleosts include macrocytic anemia with abnormal cell nucleus segmentation of the blood cells (Smith 1968; Waagbø et al. 2001). The quantitative requirement for folate has been established in channel catfish (Duncan et al. 1993; Robinson and Li 2002) and rainbow trout (Cowey and Woodward 1993). In the latter study, the authors suggested a dietary requirement between 0.6 and 1.1 mg/kg, depending on the response criteria used (Cowey and Woodward 1993).

In developing high-quality Atlantic halibut larvae, 50% of the yolk sac folate is retained in the larval body compartment, while the other 50% is a net loss (Mæland et al. 2003). Based on this study, the authors suggested a need for 2 µg folate/g weight gain, which is in line with suggested requirements from the NRC (1993), equivalent to 2 mg/kg dry diet. The study served not only as a basis for folate recommendations for first-feed formulations for halibut larvae but also as a guideline for feeds for halibut broodstocks since a screening of several batches of eggs from Atlantic halibut broodfishes showed lower folate contents compared with high-quality eggs (Mæland et al. 2003). Later screenings of commercially produced eggs in Norway and Iceland during 1998–2000 confirmed large variations in egg folate concentrations, however, with improvement over the sampling period due to generally improved halibut broodstock vitamin nutrition during the sampling period (see review (Waagbø 2009). So far, no studies have demonstrated a direct relation between egg folate status and reliable quality markers or parameters of the offspring. Folate is one of the most unstable B vitamins in production and storage of formulated fish feeds (Marchetti et al. 1999), but currently used live feed organisms seem to supply sufficient folate to the developing larvae (Table 4.1).

4.4.2.8 Vitamin B_{12}

Despite differences in chemical structure and biochemical actions of vitamin B_{12} and folate, a close functional relationship exists between the two vitamins, especially in cell division. Requirements and interactions between vitamin B_{12} and folate, and bioavailability of these vitamins from practical diets, were studied in two experiments with Atlantic salmon (Sandnes and Mæland 1994). Difficulties in determining quantitative requirements have been related to possible intestinal microbial vitamin synthesis (Sugita et al. 1990). Feeding a purified diet without supplementation of folic acid showed growth reduction, reduced levels of folate in liver and muscle tissue, and anemia characterized by reduced blood hemoglobin concentration, enlarged immature erythrocytes with fragmented nuclei, and reduced erythrocyte hemoglobin content (MCH). Vitamin B_{12}–deficient fish also showed anemia with immature erythrocytes. Anemia was most severe in fish fed a diet without both vitamin B_{12} and folate supplementation. Dietary supplementation of 0.014 mg vitamin B_{12}/kg and 3.6 mg folate/kg in purified diets prevented vitamin deficiency signs in Atlantic salmon (Sandnes and Mæland 1994). For vitamin B_{12}, qualitative requirement studies exist for Japanese eel (*Anguilla japonica*) and red sea bream (*Pagrus major*) (Koshio 2002), yellowtail (*Seriola quinqueradiata*; Hosokawa 1999, cited by Masumoto 2002), and European sea bass (*Dicentrarchus labrax*; Kaushik et al. 1998). Fish meals contain marginal levels of folate, 0.3–1.0 mg/kg, whereas the concentration of vitamin B_{12} has been found to be 10 times higher than the recommendation of the NRC (1993). Vitamin B_{12} levels in rotifers, *Artemia*, and copepods are more than 30 times higher than the requirements in fish (NRC 1993). As is true for folate, vitamin B_{12} is among the most unstable vitamins in fish feed, and considerable losses occur during ingredient processing, fish feed production (50–65%), and storage (Gabaudan and Hardy 2000).

4.4.2.9 Vitamin C

Vitamin C, or AA, is the most studied vitamin in fish and fish larvae (see reviews by Sandnes 1991; Dabrowski 2001). This is due to the extreme consequences of deficiency arising from improper broodstock (Sandnes 1984; Blom and Dabrowski 1995; Izquierdo et al. 2001), larvae (Terova et al. 2001), and juvenile (Sandnes et al. 1992) vitamin C nutrition. Vitamin C has also been considered important in nutrition of marine fish larvae based on high concentrations of AA in the ovaries and eggs of marine fish (Mangor-Jensen et al. 1994; Nortvedt et al. 2001, 2003) and the high levels analyzed in natural live prey (Rønnestad et al. 1999; Hamre et al. 2008a)

Vitamin C has no coenzyme function but acts as cofactor in hydroxylation reactions in the fish body. One classical role is related to posttranslational hydroxylation of proline and lysine moieties in the collagen subunits, essential for cross-linking of the collagen triple-helix structure and thereby its strength. Consequently, AA nutrition affects collagen synthesis in connective tissues and structural supportive organs such as the skin, cartilage, and bone in the developing marine larvae (Terova et al. 2001).

There is an extensive body of literature on the qualitative (Mæland and Waagbø 1998) and quantitative vitamin C requirements in fish (Sandnes 1991; Dabrowski 2001), including the use of excess dietary vitamin C under stressful and unfavorable farming conditions. The minimum requirement seems to vary

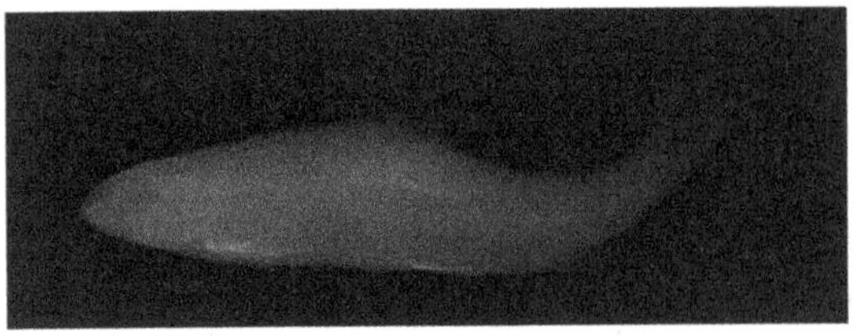

Figure 4.4 Atlantic salmon start-feeding fry fed an ascorbic acid (AA)–deficient diet for 24 weeks, showing a severe incidence of scoliosis and lordosis (Waagbø and Maage, unpublished data). (Photo by Drs. P.G. Fjelldal and R. Waagbø.)

with the size of the fish, with declining requirements at increasing size. The minimum requirement approximates 10–20 mg AA equivalents/kg feed in Atlantic salmon (*Salmo salar*) fry (Sandnes et al. 1992) using formulated diets with stable AA phosphate (AP) derivatives. In a similar study, groups of Atlantic salmon start-feeding fry fed an AA-deficient diet for 24 weeks showed 30% less vertebra hydroxyproline (in % of protein) and a severe incidence of scoliosis and lordosis (57% visually observed) as compared with groups fed marginal (10 mg/kg) and sufficient (100 mg/kg) AA in the form of AP derivatives (Waagbø and Maage, unpublished data; Figure 4.4). This illustrates the structural importance of high-quality collagen with an optimal cross-linking and that it may take considerable time for fish fry to develop visual signs of deficiency. The salmon fry fed suboptimal (10 mg AA equivalents/kg) AA showed optimal vertebra hydroxyproline and normal bone health compared with sufficiently fed fish, while other deficiency signs such as anemia were apparent. Gouillou-Coustans and Kaushik (2001) critically reviewed the AA requirements of freshwater and marine fish species relative to different response indicators (weight gain, absence of deficiency symptoms, OH-proline/collagen synthesis). They concluded that the reported AA requirement was considerably lower when using stable AP derivatives than crystalline AA. Relative to the discussion above, the authors pointed out that there may be more stringent response markers that were not included in all the reviewed feeding experiments, such as tissue AA saturation. For marine species that depend on live feed at start-feeding, there are no requirement studies in early larvae. Most studies start when formulated diets can be applied (Merchie et al. 1996b; Mæland et al. 1999). Weaning of Atlantic halibut with formulated feed with increasing doses of AA in the form of AP suggested that AA from *Artemia* (control) was retained more efficiently than AP from the formulated feeds, but this changed in the course of the experiment (Mæland et al. 1999).

Studies with common carp (*Cyprinio carpio*) larvae fed semipurified diets with graded levels of a stable and bioavailable AP demonstrated that the level for tissue saturation was six times higher than that of maximum growth (45 mg AA equivalents/kg) (Gouillou-Coustans et al. 1998). Thus, body vitamin saturation or retention efficacies may not necessarily be a useful indicator of the requirement in fish larvae, related to the daily need and limited storage capacity for water-soluble vitamins. For fish larvae, intestinal hydrolysis of stable AA derivatives (AP and AA palmitate derivatives), and subsequently AA uptake, may add to the insecurity of the requirement estimates (Dabrowski et al. 1996). In Atlantic halibut larvae followed from hatching until start-feeding, major parts of the yolk sac AA were retained in the larval body during development (Rønnestad et al. 1999), indicating minor losses of AA. As a participant in the integrated antioxidant defense system in the fish body, AA may be effectively spared and regenerated by other antioxidants (Figure 4.3). Such interactions were observed in a dietary vitamin C and E interaction study on start-feeding salmon fry (Hamre et al. 1997).

Many studies have indicated that the AA requirement decreases with age, probably related to a decreasing metabolic rate with increasing size and increased AA storage capacity. Further, temperate species such as the red sea bream (*Pagrus major*) may need more AA than cold-water species (Ren et al. 2010). For practical solutions, considerable efforts have been related to enrichment of AA through the live food chain. In enrichment procedures to boost live feed with vitamin C for fish larvae, microalgae (Lie et al. 1997), AA palmitate (Merchie et al. 1996a), and AP have been used successfully.

Besides the above-mentioned classical roles of vitamin C and its established minimum requirement in many species (Dabrowski 2001), AA has been shown to interfere with mineral metabolism (Sandnes 1991), stress response (Fletcher 1997; Ren et al. 2010), immunity (Sealy and Gatlin 2001; Waagbø 2006), wound repair (Wahli et al. 2003), and detoxification reactions (Norrgren et al. 2001) in many farmed fish species, however at considerably higher concentrations than the minimum requirement for optimal growth and survival.

In summary, wide requirement ranges have been suggested for vitamin C among marine fish species, mostly reflecting imperfect experimental conditions with variable individual feed intake, maturation of the gastrointestinal tract, bioavailability of inherent and added vitamin forms, and tissue retention and storage capacity in the developing larvae (Dabrowski 1986; Segner et al. 1993). Also included are the methodological difficulties in the analysis of AA and its chemical forms (Halver and Felton 2001). Consequently, the exact minimum requirements are difficult to estimate and establish, and future recommendations need to consider these concerns.

4.5 Minerals

Shearer (1991) suggested a model for dietary mineral requirement in fish, considering the requirement for growth, endogenous loss, mineral availability, feed efficiency, and finally, uptake of waterborne minerals:

Element requirement = (Requirement for growth + Endogenous loss)/ (Mineral availability/feed efficiency) – Waterborne minerals.

The exact contribution of dietary and waterborne elements to the requirements has not been examined in fish larvae. Excess in both water and diet mineral concentrations constitute risks for toxicity in fish (Waagbø 2006); water exposure has proved to represent more acute toxicity (Handy 1996). Increased uptake over the gill epithelium of elevated waterborne element concentrations may therefore represent an additional stressor to fish (Wendelar Bonga 1997). Since uptake is well regulated in the intestine, toxic concentration of elements in the diet may lead to more chronic or sublethal toxic effects even at prolonged feeding of high concentrations. This section of the chapter will cover the requirements of elements for fish larvae, while consideration of toxicity will receive less space.

4.5.1 Macrominerals

4.5.1.1 Calcium

Most of the calcium (Ca) in vertebrates is located in the skeletal tissue in the form of hydroxyapatite, a hydroxylated polymer of calcium phosphate ($Ca_{10}(PO_4)_6(OH)_2$). The presence of minerals in skeletal tissues strengthens the structures. Free Ca, Ca^{2+}, has a vital role in the process of muscle contractions. Ca^{2+} is released from the sarcoplasmatic reticulum upon nerve stimulation and causes the muscle fibers to contract (Murphy 1993). While terrestrial animals maintain Ca homeostasis by the release and deposit of Ca from the skeleton, fish regulate their Ca mainly through the gills (reviewed by Lall 2002). Although Ca can be released from skeletal tissues, the importance of this Ca source in fish is not agreed upon, and it appears that Ca resorption from the endoskeleton only occurs under extreme conditions (Witten 1997). In contrast to what we know for mammals and birds, vitamin D is considered to have relatively little effect on the Ca and P homeostasis in fish (Vielma et al. 1998; Vielma and Lall 1998a; see Section 4.4.1.2), but the observed effects differ between studies (reviewed by Lock et al. 2010). Calcium homeostasis-regulating hormones such as stanniocalcin, parathyroid hormone, and parathyroid hormone–related protein (PTHrP) primarily target the gills where Ca uptake is regulated. However, the intestine is also an important target organ for these hormones (Verbost et al. 1993; Abbink et al. 2007), where calcium from both water and diet can be absorbed. Some studies point to the diet as an important source of calcium, particularly in freshwater species; for example, when Ca concentration ranged between 27.1 and 33.3 mg/L in the water, tilapia larvae needed a diet with 3.5–4.2 g Ca/kg to gain maximum growth and calcium deposits in scales and bones (Shiau and Tseng 2007). Some results show that the ratio between Ca and phosphorus (P; see below for more details) affects the uptake of calcium. The optimum Ca-to-P ratio for red sea bream was 1:2 (Sakamoto and Yone 1978). On the other hand, Ca levels do not seem to affect the uptake of P (Vielma and Lall 1998b).

4.5.1.2 Phosphorus

Although a major part of body phosphorus is bound to bone structures, it also serves important roles in cellular phospholipids, energy metabolism, and cell signaling. The standing dietary requirement for available phosphorus in freshwater and seawater fish is ~6 g/kg (Lall 2002). Lower P bioavailability and reduced feed efficiencies increase the requirements. Due to inconsistencies in results and measures of P status, whole-body elemental concentrations have been suggested as a standard and comparable response parameter among experiment and fish species.

While many elements may fluctuate with changes in water quality, water phosphorus is most often marginal and fails to fulfill the fish requirement. Phosphorus therefore needs to be supplied through the diet (Lall 2002). Bone deformities are often observed as production-related disorders in marine as well as salmonid fish farming, related to a deficiency or macromineral imbalance (Lall and Lewis-McCrea 2007; Waagbø 2008). Bone deformations such as vertebral compressions were recently suggested to be late effects of previous mineral or P malnutrition in Atlantic salmon (Fjelldal et al. 2009).

The P concentration in live feed alternatives are all in excess of the requirements (Table 4.2), given a reasonable availability. The P concentration in rotifers was influenced only to a minor extent by rearing water quality (Yamamoto et al. 2009). The immature intestine of marine fish larvae may, however, represent a risk for P deficiency.

4.5.1.3 Electrolytes

Potassium (K^+), sodium (Na^+), and chlorine (Cl^-) are the most important inorganic ions active in the regulation of water and ion balance of living organisms. Na^+ and Cl^- account for more than 90% of the extracellular electrolytes (reviewed by Kaneco and Hiroi 2008), while K^+, balanced with a range of anions (reviewed by Lall 2002), is present intracellularly. Fish that live in freshwater meet the challenge of a hypotonic environment, with a potential loss of ions and inflow of water, by active uptake of ions and production of ample amounts of dilute urine. In seawater, fish drink water and excrete excess ions over the gills and kidneys. The osmoregulatory organs develop during the larval stage. In Japanese flounder, gills are not distinguishable at hatching but appear to be functional with mature chloride cells at 18 dph, for example, in premetamorphic larvae. The euryhaline Mozambique tilapia (*Oreochromis mossambicus*) larvae drink more in seawater than in freshwater, indicating that their gut functions as an osmoregulatory organ already in early larvae. Furthermore, the kidney in chum salmon is underdeveloped during the late embryonic stages (reviewed by Kaneco and Hiroi 2008). Chloride cells are cells specialized in the exchange of Na^+, K^+, and Cl^- between the organism and the environment. There are four different types of chloride cells in fish, adapted to different environments. During ontogeny, chloride cells appear first on the yolk sac membrane. After yolk absorption, chloride cells are widely distributed in the skin, and in the course of metamorphosis, the cells become concentrated to the gills (reviewed by Kaneco and Hiroi 2008). Na^+K^+ ATPase is also located in the plasma membrane of all cells, keeping the intracellular concentration of K^+ high and Na^+ low. This electrochemical gradient is the driving force of active transport over the cell membrane.

The surrounding water is the most important source of electrolytes for fish. Sodium concentration in seawater is more than 3 g/L, while potassium concentrations in sea- and freshwater may be approximately 380 and 10 mg/L, respectively. However, potassium in freshwater can drop to below 1 mg/L (Steffens 1989). In addition, fish larvae will obtain ions both from the live feed and from the formulated diets fed in intensive culture, which are commonly based on fish meal. Fish meal is a good source of sodium but a relatively poor source of potassium (Lall 2002). Both elements are present as freely soluble salts and have good bioavailability.

In freshwater, there may thus be a possibility for potassium deficiency as shown by Shearer (1988). When the water potassium concentration dropped below 1 mg/L, a dietary potassium concentration of 8 g/kg was necessary to maintain growth in king salmon in this study. Dietary requirements of the other electrolytes have not been measured in fish due to interference from the dietary and waterborne high concentrations of these elements.

4.5.2 Microminerals

4.5.2.1 Iodine

Fish accumulate iodide from the surrounding water by active transport at the gills, and through absorption in the gut (Hunn and Fromm 1966; Eales 1997). Iodine is essential for the production of THs thyroxin (T_4) and tri-iodothyronine (T_3). T_4 is produced in the thyroid follicles of the subpharyngeal region in larvae, juveniles, and adult fish species, although T_4 production is also found in the kidney in adult carp (Geven et al. 2007), which may be the case for several species. The conversion of T_4 to T_3, the active form, takes place in peripheral tissues by enzymatic deiodination of T_4 (see below on selenium-dependent deiodinase enzymes). THs regulate growth and development in teleost fish; for example, metamorphosis in flatfish and parr–smolt transformation in salmonids are regulated by THs (Power et al. 2001). In general, both seawater and marine plankton contain iodine that fish larvae utilize (Solbakken et al. 2002; Moren et al. 2008). Most of the iodine found in the oceans is organically bound in the marine biota, mainly in plankton and kelp, while the iodine found in the seawater is primarily found as inorganic forms such as iodide (the biologically accessible form) and iodate (reviewed by Wong 1991). The concentration of iodide varies with depth and latitude, although the total amount of inorganic iodine in the seawater is relatively constant (400–500 nM). Iodide (I^-) concentration can be as high as 230 nM in the upper surface level in tropical and subtropical areas and down to 20 nM more poleward (Schwehr and Santschi 2003; Waite et al. 2006; Moren et al. 2008). Witt et al. (2009) discovered that fish reared in seawater from a well with 78 nM iodide had less T_4 production and lacked a T_4 peak compared with fish reared in fresh seawater that contained 160 nM iodide. The fish in the fresh seawater also had better growth and survival than the fish reared in well water. The use of protein skimmers and ozone injection in recirculation systems alters the original level of iodide. An oxidation occurs, and iodide is converted to iodate. Ribeiro et al. (2009) found that postmetamorphic Senegalese sole larvae reared in a recirculation system, without iodine added to the diet, developed goiter within 15 days, while larvae fed extra iodine had normal thyroid follicles and grew better as well. Commercially enriched live prey used in aquaculture, that is, rotifers and *Artemia*, contain very low levels of iodine (0.5–1.1 μg I/g DW) compared with marine copepods (60–300 μg I/g DW) (Solbakken et al. 2002; Moren et al. 2006; Hamre et al. 2008a; Ribeiro et al. 2009). Better growth and survival was observed by Hamre et al. (2008b) when cod were fed rotifers and *Artemia* enriched with a combination of selenium and iodide. The water in this facility has not been analyzed for iodide, but water taken from nearby areas ranged from 12 to 22 nM (Moren et al. 2008). Mechanisms of uptake have not been described in many fish species, but there is evidence that both marine and freshwater fish possess the natrium iodide symporter (NIS) not only in the thyroid follicles (Alt et al. 2006) but also possibly in the digestive system and surface areas since perchlorate, a known inhibitor, blocks the uptake of iodide and gives lower levels of TH (Mukhi and Patino, 2007; Moren et al. 2008). Since rotifers and *Artemia* are low in iodine and there is an increase in the use of recirculation systems for larval rearing, iodine should be supplemented in the enrichments used. The requirement is, however, not determined.

4.5.2.2 Selenium

Approximately 30 selenoproteins have been detected in mammals by feeding radiolabeled selenium, but the sequence and function are known for only 12 of them (Brigelius-Flohe 1999). Selenium is an integral part of several enzymes with antioxidant activity, the GPxs, which reduce hydroperoxides at the expense of reduced glutathione (GSH) (Brigelius-Flohe 1999; Arteel and Sies 2001). Four GPxs have been described in mammals: cytosolic GPx (cGPx), which is the classical one that neutralizes water-soluble and fatty acid hydroperoxides; plasma GPx (pGPx), which has similar properties as the cGPx; gastrointestinal GPx (GIGPx), which is exclusively expressed in the gastrointestinal tract; and phospholipid hydroperoxide GPx (PHGPx), which is active in biological membranes and reduces lipid hydroperoxides (Brigelius-Flohe 1999; Arteel and Sies 2001). The active sites of these enzymes contain selenocysteine residues. In addition, selenoprotein P (SeP), present in the plasma of mammals and expressed in cellular membranes, contains 10 selenocysteine residues and is not only regarded as a transport protein for selenium but also has antioxidative properties (Steinbrenner et al. 2006). The term "hierarchy of selenoproteins" is used to describe the differences in response of the different variants of GPxs to selenium deficiency, where the stability of the enzymes decreases in the order GI-GPx > PHGPx > pGPx = cGPx (Brigelius-Flohe et al. 2002). The GPxs are an integral part of the fish's antioxidant defense system as shown in Figure 4.5.

Selenium also plays a role in TH metabolism. TH is excreted from the thyroid gland mainly as T_4 into the circulation, taken up by peripheral organs, and converted to the more potent form, T_3, by type I or type II deiodinases (outer ring deiodinases [ORDs]), which contain selenium (Yen 2001). The dominant organs for T_3 synthesis are the liver and head kidney where deiodinase I is the principle factor (Eales and Brown 1993; Geven et al. 2007). Deiodinase II is more important for conversion of T_4 to T_3 in the target tissues (Watanabe et al. 2006). Deiodinase III is the

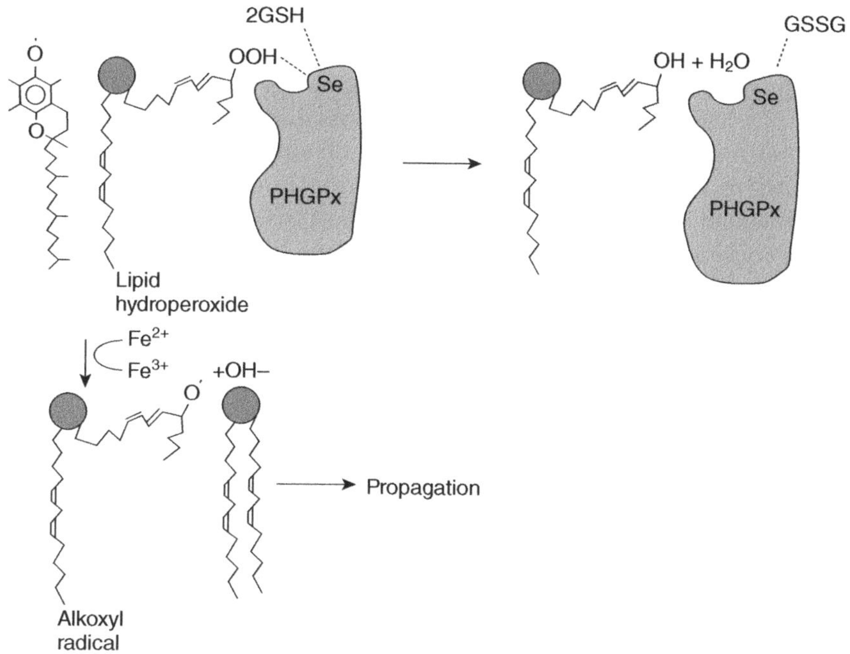

Figure 4.5 Model of the proposed role of phospholipid hydroperoxide glutathione peroxidase (PHGPx) in protecting membranes against lipid oxidation. PHGPx decomposes lipid hydroperoxides to alcohol and water, oxidizing 2GSH in the process. This prevents the formation of the lipid alkoxyl radical, which will stimulate lipid oxidation even in the presence of α-TOH (Maiorino et al. 1989; Ursini 1993).

inner ring deiodification resulting in the deactivation of either T_4 or T_3; this takes place in tissues that need to be protected from TH actions, for example, parts of the brain during development (Mol et al. 1997).

In fish, studies of the interactions between selenium and vitamin E have shown that selenium deficiency may lead to reduced levels of tissue α-tocopherol. Combined selenium and vitamin E deficiency signs are muscular dystrophy, muscle-specific proteins in plasma, and anemia (Poston 1976; Gatlin et al. 1986; Bell et al. 1985, 1986, 1987). The symptoms of selenium deficiency alone were reported to be disintegration of membranes and condensed nuclei in liver cells, pathological changes in nerve cells, and anemia, while no symptoms were visible on the macroscopic scale (Bell et al. 1986). Selenium and iodine deficiency in rats affect TH metabolism and deiodinase activities, while selenium deficiency affects both deiodinase and GPx activities (Meinhold et al. 1993; Brigelius-Flohe 1999; Arteel and Sies 2001).

Decreased GPx activity caused by low dietary selenium has been measured in rainbow trout, Atlantic salmon, and Channel catfish (Hilton et al. 1980; Bell et al. 1986, 1987; Gatlin at al., 1986), and Hilton et al. (1980) found that the selenium requirement in rainbow trout based on GPx activity was 0.35 mg/kg dry diet. This is above the levels in rotifers found by Hamre et al. (2008b; Table 4.2). Rotifers enriched with iodine and inorganic selenium to match levels found in copepods gave an increase in survival and a near-to-significant increase in GPx activity in cod larva (Hamre et al. 2008b). Penglase (2009) reared cod larvae on rotifers enriched with selenium-containing yeast, with iodine enrichment in both the control and experimental groups. The survival in both groups was nearly 100%, and there was an increase in GPx in response to enrichment with selenium. The data thus indicate that rotifers not enriched with extra selenium do not cover the requirement for selenium in cod larvae. On the other hand, *Artemia*-fed Atlantic halibut larvae not supplemented with extra selenium had higher tissue concentrations of selenium than larvae fed copepods, indicating that the supply of selenium from *Artemia* is sufficient for fish larvae (Solbakken 2003).

4.5.3 Other trace elements

The requirements of trace elements in marine animals are covered by input from diet and water (Shearer 1991). However, for most marine species such as European sea bass, Asian sea bass, red sea bream, gilthead sea bream, Atlantic halibut, Japanese flounder, North American flounder, yellowtail, red drum, and Southern bluefin tuna, trace element requirements have not been adequately evaluated (Webster and Lim, 2002).

In Table 4.2, element composition of live feed alternatives such as rotifers and *Artemia* are listed, together with natural zooplankton (copepods) regarded as providing the target concentration for optimal growth and health of marine larvae. Several elements are low in rotifers and *Artemia*, as compared with copepods, but higher than that extrapolated from fish species with known element requirements (NRC 1993). As discussed for iodine above, high levels of elements found in copepods may be necessary to compensate for the developmental immaturity of the larvae and low retention efficacies of the elements. Accordingly, manganese and zinc have been suggested to be marginal for marine fish larvae compared with the content of natural copepods and established requirements for larger fish (Hamre et al. 2008b).

The intestines of marine larvae have the ability to absorb iron from seawater, as observed in European flounder (*Platichthys flesus*), and the uptake was significantly enhanced in the presence of ascorbate as a reducing agent (Bury and Grosell 2003). The authors discuss the growing evidence that the aquatic uptake route for essential metals may contribute considerably to overall metal homeostasis, particularly when the dietary elements are low. A significant relation between liver iron and blood Hb was seen among salmon groups fed marginal vitamin C levels (10 mg AA equivalents/kg) and not among the groups fed 100 mg AA equivalents/kg, which also indicate *in vivo* interactions between AA status and available iron for Hb synthesis. Vitamin C–deficient Atlantic salmon seem to suffer from an iron-deficiency-like anemia despite the excess iron storage (Sandnes et al. 1990). Nutrient interactions that may influence element requirements should be taken into account.

4.6 Future challenges

The micronutrient requirements are probably relatively uniform among fish species and are related to growth and metabolism. There may, however, be species specificity in practical requirements among farmed species during early development due to different degrees of biological immaturity and currently used imperfect diets. Since fish larvae, for many reasons, are more susceptible to nutrient deficiencies, larval diets should contain safe and surplus micronutrient levels to compensate for such varying biological and technical conditions.

Literature cited

Abbink, W., Hang, X.M., Guerreiro, P.M., Spanings, F.A.T., Ross, H.A., Canario, A.V.M., and Flik, G. 2007. Parathyroid hormone–related protein and calcium regulation in vitamin D–deficient sea bream (*Sparus auratus*). *Journal of Endocrinology* 193:473–480.

Albrektsen, S. 1994. Studies on vitamin B6 nutrition in Atlantic salmon (*Salmo salar*). Department of Fisheries and Marine Biology, University of Bergen, Bergen, Norway.

Albrektsen, S., Waagbø, R., and Sandnes, K. 1993. Tissue vitamin B_6 concentrations and aspartate aminotransferase (AspT) activity in Atlantic salmon (*Salmo salar*) fed graded dietary levels of vitamin B_6. *Fiskeridirektoratets Skrifter, Serie Ernæring* 6:21–34.

Albrektsen, S., Waagbø, R., Lie, Ø., and Sandnes, K. 1994. Contents and organ distribution of vitamin B_6 in Atlantic salmon (*Salmo salar*) and turbot (*Psetta maxima*) during the reproductive cycle. *Comparative Biochemistry and Physiology* 109A:705–712.

Alt, B., Reibe, S., Feitosa, N.M., Elsalini, O.A., Wendl, T., and Rohr, K.B. 2006. Analysis of origin and growth of the thyroid gland in zebrafish. *Developmental Dynamics* 235: 1872–1883.

Amcoff, P., Akerman, G., Tjarnlund, U., Borjeson, H., Norrgren, L., and Balk, L. 2002. Physiological, biochemical and morphological studies of Baltic salmon yolk-sac fry with an experimental thiamine deficiency: relations to the M74 syndrome. *Aquatic Toxicology* 61:15–33.

Amezaga, M.R., and Knox, D. 1990. Riboflavin requirements in on-growing rainbow trout, *Oncorhynchus mykiss*. *Aquaculture* 88:87–98.

Aoe, H., Saito, T., and Takada, T. 1957. Water soluble requirements of carp: V. Requirements for folic acid. *Nippon Suisan Gakkaishi* 33:1068–1071.

Arteel, G.E., and Sies, H. 2001. The biochemistry of selenium and the glutathione system. *Environmental Toxicology and Pharmacology* 10:153–158.

Atalah, E., Hernandez-Cruz, C.M., Montero, D., Ganuza, E., Benitez-Santana, T., Ganga, R., Roo, J., Fernandez-Palacios, H., and Izquierdo, M.S. 2008. Enhancement of gilthead seabream and sea bass larval growth by dietary vitamin E in relation to different levels of essential fatty acids. XIII International Symposium on Fish Nutrition and Feeding, June 1–5, Florianopolis, Brazil.

Azzi, A.M., Bartoli, G., Boscoboinic, D., Hensey, C., and Szewczyk, A. 1993. Alpha-tocopherol and protein kinase C regulation of intracellular signalling. In: Packer, L., and Fuchs, J. (eds.) *Vitamin E in Health and Disease*. Marcel Dekker, Inc., New York, Basel, Hong Kong, pp. 371–384.

Balmer, J.E., and Blomhoff, R. 2002. Gene expression regulation by retinoic acid. *Journal of Lipid Research* 43:1773–1808.

Barrows, F.T., Gaylord, G.T., Sealy, W.M., Porter, L., and Smith, C.E. 2008. The effect of vitamin premix in extruded plant-based and fish meal based diets on growth efficiency and health of rainbow trout, *Oncorhynchus mykiss*. *Aquaculture* 283:148–155.

Bell, J.G., Cowey, C.B., Adron, J.W., and Shanks, A.M. 1985. Some effects of vitamin E and selenium deprivation on tissue enzyme levels and indices of tissue peroxidation in rainbow trout (*Salmo gairdneri*). *The British Journal of Nutrition* 53:149–157.

Bell, J.G., Pirie, B.J.S., Adron, J.W., and Cowey, C.B. 1986. Some effects of selenium deficiency on glutathione peroxidase (EC 1.11.1.9) activity and tissue pathology in rainbow trout (*Salmo gairdneri*). *The British Journal of Nutrition* 55:305–311.

Bell, J.G., Cowey, C.B., Adron, J.W., and Pirie, B.J.S. 1987. Some effects of selenium deficiency on enzyme activities and indices of tissue peroxidation in Atlantic salmon parr (*Salmo salar*). *Aquaculture* 65:43–54.

Betancor, M.B., Izquierdo, M.J., Benitez-Santana, T., Quesada, O., Atalah, E., Montero, D., and Izquierdo, M.S. 2008. Dystrophic alterations in skeletal muscle of sea bass (*Dicentrarchus labrax*) larvae in relation to the dietary DHA/vitamin E ratio. XIII International Symposium on Fish Nutrition and Feeding, June 1–5, Florianopolis, Brazil.

Blom, J.H., and Dabrowski, K. 1995. Reproductive success of female rainbow trout (*Oncorhynchus mykiss*) in response to graded dietary ascorbyl monophosphate levels. *Biology of Reproduction* 52:1073–1080.

Blomhoff, R., Green, M.H., Berg, T., and Norum, K.R. 1990. Transport and storage of vitamin-A. *Science* 250:399–404.

Bowry, V.W., Ingold, K.U., and Stocker, R. 1992. Vitamin E in human low-density lipoprotein. When and how this antioxidant becomes a pro-oxidant. *The Biochemical Journal* 288: 341–344.

Brækkan, O.R. 1955. Role of pantothenic acid in the reproductive cycle of ovaries in fish. *Nature* 176:1.

Brækkan, O.R. 1958. Vitamins and the reproductive cycle of ovaries in cod. *Fiskeridirektoratets Skrifter, Serie Teknologiske Undersøkelser* 3:1–19.

Brækkan, O.R. 1959. *Comparative Studies of Vitamins in Fishes*. John Griegs Boktrykkeri Bergen, Bergen, Norway.

Brigelius-Flohe, R. 1999. Tissue-specific functions of individual glutathione peroxidases. *Free Radical Biology & Medicine* 27:951–965.

Brigelius-Flohe, R., Wingler, K., and Muller, C. 2002. Estimation of individual types of glutathione peroxidases. *Methods in Enzymology* 347:101–112.

Brønstad, I., Bjerkås, I., and Waagbø, R. 2002. The need for riboflavin supplementation in high and low energy diets for Atlantic salmon *Salmo salar* L. parr. *Aquaculture Nutrition* 8: 209–220.

Brown, M.R., Jeffrey, S.W., Volkman, J.K., and Dunstan, G.A. 1997. Nutritional properties of microalgae for mariculture. *Aquaculture* 262:315–331.

Bury, N., and Grosell, M. 2003. Iron acquisition by teleost fish. *Comparative Biochemistry and Physiology Part C* 135:97–105.

Bylund, G., and Lerche, O. 1995. Thiamine therapy of M 74 affected fry of Atlantic salmon *Salmo salar*. *Bulletin of the European Association of Fish Pathologists* 15:93–97.

Casteldine, A.J., Cho, C.Y., Slinger, S.J., Hicks, B., and Bayley, H.S. 1978. Influence of dietary biotin level on growth, metabolism and pathology of rainbow trout. *The Journal of Nutrition* 108:698–711.

Chuang, J.L. 1991. 5.7. Fish and shrimp. In: Fenster, R., and Blum, R.A. (eds.) *Niacin in Animal Nutrition*. F. Hoffmann-La Roche Ltd., Basel, Switzerland, pp. 34–37.

Cornwell, D.G., and Panganamala, R.V. 1993. Vitamin E action in modulating the arachidonic acid cascade. In: Packer, L., and Fuchs, J. (eds.) *Vitamin E in Health and Disease*. Marcel Dekker, Inc., New York, Basel, Hong Kong, pp. 385–410.

Cowey, C.B., and Woodward, B. 1993. The dietary requirement of young rainbow trout (*Oncorhynchus mykiss*) for folic acid. *The Journal of Nutrition* 123:1594–1600.

Dabrowski, K. 1986. Ontogenetical aspects of nutritional requirements in fish. *Comparative Biochemistry and Physiology* 85A:639–655.

Dabrowski, K. 2001. *Ascorbic Acid in Aquatic Organisms—Status and Perspectives*. CRC Press, Boca Raton, FL.

Dabrowski, K., Moreau, R., and El-Saidy, D. 1996. Ontogenetic sensitivity of channel catfish to ascorbic acid deficiency. *Journal of Aquatic Animal Health* 8:22–27.

Dakshinamurti, K., and Cauhan, J. 1989. Biotin. *Vitamins & Hormones* 45:337–384.

Duncan, P.L., Lovell, R.T., Butterworth, C.E., Freeberg, L.E., and Tamura, T. 1993. Dietary folate requirement determined for channel catfish, *Ictalurus punctatus*. *The Journal of Nutrition* 123:1888–1897.

Eales, J.G. 1997. Iodine metabolism and thyroid-related functions in organisms lacking thyroid follicles: are thyroid hormones also vitamins. *Proceedings of the Society for Experimental Biology and Medicine* 214:302–317.

Eales, J.G., and Brown, S.B. 1993. Measurement and regulation of thyroidal status in teleost fish. *Reviews in Fish Biology and Fisheries* 3:299–347.

Estevez, A., and Kanazawa, A. 1995. Effect of (n-3) PUFA and vitamin A *Artemia* enrichment on pigmentation success of turbot, *Scophthalmus maximus*. *Aquaculture Nutrition* 1:159–168.

Fernandez, I., Hontoria, F., Ortiz-Delgado, J.B., Kotzamanis, Y., Estevez, A., Zambonino-Infante, J.L., and Gisbert, E. 2008. Larval performance and skeletal deformities in farmed gilthead sea bream (*Sparus aurata*) fed with graded levels of vitamin A enriched rotifers (*Brachionus plicatilis*). *Aquaculture* 283:102–115.

Finn, R.N., Rønnestad, I., van der Meeren, T., and Fyhn, H.J. 2002. Fuel and metabolic scaling during the early life stages of Atlantic cod *Gadus morhua*. *Marine Ecology Progress Series* 243:217–234.

Fisher, J.P., Spitsbergen, J.M., Iamonte, T., Little, E., and DeLonay, A. 1995. Pathological and behavioral manifestations of the "Cayuga syndrome," a thiamine deficiency in larval landlocked Atlantic salmon. *Journal of Aquatic Animal Health* 7:269–282.

Fitzsimons, J.D., Williston, B., Amcoff, P., Balk, L., Pecor, C., Ketola, H.G., Hinterkopf, J.P., and

Honeyfield, D.C. 2005. The effect of thiamine injection on upstream migration, survival, and thiamine status of putative thiamine-deficient coho salmon. *Journal of Aquatic Animal Health* 17:48–58.

Fjelldal, P.G., Hansen, T., Breck, B., Sandvik, R., Waagbø, R., Berg, A., and Ørnsrud, R. 2009. Supplementation of dietary minerals during the early seawater phase increase vertebral strength and reduce the prevalence of vertebral deformities in fast growing under-yearling Atlantic salmon (*Salmo salar* L.) smolt. *Aquaculture Nutrition* 15:366–378.

Fletcher, T.C. 1997. Dietary effects on stress and health. In: Iwama, G.K., Pickering, A.D., Sumpter, J.P., and Schreck, C.B. (eds.) *Fish Stress and Health in Aquaculture*. Cambridge University Press, Cambridge, UK, pp. 223–246.

Frankel, E.N. 1998. *Lipid Oxidation*. The Oily Press, Ltd., Dundee, Scotland.

Frigg, M. 1976. Bio-availability of biotin in cereals. *Poultry Science* 55:2310–2318.

Frigg, M. 1984. Available biotin content of various feed ingredients. *Poultry Science* 63:750–753.

Gabaudan, J., and Hardy, R.W. 2000. Vitamin sources for fish feed. In: Stickney, R.R. (ed.) *Encyclopedia of Aquaculture*. John Wiley & Sons, New York, pp. 961–964.

Gatlin, D.M., Poe, W.E., and Wilson, R.P. 1986. Effect of singular and combined dietary deficiencies of selenium and vitamin E on fingerling channel catfish (*Ictalurus punctatus*). *The Journal of Nutrition* 116:1061–1067.

Geven, E.J.W., Nguyen, N.K., van den Boogaart, M., Spanings, F.A.T., Flik, G., and Klaren, P.H.M. 2007. Comparative thyroidology: thyroid gland location and iodothyronine dynamics in Mozambique tilapia (*Oreochromis mossambicus* Peters) and common carp (*Cyprinus carpio* L.). *The Journal of Experimental Biology* 210:4005–4015.

Gouillou-Coustans, M.F., and Kaushik, S.J. 2001. Ascorbic acid requirement in freshwater and marine fish: is there a difference? In: Dabrowski, K. (ed.) *Ascorbic Acid in Aquatic Organisms—Status and Perspectives*. CRC Press, Boca Raton, FL, pp. 49–68.

Gouillou-Coustans, M.F., Bergot, P., and Kaushik, S.J. 1998. Dietary ascorbic acid needs of common carp (*Cyprinus carpio*) larvae. *Aquaculture* 161:451–459.

Graff, I.E., Hoie, S., Totland, G.K., and Lie, O. 2002a. Three different levels of dietary vitamin D-3 fed to first-feeding fry of Atlantic salmon (*Salmo salar* L.): effect on growth, mortality, calcium content and bone formation. *Aquaculture Nutrition* 8:103–111.

Graff, I., Waagbø, R., Fivelstad, S., Vermeer, C., Lie, Ø., and Lundebye, A. 2002b. A multivariate study on the effects of dietary vitamin K, vitamin D3 and calcium, and dissolved carbon dioxide on growth, bone minerals, vitamin status and health performance in smolting Atlantic salmon *Salmo salar* L. *Journal of Fish Diseases* 25:599–614.

Grahl-Madsen, E., and Lie, Ø. 1997. Effects of different levels of vitamin K in diets for cod (*Gadus morhua*). *Aquaculture* 151:269–274.

Haga, Y., Suzuki, T., and Takeuchi, T. 2002. Retinoic acid isomers produce malformations in postembryonic development of the Japanese flounder, *Paralichthys olivaceus*. *Zoological Science* 19:1105–1112.

Halver, J.E. 2002. The vitamins. In: Halver, J.E., and Hardy, R.W. (eds.) *Fish Nutrition*. Academic Press, London, pp. 61–141.

Halver, J.E., and Felton, S.P. 2001. Analytical enigmas in assaying for vitamin C. In: Dabrowski, K. (ed.) *Ascorbic Acid in Aquatic Organisms*. CRC Press, Boca Raton, FL.

Halver, J.E., and Hardy, R.W. 2002. *Fish Nutrition*, 3rd edition. Academic Press, London.

Hamre, K. 2006. Nutrition in cod (*Gadus morhua*) larvae and juveniles. *ICES Journal of Marine Science* 63:267–274.

Hamre, K., and Mangor-Jensen, A. 2006. A multivariate approach to optimization of macronutrient composition in weaning diets for cod (*Gadus morhua*). *Aquaculture Nutrition* 12:15–24.

Hamre, K., Hjeltnes, B., Kryvi, H., Sandberg, S., Lorentzen, M., and Lie, Ø. 1994. Decreased concentration of hemoglobin, accumulation of lipid oxidation products and unchanged skeletal muscle in Atlantic salmon (*Salmo salar*) fed low dietary vitamin E. *Fish Physiology and Biochemistry* 12:421–429.

Hamre, K., Waagbø, R., Berge, R.K., and Lie, Ø. 1997. Vitamins C and E interact in juvenile Atlantic salmon (*Salmo salar*, L.). *Free Radical Biology & Medicine* 22:137–149.

Hamre, K., Holen, E., and Moren, M. 2007. Pigmentation and eye-migration in Atlantic halibut (*Hippoglossus hippoglossus* L.) larvae: new findings and hypotheses. *Aquaculture Nutrition* 13:65–80.

Hamre, K., Srivastava, A., Ronnestad, I., Mangor-Jensen, A., and Stoss, J. 2008a. Several micronutrients in the rotifer *Brachionus* sp. may not fulfil the nutritional requirements of marine fish larvae. *Aquaculture Nutrition* 14:51–60.

Hamre, K., Mollan, T.A., Sæle, Ø., and Erstad, B. 2008b. Rotifers enriched with iodine and selenium increase survival in Atlantic cod (*Gadus morhua*) larvae. *Aquaculture* 284:190–195.

Handy, R.D. 1996. Dietary exposure to toxic metals in fish. In: Taylor, E.W. (ed.) *Toxicology of Aquatic Pollution—Physiological, Cellular and Molecular Approaches.* Cambridge University Press, Cambridge, UK, pp. 29–61.

Hashimoto, Y. 1953. Effect of antibiotics and vitamin B_{12} supplement on carp growth. *Bulletin of the Japanese Society of Scientific Fisheries* 19:899–904.

Hernandez, L.H.H., Teshima, S.I., Ishikawa, M., Alam, S., Koshio, S., and Tanaka, Y. 2005. Dietary vitamin A requirements of juvenile Japanese flounder *Paralichthys olivaceus. Aquaculture Nutrition* 11:3–9.

Heyman, R.A., Mangelsdorf, D.J., Dyck, J.A., Stein, R.B., Eichele, G., Evans, R.M., and Thaller, C. 1992. 9-Cis retinoic acid is a high-affinity ligand for the retinoid-X receptor. *Cell* 68:397–406.

Hilton, J.W., Hodson, P.V., and Slinger, S.J. 1980. The requirement and toxicity of selenium in rainbow trout (*Salmo gairdneri*). *The Journal of Nutrition* 110:2527–2535.

Horvli, O., and Lie, Ø. 1994. Determination of vitamin D_3 in fish meals by HPLC. *Fiskeridirektoratets Skrifter, Serie Ernæring* 6:163–175.

Hosokawa, H. 1999. Studies on the vitamin nutrition in yellowtail. PhD thesis, United Graduate School of Agriculture Sciences, Ehime University, Ehime, Japan.

Hughes, S.G., Riis, R.C., Nickum, J.G., and Rumsey, G.L. 1981. Biomicroscopic and histologic pathology of the eye in riboflavin deficient rainbow trout (*Salmo gairdneri*). *The Cornell Veterinarian* 71:269–279.

Hunn, J.B.F., and Fromm, P.O. 1966. *In vivo* uptake of radioiodide by rainbow trout. *Journal (Water Pollution Control Federation)* 38: 1981–1985.

Ingold, K.U., Burton, G.W., Foster, D.O., Hughes, L., Lindsay, D.A., and Webb, A. 1987. Biokinetics of and discrimination between dietary RRR- and SRR-alpha-tocopherols in the male rat. *Lipids* 22:163–172.

Izquierdo, M.S., Fernandez-Palacios, H., and Tacon, A.G.J. 2001. Effect of broodstock nutrition on reproductive performance of fish. *Aquaculture* 197:25–42.

Kanazawa, A. 1995. Nutrition of larval fish. In: Lim, C.E., and Sessa, D.J. (eds.) *Nutrition and Utilization Technology in Aquaculture.* AOAC Press, Champaign, IL, pp. 50–59.

Kaneco, T., and Hiroi, J. 2008. Osmo- and iono-regulation. In: Finn, R.N., and Kapoor, P.G. (eds.) *Fish Larval Physiology.* Science Publishers, Enfield, NH, pp. 163–184.

Kaushik, S.J., Gouillou-Coustans, M.F., and Cho, C.Y. 1998. Application of the recommendations on vitamin requirements of finfish by NRC (1993) to salmonids and sea bass using practical and purified diets. *Aquaculture* 161:463–474.

Koppe, W. 1993. Auswirkungen einer unterschiedlichen Biotinversorgung auf Wachstum, Biotinstatus sowie Parameter der Immun- und swtressrespons bei Regenbogenforellen (*Oncorhynchus mykiss*). Institut für Tierphysiologie und Tierernährung, Georg-August-Universität zu Göttingen, Göttingen, Germany.

Koshio, S. 2002. Red sea bream, *Pagrus major*. In: Webster, C.D., and Lim, C.E. (eds.) *Nutrient Requirements and Feeding of Finfish for Aquaculture.* CABI Publishing, Wallingford, Oxon, UK, pp. 51–63.

Krossøy, C. 2009. Vitamin K in farmed Atlantic salmon (*Salmo salar* L.) with emphasis on bone. PhD thesis, Institute of Biology, University of Bergen, Bergen, Norway.

Krossoy, C., Waagbø, R., Fjelldal, P.G., Wargelius, A., Lock, E.J., Graff, I.E., and Ørnsrud, R. 2009. Dietary menadione nicotinamide bisulphite (vitamin K_3) does not affect growth or bone health in first-feeding fry of Atlantic salmon (*Salmo salar* L.). *Aquaculture Nutrition* 15:638–649.

Kurlandsky, S.B., Xiao, J.H., Duell, E.A., Voorhees, J.J., and Fisher, G.J. 1994. Biological-activity of all-trans retinol requires metabolic conversion to all-trans-retinoic acid and is mediated through activation of nuclear retinoid receptors in human keratinocytes. *Journal of Biological Chemistry* 269:32821–32827.

Lall, S.P. 2002. The minerals. In: Halver, J.E., and Hardy, R.W. (eds.) *Fish Nutrition*. Academic Press, San Diego, CA, pp. 259–308.

Lall, S.P., and Lewis-McCrea, L.M. 2007. Role of nutrients in skeletal metabolism and pathology in fish—an overview. *Aquaculture* 267:3–19.

Lampert, J.M., Holzschuh, J., Hessel, S., Driever, W., Vogt, K., and von Lintig, J. 2003. Provitamin A conversion to retinal via the b,b-carotene-15,15′-oxygenase (bcox) is essential for pattern formation and differentiation during zebrafish embryogenesis. *Development* 130:2173–2186.

Langdon, C. 2003. Microparticle types for delivering nutrients to marine fish larvae. *Aquaculture* 227:259–275.

LD/FD. 2002. Forskrift om fôrvarer av 7. november 2002. In: Landbruksdepartementet & Fiskeridepartementet (eds.) Informasjonsforvatning, Norwegian Feed Regulation, p. 256 (in Norwegian). Available at www-lovdata.no (accessed December 7, 2010).

Levi, L., Levavi-Sivan, B., and Lubzens, E. 2008. Expression of genes associated with retinoid metabolism in the trout ovarian follicle. *Biology of Reproduction* 79:570–577.

Lie, Ø., Haaland, H., Hemre, G.I., Maage, A., Lied, E., Rosenlund, G., Sandnes, K., and Olsen, Y. 1997. Nutritional composition of rotifers following a change in diet from yeast and emulsified oil to microalgae. *Aquaculture International* 5:427–438.

Lim, C., Leamaster, B., and Brock, J.A. 1993. Riboflavin requirement of fingerling red hybrid tilapia grown in seawater. In: Kaushik, S.J., and Luquet, P. (eds.) *Fish Nutrition and Practice*. Proceedings of the IV International Symposium on Fish Nutrition and Feeding 1991. INRA Editions, Paris, pp. 743–752.

Lock, E.J., Waagbø, R., Wendelaar Bonga, S., and Flik, G. 2010. The significance of vitamin D for fish: a review. *Aquaculture Nutrition* 16: 100–116.

Lopez-Alvarado, J., Langdon, C.J., Teshima, S.-I., and Kanazawa, A. 1994. Effect of coating and encapsulation of crystalline amino acids on leaching in larval feeds. *Aquaculture* 122: 335–346.

Lorentzen, M., Berntssen, M.H.G., and Måge, A. 2001. Mineraler og sporelementer. In: Waagbø, R., Espe, M., Hamre, K., and Lie, Ø. (eds.) *Fiskeernæring*. Kystnæringen Forlag & Bokklubb AS, Bergen, Norway.

Lovell, R.T., and Buston, J.C. 1984. Biotin supplementation of practical diets for channel catfish. *Journal of Nutrition* 114:1092–1096.

Mæland, A. 2000. Water-soluble vitamins in the nutrition of Atlantic halibut *Hippoglossus hippoglossus* L. larvae and juveniles. D.Scient. thesis, Department of Fisheries and Marine Biology, University of Bergen, Bergen, Norway.

Mæland, A., and Waagbø, R. 1998. Examination of the qualitative ability of some cold water marine teleosts to synthesise ascorbic acid. *Comparative Biochemistry and Physiology* 121A:249–255.

Mæland, A., Sandnes, K., Hjeltnes, B., and Waagbø, R. 1998. Biotin in practical fish-meal based diet for Atlantic salmon, *Salmo salar* L. fry. *Aquaculture Nutrition* 4:241–247.

Mæland, A., Rosenlund, G., Stoss, J., and Waagbø, R. 1999. Weaning of Atlantic halibut (*Hippoglossus hippoglossus* L.) using formulated diets with various levels of ascorbic acid. *Aquaculture Nutrition* 5:211–219.

Mæland, A., Rønnestad, I., and Waagbø, R. 2003. Short communication—folate in eggs and developing larvae of Atlantic halibut, *Hippoglossus hippoglossus* L. *Aquaculture Nutrition* 9:185–188.

Maiorino, M., Coassin, M., Roveri, A., and Ursini, F. 1989. Microsomal lipid peroxidation: effect of vitamin E and its functional interaction with phospholipid hydroperoxide glutathione peroxidase. *Lipids* 24:721–726.

Mangor-Jensen, A., Holm, J.C., Rosenlund, G., Lie, Ø., and Sandnes, K. 1994. Effects of dietary vitamin C on maturation and egg quality of cod *Gadus morhua* L. *Journal of the World Aquaculture Society* 25:30–40.

Marchetti, M., Tassinari, M., and Bauce, G. 1995. Tolerance of high dietary levels of menadione bisulfite-nicotinamide by rainbow trout, *Onchorynchus mykiss*. *Aquaculture* 134:137–142.

Marchetti, M., Tossani, N., Marchetti, S., and Bauce, G. 1999. Stability of crystalline and coated vitamins during manufacture and storage of fish feeds. *Aquaculture Nutrition* 5:115–120.

Mårtensson, J., and Meister, A. 1991. Glutation deficiency decreases tissue ascorbate levels in newborn rats: ascorbate spares glutathion and protects. *Proceedings of the National Academy of Sciences of the United States of America.* 88:4656–4660.

Martinez, G.M., Baron, M.P., and Bolker, J.A. 2007. Skeletal and pigmentation defects following retinoic acid exposure in larval summer flounder, *Paralichthys dentatus*. *Journal of the World Aquaculture Society* 38:353–366.

Masumoto, T. 2002. Yellowtail, *Seriola quinqueradiata*. In: Webster, C.D., and Lim, C. (eds.) *Nutrient Requirements and Feeding of Finfish for Aquaculture*. CABI Publishing, Wallingford, Oxon, UK, pp. 131–146.

van der Meeren, T., Olsen, R.E., Hamre, K., and Fyhn, H.J. 2008. Biochemical composition of copepods for evaluation of feed quality in production of juvenile marine fish. *Aquaculture* 274:375–397.

Meinhold, H., Campos-Barros, A., Walzog, B., Köhler, R., Müller, R., and Behne, D. 1993. Effects of iodine deficiency on type I, type II and type III iodothyronine deiodinases and circulating thyroid hormones in the rat. *Experimental and Clinical Endocrinology* 101:87–93.

Merchie, G., Lavens, P., Dhert, P., Garcia Ulloa Gómez, M., Nelis, H., De Leenheer, A., and Sorgeloos, P. 1996a. Dietary ascorbic acid requirement during the hatchery production of turbot larvae. *Journal of Fish Biology* 49:573–583.

Merchie, G., Lavens, P., Storch, V., Übel, U., Nelis, H., De Leenheer, A., and Sorgeloos, P. 1996b. Influence of dietary vitamin C dosage on turbot (*Scophthalmus maximus*) and European sea bass (*Dicentrarchus labrax*) nursery stages. *Comparative Biochemistry and Physiology* 114A:123–133.

Mock, D.M. 1999. Biotin. In: Shils, M.E., Olson, J.A., Shike, M., and Ross, A.C. (eds.) *Modern Nutrition in Health and Disease*. Williams & Wilkins, Baltimore, MD, pp. 459–466.

Mol, K.A., VanderGeyten, S., Darras, V.M., Visser, T.J., and Kuhn, E. 1997. Characterization of iodothyronine outer ring and inner ring deiodinase activities in the blue tilapia, *Oreochromis aureus*. *Endocrinology* 138: 1787–1793.

Moren, M., Opstad, I., and Hamre, K. 2004a. A comparison of retinol, retinal and retinyl ester concentrations in larvae of Atlantic halibut (*Hippoglossus hippoglossus* L.) fed *Artemia* or zooplankton. *Aquaculture Nutrition* 10:253–259.

Moren, M., Opstad, I., Berntssen, M.H.G., Zambonino Infante, J.-L., and Hamre, K. 2004b. An optimum level of vitamin A supplements for Atlantic halibut (*Hippoglossus hippoglossus* L.) juveniles. *Aquaculture* 235: 587–599.

Moren, M., Gundersen, T.E., and Hamre, K. 2005. Quantitative and qualitative analysis of retinoids in *Artemia* and copepods by HPLC and diode array detection. *Aquaculture* 246: 359–365.

Moren, M., Opstad, I., van der Meeren, T., and Hamre, K. 2006. Iodine enrichment of *Artemia* and enhanced levels of iodine in Atlantic halibut larvae (*Hippoglossus hippoglossus* L.) fed the enriched *Artemia*. *Aquaculture Nutrition* 12: 97–102.

Moren, M., Sloth, J.J., and Hamre, K. 2008. Uptake of iodide from water in Atlantic halibut larvae (*Hippoglossus hippoglossus* L.). *Aquaculture* 285:174–178.

Morito, C.L.H., Conrad, D.H., and Hilton, J.W. 1986. The thiamin deficiency signs and requirement of rainbow trout (*Salmo gairdneri*, Richardson). *Fish Physiology and Biochemistry* 1:93–104.

Morris, P.C., and Davies, S.J. 1995a. The requirement of the gilthead sea bream (*Sparus aurata* L.) for nicotinic acid. *Animal Science* 61:437–443.

Morris, P.C., and Davies, S.J. 1995b. Thiamine supplementation of diets containing varied lipid : carbohydrate ratios given to gilthead seabream (*Sparus aurata* L.). *Animal Science* 61: 597–603.

Mukhi, S., and Patino, R. 2007. Effects of prolonged exposure to perchlorate on thyroid and reproductive function in zebrafish. *Toxicological Sciences* 96:246–254.

Murphy, A.R. 1993. Muscle. In: Berne, R.M., and Levy, M.N. (eds.) *Physiology*. Mosby Year Book, St. Louis, MO, pp. 281–324.

National Research Council (NRC). 1993. Nutrient requirements of fish. In: Committee on Animal

Nutrition, B.o.a. (ed.) National Academy Press, Washington, DC, p. 114.

Ng, W.-K., Serrini, G., Zhang, Z., and Wilson, R.P. 1997. Niacin requirement and inability of tryptophan to act as a precursor of NAD^+ in channel catfish, *Ictalurus punctatus*. *Aquaculture* 152:273–285.

Nordgreen, A., Yufera, M., and Hamre, K. 2008. Evaluation of cross-linked protein capsules for delivering nutrients to marine fish larvae and suspension-feeders. *Aquaculture* 285:159–166.

Norrgren, L., Börjeson, H., Förlin, L., and Åkerblom, N. 2001. The role of ascorbic acid and its derivatives in resistance to environmental and dietary toxicity of aquatic organisms. In: Dabrowski, K. (ed.) *Ascorbic Acid in Aquatic Organisms—Status and Perspectives*. CRC Press, Boca Raton, FL, pp. 133–147.

Nortvedt, R., Mangor-Jensen, A., Waagbø, R., and Norberg, B. 2001. Variability in egg composition in captive broodstock of Atlantic halibut (*Hippoglossus hippoglossus*) from Iceland and Norway. In: Hendry, C.I., Van Stappen, G., Wille, M., and Sorgeloos, P. (eds.) *Larvi'01*. European Aquaculture Society, Oostende, Belgium, pp. 1–3.

Nortvedt, R., Mangor-Jensen, A., Waagbø, R., and Norberg, B. 2003. *Ernæringsbetinget Eggkvalitet Hos Kveite, Hippoglossus hippoglossus*. NIFES, Bergen, Norway.

O'Connell, J.P., and Gatlin, D.M. 1994. Effects of dietary calcium and vitamin D_3 on weight gain and mineral composition of the blue tilapia (*Oreochromis aureus*) in low-calcium water. *Aquaculture* 125:107–117.

Önal, U., and Langdon, C. 2000. Characterization of two microparticle types for delivery of food to altricial fish larvae. *Aquaculture Nutrition* 6:159–170.

Opstvedt, J., Knudsen, G., and Asbjørnsen, B. 1997. Innhold av fettløselige vitaminer i fiskemel og fiskeolje produsert fra sild, lodde, tobis, kolmule og brisling (in Norwegian). *Sildeolje- og Sildemelindustriens Forskningsinstitutt* C-295:10.

Otterlei, E., Nyhammer, G., Folkvord, A., and Stefansson, S.O. 1999. Temperature- and size-dependent growth of larval and early juvenile Atlantic cod (*Gadus morhua*): a comparative study of Norwegian coastal cod and northeast Arctic cod. *Canadian Journal of Fisheries and Aquatic Sciences* 56:2099–2111.

Packer, J.E., Slater, T.F., and Willson, R.L. 1979. Direct observation of a free radical interaction between vitamin E and vitamin C. *Nature* 278:737–738.

Penglase, S.J. 2009. Enrichment of rotifers (*Brachionus* sp.) with selenium and the effect of feeding selenium-enriched rotifers to cod larvae (*Gadus morhua*). Master thesis, Institute of Biology, University of Bergen, Norway.

Phillips, A.M., Brockway, D.R., Kolb, A.J.J., and Maxwell, J.M. 1950. Further studies upon biotin and trout diets. *Fisheries Research Bulletin* 19:9–12.

Poston, H.A. 1976. Optimum level of dietary biotin for growth, feed utilization, and swimming stamina of fingerling lake trout (*Salvelinus namaycush*). *Canadian Journal of Fisheries and Aquatic Sciences* 33:1803–1806.

Poston, H.A., and Wolfe, M.J. 1985. Niacin requirement for optimum growth, feed conversion and protection of rainbow trout, *Salmo gairdneri* Richardson, from ultraviolet-B irradiation. *Journal of Fish Diseases* 8:451–460.

Power, D.M., Llewellyn, L., Faustino, M., Nowell, M.A., Bjornsson, B.T., Einarsdottir, I.E., Canario, A.V.M., and Sweeney, G.E. 2001. Thyroid hormones in growth and development of fish. *Comparative Biochemistry and Physiology C: Toxicology & Pharmacology* 130:447–459.

Power, D.M., Einarsdottir, I.E., Pittman, K., Sweeney, G.E., Hildahl, J., Campinho, M.A., Silva, N., Saele, O., Galay-Burgos, M., Smaradottir, H., and Bjornsson, B.T. 2008. The molecular and endocrine basis of flatfish metamorphosis. *Reviews in Fisheries Science* 16(Suppl. 1):95–111.

Rando, R.R. 1994. Vision. In: Blomhoff, R. (ed.) *Vitamin A in Health and Disease*. Marcel Dekker, Inc., New York, Basel, Hong Kong, pp. 677.

Rao, D.S., and Raghuramulu, N. 1999. Vitamin D_3 and its metabolites have no role in calcium and phosphorus metabolism in *Tilapia mossambica*. *Journal of Nutritional Science and Vitaminology* 45:9–19.

Ren, T., Koshio, S., Jiang, Z.-Q., Yokoyama, S., Komilus, C.F., Gao, J., and Ishikawa, M. 2010. Interactive effects of dietary vitamin C and phospholipid in micro-bound diet for growth, survival, and stress resistance of larval red sea bream, *Pagrus major*. *Aquaculture Nutrition* 16:475–482.

Ribeiro, A.R.A., Ribeiro, L., Sæle, Ø., Hamre, K., Dinis, M.T., and Moren, M. 2009. Iodine enriched rotifers and *Artemia* prevents goitre in Senegalese sole (*Solea senegalensis*) larvae reared in a recirculation system. *Aquaculture Nutrition* doi:10.1111/j.1365-2095.2009.00740.x.

Robinson, E.H., and Li, M.H. 2002. Channel catfish, *Ictalurus punctatus*. In: Webster, C.D., and Lim, C.E. (eds.) *Nutrient Requirements and Feeding of Finfish for Aquaculture*. CABI Publishing, Wallingford, Oxon, UK, pp. 293–318.

Roem, A.J., Stickney, R.R., and Kohler, C.C. 1991. Dietary pantothenic-acid requirement of the blue tilapia. *Progressive Fish-Culturist* 53:216–219.

Rønnestad, I., Lie, Ø., and Waagbø, R. 1997. Vitamin B_6 in Atlantic halibut *Hippoglossus hippoglossus*—endogenous utilization and retention in larvae fed natural zooplankton. *Aquaculture* 157:337–345.

Rønnestad, I., Helland, S., and Lie, O. 1998. Feeding *Artemia* to larvae of Atlantic halibut (*Hippoglossus hippoglossus* L) results in lower larval vitamin A content compared with feeding copepods. *Aquaculture* 165:159–164.

Rønnestad, I., Hamre, K., Lie, Ø., and Waagbø, R. 1999. Ascorbic acid and α-tocopherol in larvae of Atlantic halibut before and after exogenous feeding. *Journal of Fish Biology* 55:720–731.

Sakamoto, S., and Yone, Y. 1978. Effect of dietary phosphorus level on chemical composition of red sea bream. *Bulletin of the Japanese Society of Scientific Fisheries* 44:227–229.

Sandnes, K. 1984. Some aspects of ascorbic acid and reproduction in fish. Proceedings of Ascorbic Acid in Domestic Animals, Skjoldn'sholm, September 1983. The Royal Danish Agriculture Society, Copenhagen.

Sandnes, K. 1991. Vitamin C in fish nutrition—a review. *Fiskeridirektoratets Skrifter, Serie Ernæring* 4:3–32.

Sandnes, K., and Mæland, A. 1994. Vannløselige vitaminer—Folinsyre og vitamin B12 i fôr til laks. Norwegian Research Council Research Report, Institute of Nutrition, Directorate of Fisheries, Bergen, Norway, p. 5.

Sandnes, K., Hansen, T., Killie, J.E.A., and Waagbø, R. 1990. Ascorbate-2-sulfate as a dietary vitamin C source for Atlantic salmon (*Salmo salar*): 1. Growth, bioactivity, haematology and immune response. *Fish Physiology and Biochemistry* 8:419–427.

Sandnes, K., Torrissen, O.J., and Waagbø, R. 1992. The minimum dietary requirement of vitamin C in Atlantic salmon (*Salmo salar*) fry using Ca ascorbate-2-monophosphate as dietary source. *Fish Physiology and Biochemistry* 10:315–319.

Sandnes, K., Rosenlund, G., Mangor-Jensen, A., and Lie, Ø. 1998. Contents and organ distribution of pantothenic acid in maturing turbot (*Psetta maxima*). *Aquaculture Nutrition* 4:285–286.

Sato, M., Yoshinaka, R., Kuroshima, R., Morimoto, H., and Ikeda, S. 1987. Changes in water soluble vitamin contents and transaminase activity of rainbow trout egg during development. *Nippon Suisan Gakkaishi* 53:795–799.

Schaeffer, M.C., Gretz, D., Mahuren, J.D., and Coburn, S.P. 1995. Tissue B-6 vitamin concentrations in rats fed excess vitamin B-6. *Journal of Nutrition* 125:2370–2378.

Schwehr, K.A., and Santschi, P.H. 2003. Sensitive determination of iodine species, including organo-iodine, for freshwater and seawater samples using high performance liquid chromatography and spectrophotometric detection. *Analytica Chimica Acta* 482:59–71.

Sealy, W.M., and Gatlin, D.M. 2001. Overview of nutritional strategies affecting the health of marine fish. In: Lim, C., and Webster, C.D. (eds.) *Nutrition and Fish Health*. Food Products Press, Birminghamton, NY, pp. 103–118.

Segner, H., Rösch, R., Verreth, J., and Witt, U. 1993. Larval nutritional physiology: studies with *Clarias gariepinus*, *Coregonus lavaretus* and *Scophthalmus maximus*. *Journal of the World Aquaculture Society* 24:123–135.

Serrini, G., Zhang, Z., and Wilson, R.P. 1996. Dietary riboflavin requirement of fingerling channel catfish (*Ictalurus punctatus*). *Aquaculture* 139:285–290.

Shearer, K.D. 1988. Dietary potassium requirement of juvenile chinook salmon. *Aquaculture* 73:119–129.

Shearer, K.D. 1991. Determination of the dietary requirements for essential elements in fish. PhD thesis, Department of Fisheries and Marine Biology, University of Bergen, Bergen, Norway.

Shiau, S.Y., and Chin, Y.H. 1999. Estimation of the dietary biotin requirement of juvenile hybrid tilapia, *Oreochromis niloticus* × *O. aureus*. *Aquaculture* 170:71–78.

Shiau, S.Y., and Tseng, H.C. 2007. Dietary calcium requirements of juvenile tilapia, *Oreochromis niloticus* × *O. aureus*, reared in fresh water. *Aquaculture Nutrition* 13:298–303.

Smith, C.E. 1968. Haematological changes in coho salmon fed a folic acid deficient diet. *Journal of the Fisheries Research Board of Canada* 25:151–156.

Smith, C.M., and Song, W.O. 1996. Comparative nutrition of pantothenic acid. *The Journal of Nutritional Biochemistry* 7:312–321.

Solbakken, J.S. 2003. Plasticity of metamorphosis in Atlantic halibut (*Hippoglossus hippoglossus* L.); responses to internal and external cues. Dr.Scient., Department of Fisheries and Marine Biology, University of Bergen, Bergen, Norway.

Solbakken, J.B., Berntssen, M.H.G., Norberg, B., Pittman, K., and Hamre, K. 2002. Different iodine and thyroid hormone levels between Atlantic halibut larvae fed wild zooplankton or *Artemia* from first exogenous feeding until post metamorphosis. *Journal of Fish Biology* 60:1345–1362.

Soliman, A.K., and Wilson, R.P. 1992. Water-soluble vitamin requirements of tilapia. 2. Riboflavin requirement of blue tilapia, *Oreochromis aureus*. *Aquaculture* 104:309–314.

Steffens, W. 1989. *Principles of Fish Nutrition*. Ellis Horwood Limited, Chichester, West Sussex, England, pp. 290–291.

Steinbrenner, H., Alili, L., Bilgic, E., Sies, H., and Brenneisen, P. 2006. Involvement of selenoprotein P in protection of human asterocytes from oxidative damage. *Free Radical Biology & Medicine* 40:1513–1523.

Sugita, H., Miyajima, C., and Deguchi, Y. 1990. The vitamin B_{12}–producing ability of intestinal bacteria isolated from tilapia and channel catfish. *Nippon Suisan Gakkaishi* 56:701.

Sugita, H., Miyajima, C., and Deguchi, Y. 1991. The vitamin B_{12}–producing ability of the intestinal microflora of freshwater fish. *Aquaculture* 92:267–276.

Sugita, H., Takahashi, J., and Deguchi, Y. 1992. Production and consumption of biotin by the intestinal microflora of cultured freshwater fishes. *Bioscience Biotechnology & Biochemistry* 56:1678–1679.

Tacon, A.G.J. 1992. Nutritional fish pathology. Morphological signs of nutrient deficiency and toxicity in farmed fish. FAO Fish Technical Paper, Rome, pp. 75.

Takeshi, M., and Andrews, J.W. 1977. Pantothenic acid requirements of channel catfish fingerlings. *The Journal of Nutrition* 109:1140–1142.

Takeuchi, L., Takeuchi, T., and Ogino, C. 1980. Riboflavin requirements in carp and rainbow trout. *Bulletin of the Japanese Society of Scientific Fisheries* 46:733–737.

Takeuchi, T., Dedi, J., Ebisawa, C., Watanabe, T., Seikai, T., Hosoya, K., and Nakazoe, J.-I. 1995. The effect of α-carotene and vitamin A enriched *Artemia* nauplii on the malformation and color abnormality of larval Japanese flounder. *Fisheries Science* 61:141–148.

Takeuchi, T., Dedi, J., Haga, Y., Seikai, T., and Watanabe, T. 1998. Effect of vitamin A compounds on bone deformity in larval Japanese flounder (*Paralichthys olivaceus*). *Aquaculture* 169:155–165.

Takitani, K., Zhu, C.L., Inoue, A., and Tamai, H. 2006. Molecular cloning of the rat beta-carotene 15,15′-monooxygenase gene and its regulation by retinoic acid. *European Journal of Nutrition* 45:320–326.

Tappel, A.L. 1962. Vitamin E as the biological lipid antioxidant. *Vitamins & Hormones* 20:493–510.

Taveekijakarn, P., Miyazaki, T., Matsumoto, M., and Arai, S. 1996a. Histopathological and haematological changes in amago salmon, *Oncorhynchus rhodurus* (Jordan & McGregor), fed a vitamin-D-free diet. *Journal of Fish Diseases* 19:289–294.

Taveekijakarn, P., Miyazaki, T., Matsumoto, M., and Arai, S. 1996b. Studies on vitamin K deficiency in amago salmon, *Oncorhynchus rhodurus* (Jordan & McGregor). *Journal of Fish Diseases* 19:209–214.

Terova, G., Cecchini, S., Saroglia, M., Caricato, G., and Jeney, Z. 2001. Live food–mediated vitamin C transfer in sea bass (*Dicentrarchus labrax*, L.) during first feeding. In: Dabrowski, K. (ed.) *Ascorbic Acid in Aquatic Organisms—Status and Perspectives*. CRC Press, Boca Raton, FL, pp. 191–210.

Truesdale, V.W., Bale, A.J., and Woodward, E.M.S. 2000. The meridional distribution of dissolved iodine in near-surface waters of the Atlantic Ocean. *Progress in Oceanography* 45:387–400.

Udagawa, M. 2000. Physiological role of vitamin K in fish—review. *Japan Agricultural Research Quarterly* 34:279–284.

Ursini, F. 1993. Antioxidant activity of selenium dependent peroxidases: the molecular basis of the synergistic effect between selenium and vitamin E. In: Diplock, A.T., Gutteridge, J.M.C., and Shukla, V.K.S. (eds.) *Antioxidants, Free Radicals and Polyunsaturated Fatty Acids in Biology and Medicine*. International Food Science Centre A/S, Lystrup, Denmark.

Verbost, P.M., Flik, G., Fenwick, J.C., Greco, A.M., Pang, P.K.T., and Bonga, S.E.W. 1993. Branchial calcium-uptake—possible mechanisms of control by stanniocalcin. *Fish Physiology and Biochemistry* 11:205–215.

Vielma, J., and Lall, S.P. 1998a. Control of phosphorus homeostasis of Atlantic salmon (*Salmo salar*) in fresh water. *Fish Physiology and Biochemistry* 19:83–93.

Vielma, J., and Lall, S.P. 1998b. Phosphorus utilization by Atlantic salmon (*Salmo salar*) reared in freshwater is not influenced by higher dietary calcium intake. *Aquaculture* 160:117–128.

Vielma, J., Lall, S.P., Koskela, J., Schöner, F.J., and Mattila, P. 1998. Effects of dietary phytase and cholecalciferol on phosphorus bioavailability in rainbow trout (*Oncorhynchus mykiss*). *Aquaculture* 163:309–323.

Villeneuve, L., Gisbert, E., Cahu, C.L., Le Gall, M.M., and Zambonino-Infante, J.L. 2004. Expression and localization of some retinoid receptors during European sea bass (*Dicentrarchus labrax*) larvae development. *Aquaculture* 242:537–551.

Villeneuve, L., Gisbert, E., Delliou, H.L., Cahu, C.L., and Zambonino-Infante, J.L. 2005a. Dietary levels of all-trans retinol affect retinoid nuclear receptor expression and skeletal development in European sea bass larvae. *British Journal of Nutrition* 93:791–801.

Villeneuve, L., Gisbert, E., Zambonino-Infante, J.L., Quazuguel, P., and Cahu, C.L. 2005b. Effect of nature of dietary lipids on European sea bass morphogenesis: implication of retinoid receptors. *British Journal of Nutrition* 94: 877–884.

Waagbø, R. 2006. Feeding and disease resistance in fish. In: Mosenthin, J., Zentek, R., and Zebrowska, T. (eds.) *Biology of Growing Animal*. Elsevier Limited, London, UK, pp. 387–415.

Waagbø, R. 2008. Reducing production related diseases in farmed fish. In: Lie, Ø. (ed.) *Improving Farmed Fish for the Consumer*. VS Woodhead Publishing, Cambridge, UK, pp. 363–398.

Waagbø, R. 2009. Water soluble vitamins in fish ontogeny. *Aquaculture Research* doi: 10.1111/j.1365-2109.2009.02223.x.

Waagbø, R., Sandnes, K., and Mæland, A. 1994. Biotinbehov hos smolt. *Norsk Fiskeoppdrett* 2A:24–25.

Waagbø, R., Graff, I.E., and Hamre, K. 2001. Vitaminer. In: Waagbø, R., Espe, M., Hamre, K., and Lie, Ø. (eds.) *Fiskeernæring*. Kystnæringen, Bergen, pp. 93–124.

Waagbø, R., Kryvi, H., Breck, O., and Ørnsrud, R. 2005. Final report—workshop on bone disorders in intensive aquaculture of salmon and cod. National Institute of Nutrition and Seafood Research, Bergen, Norway.

Wahli, T., Verlhac, V., Girling, P., Gabaudan, J., and Aebischer, C. 2003. Influence of dietary vitamin C on the wound healing process in rainbow trout (*Oncorhynchus mykiss*). *Aquaculture* 225:371–386.

Waite, T.J., Truesdale, V.W., and Olafsson, J. 2006. The distribution of dissolved inorganic iodine in the seas around Iceland. *Marine Chemistry* 101:54–67.

Watanabe, M., Houten, S.M., Mataki, C., Christoffolete, M.A., Kim, B.W., Sato, H., Messaddeq, N., Harney, J.W., Ezaki, O., Kodama, T., Schoonjans, K., Bianco, A.C., and Auwerx, J. 2006. Bile acids induce energy expenditure by promoting intracellular thyroid hormone activation. *Nature* 439:484–489.

Webster, C.D., and Lim, C. 2002. *Nutrient Requirements and Feeding of Finfish for Aquaculture*. CABI Publishing, Wallingford, Oxon, UK.

Wendelar Bonga, S.E. 1997. The stress response in fish. *Physiological Reviews* 77:591–625.

Witt, E.M., Laidley, C.W., Liu, K.K.M., Hirano, T., and Grau, E.G. 2009. Correlation between environmental iodide concentrations and larval growth, survival, and whole body concentrations of thyroid hormones and cortisol in Pacific threadfin (*Polydactylus sexfilis*). *Aquaculture* 289:357–364.

Witten, P. E. 1997. Enzyme histochemical characteristics of osteoblasts and mononucleated osteoclasts in a teleost fish with acellular bone (*Oreochromis niloticus*, Cichlidae). *Cell and Tissue Research* 287:591–599.

Wong, G.T.F. 1991. The marine geochemistry of iodine. *Reviews in Aquatic Sciences* 4:45–73.

Woodward, B. 1984. Symptoms of severe riboflavin deficiency without ocular opacity in rainbow trout (*Salmo gairdneri*). *Aquaculture* 37: 275–281.

Woodward, B. 1985. Riboflavin requirement for growth, tissue saturation and maximal flavin-dependent enzyme activity in young rainbow trout (*Salmo gairdneri*) at two temperatures. *Journal of Nutrition* 115:78–84.

Woodward, B. 1994. Dietary vitamin requirements of cultured young fish, with emphasis on quantitative estimates for salmonids. *Aquaculture* 124:133–168.

Woodward, B., and Frigg, M. 1989. Dietary biotin requirements of young rainbow trout (*Salmo gairdneri*) determined by weight gain, hepatic biotin concentration and maximal biotin-dependent enzyme activities in liver and white muscle. *Journal of Nutrition* 119:54–60.

Yamamoto, T., Teruya, K., Hara, T., Hokazono, H., Kai, I., Hashimoto, H., Furuita, H., Matsunari, H., and Mushiake, K. 2009. Nutritional evaluation of rotifers in rearing tanks without water exchange during seed production of amberjack *Seriola dumerili*. *Fisheries Science* 75:697–705.

Yen, P.M. 2001. Physiological and molecular basis of thyroid hormone action. *Physiological Reviews* 81:1097–1142.

Section 2

Nutritional Physiology

Chapter 5

Effects of broodstock diet on eggs and larvae

Hipólito Fernández-Palacios, Birgitta Norberg, Marisol Izquierdo, and Kristin Hamre

5.1 Introduction

The success of early larval rearing is markedly affected by the quality of each spawning batch and the larvae subsequently obtained, which in turn is strikingly dependent on broodstock nutrition. However, despite the long-recognized importance of adequate diets for reproduction and spawning quality in teleosts (Luquet and Watanabe 1986), the nutritional requirements of broodstock remains one of the least explored areas in the field of fish nutrition. Few studies are available and they are limited to only a few species in comparison to the large number of fishes produced by aquaculture worldwide (Brooks et al. 1997; Izquierdo et al. 2001). This is due to the need for huge facilities to keep substantial groups of large broodstock fish, and the high costs both in construction and maintenance of these facilities that are required to carry out long-term nutritional experiments on broodstock. Moreover, reproductive biology varies greatly among the many fish species produced by aquaculture, being closely related to the feeding strategy and nutritional requirements of each species. As also happens in other vertebrates, many of the problems that appear during early development in fish are directly related to broodstock nutrition. Thus, the nutritional components of the diet, the feed intake rate, or the feeding period can all directly or indirectly affect spawning, egg, and larval quality. There is a wide range of reproduction-related parameters being used by commercial or experimental hatcheries in the diverse farmed fish, their sensitivity to the different nutrients varying among them. The present chapter, after a brief introduction to teleost reproductive biology and description of the quality parameters employed, gathers the current

Larval Fish Nutrition, First Edition. Edited by G. Joan Holt.

information on nutritional requirements of broodstock fish and the consequences of feeding on spawning, egg, and larval quality.

5.2 Reproductive biology and ovarian development

Teleost fish display a variety of reproductive strategies that result from adaptations to their respective environments. Semelparous fish, such as many species of Pacific salmon (*Oncorhynchus* spp.) and freshwater eels (*Anguilla* spp.), spawn only once in their lifetime and then die, while most other species are iteroparous and go through repeated spawning cycles. Anadromous fish, for example, most salmonids, spawn and hatch in freshwater, migrate to the sea as juveniles, and return to spawn in freshwater as they become sexually mature. Catadromous fish, for example, eels, spawn in the sea but spend the immature stages in freshwater. Reproduction is usually seasonal in temperate waters but can be continuous in tropical environments.

Ovarian development in female teleosts has been described as follows: (1) synchronous—all oocytes develop at the same rate and are spawned simultaneously, as in many *Oncorhynchus* species; (2) group synchronous—several populations of oocytes are present during the reproductive season and clusters of oocytes develop, mature, and are spawned simultaneously; and (3) asynchronous—oocytes in all stages of development are present at the same time and often independent of season, as is often the case in tropical species (Wallace and Selman 1981). Species that spawn only one batch of eggs may display some degree of parental care such as nest building or guarding of eggs. On the other hand, many marine species with very high fecundity and no parental care are group synchronous batch spawners that release several batches of eggs each reproductive season in order to optimize survival of the offspring.

Reproductive development in female teleosts starts with mitotic production of primordial germ cells that migrate to the ovary and are differentiated into oogonia. These are further transformed into primary oocytes that enter meiosis, which is arrested during the diplotene stage of the first prophase, and the oocyte is prepared for fertilization and embryonic development (Yoshizaki et al. 2002; Strussman and Nakamura, 2002; Lubzens et al. 2010).

During the primary growth phase, the surrounding granulosa and theca cells envelop the oocyte to form the ovarian follicle (Wallace and Selman 1981, 1990; Patiño and Sullivan 2002). Follicle cell synthesis of estrogens starts at this stage. During the perinucleolus stage, the cell nucleus increases in size, with numerous ribosome-producing nucleoli surrounding its periphery. A large part of the RNA stored for use by the embryo is believed to be synthesized during this period (Pelegri 2003; Cerdá et al. 2008).

At the early stages of secondary growth, cortical alveoli appear and accumulate in the periphery of the oocyte. The cortical alveoli are secretory granules formed from the Golgi complex and contain large amounts of glycoprotein. At fertilization, the glycoproteins are secreted into the perivitelline space from the cortical alveoli in order to prevent polyspermy (Wallace and Selman 1981; Wessel et al. 2001). The oocyte is surrounded by an acellular protein envelope, the vitelline envelope, which hardens to form the eggshell at oviposition. Hepatic and intraovarian synthesis of the zona pellucida (ZP) proteins (also known as zona radiata proteins, vitelline envelope proteins, eggshell proteins, or choriogenins) is also initiated during the early secondary growth stages (Hyllner et al. 1994; see also review by Lubzens et al. 2010).

Even though the oocyte may increase in size several thousandfold during primary and early secondary growth, the most conspicu-

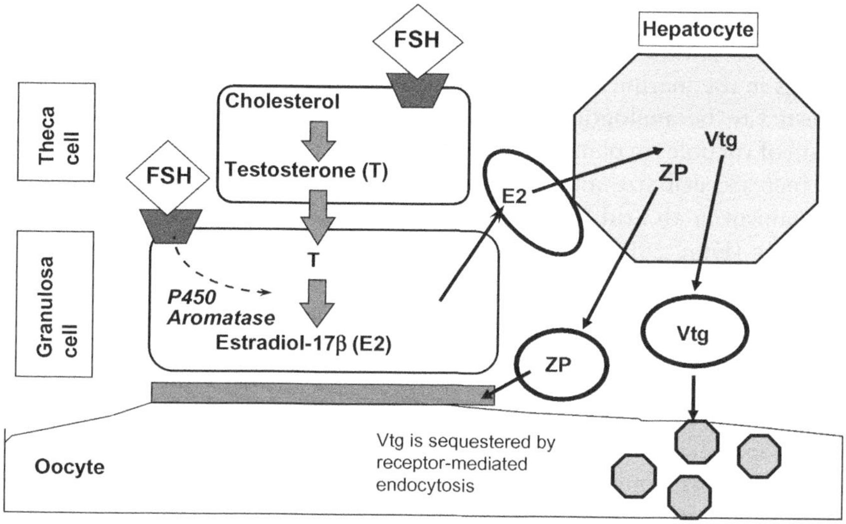

Figure 5.1 Vitellogenesis. FSH = follicle stimulating hormone, Vtg = vitellogenin, ZP = zona pellucida proteins.

ous size increase occurs during the last part of secondary growth, vitellogenesis (Figure 5.1). During this stage, lipids and the yolk proteins, lipovitellins, phosvitins, phosvettes, and β′-component, are incorporated and stored in large amounts in yolk vesicles or granules. Incorporation of maternal RNA continues, together with vitamins and hormones that are necessary for early embryonic development (cf. Lubzens et al. 2010).

Vitellogenesis is the process of yolk formation and incorporation in the growing oocytes. The yolk protein precursors, vitellogenins (Vtgs) are high-molecular-weight lipoproteins that are synthesized in the liver and secreted into the blood after posttranslational lipidation, glycosylation, and phosphorylation (reviewed, e.g., by Wallace 1985; Tyler and Sumpter 1996). Recent studies on the molecular regulation of vitellogenesis have revealed three forms of Vtgs—VtgA, VtgB, and VtgC (also named VtgAa, VtgAb, and VtgC)—that are encoded by separate genes (Matsubara et al. 2003; Finn 2007a). VtgA and VtgB are constituted of heavy-chain lipovitellin, phosvitin, light-chain lipovitellin, and β′-component domains, while the phosvitin domain is not present in VtgC. Vtg is sequestered by receptor-mediated endocytosis, and is enzymatically processed by cathepsin D (reviewed by Wallace 1985; Carnevali et al. 2006). Vertebrate Vtgs contain ca. 20% lipids, and constitute the most important lipid source for the growing embryo, although part of the egg lipids are also derived from plasma very low-density lipoproteins (VLDLs) (Babin et al. 2007). The fatty acid composition of Vtg appears to be conserved but may be affected by long-term imbalances in broodstock diet (Silversand 1996).

At final maturation, meiosis is resumed, the germinal vesicle migrates from the center to the periphery of the oocyte and is broken down, and the first polar body is expelled. Meiosis is again arrested, in metaphase II (cf. Suwa and Yamashita 2007). Concurrent with final maturation, the oocyte is hydrated (Wallace and Selman 1978). Hydration is especially pronounced in marine teleosts spawning pelagic eggs, where a three- to eight-fold increase in volume may occur. The mechanism behind this large volume increase is secondary processing of yolk proteins into free amino acids, thus creating an osmotic gradient (cf. Cerdà et al. 2007). Molecular water channels, aquaporins, appear in the oocyte membrane and mediate the influx of water into the cell (Fabra et al. 2005). Hydration of eggs in the marine environment has been suggested to be analogous to lysosomal generation of vacuoles in plants, as both processes will increase cell size and provide water to the organism in an arid or hyperosmotic environment (Finn 2007b). The free amino acids generated by secondary cleavage appear to be important nutrients in early embryonic development, as the yolk protein stores do not seem to be utilized before hatching (Rønnestad et al. 1999; Ohkubo and Matsubara 2002; see Chapter 3).

The increase in cell size caused by oocyte hydration also facilitates the ovulation process, when the ovarian follicle is ruptured and the mature egg is expelled. Depending on the reproductive strategy, the ovulated eggs may be kept in the body cavity and undergo postovulatory maturation for a few days, or they may be released into the surrounding water where they are fertilized. A few species also have internal fertilization, for example, the wolffishes (*Anarhicas* spp.).

Sexual maturation in teleost fish is under endocrine regulation via the brain–pituitary–gonad axis. Gonadotropin-releasing hormone(s) (GnRHs) from the brain induce pituitary synthesis and release of the gonadotropins follicle-stimulating hormone (FSH) and luteinizing hormone (LH). The gonadotropins act on receptors on the ovarian follicle cells through the synthesis of steroid hormones to stimulate gonadal differentiation and development at different stages in the life cycle (reviewed by Nagahama 1994, 2000). FSH is presumed to act during ovarian growth to stimulate synthesis of testosterone (T) and estradiol-17β (E2). E2 stimulates liver synthesis of ZP proteins and Vtgs during secondary oocyte growth. The role of FSH during ovarian maturation is not well understood but is relevant to follicular maturation and granulosa cell proliferation (e.g., Zhou et al. 1997). The transition of the ovarian follicle from vitellogenesis to final maturation involves an LH-controlled switch in the steroid biosynthesis from E2 to maturation-inducing steroid, MIS. Two different MIS have been identified in fish; the progestines 17α,20β-dihydroxy-4-pregnen-3-one (17,20β-P), and 17α,20β,21-trihydroxy-4-pregnen-3-one, with 17,20β-P as the most common (cf. Nagahama and Yamashita 2008). The MIS acts via a membrane-bound receptor to induce final maturation (see Thomas et al. 2002). Ovulation and oviposition are further controlled by prostaglandins (PGs) synthesized within the ovary (Goetz et al. 1989; 1991). PGs are eicosanoids produced from specific fatty acids that are liberated from cell membranes by the action of phospholipases. They are highly biologically active, and act both to control ovulation and oviposition and to stimulate reproductive behavior in fish (Stacey and Goetz 1982; Sorensen et al. 1988). Arachidonic acid (ARA) is generally considered to be the preferred eicosanoid substrate, at least in mammals (Wada et al. 2007), but very few studies have investigated the importance of eicosanoids produced by other fatty acids in fish. There is evidence in both fish and mammals that the ratio of various fatty acids can alter the production of eicosanoids (Farndale et al. 1999; Wada et al. 2007). This is most likely the result of competition by different fatty acids for key converting enzymes such as PG endoperoxide synthase (cyclooxygenase). Although different fatty acids are not converted at the same efficiency, they can still bind the active site of the enzyme, and may act to decrease the overall production of eicosanoids depending on their relative levels. However, further studies are needed on the effect of dietary fatty acids on reproductive physiology and broodstock health.

This chapter addresses the effects of broodstock diet on eggs and larvae. Reproductive success also depends on the paternal contribution, and spermatogenesis as well as sperm quality may be affected by diet. For a detailed account of spermatogenesis and factors affecting sperm quality, the reader is referred to recent reviews by Schulz et al. (2010) and Bobe and Labbé (2010).

5.3 Criteria for egg and larval quality

Nutritional studies with fish broodstock most often use fecundity and egg and larval quality to measure the effects of different treatments, whereas effects on the parent fish themselves receive less attention. Fecundity is defined as the number of mature eggs that are produced by a female, or per kilogram body weight of females, during the annual spawning cycle (Bromage et al. 1992). Good-quality eggs are those that exhibit low levels of mortality at fertilization, hatch, and first feeding and those that produce the fastest-growing and healthiest fry and older fish (Bromage et al. 1992). Egg quality in nutritional studies is most often indicated by fertilization rate, survival during the egg stage, and hatching rate. Cleavage symmetry of the 8- and 32-cell stage has been proposed as an additional quality parameter (Kjørsvik et al. 1990). Larval quality is most often assessed in yolk sac larvae since keeping the larvae after first feeding involves large costs and introduces appreciable amounts of noise into the experiment.

Kjørsvik et al. (2003) conducted an experiment where the quality criteria (fertilization rate and blastomere morphology) were correlated with hatching rate and later rearing success. The results indicated that these parameters were good predictors of egg quality. However, blastomere morphology is a difficult parameter to measure since the embryonic cells may be in different stages of cleavage and therefore have different morphologies, even though the eggs are fully viable (unpublished data).

The assembly of fish eggs is a very complicated process. The ovulated egg contains the genetic code for the new individual, and in addition, nutrients, maternal RNA, and maternally deposited hormones, which all may affect the egg quality (Brooks et al. 1997; Lubzens et al. 2010). Recent and future advances in ovary and egg transcriptome and proteome profiling may provide more specific markers of egg quality than those used at present (Lubzens et al. 2010).

5.4 Effective feeding periods for optimum broodstock performance

According to the differences in reproductive biology discussed above, the length of the gametogenetic and, particularly, vitellogenetic periods, feeding history during reproduction, the time period needed to change the nutrient composition of the developing gonads, and hence the reproductive success varies among the different fish species. In gilthead sea bream (*Sparus aurata*) and red sea bream (*Pagrus major*), the egg composition is quickly affected by the diet, even after only few weeks of feeding (Watanabe et al. 1985b; Fernández-Palacios et al. 1995; Tandler et al. 1995). These species are batch spawners with group synchronic ovaries and a short vitellogenesis period where it is possible to improve the spawning quality by modifying the nutritional quality of the diet even during the spawning season (Tandler et al. 1995; Fernández-Palacios et al. 1995, 1997, 1998, 2005). In sea bass (*Dicentrarchus labrax*), vitellogenesis lasts slightly longer than in sea breams, and a longer period is needed for obtaining appropriate levels of n-3 highly unsaturated fatty acids (HUFA) in the eggs (Navas et al. 1997). The feeding period for red snapper *Lutjanus campechanus* to obtain good n-3 HUFA levels is at least 2 months before the spawning period starts (Papanikos 2005). On the contrary, in fish with more than 6 months of vitellogenesis (Frémont et al. 1984) and synchronic ovaries as in salmonids, the broodstock must be fed with a good-quality diet for several months before the spawning season (Watanabe et al. 1984d; Corraze et al. 1993). Thus, in coho salmon, *Oncorhynchus kisutch*, the fatty acid profiles of the muscle and developing eggs (Hardy et al. 1990) start to reflect the dietary fatty acids only after 2 months of feeding, while in gilthead sea bream, the fatty acid composition of the female gonads is affected after just 15 days (Harel et al. 1992; Tandler et al. 1995).

Many fish species tend to decrease their feed intake rate during sexual maturation. The mechanism responsible for this behavior is not clear, but it also affects fish in the natural environment (Trippel et al. 1995; Link and Burnett 2001). The swelling of the females when egg hydration starts in the ovary may restrict space in the body cavity and, as a consequence, the food volume that can be ingested. Endocrine changes in connection with sexual maturation and spawning may also interfere with appetite regulation (Thorsen et al. 2003). As a consequence, the energy and nutrients needed for gonadal development may be taken from the body reserves. Rainbow trout, *Oncorhynchus mykiss*, mobilizes its lipid reserves from the carcass and viscera (Nassour and Leger 1989). African catfish, *Clarius batrachus*, uses the abdominal fat as the major energy source for sexual maturation (Lall and Singh 1987). Atlantic salmon, *Salmo salar*, uses the proteins and lipids from the muscle (Aksnes et al. 1986), and the cod *Gadus morhua* uses the accumulated lipid reserves of the liver (Tocher and Harvie 1988). The gilthead sea bream broodstock continues to feed during sexual maturation and during the spawning period, producing an egg biomass equivalent to its own body weight. Under these circumstances, the nutrients deposited in the ovaries may come from the broodstock diets, and the nutritional composition of the eggs can be modified during the spawning season and have an effect on the spawning quality (Almansa et al. 1999). The turbot *Scophthalmus maximus* could be an exception to this observation since the ovary composition is more affected by diet during the early stages of gonad development (Lie et al. 1993).

5.5 Importance of adequate feed intake on spawning performance

Food availability seems to be a main factor determining fecundity in wild fish. Therefore, geographic areas with low food availability, such as certain river ecosystems, in comparison with areas with high food productivity, denote differences in fecundity in species such as the rutile *Rutilus rutilus* (McKay and Mann 1969; Kuznetzov and Khalitov 1978); the bream *Abramis brama* (Brylinska and Brylinski 1972); *Poeciliopsis occidentalis* (Constanz 1975); *Gasterosteus acuelatus* (Ali and Wootton 2000); river trout *Salmo trutta* (Fry 1949; McFadden et al. 1965); lake trout *Salvelinus namaycush* (Martin 1970); or coho salmon, *Oncorhynchus kisuth* (Stauffer 1976). This type of study has also been conducted with marine species such as plaice *Pleuronectes platessa* (Bagenal 1966); haddock *Melanogrammus aeglefinus* (Hodder 1963); Norway pout, *Gadus esmarkii* (Raitt 1968); and Baltic cod, *Gadus morhua callaris* (Kraus et al. 2000; Kraus, 2002), where a relation between fecundity of the different populations and food intake was found.

In cultured fish, fecundity and other spawning quality parameters are also affected by ration size. An insufficient feeding rate inhibits gonadal maturation in several species, such as carp *Carassius auratus* (Sasayama and Takahashi 1972), sea bass *Dicentrarchus labrax* (Cerdá et al. 1994a), and Atlantic salmon (Berglund 1995). After 6 months of feeding sea bass broodstock with half the food ration, growth rates decreased, the spawning period was delayed, and the eggs and newly hatched larvae were smaller than those obtained from fish eating the full ration

(Cerdá et al. 1994a). These negative effects of food restriction have been associated with a decrease in plasma estradiol levels (Cerdá et al. 1994b). However, gene expression of gonadotropins was not affected by decreased food intake in mature carp females (Sohn et al. 1998). Food restriction resulted in a decrease in total fecundity in the following species: rainbow trout, *Oncorhynchus mykiss* (Springate et al. 1985; Jones and Bromage 1987); haddock *Melanogrammus aeglefinus* (Hislop et al. 1978; Robb 1982); herring *Clupea harengus* (Ma et al. 1998); the viviparous guppy *Poecilia reticulata* (Hester 1964; Dahlgren 1980); tilapia *Tilapia mossambica* and *Tilapia zillii* (Mironova 1977; Coward and Bromage 1999); rutile *Rutilus rutilus* (Kuznetzov and Khalitov 1978); medaka *Oryzias latipes* (Hirshfield 1980); the cycle zebra *Cichlasoma nigrofasciatum* (Townshend and Wootton 1984); flounder *Pleuronectes platessa* (Horwood et al. 1989); and milkfish, *Chanos chanos* (Emata et al. 1996). In cod *Gadus morhua*, there was also a positive correlation between feed intake and fecundity (Karlsen et al. 1995; Kjesbu et al. 1998). On the contrary, Bromley et al. (2000) did not find differences in fecundity of turbot broodstock fed different food rations. Yoneda and Wright (2005) maintained cod at different temperatures and feeding rations and observed that fecundity was positively correlated with body size, but neither the feeding rations nor the temperature affected this relation. In hybrid tilapia (*Oreochromis niloticus* × *Oreochromis aureus*), differences were also found in the broodstock fecundity when they were fed with different feeding rations, but this factor did not affect the hatching percentage or the egg diameter (Siddiqui et al. 1998). In espinosillo, *Gasterosteus aculeatus*, high feed intake not only increased fecundity but also the egg size and female body size (Fletcher and Wootton 1995). Furthermore, the length of time between successive spawns was lower in fish with larger food intakes (Ali and Wootton 1999).

5.6 Nutrient requirements and their effects on reproductive physiology and egg quality

5.6.1 Protein and amino acid requirements

Protein constitutes the most abundant nutrient in fish eggs (Watanabe and Kiron 1994) and is a main energy source during embryonic development of most teleost species (Fyhn and Serigstad 1987; Rønnestad et al. 1992; Sivaloganathan et al. 1998). Furthermore, growth of the embryo mainly involves deposition of protein. Proteins have an important role in fertilization and embryonic development (Fyhn and Serigstad 1987; Srivastava and Brown 1992; Srivastava et al. 1995); for example, proteins surrounding the yolk are significant for fertilization (Hart 1990), their amino acid composition being high in proline and glutamic acid and low in cystine (Hyllner et al. 1995). Vtgs comprise the main protein precursor of yolk in teleosts; their amino acid composition being characterized by high contents of alanine, glutamic acid, and leucine, and low levels of serine. Free amino acids appear in high levels in the eggs of pelagic fish; for example, more than 43 nmoles per egg in gilthead sea bream. Abundant free amino acids are leucine, lysine, valine, isoleucine, alanine, and serine (Rønnestad 1992).

A sufficient amount of dietary protein with a balanced composition of amino acids is vital for the supplementation of protein and amino acids for the embryo, through the yolk. It is therefore not surprising that the protein level and composition of broodstock diets has an influence on the spawning quality. Thus, a well-balanced diet in essential amino acids for gilthead sea bream (*Sparus aurata*) broodstock improved the synthesis of Vtgs (Tandler et al. 1995). In studies with Japanese sea bream (*Pagrus major*) broodstock, the optimal dietary protein level in diets with fish meal as the main protein source was around 45%

(Watanabe et al. 1984a, 1984b, 1984d, 1984e); broodstock that were fed below this level produced approximately 30% fewer eggs. In sea bass (*Dicentrarchus labrax*) fed 510 g/kg protein, fecundity was 1.5 times higher than in fish fed 340 g/kg protein (Cerdá et al. 1994b). The high protein diets also gave a reduced rate of deformed larvae. The protein was exchanged with carbohydrate in this experiment. During the spawning of sea bass broodstock, diets low in protein altered secretion of GnRHs (Kah et al. 1994) and LH (Navas et al. 1996), which play important roles in regulation of oocyte maturation and ovulation regulation as discussed above. Labeo roho (*Labeo rohita*) broodstock fecundity increased with an increase in dietary protein levels from 20 to 25 and 30%, while an increase over 35 or 40% reduced fecundity (Khan et al. 2005). Optimum dietary protein levels in catfish *Mystus nemurus* broodstock diets for maximum fecundity were around 35% (Abidin et al. 2006). In the case of turbot (*Scophthalmus maximus*), it is necessary to increase the protein and fat content above 45 and 10%, respectively, with 2% HUFA to obtain the highest fecundity (Aijun et al. 2005). In summary, requirements of dietary protein for broodstock fish range between 30 and 45% (Table 5.1).

Table 5.1 Requirements of dietary protein for broodstock fish.

Species	Requirements (% dry weight)	References
Pagrus major	45	Watanabe et al. (1984a, 1984b, 1984d, 1984e)
Dicentrarchus labrax	>34	Cerdá et al. (1994b)
Labeo rohita	30	Khan et al. (2005)
Mystus nemurus	35	Abidin et al. (2006)
Scophthalmus maximus	45	Aijun et al. (2005)

Tryptophan and taurine have been found to be particularly important for fish reproduction. Tryptophan is a precursor of serotonin and can affect gonadal maturation in both males and females. Broodstock diets for ayu (*Plecoglossus altivelis*) supplemented with 0.1% of tryptophan gave a significant increase in testosterone levels favoring spermiation in males and induced female maturation (Akiyama et al. 1996). Taurine is one of the most abundant free amino acids in tissues of fish and is involved in antioxidant processes, osmoregulation, neurotransmitter modulation, calcium regulation in cells, hormone release, and formation of bile salts (Huxtable 1992). Supplementation of at least 1% of taurine in diets for yellowtail (*Seriola quinqueradiata*) broodstock improved fecundity, percentage of viable eggs, and fertilization rate (Matsunari et al. 2006).

5.6.2 Lipids and fatty acids

Lipid is the most studied dietary component in broodstock nutrition. The content of total dietary lipids and essential fatty acids greatly influences the spawning quality in fish (Watanabe et al. 1984a; Mourente et al. 1989; Dhert et al. 1991; Bruce et al. 1993; Fernández-Palacios et al. 1995; Navas et al. 1997; Rodriguez et al. 1998; Lavens et al. 1999; Furuita et al. 2002, 2003b; Mazorra et al. 2003; Aijun et al. 2005). Dietary contents of total lipids, energy, essential fatty acids, and the ratios among them may drastically affect fish reproductive performance, but since altering the dietary lipid source can affect all these factors, the importance of each of them may be difficult to determine.

5.6.2.1 Total dietary lipid contents

Few studies have shown an effect of total dietary lipid on broodstock performance, but increased dietary energy seems to be responsible for an increase in the gonadosomatic index, as has been seen in cod (*Gadus morhua*) (Rosenlund, personal communication). This may be a result of increased fecundity or increased egg size. Furthermore, increasing fat in broodstock diets increases fecundity and survival of 14 days posthatch larvae of rabbit fish (*Signatus guttatus*) (Hara et al. 1996), whereas in gilthead sea bream, high fat leads to an increase in larval weight and length even at 28 days posthatch (Bueno 2001). Since the lipid source in both studies contained essential fatty acids, it is difficult to determine if the improvement was related to an increase in fat, energy, or essential fatty acid content.

5.6.2.2 Essential fatty acids

The fatty acid composition of fish eggs is directly affected by the fatty acid content of the broodstock diets. Fatty acids of the n-3 series and the n-3 HUFA found in gilthead sea bream eggs and cod (*Gadus morhua*) gonads increased when n-3 HUFA increased in the diet (Fernández-Palacios et al. 1995; Lie et al. 1993). However, the fatty acid composition of the lipids in fish eggs may vary with the species and even with different batches of the same species (Pickova et al. 1997), or with the environmental conditions in which gametogenesis occurs (Dantagnan et al. 2007). In rainbow trout (*Oncorhynchus mykiss*) fed a diet deficient in fatty acids of the series n-3 during the last 3 months of vitellogenesis, there was only a slight reduction of docosahexaenoic acid (DHA, 22:6n-3) content in the egg, while eicosapentaenoic acid (EPA, 20:5n-3) concentration declined by 50% (Frémont et al. 1984). This selective retention of DHA has also been found during embryogenesis (Izquierdo 1996) and during starvation in the larval stages (Tandler et al. 1989), showing the importance of this fatty acid for embryonic and larval development (Watanabe et al. 1989). Many fish species also retain tissue DHA to a larger extent than EPA (Watanabe et al. 1989; Watanabe 1993).

The fatty acids are important sources of energy during early embryonic development (Tocher et al., 1985a, 1985b; Falk-Petersen et al. 1986, 1989; Rainuzzo 1993; Sargent 1995). They have an important structural function as components of the phospholipids in fish biomembranes. The ratio between saturated and unsaturated fatty acids regulates the fluidity of cell and organelle membranes, and thereby their function (Bell et al. 1986, 1997; Sargent 1995; Takeuchi 1997; Sargent et al. 1999). DHA is especially important in the neural tissue, the retina, and the optic nerve and is related to other structures of the sensory organs (Sargent et al. 1993; Benítez et al. 2007) that develop during early larval stages.

The 20-carbon fatty acids ARA (20:4n-6) and EPA are eicosanoid precursors, producing PGs of the II and III series, respectively, and leukotrienes LTB4 and LTB5 (Stacey and Goetz 1982; Bell et al. 1994; Knight et al. 1995; Ganga et al. 2005). The key enzymes for production of PGs and leukotrienes from fatty acids are the cyclooxygenases and lipoxygenases, respectively. The PGs have been shown to be involved in numerous reproductive processes, including the production of steroid hormones, gonadal development, and ovulation (Moore 1985). The inhibitors of lipoxygenases reduce the maturation of sea bass (*Dicentrarchus labrax*) oocytes induced by gonadotropin (Asturiano 1999), suggesting that the leukotrienes may also be involved in reproduction in fish.

ARA, EPA, and DHA compete with each other for the enzymes that regulate the synthesis of eicosanoids. The relationship between them in the diet (the EPA/DHA and the ARA/EPA ratios) will be decisive, for example, for the presence of PGs of the II and III series in the tissues. These PGs have very different activities in the regulation of various physiological functions, including reproduction. Therefore, a deficiency or imbalance of

ARA, EPA, and DHA in the broodstock diet can have large effects on reproduction. These fatty acids also compete in other metabolic pathways, such as glycerophosphate synthesis, but the consequences of this for spawning performance have not been determined.

ARA supplementation has a profound effect on spawning performance in several fish species. By supplementation with ARA, the ratio of ARA to the other fatty acids and the ARA/EPA ratio will increase. It is therefore not possible to decide if it is the amount of ARA in itself or the change in ratios that causes the effects. There is an optimal level of ARA in broodstock diets, where addition below or above this level gives reduced performance (Furuita et al. 2003b). Optimal ARA may increase fecundity, egg viability, hatching rate, and larval survival (Bruce et al. 1999; Fernández-Palacios et al. 1995, 2005; Navas et al. 2001). The optimal ratio of ARA to EPA is species-dependent in fish larvae; Sargent et al. (1999) observed a ratio of 1:1 for sea bass, and 10:1 or higher for halibut and turbot. It is not known if a similar dependency on species is the case for the optimal ARA/EPA ratio in broodstock diets.

The fatty acid composition of the male gonad and the sperm depends on the composition of essential fatty acids in the broodstock diets (Leray and Pelletier 1985). The dietary fatty acid composition therefore affects sperm production and quality. *In vitro*, ARA, but not EPA or DHA, stimulates the release of testosterone in carp (*Carassius auratus*) testicles through its conversion into prostaglandin PGE2, while EPA and DHA block the steroidogenic action of ARA and the PGE2 (Wade et al. 1994). ARA in broodstock diets may therefore increase the fertilization rate. Moreover, PGs are also well known to act as pheromones in some teleosts (Mustafa and Srivastava 1989; Sorensen and Goetz 1993; Rosenblum et al. 1995), stimulating male sexual behavior and synchronizing spawning of the females and males, directly affecting the success of fertilization (Sorensen et al. 1988). Accordingly, ARA content in broodstock diets is directly related to the fertilization rate in halibut (*Hippoglossus hippoglossus*) and gilthead sea bream, *Sparus aurata* (Fernández-Palacios et al. 1995, 1997, 2005; Mazorra et al. 2003).

As for ARA, there seems to be an optimal level of n-3 fatty acids in broodstock diets. Increasing these fatty acids may lead to improvement of fecundity and egg and larval quality in terms of survival, stress resistance, larval weight, and inflation of the swim bladder (Watanabe et al. 1984a, 1984b, 1984c, 1985a, 1985b; Tandler et al. 1995; Fernández-Palacios et al. 1995, 2005; Abyayad et al. 1997). Diets deficient in essential fatty acids also cause an increase in the number of fat droplets in fish eggs (Watanabe et al. 1984a; Fernández-Palacios et al. 1997, 2005). Makino et al. (1999) observed that in Japanese snook (*Lateolabrax japonicus*), fusion of fat droplets into a single lipid droplet led to increased percentages of hatching and normal larvae. On the other hand, very high levels of n-3 fatty acids caused a decrease in the total number of eggs produced and egg quality in gilthead sea bream, swordfish, (*Xiphophorus helleri*), and Pacific sole (*Paralichthys olivaceus*) (Fernández-Palacios et al. 1995, 2005; Furuita et al. 2002; Ling et al. 2006). Diets with excess or deficiency in n-3 HUFA for crescent sweetlips (*Plectorhynchus cinctus*) also had negative effects on egg and larval quality (Li et al. 2005); n-3 HUFA also affects the fertilization rate (Fernández-Palacios et al. 1995; Izquierdo et al. 2001). In gilthead sea bream, the fertilization rate was lowered when the diet was deficient in n-3 HUFA (Izquierdo et al. 2001). This may have been due to a decrease in sperm motility (Vassallo-Agius et al. 2001c).

The requirement of n-3 fatty acids in sea bream broodstock varies between 1.5 and 2.5% in the diet (Watanabe et al. 1984a, 1984b, 1984c, 1985b; Fernández-Palacios et al. 1995, 2005), being higher than that established for juveniles, ranging between 0.5

and 0.8% (Izquierdo 1996). The value is also higher than that established for salmonids of approximately 1% n-3 HUFA in the diet (Watanabe 1990).

The positive effect of n-3 fatty acids on spawning performance may be explained by their roles as structural components in biological membranes, where they increase fluidity. Overly high levels of these fatty acids may, on the other hand, give membranes with too low melting points. The levels of n-3 fatty acids will also affect the ratio of ARA to EPA and can influence eicosanoid production and fat metabolism in general, calling for a balanced supplementation of both n-3 fatty acids and ARA. Finally, the n-3 fatty acids are susceptible to lipid oxidation, which may be a great challenge to the health of the organism. Increasing the level of polyunsaturated fatty acids in broodstock diets should therefore be accompanied by an increase in dietary antioxidants (Fernández-Palacios et al. 2005).

5.6.3 Carbohydrates

Although the carbohydrates are not essential nutrients, they have important biological functions in fish, and constitute a basic source of energy in some tissues. Therefore, carbohydrate inclusion in broodstock diets at levels that can be utilized by fish has been assessed by a few authors. Washburn et al. (1990) found a decrease in rainbow trout (*Oncorhynchus mykiss*) fecundity when the broodstock were fed diets with low levels of carbohydrates. However, feeding cod (*Gadus morhua*) broodstock with increasing levels of carbohydrates slightly reduced spawning quality (Mangor-Jensen and Birkeland 1993).

5.6.4 Vitamins

There are 15 vitamins recognized as essential for fish, the same as for terrestrial vertebrates (Woodward 1994), that will be necessary in broodstock diets. However, deficiencies are most often encountered in those vitamins that are unstable or present at low concentrations in the feed ingredients. Therefore, vitamins C, E, and thiamine are the vitamins most often studied in fish broodstock diets. Vitamins C and E are antioxidants and are degraded as they fulfill their function. Broodstock fed trash fish are most often low in these vitamins due to degradation during storage of the feed (Figure 5.2). Thiamine is also often low in trash fish due to activity of the enzyme thiaminase, which is present in most fish species and which destroys thiamine. The low availability of thiamine from the diet will then eventually lead to thiamine deficiency in the broodstock. In modern commercial diets, vitamins C and E are, however, added as stable compounds and thiaminase is inactivated by heat treatment of the feed ingredients.

The requirements of vitamin A (retinol), vitamin D (cholecalciferol), vitamin E (tocopherol), vitamin K (menadione), and vitamin C (ascorbic acid) have been determined for some teleost species (National Research Council, 1993; see Chapter 4), and there are a few studies of their effects on reproduction and spawning quality (Table 5.2). However, direct requirement studies with broodstock are relatively rare due to the high costs involved in running the many groups necessary for dose–response experiments.

5.6.4.1 Vitamin E

Vitamin E acts as a powerful natural antioxidant, preventing peroxidation of lipids in animal cells (Huber 1988). Although the negative effects of vitamin E deficiency in the reproduction of other vertebrates has been known since the beginning of last century, the importance of vitamin E in fish reproduction was not verified until 1990. According to Watanabe (1990), deficiencies of vitamin E in common carp (*Cyprinus carpio*) and ayu

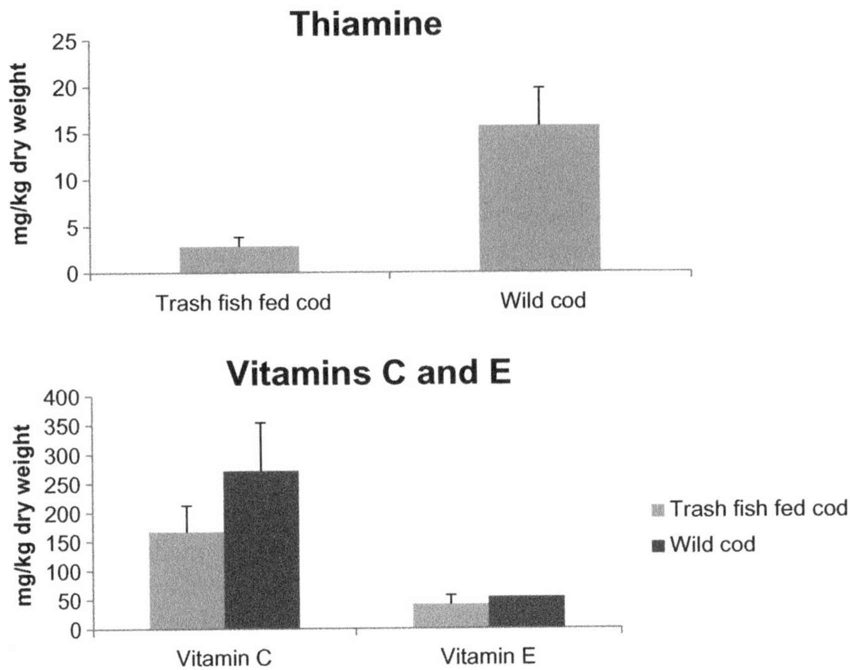

Figure 5.2 Vitamin levels in the ovaries of cod broodstock fed whole capelin, compared with wild cod and cod fed a formulated broodstock diet.

Table 5.2 Requirements of vitamins A, D, E, K, and C in broodstock of five teleost species.

	Species				
Vitamins	**Catfish**	**Rainbow trout**	**Pacific salmon**	**Carp**	**Tilapia**
A (IU/kg)	1,000–2,000	2,500	2,500	4,000	NT
D (IU/kg)	500	2,400	NT	NT	NT
E (IU/kg)	50	50	50	100	50
K (mg/kg)	N	NT	N	NT	NT
C (mg/kg)	25–50	50	50	N	50

N = necessary in the diets but quantity not determined, NT = not tested.

(*Plecoglossus altivelis*) inhibited gonadal maturation and decreased the hatching and survival rates of larvae. Also, in Japanese sea bream (*Pagrus major*), increasing the levels of dietary vitamin E (above 2,000 mg/kg) improved the percentage of buoyant eggs with normal development, the ratio of hatched eggs, the ratio of larvae with normal development, and larval survival (Watanabe et al. 1991a). This is in agreement with the results obtained in diabetic rats, where dietary vitamin E increased the α-tocopherol concentrations in the mothers' tissue and in the tissues of the embryo and fetus, and reduced congenital malformations (Siman and Erikkson 1997). The vitamin E content is generally high in fish eggs and low in broodstock tissues after the spawning period (Mukhopadhyay et al. 2003). This may be a result of mobilization of vitamin E from peripheral tissues to the ovary during vitellogenesis as has been shown in turbot and Atlantic salmon (Hemre et al. 1994; Lie et al. 1994).

Vitamin E improves spawning quality in a wide range of species, for example, common carp, *Cyprinus carpio* (Watanabe and Takashima 1977; Watanabe 1990); ayu, *Plecoglossus altivelis* (Takeuchi et al. 1981a); carp *Carassius auratus* (Sutjaritvongsanon 1987); Japanese sea bream, *Pagrus major* (Watanabe et al. 1985b, 1991a, 1991b); catfish *Heteropneustes fossilis* (Dube 1994); tilapia *Oreochromis niloticus* (Schimittou 1993); yellowtail *Seriola quinqueradiata* (Mushiake et al. 1993); the pearlspot, *Etroplus suratensis* (Shiranee and Natarajan 1996); the gilthead sea bream, *Sparus aurata* (Izquierdo et al. 2001); and the grouper *Epinephelus coioides* (Xiao et al. 2003), as well as other marine species (Verakunpiriya et al. 1996). Likewise, its exclusion, together with vitamin C, in diets for the Australian trumpeter, *Latris lineata*, significantly reduced spawning quality (Morehead et al. 2001). In gilthead sea bream, diets deficient in vitamin E decreased the percentage of fertilized egg (Fernández-Palacios et al. 2005). This may have been related to the decrease in the number and motility of the spermatozoids, as has been described for other vertebrates (Donnelly et al. 1999; Danikowski et al. 2002) and in fish such as ayu, *Plecoglossus altivelis* (Hsiao and Mak 1978).

Lee and Dabrowski (2004) found that the level of sperm plasma tocopherol decreased significantly and sperm viability was seriously compromised in American perch (*Perca flavescens*) broodstock fed with diets deficient in vitamin E. Insufficient vitamin E also decreased the percentage of viable eggs with normal morphology in several species: for example, rainbow trout, *Oncorhynchus mykiss* (King 1985), Japanese sea bream (Watanabe et al. 1991a, b), gilthead sea bream (Fernández-Palacios et al. 2005), and milkfish, *Chanos chanos* (Emata et al. 2000). Furthermore, larval survival significantly improved with vitamin E inclusion in broodstock diets of ayu (*Plecoglossus altivelis*; Takeuchi et al. 1981a), rainbow trout (King 1985), and seriola (Mushiake et al. 1993). Vitamin E deficiencies also caused loss of sexual coloration in tilapia (*Oreochromis niloticus*) and decreased the reproductive activity (Schimittou 1993).

The vitamin E requirement is dependent on the dietary content of polyunsaturated fatty acids, considered to be essential for teleosts (Watanabe et al. 1991a). An increase in n-3 HUFA levels at a fixed level of vitamin E improved the spawning quality in gilthead sea bream, but it also caused a higher percentage of deformed larvae with hypertrophy of the yolk sac (Fernández-Palacios et al. 1995, figure 10A). However, elevation of both n-3 HUFA and vitamin E prevented deformities in the larvae (Fernández-Palacios et al. 2005). Similar interactions between DHA and vitamin E have been found in cod (Takeuchi et al. 1994).

The optimal level of dietary vitamin E for broodstock has been investigated for several fish species. Guppy (*Poecilia reticulata*), swordfish (*Xiphophorus helleri*), and catfish have an optimum dietary vitamin E level of 150 mg/kg (Dube 1994; Koprücü and Seker 2003), while in gilthead sea bream (*Sparus aurata*), vitamin E up to 190 mg/kg α-tocopherol greatly improved spawning quality (Fernández-Palacios et al. 1998, 2005). These levels still appear to be suboptimal for turbot (*Scophthalmus maximus*) broodstock (Hemre et al. 1994).

5.6.4.2 Vitamin C

Ascorbic acid also plays an important role in fish reproduction. In salmonids, vitamin C has been shown to affect steroidogenesis and vitellogenesis (Eskelinen 1989; Sandnes 1991; Blom and Dabrowski 1995). The ascorbic acid levels in broodstock diets affected the concentration in the seminal fluid, and seminal vitamin C concentration was directly related to sperm motility at the end of the spawning season (Ciereszco and Dabrowski 1995). Also, the level of vitamin C in rainbow trout eggs reflected the dietary concentration and was associated with the improvement of egg quality (Sandnes et al. 1984). An increase in dietary vitamin C up to 1,200 mg/kg for rainbow trout broodstock increased egg hatching rates (Ridelman 1981). The hatching percentage and rate of normal larvae in Nile tilapia (*Oreochromis niloticus*) was also increased by higher dietary vitamin C (Soliman et al. 1986).

Studies on specific Vitamin C requirements are scarce and show large differences for the different species studied. In rainbow trout (*Oncorhynchus mykiss*), the vitamin C requirement of broodstock appears to be eight times higher than that of juveniles (Blom and Dabrowski 1995). In other species, such as cod, lower requirements have been suggested (Mangor-Jensen et al. 1993). Fecundity and egg quality are affected not only by the dietary vitamin C content (Blom and Dabrowski 1995) or vitamin E (Izquierdo and Fernández-Palacios 1997; Fernández-Palacios et al. 1998, 2005) but also by the interaction between them (Silveira et al. 1996; Emata et al. 2000).

5.6.4.3 Vitamin A

Vitamin A is necessary for growth, reproduction, and embryonic development of fish and must be obtained from the diet (Craik 1985; Madden 2001). Vitamin A is present in fish as vitamin A_1 or retinol, vitamin A_2 or didehydroretinol, and retinal (Palace and Werner 2006). Recently, it was shown that vitamin A is transported to the ovary as retinal bound as a Schiff base to Vtg in zebrafish (Lubzens et al. 2010). This makes regulation of the vitamin A concentration in the egg possible, in contrast to the situation for vitamin E and carotenoids, which to a greater extent are transported in a dose-dependent manner by other lipoproteins (Palace and Werner 2006; Lubzens et al. 2010).

The vitamin A requirement during gonadal maturation and spawning in fish is not well known. However, Furuita et al. (2003a) found that feeding broodstock of the Pacific sole (*Paralichthys olivaceus*) with vitamin A increased fecundity, viable egg percentage, and normal larvae percentage. Furthermore, vitamin A, along with vitamin C, increased the number of 3-day-old larvae produced from largehead carp (*Aristichtys nobilis*) broodstock (Santiago and Gonzalo 2000).

5.6.4.4 Thiamine

Another dietary component that may sometimes be deficient for normal development of embryo and larvae is thiamine (vitamin B_1). For example, thiamine injections in pregnant females of Atlantic salmon (*Salmo salar*) may reduce their offspring mortality (Ketola et al. 1998). Furthermore, high thiamine concentration in eggs and yolk sac fry of lake trout *Salvelinus namaycush*; rainbow trout, *Oncorhynchus mykiss*; coho salmon, *Oncorhynchus kisutch*; and Atlantic salmon, *Salmo salar* is related to a reduction in early mortality syndrome (Brown et al. 1998; Hornung et al. 1998; Wooster and Bowser 2000).

5.6.5 Minerals

There are very few published papers on the effect of minerals on the spawning quality of fish. Takeuchi et al. (1981b) suggest that rainbow trout (*Oncorhynchus mykiss*) broodstock fed diets not supplemented with minerals have poor-quality spawns in terms of low egg viability and hatching rates. Phosphorus deficiency in Japanese sea bream (*Pagrus *major*) broodstock was related to a decrease in fecundity, percentage of viable eggs, and hatching rate, and an increase in the number of abnormal larvae (Watanabe et al. 1984a, 1984b). A lower fecundity in ayu (*Plecoglossus altivelis*) was also correlated to a deficiency of phosphorus in the diet (Luquet and Watanabe 1986). However, in studies by Hardy et al. (1984) and Ketola (1985) with coho salmon (*Oncorhynchus kisutch*) and Atlantic salmon (*Salmo salar*), there were no changes in spawning quality in response to mineral supplementation.

5.6.6 Carotenoids

The carotenoids are orange to yellow pigments with a wide variety of functions in fish that include protection against adverse light conditions, provitamin A, and antioxidant functions. Astaxanthin is the most abundant carotenoid in the marine environment. There has been great controversy about the relationship between the carotene content of the egg and egg quality in salmonids (Craik 1985; Choubert 1986; Torrissen, 1990; Torrissen and Christiansen 1995), and studies published on this issue have often been contradictory. Some authors observe a positive relation between egg pigmentation and egg quality in rainbow trout, *Oncorhynchus mykiss* (Harris 1984; Craik 1985), while others have not found evidence of this association (Torrissen 1984; Craik and Harvey 1986; Torrissen and Christiansen 1995). These contradictions appear to be due to differences in the methodology used by the different authors, including broodstock age, total carotenoid content in the eggs, carotenoid type (astaxanthin, canthaxanthin, etc.) used in the diet or determined in the eggs, sample size, and differences in the indicator used in determining egg quality.

However, positive effects of carotenoids in reproduction of salmonids have been found in several studies (Harris 1984; Choubert and Blanc 1993; Watanabe and Kiron 1995; Ahmadi et al. 2006). In Japanese sea bream (*Pagrus major*), the inclusion of purified astaxanthin in broodstock diets improved the percentage of floating and hatched eggs, and the percentage of normal larvae (Watanabe and Kiron 1995). Dietary astaxanthin increased fecundity but did not improve the egg quality in mackerel (*Pseudocaranx dentex*) (Vassallo-Agius et al. 2001a). Similar results were obtained in yellowtail (*Seriola quinqueradiata*) by Verakunpiriya et al. (1997). Interactions of carotenoids with other nutrients have been studied in gilthead sea bream (*Sparus aurata*), where the combined increase in levels of n-3 HUFA and carotenoids significantly improved spawning quality in terms of fecundity and percentage of viable eggs, hatching rate, and larval survival (Scabini et al. 2006).

Paprika as a source of carotenoids was supplemented in broodstock diets for striped jack and improved fecundity, fertilization, hatching, and larval survival (Vassallo-Agius et al. 2001b). Supplementation of β-carotene in broodstock diets does not appear to exert a significant effect on reproduction of red sea bream (Watanabe, personal communication). This may be due to the poor intestinal absorption of β-carotene compared with canthaxanthin or astaxanthin (Torrissen and Christiansen 1995). Miki et al. (1984) found that canthaxanthin and astaxanthin in broodstock diets were incorporated in the eggs of red sea bream but were not converted into β-carotene

5.6.7 Nucleotides

Other nutrients that can affect spawning quality are the nucleotides. Diets enriched with nucleotides produced an improvement in larvae survival of haddock (*Melanogrammus aeglefinus*) determined 10 days after hatching. This may have been caused by better development of the intestine of the nucleotide-supplemented larvae, and as a consequence, better utilization of the first exogenous feed

(González-Vecino et al. 2004). Furthermore, both halibut (*Hippoglossus hippoglossus*) and haddock (*Melanogrammus aeglefinus*) broodstock fed diets enriched with nucleotides had a higher fecundity than control fish (Gonzalez-Vecino 2005).

5.7 Specific feed ingredients

5.7.1 Squid

Several studies have shown the beneficial effects of feeding sea bream broodstock with cuttlefish, squid, or meals made from these cephalopods (Watanabe et al. 1984a, 1984b; Mourente et al. 1989; Zohar et al. 1995), suggesting that they contain nutritional components essential for reproduction. Japanese sea bream (*Pagrus major*) broodstock fed a diet based on cuttlefish meal instead of fish meal increased their total egg production and viability of eggs (Watanabe et al. 1984a, 1984b). Even partial replacement of fish meal by cuttlefish meal (50%; Watanabe et al. 1984a, 1984b) improved egg viability. Replacement of 50% of the fish meal by squid meal in diets for striped jack (*Pseudocaranx dentex*) did not increase fecundity but improved the percentage of fertilized and hatched eggs (Vassallo-Agius et al. 2001b). Furthermore, high rates of hatching and larval survival from *Lutjanus argentimaculatus* spawns were related to feeding with squid meal and oil, substituting fish meal and fish oil (Emata and Borlongan 2003).

The beneficial effects of squid and cuttlefish may have several causes. It may simply be that these feed ingredients make the diet more attractive and therefore increase feed intake. Squid and cuttlefish also contain high levels of essential fatty acids that may have given the beneficial effect (Mourente et al. 1989). However, the high dietary value of squid and cuttlefish for gilthead sea bream was shown to be present in the water-soluble fraction of the meal (Fernández-Palacios et al. 1997).

5.7.2 Soybean

The dietary protein or fat substitution of squid meal by protein or fat derived from soybeans produced a significant decrease in the hatching and survival rates of 3-day-old gilthead sea bream larvae (Zohar et al. 1995). This decrease in spawning quality could be due not only to the reduced supplementation of squid meal but also to the negative effects of soybean meal. Soybean meal is extensively used in commercial feeds for ongrowing, but the presence of antinutritional factors can limit its inclusion at high levels. Moreover, an imbalance of the polyunsaturated fatty acid n-3/n-6 ratio, along with a decrease in phosphorus availability in diets based on soybean meal, could directly reduce spawning quality (Watanabe et al. 1984a; Watanabe and Kiron 1995). Soybean meal and fish meal also differ in other essential components such as vitamin B_6, vitamin B_{12}, and minerals (Hansen et al. 2007). The high concentrations of phytoestrogens in soybean meal, such as genestein, may also be of concern since these compounds have been shown to affect sex determination in rainbow trout embryos and inhibit the cytochrome P450 CYP1A detoxification system in rats. In zebrafish larvae, phytoestrogens have a number of teratogenic effects that may correspond to CYP1A inhibition (Ronis et al. 1994; Green and Kelly 2009; Kim et al. 2009). In broodstock diets, which are needed in minor quantities compared with ongrowing diets, and where the nutritional quality is very important, one can argue for the use of high-quality marine ingredients, even though the costs may be higher than for diets based on plant ingredients.

5.7.3 Krill

Krill inclusion in broodstock diets greatly improves the spawning quality of Japanese sea bream (*Pagrus major*) (Watanabe et al. 1991a). The nutritional factors responsible for this improvement appear to be present in the lipid phase, in both the polar and nonpolar fractions, as phosphatidylcholine and astaxanthin, respectively (Watanabe et al. 1991a). Other studies confirm that dietary phospholipids improve the egg quality in Japanese sea bream (*Pagrus major*) (Watanabe et al. 1991b). The beneficial effects of phospholipids have been attributed to their ability to stabilize free radicals (Watanabe and Kiron 1995). The positive effect of krill may also be explained by increased feed intake since this feed ingredient contains components that function as attractants for fish.

5.8 Summary and conclusions

The studies summarized in this chapter show that broodstock nutrition has clear effects on fecundity and egg and larval quality in fish despite the small number of studies on nutrients and species. Development of balanced diets for broodstock should be a priority in fish culture since egg and larval quality are the basis for the successful culture of later stages of fish. Many of the experiments referred to in this chapter show that nutrient deficiencies give poor spawning performance, but optimal dietary levels remain undetermined. Furthermore, the requirements for many of the nutrients have not been examined yet in broodstock of all commercially important species.

Fish nutritionists most often measure end points, such as fecundity and egg and larval quality, but the effects of nutrition on the biological processes that produce gametes and the transport and deposition of nutrients in these gametes deserve more attention. Therefore, further studies on broodstock nutrition should aim at an integration of nutrition and reproductive biology in order to obtain a better understanding of the biological mechanisms involved in the development of a healthy egg.

Literature cited

Abidin, M.Z., Hashim, R., and Chong, A.S.C. 2006. Influence of dietary protein levels on growth and egg quality in broodstock female bagrid catfish (*Mystus nemurus* Cuv. & Val.). *Aquaculture Research* 37:416–418.

Aby-ayad, S.M.E.A., Melard, C., and Kestemont, P. 1997. Effects of fatty acids in Eurasian perch broodstock diet on egg fatty acid composition and larvae stress resistance. *Aquaculture International* 5:161–168.

Ahmadi, M.R., Bazyar, A.A., Safi, S., Ytrestøyl, T., and Bjerkeng, B. 2006. Effects of dietary astaxanthin supplementation on reproductive characteristics of rainbow trout (*Oncorhynchus mykiss*). *Journal of Applied Ichthyology* 22: 388–394.

Aijun, M., Chao, C., Jilin, L., Siqing, C., and Zhimin, Z. 2005. The effect of protein and n-3 HUFA on the reproduction of turbot (*Scophthalmus maximus*). *Marine Fisheries Research (Haiyang Shuichan Yanjiu)* 26:7–12.

Akiyama, T., Shiraaishi, M., Yamamoto, T., and Unuma, T. 1996. Effect of dietary tryptophan on maturation of ayu *Plecoglossus altivelis*. *Fisheries Science* 62:776–782.

Aksnes, A., Gjerde, B., and Roald, S.O. 1986. Biological, chemical and organoleptic changes during maturation of farmed Atlantic salmon, *Salmo salar*. *Aquaculture* 53:7–20.

Ali, M., and Wootton, R.J. 1999. Effect of variable food levels on reproductive performance of breeding female three-spined sticklebacks. *Journal of Fish Biology* 55:1040–1053.

Ali, M., and Wootton, R.J. 2000. Variation in rates of food consumption and evidence for compensatory responses in the three-spined stickleback, *Gasterosteus aculeatus* L. in relation to growth and reproduction. *Ecology of Freshwater Fish* 9:103–108.

Almansa, E., Perez, M.J., Cejas, J.R., Badia, P., Villamandos, J.E., and Lorenzo, A. 1999. Influence of broodstock gilthead seabream (*Sparus aurata* L.) dietary fatty acids on egg quality and egg fatty acid composition throughout the spawning season. *Aquaculture* 170: 323–336.

Asturiano, J.F. 1999. El proceso reproductivo de la lubina europea (*Dicentrarchus labrax* L.). Efectos de los ácidos grasas de la dieta: estudios "in vivo" e "in vitro". PhD thesis, Universidad de Valencia, Spain.

Babin, P.J., Carnevali, O., Lubzens, E., and Schneider, W.J. 2007. Molecular aspects of oocyte vitellogenesis in fish. In: Babin, P.J., Cerdá, J., and Lubzens, E. (eds.) *The Fish Oocyte: From Basic Studies to Biotechnological Applications*. Springer, Dordrecht, pp. 39–76.

Bagenal, T.M. 1966. The ecological and geographical aspects of the fecundity of the plaice. *Journal of the Marine Biological Association of the United Kingdom* 46:161–186.

Bell, M.V., Henderson, R.J., and Sargent, J.R. 1986. The role of polyunsaturated fatty acids in fish. *Comparative Biochemistry and Physiology* 83 B:711–719.

Bell, J.G., Tocher, D.R., Macdonald, F.M., and Sargent, J.R. 1994. Effects of diets rich in linoleic (18:2n-6) and alpha-linolenic (18:3n-3) acids on the growth, lipid class and fatty acid compositions and eicosanoid production in juvenile turbot (*Scophthalmus maximus* L.). *Fish Physiology and Biochemistry* 13: 105–118.

Bell, J.G., Farndale, B.M., Bruce, M.P., Navas, J.M., and Carillo, M. 1997. Effects of broodstock dietary lipid on fatty acid compositions of eggs from sea bass (*Dicentrarchus labrax*). *Aquaculture* 149:107–119.

Benítez, T., Masuda, R., Juárez Carrillo, E., Ganuza, E., Valencia, A., Hernández-Cruz, C.M., and Izquierdo, M.S. 2007. Dietary n-3 HUFA deficiency induces a reduced visual response in gilthead seabream. *Aquaculture* 264:408–417.

Berglund, L. 1995. Effects of spring temperature and feeding regime on sexual maturation in Atlantic salmon (*Salmo salar* L.) male parr. In: Goetz, F.W., and Thomas, P. (eds.) *Reproductive Physiology of Fish, Fish Symposium '95*, University of Texas Press, Austin, TX, pp. 170–172.

Blom, J.H., and Dabrowski, K. 1995. Reproductive success of female rainbow trout (*Oncorhynchus mykiss*) in response to graded dietary ascorbyl monophosphate levels. *Biology of Reproduction* 52:1073–1080.

Bobe, J., and Labbé, C. 2010. Egg and sperm quality in fish. *General and Comparative Endocrinology* 165:535–548.

Bromage, N., Jones, J., Randall, C., Trush, M., Davies, B., Springate, J., Duston, J., and Barker, G. 1992. Broodstock management, fecundity, egg quality and the timing of egg production in the rainbow trout (*Oncorhynchus mykiss*). *Aquaculture* 100:141–166.

Bromley, P., Ravier, C., and Wittihame, P.R. 2000. The influence of feeding regime on sexual maturation, fecundity and atresia in first-time spawning turbot. *Journal of Fish Biology* 56: 264–278.

Brooks, S., Tyler, C.R., and Sumpter, J.P. 1997. Egg quality in fish: what makes a good egg? *Reviews in Fish Biology and Fisheries* 7:387–416.

Brown, S.B., Fitzsimons, J.O., Palace, W.T., and Vandenbillaardt, L. 1998. Thiamine and early mortality syndrome in lake trout. In: McDonald, G., Fitzsimons, J.O., and Honeyfield, O.C. (eds.) *Early Life Stage Mortality Syndrome in Fishes of the Great Lake and Baltic Sea*. American Fisheries Society Symposium 21. American Fisheries Society, Bethesda, MD, pp. 18–25.

Bruce, M.P., Shields, R.J., Bell, M.V., and Bromage, N.R. 1993. Lipid class and fatty acid composition of eggs of Atlantic halibut, *Hippoglossus hippoglossus* (L.), in relation to egg quality in captive broodstock. *Aquaculture and Fishery Management* 24:417–422.

Bruce, M., Oyen, F., Bell, G., Asturiano, J.F., Farndale, B., Ramos, J., Bromage, N., Carrillo, M., and Zanuy, S. 1999. Development of broodstock diets for the European sea bass (*Dicentrarchus labrax*) with special emphasis on the importance of n-3 and n-6 HUFA to reproductive performance. *Aquaculture* 117:85–97.

Brylinska, M., and Brylinski, E. 1972. Methods for estimation of fish fecundity on the example of bream (*Abramis brama* L.). *Roczniki Nauk Rolniczych* 94–H–2:7–40.

Bueno, D. 2001. Evaluación del cultivo larvario de la dorada (*Sparus aurata* L.) en funcion de la calidad de puesta. Master thesis, Universidad de Las Palmas de Gran Canaria, Spain.

Carnevali, O., Cionna, C., Tosti, L., Lubzens, E., and Maradonna, F. 2006. Role of cathepsins in ovarian follicle growth and maturation. *General and Comparative Endocrinology* 146: 195–203.

Cerdá, J., Carrillo, M., Zanuy, S., and Ramos, J. 1994a. Effect of food ration on estrogen and vitellogenin plasma levels, fecundity and larval survival in captive sea bass, *Dicentrarchus labrax*: preliminary observations. *Aquatic Living Resources* 7:255–266.

Cerdá, J., Carrillo, M., Zanuy, S., Ramos, J., and De La Higuera, M. 1994b. Influence of nutritional composition of diet on sea bass *Dicentrarchus labrax* L., reproductive performance and egg and larvae quality. *Aquaculture* 128:345–361.

Cerdà, J., Fabra, M., and Raldúa, D. 2007. Physiological and molecular basis for fish oocyte hydration. In: Babin, P.J., Cerdá, J., and Lubzens, E. (eds.) *The Fish Oocyte: From Basic Studies to Biotechnological Applications*. Springer, Dordrecht, pp. 349–396.

Cerdá, J., Bobe, J., Babin, P.J., Admon, A., and Lubzens, E. 2008. Functional genomics and protemic approaches for the study of gamete formation and viability in farmed finfish. *Reviews in Fisheries Science* 16:54–70.

Choubert, G. 1986. Pigments caroténoides et reproduction des poissons. *Bulletin Francais de la Peche et de la Pisciculture* 300:25–32.

Choubert, G., and Blanc, J.M. 1993. Muscle pigmentation changes during and after spawning in male and female rainbow trout, *Oncorhynchus mykiss*, fed dietary carotenoids. *Aquatic Living Resources* 6:163–168.

Ciereszco, A., and Dabrowski, K. 1995. Sperm quality and ascorbic acid concentration in rainbow trout semen are affected by dietary vitamin C: an across season study. *Biology of Reproduction* 52:982–988.

Constanz, G.D. 1975. Behavioral ecology of mating in the male Gila topminnow, *Poeciliopsis occidentalis* (Cyprinodontiformes: Poeciliidae). *Ecology* 56:966–973.

Corraze, G., Larroquet, L., Maisse, G., Blanc, O., and Kaushik, S. 1993. Effect of temperature and of dietary lipid source on female broodstock performance and fatty acid composition of the eggs of rainbow trout. *Fish Nutrition in Practice*, Biarritz (France), Ed. INRA, Paris (Les Colloques, no. 61), pp. 61–66.

Coward, K., and Bromage, N.R. 1999. Spawning frequency, fecundity, egg size and ovarian histology in groups of *Tilapia zillii* maintained upon two distinct food ration sizes from first feeding to sexual maturity. *Aquatic Living Resources* 12:11–22.

Craik, J.C.A. 1985. Egg quality and egg pigment content in salmonid fishes. *Aquaculture* 47:61–88.

Craik, J.C.A., and Harvey, S.M. 1986. Egg quality in the Atlantic salmon. *ICES Council Meeting 1986* (Collected Papers), ICES, Copenhagen, Denmark.

Dahlgren, B.T. 1980. The effects of three different dietary protein levels on the fecundity in the guppy, *Poecilia reticulata* (Peters). *Journal of Fish Biology* 16:83–97.

Danikowski, S., Sallmann, H.P., and Flachowsky, G. 2002. Influence of high levels of vitamin E on sperm parameters of cock. *The Journal of Animal Physiology and Animal Nutrition* 86:376–382.

Dantagnan, H.P., Borquez, A.S., Valdebenito, I.N., Salgado, I.A., Serrano, E.A., and Izquierdo, M.S. 2007. Lipid and fatty acid composition during embryo and larval development of puye *Galaxias maculatus* Jenyns, 1842, obtained from estuarine, freshwater and cultured populations. *Journal of Fish Biology* 70:770–781.

Dhert, P., Lim, L.C., Lavens, P., Chao, T.M., Chou, R., Lavens, P., Sorgeloos, P., Jaspers, E., and Ollevier, F. 1991. Effect of dietary essential fatty acids on egg quality and larviculture success of the greasy grouper (*Epinephelus tauvina*). *Larvi '91. Special Publication, European Aquaculture Society* 15:58–62.

Donnelly, E.T., McClure, N., and Lewis, S.E.M. 1999. The effect of ascorbate and α-tocopherol supplementation in vitro on DNA integrity and hydrogen peroxide–induced DNA damage in human spermatozoa. *Mutagenesis* 14: 505–511.

Dube, K. 1994. Effect of vitamin E on the fecundity and maturity of *Heteropneustes fossilis* (Bloch.). In: Chou, L.M., Munro, A.D., Lam, T.J., Chen, T.W., Cheong, L.K.K., Ding, J.K., Hooi, K.K., Khoo, H.W., Phang, V.P.E., Shim, K.F., and Tan, C.H. (eds.) *Proceedings of the Third Indian Fisheries Forum*. The Asian Fisheries Science, Pant Nagar, Uttar Pradesh, India, pp. 101–103.

Emata, A.C., and Borlongan, I. 2003. A practical broodstock diet for the mangrove red snapper, *Lutjanus argentimaculatus*. *Aquaculture* 225: 83–88.

Emata, A., Borlongan, I., and Darnaso, J. 2000. Dietary vitamin C and E supplementation and reproduction of milkfish *Chanos chanos* Forsskal. *Aquacultural Engineering* 31: 557–564.

Emata, A.C., Marte, C., Borlongan, I., and Nocillado, J. 1996. The effect of dietary lipid and protein and ration size on the reproductive performance of cage reared milkfish broodstock. In: Santiago, C.B., Coloso, R.M., Millamena, O.R., Borlongan, I.G. (eds.) *Feeds for Small-Scale Aquaculture*, Proceedings of the National Seminar-Workshop on Fish Nutritíon and Feeds. Tigbauan, Iloilo, Philippines, June 1–2, 1994. Aquaculture Department, Southeast Asian Fisheries Development Center, Iloilo, Philippines, p. 122.

Eskelinen, P. 1989. Effects of different diets on egg production and egg quality of Atlantic salmon (*Salmo salar* L.). *Aquaculture* 79:275–281.

Fabra, M., Raldúa, D., Power, D.A., Deen, P.M.T., and Cerdá, J. 2005. Marine fish egg hydration is aquaporin-mediated. *Science* 307:545.

Falk-Petersen, S., Falk-Petersen, Y., Sargent, J.R., and Haug, T. 1986. Lipid class and fatty acid composition of eggs from the Atlantic halibut (*Hippoglossus hippoglossus*). *Aquaculture* 52:207–211.

Falk-Petersen, S., Sargent, J.R., Fox, C., Falk-Petersen, L.B., Haug, T., and Kjørsvik, E. 1989. Lipids in Atlantic halibut (*Hippoglossus hippoglossus*) eggs from planktonic samples in Northern Norway. *Marine Biology* 101: 553–556.

Farndale, B., Bell, J., Bruce, M., Bromage, N., Oyen, F., Zanuy, S., and Sargent, J. 1999. Dietary lipid composition affects blood leukocyte fatty acid compositions and plasma eicosanoid concentrations in European sea bass (*Dicentrarchus labrax*). *Aquaculture* 179: 330–350.

Fernández-Palacios, H., Izquierdo, M.S., Robaina, L., Valencia, A., Salhi, M., and Vergara, J.M. 1995. Effect of n-3 HUFA level in broodstock diets on egg quality of gilthead sea bream (*Sparus aurata* L.). *Aquaculture* 132:325–337.

Fernández-Palacios, H., Izquierdo, M.S., Robaina, L., Valencia, A., Salhi, M., and Montero, D. 1997. The effect of dietary protein and lipid from squid and fish meals on egg quality of broodstock for gilthead seabream (*Sparus aurata*). *Aquaculture* 148:233–246.

Fernández-Palacios, H., Izquierdo, M.S., Gonzalez, M., Robaina, L., and Valencia, A. 1998. Combined effect of dietary α-tocopherol and n-3 HUFA on egg quality of gilthead seabream broodstock (*Sparus aurata*). *Aquaculture* 161: 475–476.

Fernández-Palacios, H., Izquierdo, M.S., and Robaina, L. 2005. Efecto de distintas dietas para reproductores de dorada (*Sparus aurata*) sobre la calidad de sus puestas. Informes Técnicos del Instituto Canario de Ciencias Marinas, No. 12.

Finn, R.N. 2007a. Vertebrate yolk complexes and the functional implications of phosvitins and other subdomains in vitellogenins. *Biology of Reproduction* 76:926–935.

Finn, R.N. 2007b. The maturational disassembly and differential proteolysis of paralogous vitellogenins in a marine pelagophil teleost: a conserved mechanism of oocyte hydration. *Biology of Reproduction* 76:936–948.

Fletcher, D.A., and Wootton, R.J. 1995. A hierarchical response to differences in ration size in the reproductive performance of female three spined sticklebacks. *Journal of Fish Biology* 46:657–668.

Frémont, L., Léger, C., Petridou, B., and Gozzelino, M.T. 1984. Effects of a polyunsaturated fatty acid deficient diet on profiles of serum vitellogenin and lipoprotein in vitellogenic trout (*Salmo gairdneri*). *Lipids* 19:522–528.

Fry, F.J. 1949. Statistics of a lake trout fishery. *Biometrics* 5:27–67.

Furuita, H., Tanaka, H., Yamamoto, T., Suzuki, N., and Takeuchi, T. 2002. Effects of high levels of n-3 HUFA in broodstock diet on egg quality and egg fatty acid composition of Japanese flounder, *Paralichthys olivaceus*. *Aquaculture* 210:323–333.

Furuita, H., Tanaka, H., Yamamoto, T., Suzuki, N., and Takeuchi, T. 2003a. Supplemental effect of vitamin A in diet on the reproductive performance and egg quality of the Japanese flounder *Paralichthys olivaceus* (T&S). *Aquacultural Engineering* 34:461–468.

Furuita, H., Yamamoto, T., Shima, N., and Takeuchi, T. 2003b. Effect of arachidonic acid levels in broodstock diet on larval and egg quality of Japanese flounder *Paralichthys olivaceus*. *Aquaculture* 220:725–735.

Fyhn, H.J., and Serigstad, B. 1987. Free amino acids as energy substrate in developing eggs and larvae of the cod *Gadus morhua*. *Marine Biology* 96:335–341.

Ganga, R., Bell, J.G., Montero, D., Robaina, L., Caballero, M.J., and Izquierdo, M.S. 2005. Effect of feeding gilthead seabream (*Sparus aurata*) with vegetable lipid sources on two potential immunomodulator products: prostanoids and leptins. *Comparative Biochemistry and Physiology* 142:410–418.

Goetz, F.W., Ranjan, M., Duman, P., and Herman, C. 1989. PGF and PGE synthesis by specific tissue components of the brook trout (*Salvelinus fontinalis*) ovary. *The Journal of Experimental Zoology* 250:196–205.

Goetz, F.W., Berndtson, A., and Ranjan, M. 1991. Ovulaton: mediators at the ovarian level. In: Pang, P., and Schreibman, M. (eds.) *Vertebrate Endocrinology, Fundamentals and Biomedical Implications, Vol. 4(A)—Reproduction.* Academic Press, New York, pp. 127–203.

Gonzalez-Vecino, J.L. 2005. Nucleotide enhancement of diets, fish reproduction and egg quality (*Hippoglossus hippoglossus, Melanogrammus aeglefinus*). *Dissertation Abstracts International Part C: Worldwide* 66(2):310.

González-Vecino, J.L., Cutis, C.J., Batty, R.S., Mazorra de Quero, C., Greenhaff, P.L., and Wadsworth, S. 2004. Short and long term effects of a nucleotide enriched broodstock diet on the reproductive performance of haddock (*Melanogrammus aeglefinus* L.). 11th International Symposium on Nutrition and Feeding in Fish, Phuket, Thailand, p. 99.

Green, C.C., and Kelly, A.M. 2009. Effects of the estrogen mimic genistein as a dietary component on sex differentiation and ethoxyresorufin-O-deethylase (EROD) activity in channel catfish (*Ictalurus punctatus*). *Fish Physiology and Biochemistry* 35:377–384.

Hansen, A.C., Rosenlund, G., Karlsen, Ø., Koppe, W., and Hemre, G.I. 2007. Total replacement of fish meal with plant proteins in diets for Atlantic cod (*Gadus morhua*). I: effects on growth and protein retention. *Aquaculture* 272:599–611.

Hara, S., Duraya, M.N., Parazoa, M., and Taki, Y. 1986. Year-round spawning and seed production of the rabbitfish, *Siganus guttatus*. *Aquaculture* 59:259–272.

Hardy, R.W., Shearer, K.D., and King, I.B. 1984. Proximate and elemental composition of developing eggs and maternal soma of pen-reared coho salmon (*Oncorhynchus kisutch*) fed production and trace element fortified diets. *Aquaculture* 43:147–165.

Hardy, R.W., Matsumoto, T., Fairgrieve, W.T., and Stickney, R.R. 1990. The effects of dietary lipid source on muscle and egg fatty acid composition and reproductive performance of Coho Salmon (*Oncorhynchus kisutch*). In: Takeda, M., and Watanabe, T. (eds.) *The Current Status of Fish Nutrition in Aquaculture.* Proceedings of the Third International Symposium on Feeding and Nutrition in Fish. Japan Translation Center, Tokyo, pp. 347–356.

Harel, M., Tandler, A., and Kissil, G.W. 1992. The kinetics of nutrient incorporation into body tissues of gilthead sea bream *S. aurata* females and subsequent effects on egg composition and egg quality. *The Israeli Journal of Aquaculture—Bamidgeh* 44:127.

Harris, L.E. 1984. Effects of a broodfish diet fortified with canthaxanthin on female fecundity and egg color. *Aquaculture* 43:179–183.

Hart, N.F. 1990. Fertilization in teleost fishes: mechanism of sperm–egg interactions. *International Review of Cytology* 121:1–66.

Hemre, G.I., Mangor-Jensen, A., and Lie, Ø. 1994. Broodstock nutrition in turbot (*Scophthalmus maximus*) effect of dietary vitamin E. *Fiskeridirektoratets Skrifter, Serie Ernæring* 8:21–29.

Hester, F.J. 1964. Effects of food supply on fecundity in the female guppy, *Lebistes reticulatus* (Peters). *Journal of the Fisheries Research Board of Canada* 21:757–764.

Hirshfield, M.F. 1980. An experimental analysis of reproductive effort and cost in the Japanese medaka, *Oryzias latipes*. *Ecology* 61: 282–292.

Hislop, J.R.G., Robb, A.P., and Gauld, J.A. 1978. Observations on effects of feeding level on growth and reproduction in haddock, *Melanogrammus aeglefinus* (L.) in captivity. *Journal of Fish Biology* 13:85–98.

Hodder, V.M. 1963. Fecundity of Grand Bank haddock. *Journal of the Fisheries Research Board of Canada* 20:1465–1487.

Hornung, M.W., Miller, L., Peterson, R.E., Marcquenski, S., and Brown, S.B. 1998. Efficacy of thiamine, astaxanthin, beta-carotene, and thyroxine treatments in reducing early mortality syndrome in Lake Michigan salmonid embryos. *American Fisheries Society Symposium* 21:124–134.

Horwood, J.W., Walker, M.G., and Witthames, P. 1989. The effect of feeding levels on the fecundity of plaice (*Pleuronectes platessa*). *Journal of the Marine Biological Association of the United Kingdom* 69:81–92.

Hsiao, S.M., and Mak, W.C. 1978. Artificial fertilization and incubation of fertilized eggs pond reared ayu. *China Fisheries Monthly* 305: 2–11.

Huber, J.T. 1988. Vitamins in ruminant nutrition. In: Church, D.C. (ed.) *The Ruminant Animal: Digestive Physiology and Nutrition*. Prentice Hall, Englewood Cliffs, NJ, pp. 313–325.

Huxtable, R.J. 1992. Physiological actions of taurine. *Physiological Reviews* 72:101–163.

Hyllner, S.J., Silversand, C., and Haux, C. 1994. Formation of the vitelline envelope precedes the active uptake of vitellogenin during oocyte development in the rainbow trout, *Oncorhynchus mykiss*. *Molecular Reproduction and Development* 39:166–175.

Hyllner, S.J., Fernández-Palacios, H., Larsson, D.O.J., and Haux, C. 1995. Amino acid composition and endocrine control of vitelline envelope proteins in European sea bass (*Dicentrarchus labrax*) and gilthead seabream (*Sparus aurata*). *Molecular Reproduction and Development* 41:339–347.

Izquierdo, M.S. 1996. Essential fatty acid requirements of cultured marine fish larvae. *Aquaculture Nutrition* 2:183–191.

Izquierdo, M.S., and Fernández-Palacios, H. 1997. Nutritional requirements of marine fish larvae and broodstock. *Cahiers Options Mediterraneennes* 22:243–264.

Izquierdo, M.S., Fernández-Palacios, H., and Tacon, A.G.J. 2001. Effect of broodstock nutrition on reproductive performance of fish. *Aquaculture* 197:25–42.

Jones, J., and Bromage, N.R. 1987. The influence of ration size in the reproductive performance of female rainbow trout (*Salmo gairdneri*). In: Scott, A.P., Canario, A.V.M., Idler, D.R., and Crim, L.M. (eds.) *Proceedings of the Third International Symposium on Reproductive Physiology of Fish*, Memorial University Press, St. John's, Newfoundland, Canada, p. 202.

Kah, O., Zanuy, S., Pradelles, P., Cerdá, J., and Carrillo, M. 1994. An enzyme inmunoassay for salmon gonadotropin-releasing hormone and its application to the study of the effects of diet on brain and pituitary GnRH in the sea bass, *Dicentrarchus labrax*. *General and Comparative Endocrinology* 95:464–474.

Karlsen, O., Holm, J.C., and Kjesbu, O.S. 1995. Effects of periodic starvation on reproductive investment in first time spawning Atlantic cod (*Gadus morhua* L.). *Aquaculture* 133: 159–170.

Ketola, H.G. 1985. Mineral supplementation of Atlantic salmon broodstock diets. In: Iwamoto, R.N., and Sower, S. (eds.) *Salmonid Reproduction: An International Symposium*. Washington Sea Grant Program. University of Washington, Seattle, WA, p. 111.

Ketola, H.G., Bowser, P.R., Wooster, L.R., Wedge, L.R., and Hurst, S. 1998. Thiamine remediation of early mortality in fry of Atlantic salmon from Cayuga Lake. *Great Lakes Research Review* 3:21–26.

Khan, M.A., Jafri, A.K., and Chadha, N.K. 2005. Effects of varying dietary protein levels on growth, reproductive performance, body and egg composition of rohu, *Labeo rohita* (Hamilton). *Aquaculture Nutrition* 11:11–17.

Kim, D.J., Seok, S.H., Baek, M.W., Lee, H.Y., Na, Y.R., Park, S.H., Lee, H.K., Dutta, N.K., Kawakami, K., and Park, J.H. 2009. Developmental toxicity and brain aromatase induction by high genistein concentrations in zebrafish embryos. *Toxicology Mechanisms and Methods* 19:251–256.

King, I.B. 1985. Influence of vitamin E in reproduction in rainbow trout (*Salmo gairdneri*). PhD dissertation, University of Washington, Seattle, WA.

Kjesbu, O.S., Witthames, P.R., Solemdal, P., and Walker, M.G. 1998. Temporal variations in the fecundity of Arcto-Norwegian cod (*Gadus morhua)* in response to natural changes in food and temperature. *Journal of Sea Research* 40:303–321.

Kjørsvik, E., Mangor-Jensen, A., and Holmefjord, I. 1990. Egg quality in fishes. *Advances in Marine Biology* 26:71–113.

Kjørsvik, E., Hoehne-Reitan, K., and Reitan, K.I. 2003. Egg and larval quality criteria as predictive measures for juvenile production in turbot (*Scophthalmus maximus* L.). *Aquaculture* 227:9–20.

Knight, J., Holland, J.W., Bowden, L.A., Halliday, K., and Rowley, A.F. 1995. Eicosanoid generating capacities of different tissues from the rainbow trout, *Oncorhynchus mykiss*. *Lipids* 30:451–458.

Koprücü, K., and Seker, E. 2003. Effect of supplemental dietary vitamin E on the fecundity of guppy (*Poecilia reticulata* Peters, 1895) and swordtail (*Xiphophorus helleri* Heckel, 1848). *Fırat Üniversitesi Fen ve Mühendislik Bilimleri Dergisi* 15:83–88.

Kraus, G. 2002. Variability in egg production of cod (*Gadus morhua callarias* L.) in the Central Baltic Sea. PhD thesis, University of Kiel, Germany.

Kraus, G., Mueller, A., Trella, K., and Koester, F.W. 2000. Fecundity of Baltic cod: temporal and spatial variation. *Journal of Fish Biology* 56:1327–1341.

Kuznetzov, V.A., and Khalitov, N.K. 1978. Alterations in the fecundity and egg quality of the roach, *Rutilus rutilus*, in connection with different feeding conditions. *Journal of Ichthyology* 18:63–70.

Lall, B., and Singh, T.P. 1987. Changes in tissue lipid levels in the freshwater catfish *Clarias batrachus* associated with the reproductive cycle. *Fish Physiology and Biochemistry* 3:191–201.

Lavens, P., Lebegue, E., Jaunet, H., Brunel, A., Dhert, P., and Sorgeloos, P. 1999. Effect of dietary essential fatty acids and vitamins on egg quality in turbot broodstocks. *Aquaculture International* 7:225–240.

Lee, K., and Dabrowski, K. 2004. Long-term effects and interactions of dietary vitamins C and E on growth and reproduction of yellow perch, *Perca flavescens*. *Aquaculture* 230: 377–389.

Leray, C., and Pelletier, X. 1985. Fatty acid composition of trout phospholipids: Effect of (n-3) essential fatty acid deficiency. *Aquaculture* 50:51–59.

Li, Y., Chen, W., Sun, Z., Chen, J., and Wu, K. 2005. Effects of n-3 HUFA content in broodstock diet on spawning performance and fatty acid composition of eggs and larvae in *Plectorhynchus cinctus*. *Aquaculture* 245: 263–272.

Lie, Ø., Mangor-Jensen, A., and Hemre, G.I. 1993. Broodstock nutrition in cod (*Gadus morhua*) effect of dietary fatty acids. *Fiskeridirektoratets Skrifter, Serie Ernæring* 6:11–19.

Lie, Ø., Sandvin, A., and Waagbø, R. 1994. Transport of alpha-tocopherol in Atlantic salmon (*Salmo salar*) during vitellogenesis. *Fish Physiology and Biochemistry* 13:241–247.

Ling, S., Kuah, M., Muhammad, T., Kolkovski, S., and Chong, A.S.C. 2006. Effect of dietary HUFA on reproductive performance, tissue fatty acid profile and desaturase and alongase mRNAs in female swordtail *Xiphophorus helleri*. *Aquaculture* 261:204–214.

Link, J., and Burnett, J. 2001. The relationship between stomach contents and maturity state for major northwest Atlantic fishes: new paradigms? *Journal of Fish Biology* 59: 783–794.

Lubzens, E., Young, G., Bobe, J., and Cerdá, J. 2010. Oogenesis in teleosts: how fish eggs are formed. *General and Comparative Endocrinology* 165:367–389.

Luquet, P., and Watanabe, T. 1986. Interaction "nutrition–reproduction" in fish. *Fish Physiology and Biochemistry* 2:121–129.

Ma, Y., Kjesbu, O.S., and Jørgensen, T. 1998. Effects of ration on the maturation and fecundity in captive Atlantic herring (*Clupea harengus*). *Canadian Journal of Fisheries and Aquatic Sciences* 55:900–908.

Madden, M. 2001. Vitamin A and the developing embryo. *Journal of Postgraduate Medicine* 77:489–491.

Makino, N., Uchiyama, M., Iwanami, S., Tohyama, T., and Tanaka, M. 1999. Developmental changes in multiple oil globules of Japanese sea bass eggs. *Bulletin of the Japanese Society of Scientific Fisheries* 65:268–277.

Mangor-Jensen, A., and Birkeland, R.N. 1993. Effects of dietary carbohydrate on broodstock maturation and egg quality in cod. *Milestone Report Center for Havbruk, Institute Marine Research* 9:1–4.

Mangor-Jensen, A., Birkeland, R.N., and Sandnes, K. 1993. Effects of cod broodstock dietary vitamin C on embryonic growth and survival. *Milestone Report Center for Havbruk, Institute Marine Research* 18:1–8.

Martin, N.V. 1970. Long-term effects of diet on the biology of the lake trout and the fishery in Lake Opeongo, Ontario. *Journal of the Fisheries Research Board of Canada* 27:125–126.

Matsubara, T., Nagae, M., Ohkubo, N., Andoh, T., Sawaguchi, S., Hiramatsu, N., Sullivan, C.V., and Hara, A. 2003. Multiple vitellogenins and their unique roles in marine teleosts. *Fish Physiology and Biochemistry* 28:295–299.

Matsunari, H., Hamada, K., Mushiake, K., and Takeuchi, T. 2006. Effects of taurine levels in broodstock diets on reproductive performance of yellowtail *Seriola quinqueradiata. Fisheries Science* 72:955–960.

Mazorra, C., Bruce, M., Bell, J.G., Davie, A., Alorend, E., Jordan, N., Rees, J., Papanikos, N., Portero, M., and Bromage, N. 2003. Dietary lipid enhancement of broodstock reproductive performance and egg and larval quality in Atlantic halibut (*Hippoglossus hippoglossus*). *Aquaculture* 227:21–33.

McFadden, J.T., Cooper, E.L., and Andersen, J.K. 1965. Some effects of environment on egg production in brown trout (*Salmo trutta*). *Limnology and Oceanography* 10:88–95.

McKay, I., and Mann, K.H. 1969. Fecundity of two cyprinid fishes in the River Thames, Reading, England. *Journal of the Fisheries Research Board of Canada* 26:2795–2805.

Miki, W., Yamaguchi, K., Konosu, S., and Watanabe, T. 1984. Metabolism of dietary carotenoids in eggs of red sea bream. *Comparative Biochemistry and Physiology* 77 B:665–668.

Mironova, N.V. 1977. Energy expenditure on egg production in young *Tilapia mossambica* and the influence of maintenance conditions on their reproductive intensity. *Journal of Ichthyology* 17:627–633.

Moore, P.K. 1985. *Prostanoids: Pharmacological, Physiological and Clinical Relevance.* Cambridge University Press, Cambridge.

Morehead, D.T., Hart, P.R., Dunstan, G.A., Brown, M., and Pankhurst, M.W. 2001. Differences in egg quality between wild striped trumpeter (*Latris lineata*) and captive striped trumpeter that were fed different diets. *Aquaculture* 192:39–53.

Mourente, G., Carrascosa, M.A., Velasco, C., Odriozola, J.M., Billard, R., and De Pauw, N. 1989. Effect of gilthead sea bream (*Sparus aurata* L.) broodstock diets on egg lipid composition and spawning quality. Aquaculture Europa '89. *Special Publication, European Aquaculture Society* 10:179–180.

Mukhopadhyay, P.K., Chattopadhyay, D.N., and Mitra, G. 2003. Broodstock nutrition, the key to quality seed production. *Infofish International* 3:25–23.

Mushiake, K., Arai, S., Matsumoto, A., Shimma, H., and Hasegawa, I. 1993. Artificial insemination from 2 year old cultured yellowtail fed with moist pellets. *Bulletin of the Japanese Society of Scientific Fisheries* 59:1721–1726.

Mustafa, T., and Srivastava, K.C. 1989. Prostaglandins (eicosanoids) and their role in ectothermic organisms. *Advances in Comparative and Environmental Physiology* 5:157–207.

Nagahama, Y. 1994. Endocrine regulation of gametogenesis in fish. *The International Journal of Developmental Biology* 38:217–229.

Nagahama, Y. 2000. Gonadal steroid hormones: major regulators of gonadal sex differentiation and gametogenesis in fish. In: Norberg, B., Kjesbu, O.S., Taranger, G.L., Andersson, E., and Stefansson, S.O. (eds.) *Reproductive Physiology of Fish*. John Grieg Forlag A/S, Bergen, pp. 211–222.

Nagahama, Y., and Yamashita, Y. 2008. Regulation of oocyte maturation in fish. *Development, Growth and Differentiation* 50:S195–S219.

Nassour, I., and Leger, C.L. 1989. Deposition and mobilisation of body fat during sexual maturation in female trout (*Salmo gairdneri*, Richardson). *Aquatic Living Resources* 2:153–159.

Navas, J., Thrush, M., Ramos, J., Bruce, M., Carrillo, M., Zanuy, S., and Bromage, N. 1996. The effect of seasonal alteration in the lipid composition of broodstock diets on egg quality in the European sea bass (*Dicentrarchus labrax*). In: Goetz, F.W., and Thomas, P. (eds.) *Proceedings of the Fifth International Symposium on the Reproductive Physiology of Fish*. The University of Texas Press, Austin, TX, pp. 108–110.

Navas, J.M., Bruce, M., Trush, M., Farndale, B.M., Bromage, N., Zanuy, S., Carrillo, M., Bell, J.G., and Ramos, J. 1997. The impact of seasonal alteration in the lipid composition of broodstock diets on egg quality in the European sea bass. *Journal of Fish Biology* 51:760–773.

Navas, J.M., Thrush, M., Zanuy, S., Ramos, J., Bromage, N., and Carrillo, M. 2001. Total lipid in the broodstock diet did not affect fatty acid composition and quality of eggs from sea bass (*Dicentrarchus labrax* L.). *Scientia Marina (Barcelona)* 65:11–19.

National Research Council (NRC). 1993. *Nutrient Requirements of Fish*. National Academic Press, Washington DC.

Ohkubo, N., and Matsubara, T. 2002. Sequential utilization of free amino acids, yolk proteins, and lipids in developing eggs and yolk-sac larvae of barfin flounder *Verasper moseri*. *Marine Biology* 140:187–196.

Palace, V.P., and Werner, J. 2006. Vitamins A and E in the maternal diet influence egg quality and early life stage development in fish: a review. *Scientia Marina* 70(Suppl. 2):41–57.

Papanikos, N. 2005. Egg quality of red snapper *Lutjanus campechanus*, significance of spawning method and broodfish nutrition. *Dissertation Abstracts International Part B: Science and Engineering* 65:54–72.

Patiño, R., and Sullivan, C.V. 2002. Ovarian follicle growth, maturation and ovulation: an integrated perspective. *Fish Physiology and Biochemistry* 26:57–70.

Pelegri, F. 2003. Maternal factors in zebrafish development. *Developmental Dynamics* 228: 535–554.

Pickova, J., Dutta, P.C., Larsson, P.O., and Kiessling, A. 1997. Early embryonic cleavage pattern, hatching success and egg-lipid fatty acid composition: comparison between two cod stocks. *Canadian Journal of Fisheries and Aquatic Sciences* 54:2410–2416.

Rainuzzo, J.R. 1993. Fatty acid and lipid composition of fish egg and larvae. In: Reinertsen, H., Dahle, L.A., and Jorgensen, L. (eds.) *Proceedings of the First International Conference on Fish Farming Technology*. A.A. Balkema Publishers, Trondheim, Norway, pp. 43–49.

Raitt, D.F.S. 1968. The population dynamics of the Norway pout in the North Sea. *Marine Research* 5:1–23.

Ridelman, J.M. 1981. Effects of starvation and diet formulation on ovarian development and egg viability of steelhead × rainbow trout hybrids. MSc thesis, University of Washington, Seattle, WA.

Robb, A.P. 1982. Histological observations on the reproductive biology of the haddock, *Melanogrammus aeglefinus* (L.). *Journal of Fish Biology* 20:397–408.

Rodriguez, C., Cejas, J.R., Martin, M.V., Badia, P., Samper, M., and Lorenzo, A. 1998. Influence of n-3 highly unsaturated fatty acid deficiency on the lipid composition of broodstock gilthead seabream (*Sparus aurata* L.) and on egg quality. *Fish Physiology and Biochemistry* 18:177–187.

Ronis, M.J., Ingelman-Sundberg, J., and Badger, T. 1994. Induction, suppression and inhibition of multiple hepatic cytochrome P450 isozymes in male rat and bobwhite quail (*Colinus virginianus*) by ergosterol biosynthesis inhibiting compounds. *Toxicology and Applied Pharmacology* 44:1953–1965.

Rønnestad, I. 1992. Utilization of free amino acids in marine fish eggs and larvae. PhD thesis, University of Bergen, Norway.

Rønnestad, I., Fyhn, H.J., and Gravningen, G. 1992. The importance of free amino acids to the energy metabolism of eggs and larvae of turbot (*Scophthalmus maximus*). *Marine Biology* 114:517–525.

Rønnestad, I., Thorsen, A., and Finn, R.N. 1999. Fish larval nutrition: a review of recent advances in the roles of amino acids. *Aquaculture* 177:201–216.

Rosenblum, P., Horne, H., Garwood, G., Brandt, T., and Villarreal, B. 1995. Delayed ovarian development and reduced fecundity in largemouth bass raised on a pelleted feed containing high levels of steroids and low levels of arachidonic acid. In: Goetz, F.W., and Thomas, P. (eds.) *Proceedings of the Fifth International Symposium on the Reproductive Physiology of Fish*. The University of Texas Press, Austin, TX, p. 138.

Sandnes, K. 1991. Vitamin C in fish nutrition: a review. *Fiskeridirektoratets Skrifter, Serie Ernæring* 4:3–32.

Sandnes, K., Ulgenes, Y., Braekkan, O.R., and Utne, F. 1984. The effect of ascorbic acid supplementation in broodstock feed on reproduction of rainbow trout (*Salmo gairdneri*). *Aquaculture* 43:167–177.

Santiago, C.B., and Gonzalo, A.C. 2000. Effect of prepared diet and vitamins A, E and C supplementation on the reproductive performance of cage-reared bighead carp *Aristichthys nobilis* (Richardson). *Journal of Applied Ichthyology* 16:8–13.

Sargent, J.R. 1995. Origin and functions of egg lipids: nutritional implications. In: Bromage,

N.R., and Roberts, R.J. (eds.) *Broodstock Management and Egg and Larval Quality*. Blackwell Science, London, pp. 353–372.

Sargent, J.R., Bell, J.G., Bell, M.V., Henderson, R.J., and Tocher, D.J. 1993. The metabolism of phospholipids and polyunsaturated fatty acids in fish. In: Lahlou, B., and Vitiello, P. (eds.) *Aquaculture: Fundamental and Applied Research Coastal and Estuarine Studies 43*. American Geophysical Union, Washington, DC, pp. 103–124.

Sargent, J., McEvoy, L., Estevez, A., Bell, G., Bell, M., Henderson, J., and Tocher, D. 1999. Lipid nutrition of marine fish during early development: current status and future directions. *Aquaculture* 179:217–229.

Sasayama, Y., and Takahashi, H. 1972. Effect of starvation and unilateral astration in male goldfish, *Carassius auratus*, and a design of bioassay for fish gonadotropin using starved goldfish. *Bulletin of the Faculty of Fisheries Hokkaido University* 22:267–283.

Scabini, V., Fernandez-Palacios, H., and Izquierdo, M.S. 2006. Inclusion of carotenoids in broodstock diets for gilthead sea bream (*Sparus aurata* L., 1758): effects on egg and spawning quality. Abstracts, XII International Symposium Fish Nutrition and Feeding, Biarritz, France.

Schimittou, H.R. 1993. High density fish culture in low volume cages. M.I.T.A. (P) No. 518, Vol. AQ41.

Schulz, R.W., de Franca, L.R., Lareyre, J.-J., LeGac, F., Chiarini-Garcia, H., Nobrega, R.H., and Miura, T. 2010. Spermatogenesis in fish. *General and Comparative Endocrinology* 165:390–411.

Shiranee, P., and Natarajan, P. 1996. Crude palm oil as a source of carotenoids and tocopherols to enhance reproductive potential in pearlspot *Etroplus suratensis*. *Asian Fisheries Science* 9:35–44.

Siddiqui, A.Q., Al-Hafedh, Y.S., and Ali, S.A. 1998. Effect of dietary protein level on the reproductive performance of Nile tilapia, *Oreochromis niloticus* (L.). *Aquaculture Research* 29:349–358.

Silveira, R., Perez, J., Fajer, E., and Franco, A. 1996. Effect of the vitamins supplement on the reproduction performance of *Ictalurus punctatus*. *Revista Cubana de Investigaciones Pesqueras* 20:28–30.

Silversand, C. 1996. Vitellogenesis in teleost fish. A study of vitellogenin and egg lipids. Doctoral thesis, Department of Zoophysiology, University of Gothenburg. Available at http://hdl.handle.net/2077/14128.

Siman, C.M., and Erikkson, U.J. 1997. Vitamin E decreases the occurrence of malformations in the offspring of diabetic rats. *Diabetes* 46:1054–1061.

Sivaloganathan, B., Walford, J., and Lam, T.J. 1998. Free amino acids and energy metabolism in eggs and larvae of sea bass, *Lates calcarifer*. *Marine Biology* 131:695–702.

Sohn, Y.C., Suetake, H., Yoshiura, Y., Kobayashi, M., and Aida, K. 1998. Structural and expression analysis of gonadotropin I-beta subunit genes in goldfish (*Carassius auratus*). *Gene* 222:257–267.

Soliman, A.X., Jauncey, K., and Roberts, R.J. 1986. The effect of dietary ascorbic acid supplementation on hatchability, survival rate and fry performance in *Oreochromis niloticus* (Peters). *Aquaculture* 59:197–208.

Sorensen, P.W., and Goetz, F.W. 1993. Pheromonal function of prostaglandin metabolites in teleost fish. *Journal of Lipid Mediators* 6:385.

Sorensen, P.W., Hara, T.J., Stacey, N.E., and Goetz, F.W. 1988. F prostaglandins function as potent stimulants that comprise the post-ovulatory female sex pheromone in goldfish. *Biology of Reproduction* 39:1039–1050.

Springate, J.R.C., Bromage, N.R., and Cumaranatunga, P.R.T. 1985. The effects of different ration on fecundity and egg quality in the rainbow trout (*Salmo gairdneri*). In: Cowey, C.B., Mackie, A.M., and Bell, J.G. (eds.) *International Symposium on Feeding and Nutrition in Fish*. Academic Press, London, pp. 371–393.

Srivastava, R.K., and Brown, J.A. 1992. Assessment of egg quality in Atlantic salmon, *Salmo salar*, treated with testosterone-II. *Comparative Biochemistry and Physiology* 103 A:397–402.

Srivastava, R.K., Brown, J.A., and Shahidi, F. 1995. Changes in the amino acid pool during embryonic development of cultured and wild Atlantic salmon (*Salmo salar*). *Aquaculture* 131:115–124.

Stacey, N.E., and Goetz, F.W. 1982. Role of prostaglandins in fish reproduction. *Canadian Journal of Fisheries and Aquatic Sciences* 39:92–98.

Stauffer, T.M. 1976. Fecundity of coho salmon (*Oncorhynchus kisutch*) from the Great Lakes and a comparison with ocean salmon. *Journal of the Fisheries Research Board of Canada* 33:1150–1155.

Strussman, C.A., and Nakamura, M. 2002. Morphology, endocrinology, and environmental modulation of gonadal sex differentiation in teleost fishes. *Fish Physiology and Biochemistry* 26:13–29.

Sutjaritvongsanon, S. 1987. Level of vitamin E content suitable for gonad developing and spawning of goldfish, *Carassius auratus* (Linn.). Kasetsart University, Bangkok, Abstracts of Master of Science Theses Fisheries Science, Notes from the Faculty of Fisheries, 12, p.2.

Suwa, K., and Yamashita, M. 2007. Regulatory mechanisms of oocyte maturation and ovulation. In: Babin, P.J., Cerdá, J., and Lubzens, E. (eds.) *The Fish Oocyte: From Basic Studies to Biotechnological Applications.* Springer, Dordrecht, pp. 323–347.

Takeuchi, T. 1997. Essential fatty acid requirements of aquatic animals with emphasis on fish larvae and fingerlings. *Reviews in Fisheries Science* 5:1–25.

Takeuchi, M., Ishii, S., and Ogiso, T. 1981a. Effect of dietary vitamin E on growth, vitamin E distribution, and mortalities of the fertilized eggs and fry in ayu *Plecoglossus altivelis. Bulletin of the Tokai Regional Fisheries Research Laboratory* 104:111–122.

Takeuchi, T., Watanabe, T., Ogino, T., Saito, M., Nishimura, M., and Nose, T. 1981b. Effects of low protein–high calorie diets and deletion of trace elements from a fish meal diet on reproduction of rainbow trout. *Bulletin of the Japanese Society of Scientific Fisheries* 47:645–654.

Takeuchi, T., Feng, Z., Yoseda, K., Hirokawa, J., and Watanabe, T. 1994. Nutritive value of DHA enriched rotifer for larval cod. *Nippon Suisan Gakkaishi* 60:641–652.

Tandler, A., Watanabe, T., Satoh, S., and Fukusho, K. 1989. The effect of food deprivation on the fatty acid and lipid profile of red seabream larvae (*Pagrus major*). *The British Journal of Nutrition* 62:349–361.

Tandler, A., Harel, M., Koven, W.M., and Kolkovski, S. 1995. Broodstock and larvae nutrition in gilthead seabream *Sparus aurata* new findings on its mode involvement in improving growth, survival and swimbladder inflation. *Israeli Journal of Aquaculture—Bamidgeh* 47:95–111.

Thomas, P., Zhu, Y., and Pace, M. 2002. Progestin membrane receptors involved in the meiotic maturation of teleost oocytes: a review with some new findings. *Steroids* 67:511–517.

Thorsen, A., Trippel, E.A., and Lambert, Y. 2003. Experimental methods to monitor the production and quality of eggs of captive marine fish. *Journal of Northwest Atlantic Fishery Science* 33:55–70.

Tocher, D.R., and Harvie, D.G. 1988. Fatty acid compositions of the major phosphoglycerides from fish neural tissues; (n-3) and (n-6) polyunsaturated fatty acids in rainbow trout *(Salmo gairdneri*) and cod (*Gadus morhua*) brains and retinas. *Fish Physiology and Biochemistry* 5:229–239.

Tocher, D., Fraser, A.J., Sargent, J.R., and Gamble, J.C. 1985a. Fatty acid composition of phospholipids and neutral lipids during embryonic and early larval development in Atlantic herring (*Clupea harengus* L.). *Lipids* 20:69–74.

Tocher, D., Fraser, A.J., Sargent, J.R., and Gamble, J.C. 1985b. Lipid class composition during embryonic and early larval development in Atlantic herring (*Clupea harengus* L.). *Lipids* 20:84–89.

Torrissen, O.J. 1984. Pigmentation of salmonids, effects of carotenoids in eggs and start feeding diet on survival and growth rate. *Aquaculture* 43:185–193.

Torrissen, O.J. 1990. Biological activities of carotenoids in fishes. In: Takeda, M., and Watanabe, T. (eds.) *The Current Status of Fish Nutrition in Aquaculture.* Japan Translation Center, Tokyo, Japan, pp. 387–399.

Torrissen, O.J., and Christiansen, R. 1995. Requirements for carotenoids in fish diets. *Journal of Applied Ichthyology* 11:225–230.

Townshend, T.J., and Wootton, R.J. 1984. Effects of food supply on the reproduction of the convict cichlid, *Cichlasoma nigrofasciatum. Journal of Fish Biology* 24:91–104.

Trippel, E.A., Hunt, J.J., and Buzeta, M.I. 1995. Evaluation of the cost of reproduction of Georges Bank Atlantic cod (*Gadus morhua*) using otolith back calculation. In: Secar, D.H., Dean, J.M., and Campana, S.E. (eds.) *Recent*

Developments in Fish Otolith Research. University of South Carolina Press, Columbia, SC, pp. 599–616.

Tyler, C.R., and Sumpter, J.P. 1996. Oocyte growth and development in teleosts. *Reviews in Fish Biology and Fisheries* 6:287–318.

Vassallo-Agius, R., Imaizumi, H., Watanabe, T., Yamazaki, T., Satoh, S., and Kiron, V. 2001a. The influence of astaxanthin supplemented dry pellets on spawning of striped jack. *Fisheries Science* 67:260–270.

Vassallo-Agius, R., Imaizumi, H., Watanabe, T., Yamazaki, T., Satoh, S., and Kiron, V. 2001b. Effect of squid meal in dry pellets on the spawning performance of striped jack *Pseudocaranx dentex*. *Fisheries Science* 67:271–280.

Vassallo-Agius, R., Watanabe, T., Yoshizaki, G., Satoh, S., and Takeuchi, Y. 2001c. Quality of eggs and spermatozoa of rainbow trout fed an n-3 essential fatty acid deficient diet and its effects on the lipid and fatty acid components of eggs, semen and livers. *Fisheries Science* 67:818–825.

Verakunpiriya, V., Watanabe, T., Mushiake, K., Kiron, V., Satoh, S., and Takeuchi, T. 1996. Effect of broodstock diets on the chemical components of milt and eggs produced by yellowtail. *Fisheries Science* 62:610–619.

Verakunpiriya, V., Mushiake, K., Kawano, K., and Watanabe, T. 1997. Supplemental effect of astaxanthin in broodstock diets on the quality of yellowtail eggs. *Fisheries Science* 63: 816–823.

Wada, K., DeLong, C.J., Hong, Y.H., Rieke, C.J., Song, I., Sidhu, R.S., Yuan, C., Warnock, M., Schmaier, A.H., Yokoyama, C., Smyth, E.M., Wilson, S.J., Fitzgerald, G.A., Garavito, M., Sui, D.X., Regan, J.W., and Smith, W.L. 2007. Enzymes and receptors of prostaglandin pathways with arachidonic acid–derived versus eicosapentaenoic acid–derived substrates and products. *The Journal of Biological Chemistry* 282:22254–22266.

Wade, M.G., Van Der Kraak, G., Gerrits, M.F., and Ballantyne, J.S. 1994. Release and steroidogenic actions of polyunsaturated fatty acids in fue goldfish testis. *Biology of Reproduction* 51:131–139.

Wallace, R.A. 1985. Vitellogenesis and oocyte growth in nonmammalian vertebrates. In: Browder, L.W. (ed.) *Developmental Biology*, Vol. 1. Plenum Publishing Corp., New York, pp. 127–177.

Wallace, R.A., and Selman, K. 1978. Oogenesis in *Fundulus heteroclitus*. I. Preliminary observations on oocyte maturation *in vivo* and *in vitro*. *Developmental Biology* 62:354–369.

Wallace, R.A., and Selman, K. 1981. Cellular and dynamic aspects of oocyte growth in teleosts. *American Zoologist* 21:325–343.

Wallace, R.A., and Selman, K. 1990. Ultrastructural aspects of oogenesis and oocyte growth in fish and amphibians. *Journal of Electron Microscopy Technique* 16:175–201.

Washburn, B.S., Frye, D.J., Hung, S.S.O., Doroshov, S.I., and Cante, F.S. 1990. Dietary effects on tissue composition, oogenesis and the reproductive performance of female rainbow trout (*Oncorhynchus mykiss*). *Aquaculture* 90:179–195.

Watanabe, T. 1990. Effect of broodstock diets on reproduction of fish. *IFREMER. Actes de Colloques* 9:542–543.

Watanabe, T. 1993. Importance of docosahexanoic acid in marine larval fish. *Journal of the World Aquaculture Society* 24:152–161.

Watanabe, T., and Kiron, V. 1994. Prospects in larval fish dietetics. *Aquaculture* 124:223–251.

Watanabe, T., and Kiron, V. 1995. Broodstock management and nutritional approaches for quality offsprings in the red sea bream. In: Bromage, N.R., and Roberts, R.J. (eds.) *Broodstock Management and Egg and Larval Quality*. Cambridge University Press, Cambridge, pp. 395–414.

Watanabe, T., and Takashima, F. 1977. Effect of alpha-tocopherol deficiency on carp. 6. Deficiency symptoms and changes of fatty acid and triglyceride distributions in adult carp. *Bulletin of the Japanese Society of Scientific Fisheries* 43:819–830.

Watanabe, T., Arakawa, T., Kitajima, C., and Fujita, S. 1984a. Effect of nutritional quality of broodstock diet on reproduction of red seabream. *Bulletin of the Japanese Society of Scientific Fisheries* 50:495–501.

Watanabe, T., Itoh, A., Murakami, A., Tsukashima, Y., Kitajima, C., and Fujita, S. 1984b. Effect of nutritional composition of diets on chemical components of red seabream broodstocks and eggs produced. *Bulletin of the Japanese Society of Scientific Fisheries* 50:503–515.

Watanabe, T., Itoh, A., Murakami, A., Tsukashima, Y., Kitajima, C., and Fujita, S. 1984c. Effect of nutritional quality of diets given to broodstock on the verge of spawning on reproduction of red seabream. *Bulletin of the Japanese Society of Scientific Fisheries* 50:1023–1028.

Watanabe, T., Takeuchi, T., Saito, M., and Nishimura, K. 1984d. Effect of low protein–high calorie or essential fatty acid deficiency diet on reproduction of rainbow trout. *Nippon Suisan Gakkaishi* 50:1207–1215.

Watanabe, T., Itoh, A., Kitajima, C., and Fujita, S. 1984e. Effect of protein levels on reproduction of red sea bream. *Bulletin of the Japanese Society of Scientific Fisheries* 50:1015–1022.

Watanabe, T., Itoh, A., Satoh, S., Kitajima, C., and Fujita, S. 1985a. Effect of dietary protein levels and feeding period before spawning on chemical components of eggs produced by red sea bream broodstock. *Bulletin of the Japanese Society of Scientific Fisheries* 51:1501–1509.

Watanabe, T., Koizumi, T., Suzuki, H., Satoh, S., Takeuchi, T., Yoshida, N., Kitada, T., and Tsukashima, Y. 1985b. Improvement of quality of red sea bream eggs by feeding broodstock on a diet containing cuttlefish meal or on raw krill shortly before spawning. *Bulletin of the Japanese Society of Scientific Fisheries* 51:1511–1521.

Watanabe, T., Izquierdo, M.S., Takeuchi, T., Satoh, S., and Kitajima, C. 1989. Comparison between eicosapentaenoic and docosahexaenoic acids in terms of essential fatty acid efficacy in larval red seabream. *Nippon Suisan Gakkaishi* 55:1635–1640.

Watanabe, T., Lee, M.J., Mizutani, J., Yamada, T., Satoh, S., Takeuchi, T., Yoshida, N., Kitada, T., and Arakawa, T. 1991a. Nutritional studies in the seed production of fish. 20. Effective components in cuttlefish meal and raw krill for improvement of quality of red seabream *Pagrus major* eggs. *Bulletin of the Japanese Society of Scientific Fisheries* 57:681–694.

Watanabe, T., Fujimura, T., Lee, M.J., Fukusho, K., Satoh, S., and Takeuchi, T. 1991b. Nutritional studies in the seed production of fish. 21. Effect of polar and nonpolar lipids from krill on quality of eggs of red seabream *Pagrus major*. *Bulletin of the Japanese Society of Scientific Fisheries* 57:695–698.

Wessel, G.M., Brooks, J.M., Green, E., Haley, S., Voronina, E., Wong, J., Zaydfudim, V., and Conner, S. 2001. The biology of cortical granules. *International Review of Cytology* 209:117–206.

Woodward, B. 1994. Dietary vitamin requirements of cultured young fish, with emphasis on quantitative estimates for salmonids. *Aquaculture* 124:133–168.

Wooster, G.A., and Bowser, P.R. 2000. Remediation of Cayuga syndrome in landlocked Atlantic salmon *Salmo salar* using egg and sac fry bath treatments of thiamine hydrochloride. *Journal of the World Aquaculture Society* 31:149–157.

Xiao, W., Liu, Y., Tian, L., Zhen, W., and Cao, J. 2003. Effect of vitamin E and vitamin C on spawning quality of broodstock for grouper *Epinephelus coioides*. *Acta Scientiarum Naturalium Universitatis Sunyatseni* 42(Suppl.2):214–217.

Yoneda, M., and Wright, P.J. 2005. Effects of varying temperature and food availability on growth and reproduction in first-time spawning female Atlantic cod. *Journal of Fish Biology* 67:1225–1241.

Yoshizaki, G., Takeuchi, Y., Kobayashi, T., Ihara, S., and Takeuchi, T. 2002. Primordial germ cells: the blueprint for piscine life. *Fish Physiology and Biochemistry* 26:3–12.

Zhou, J., Kumar, T.R., Matzuk, M.M., and Bondy, C. 1997. Insulin-like growth factor I regulates gonadotropin responsiveness in the murine ovary. *Molecular Endocrinology* 11:1924–1933.

Zohar, Y., Harel, M., Hassin, S., and Tandler, A. 1995. Gilthead seabream. In: Bromage, N.R., and Roberts, R.J. (eds.) *Broodstock Management and Egg and Larval Quality*. Blackwell Science, London, pp. 94–117.

Chapter 6

Utilization of yolk: transition from endogenous to exogenous nutrition in fish

Marta Jaroszewska and Konrad Dabrowski

6.1 Introduction

Fishes are a large and diverse group containing five subclasses (Holocephali, Elasmobranchii, Cladistia, Chondrostei, and Neopterygii) divided into 60 orders (Nelson 2006). Although the early ontogeny varies substantially across subclasses as well as orders, the development of embryos and larvae or "yolk sac juveniles" (the nomenclature according to Balon 1999) in oviparous and ovoviviparous fishes during endogenous nutrition is based on the utilization of materials accumulated in the yolk platelets and lipid droplets or oil globules as the primary sources of nutrients and energy.

The endogenous phase of nutrition is the first of four types of nutrient acquisition described in fish, followed by integument absorption, mixed feeding, and exogenous feeding (Balon 1986). There is considerable variability among fish species in the duration of the endogenous nutrition period. Most fish species rely on these endogenous maternal supplies as yolk sac larvae or "yolk sac juveniles" for approximately the first 3–5 days after hatching (Sanderson and Kupferberg 1999). Many exceptions, however, exist; for instance, it might be extended up to 60 days, as in silver arowana (*Osteoglossum bicirrhosum*; Jaroszewska and Dabrowski 2009a). During the time between hatching and exogenous feeding, the digestive, sensory, muscular, circulatory, and respiratory systems undergo dramatic metamorphosis in order to meet functional demands for first exogenous feeding (Sanderson and Kupferberg 1999). In the case of less developed larvae, mixed feeding occurs (Balon 1986). In some cases, prolongation of endogenous nutrition and acceleration of exogenous feeding cause extremely long intervals of mixed feeding, for

Larval Fish Nutrition, First Edition. Edited by G. Joan Holt.

example, in live-bearing coelacanth (*Latimeria chalumnae*) and mouthbrooding cichlid fish *Cyphotilapia frontosa* (Lake Tanganyika cichlid) "yolk sac juveniles" (Balon 1986). The total larval period, when the yolk sac and non-yolk-sac stages are taken together, is greatly abbreviated to several days in tropical species, but may last a few weeks in most temperate fishes, or even years in the eel (*Anguilla* sp.) (Bagarinao 1986; Sanderson and Kupferberg 1999; Webb 1999; Aoyama 2009). Additionally, larval fishes, like many invertebrates, may delay metamorphosis due to fasting or incomplete nutrition (Zhang et al. 2006) to the end of migration or until a suitable settlement site is found (Webb 1999).

In this chapter, the focus will be on the nourishing functions of the yolk during fish larval/alevin development (nomenclature is used according to cited authors), beyond its role as the driving force for blastoderm epiboly (Trinkaus 1993), and then for morphogenesis during embryonic stages (Sakaguchi et al. 2006). However, in the following subsections, information is provided on yolk formation and commencement of exogenous feeding as necessary features to understand all the processes that take place during the endogenous phase of nutrition.

6.2 Origin and yolk formation in teleosts

6.2.1 Vitellogenesis

The yolk/body ratio, its density, and its overall quality are determined during the induction of vitellogenin (Vtg) synthesis (vitellogenesis) and may vary even within the same species, depending on the female's age, feeding conditions during maturation of the ovaries, and annual climate changes (Balon 1999). The induction of vitellogenesis is triggered by environmental cues such as photoperiod and temperature changes (e.g., in percid fish; Dabrowski et al. 1996), and is associated with endocrine feedback interactions between the hypothalamus, pituitary, gonad, and liver (HPGL axis; Bemanian et al. 2004). Therefore, the increase in Vtg synthesis, responsible for the larger accumulation of the yolk, is the result of environmental changes via endocrine mechanisms.

During vitellogenesis, much of the yolk is synthesized by liver cells in the form of a protein precursor, Vtg (Fagotto 1995; Finn 2007a). Vtgs are large apolipoproteins (phosphoglycolipoprotein) synthesized under control of the hepatic estrogen receptor α (ERα) and estradiol-17β stimulation (Bemanian et al. 2004; Finn 2007a). Vtgs are apparently never synthesized by the oocyte itself (Jorgensen 2008). All vertebrate Vtgs are, before being released for transport from the liver, posttranslationally glucosylated and phosphorylated in the endoplasmic reticulum (ER) and Golgi complex of hepatocytes. Vtgs are secreted as dimers to the bloodstream (Finn 2007a, 2007b), and then incorporated by the oocyte from the plasma by receptor (clathrin)-mediated endocytosis. Internalized vesicles that contain the very high-density Vtg are transferred to early endosomes in the oocyte, where they are acidified as the result of the action of an ATP-dependent V-class proton pumps (Finn 2007a). Vtg is cleaved by cathepsin D–like protease (Ctsd) (Hiramatsu et al. 2002) into smaller derivatives, the primary yolk proteins: phosvitin (Pv), a phosphorus-containing protein; two lipid-containing proteins, lipovitellin heavy chain (LvH) and lipovitellin light chain (LvL); and β′-component (von Willebrand factor type D domain [Vwfd]; Krieger and Fleig 1999; Finn 2007a, 2007b). An additional cleavage of the Vtg-derived yolk proteins occurs during final oocyte maturation in brackish and marine fishes spawning pelagic eggs (Hiramatsu et al. 2002). This second proteolysis provides a free amino acid (FAA) pool, which was demonstrated to be an important osmotic effector for oocyte hydration. K^+ influx into maturing oocytes is thought to be another important factor in this process (Carnaveli et al. 1999; Hartling and Kunkel 1999). These proteins are stored in yolk granules (YGs) or yolk platelets (Finn 2007a, 2007b; Tingaud-Sequeira and Cerdà 2007). YGs appear to be specialized late endosomes or lysosomes with very low hydrolytic activity, and they contain the complete set of hydrolases needed for yolk degradation and utilization (Fagotto 1995; Jorgensen 2008).

There are two types of lipid substrates in the developing embryos and larvae. More than two-thirds of the lipid transported by Vtg is phosphatidylcholine (PC), while neutral lipids comprise the remainder (Finn 2007a). Neutral lipids (triacylglycerols [TAG] and wax esters) are thought to be the most important energy reserves for the development of fish that have oil droplets in the eggs, such as sea bass (*Dicentrarchus labrax*) (Rønnestad et al. 1998) and Japanese eel (*Anguilla japonica*) (review by Ohkubo et al. 2008). Thus, the longer Pv domains are mostly found in species that have oil globules. Short serine-rich Pv domains are associated with eggs of both freshwater (pike *Esox lucius*) and marine teleosts, such as Atlantic halibut (*Hippoglossus hippoglossus*), Atlantic cod (*Gadus morhua*), barfin flounder (*Verasper moseri*), walleye pollock (*Theragra chalcogramma*), that do not possess large oil globules but rather contain phospholipids as the major lipid substrate (Finn 2007a; Ohkubo et al. 2008). Finn (2007a) noticed that Vtg may have a higher neutral lipid load in marine species. The Pv-missing class of Vtg (VtgC) does transport lipids but at lower levels, that is, 13% compared with 16–21% by mass in complete Vtgs. Therefore, Finn (2007a) recognized that the Pv domain is not obligatory for lipid transport, and that most of the neutral lipids delivered to oocytes of fish that have a large oil globule is not via Vtg but via other lipoproteins. Vtg also binds hormones, vitamins, and ions, and it has been proposed that all of these nutritional functions have exerted evolutionary constraints on the Vtg sequence.

6.2.2 Segregation of ooplasm

Egg ooplasm consists of numerous organelles, yolk inclusions, maternal mRNA, proteins, and lipid droplets and pigments. After egg activation, by sperm penetration or by chemical or mechanical stimuli, the process of "ooplasmic segregation" takes place in stage V of the oocyte and leads to the separation of the organelles from the yolk. This phenomenon is necessary for normal embryogenesis to proceed (Fernández et al. 2006). Therefore, the postactivation phase, synonymous with the second polar body expulsion, results in ooplasmic segregation. As the result, the ooplasm is segregated from the yolk and forms three domains: (1) a peripheral yolk cytoplasmic layer surrounding the oocyte; (2) the central domain, endoplasm, composed of a network of irregularly shaped loci (lacunae) of ooplasm scattered among yolk globules; and (3) a polar domain, preblastodisc, as a thin cap at the top of the animal hemisphere (see figure 14 in Fernández et al. 2006). The yolk cell is formed from the yolk globules of the oocyte endoplasm.

6.3 Yolk morphology

6.3.1 Characteristics of yolk platelets and oil droplets or oil globules

The nutrients required for early development are accumulated preferentially in the form of YGs as well as platelets and the oil globules. In fully grown postvitellogenic oocytes of teleosts, the yolk consists of round or oval platelets of different sizes that vary among eggs, being larger in larger eggs. Also, yolk platelets are diverse within one egg, with the deeper, more centrally located platelets tending to be larger and more homogenous than the superficial ones. Two zones can be distinguished in yolk platelets: the membrane enveloping the yolk platelet, and a central core. The outer sheath consists of mucopolysaccharides and heterotrimeric G proteins. The inner zone contains lipovitellin (Lv) and Pv, or analogous lipoproteins and phosphoproteins (Heming and Buddington 1988; Krieger and Fleig 1999). The interesting feature of fish eggs is that the yolk may be sequestered within a membrane-bound compartment at the center of the egg, as in medaka, *Oryzias javanicus*, or it may be intermingled with ooplasm, as in the zebrafish, *Danio rerio* (Fernández et al. 2006). Yolk platelets are usually the dominant components of the fish egg yolk. Despite a diversity of yolk platelet sizes in the animal world, from 1-μm wide in the YGs of the nematode *Caenorhabditis elegans* up to 140-μm yolk spheres in chicken eggs, yolk platelets appear to be evolutionarily homologous organelles (Jorgensen 2008). In contrast to most of the other freshwater fishes, the yolk mass of European perch, *Perca fluviatilis*, appears to be completely homogenous, with no vesiculation and no yolk platelets (Krieger and Fleig 1999).

Oil droplets vary in size, appearance, and number among fish species. In demersal or viviparous fish eggs, they can appear as numerous small droplets about 1 μm in diameter located among yolk platelets (gilthead sea bream, *Sparus aurata*; pike). In European perch and medaka, among many other species, numerous small oil droplets dispersed in the interior of the oocyte fuse after fertilization to one large globule located near the upper site of the egg (Krieger and Fleig 1999; Iwamatsu et al. 2008). In sciaenids, there are multiple droplets after fertilization that usually coalesce into one before hatching, and there are other marine species that maintain numerous oil globules (Ditty 1989). In contrast, there are pelagic eggs without oil droplets as well (e.g., Gadiformes; Pleuronectiformes: winter flounder [*Pleuronectes americanus*], Atlantic halibut; Wiegand 1996b; Lubzens et al. 2010).

6.3.2 Structure and diversity of yolk syncytial layer (*ysl*) among fishes

Teleost fishes have meroblastic cleavage that results in the formation of an extraembryonic yolk sac. The functions of the yolk sac are gas exchange (Liem 1981), excretion, and, primarily, provision of nutrients to developing embryos and larvae. The most characteristic feature of this structure is the *ysl*.

The *ysl* is derived from collapsed marginal blastomeres in early cleavage stages and forms a syncytium between the yolk cell and inner cell mass (ICM), the precursor of the embryo (Betchaku and Trinkaus 1978; Trinkaus 1993; reviewed by Krieger and Fleig 1999). Research indicates that the *ysl* in the early life history of teleosts also serves as the primary motor for blastoderm epiboly (Betchaku and Trinkaus 1978; Trinkaus 1984a, 1984b, 1993; Solnica-Krezel and Driever 1994). It is widely believed that the high frequency of occurrence of mitochondria in the "periblast"[1] is the proof of its role as a driving force of epiboly (Krieger and Fleig 1999). The induction and/or patterning of anterior neural tissue, the body axes, formation of the ventrolateral mesoderm, and induction of the nodal-related genes in the ventrolateral marginal blastomeres during early embryonic development are also attributed to control of the *ysl* expressed genes (Ho et al. 1999; Hyodo et al. 1999; Chen and Kimelman 2000). The *ysl* function is connected with the morphogenesis of the orga-

[1] Review of the literature reveals that there is no consensus on the use of terms *yolk syncytial layer* and *periblast* in reference to the embryonic stage when this structure appears to function. Betchaku and Trinkaus (1978) suggested using *ysl* as a proper term, arguing against "periblast" as an inadequate one. Some authors use these two terms as synonyms (Fishelson 1995). However, in the recent literature, very often just the term *ysl* is used (Trinkaus 1984a; Ninhaus-Silveira et al. 2007). Following these descriptions, the authors of the present chapter adopted the term *ysl* for the superficial, multinuclear, and extraembryonic layer of the yolk during the embryonic and larval stages. The *ytl* is devoid of nuclei (Jaroszewska and Dabrowski 2009a).

nizer epithelium (D'Amico and Cooper 2001; Cooper and Virta 2007), regulation of cardiac tissue morphogenesis (Sakaguchi et al. 2006), and formation of the liver bud (Li et al. 2007).

In our opinion, as the embryo and larva mass increase exponentially, the function of the *ysl* as the "early embryonic patterning center" (Ho et al. 1999; Sakaguchi et al. 2006) transforms into those functions that are predominantly involved in yolk digestion, and then in the synthesis and release of nutrients from the yolk to the growing fish body (Fishelson 1995; Krieger and Fleig 1999; Poupard et al. 2000; Kunz 2004). In other words, the function of directing cell movement and the nourishing functions of the *ysl* do overlap at some developmental stages, and the role of this structure in yolk utilization becomes more profound. For this reason, Jaroszewska and Dabrowski (2009a) named the *ysl* in "yolk sac juvenile" stages of silver arowana as the yolk trophoblastic layer (*ytl*). In this species, *ytl* nutritional function parallels the absence of nuclei; therefore, the use of the term "syncytial" is not justified.

In teleost embryos, the *ysl* together with the mesoderm and ectoderm spread out to enclose the entire yolk mass. In contrast to nonteleost Semionotiformes, Amiiformes (Jaroszewska and Dabrowski 2009b), Acipenseriformes (Ostaszewska and Dabrowski 2009), and other lower vertebrates, teleosts do not have a connection between the endoderm of the presumptive gut and yolk reserves during embryonic and early larval/"yolk sac juvenile" development (Kunz 2004). Vitelline circulation, consisting of exclusively venous blood vessels, develops in the wall of the yolk sac. Generally, vitelline circulation forms in the liver portal system, with the subintenstinal vein and the reticulum of blood vessels as the part of the vena hepatica (Figure 6.1; Balon 1990; Kunz 2004). In teleost embryos and larvae, all the greater veins emanating from the body can provide blood to the yolk sac (Kunz 2004).

However, many species, such as gilthead sea bream, sea bass, pike perch (*Sander lucioperca*), or turbot (*Scophthalmus maximus*), are not provided with yolk sac circulation in the early stages of development (Mani-Ponset et al. 1996; Poupard et al. 2000). Furthermore, observations on larvae of these species provide evidence for very slow establishment of blood vessels. The most recent studies on white bass (*Morone chrysops*)[2] larvae are in agreement with the results

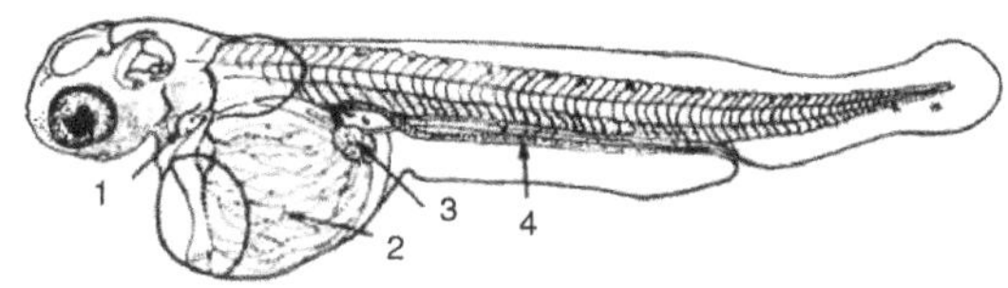

Figure 6.1 Larva (free embryo sensu Balon 1990) of *Stenodus leucichthys*, 9.4 mm of body length; 1—heart; 2—respiratory network of the vitelline subintestinal and hepatic veins; 3—liver; 4—subintestinal vein. (Figure is from the work by Balon [1990], with kind permission from *Guelph Ichthyology Reviews* and the author.)

[2]Fish were obtained from the Laboratory of Aquaculture, School of Environment and Natural Resources, the Ohio State University, Columbus, OH, where breeding under laboratory conditions was carried out. White bass males and females were obtained from Maumee River, OH, on May 24, 2008, and injected with human chorionic gonadotropin (300 IU/0.1 kg on June 3, 2008). One female was stripped 24 hours after induction and eggs were fertilized with sperm from one male and incubated in a McDonald jar with a continuous water flow at 22°C. The second female provided quality eggs in the afternoon of June 4. Hatching occurred on June 7, 2008. Larvae were kept in given condition until the third day of rearing and were then distributed into system tanks (21.5°C). To examine the ultrastructure of white bass larvae, five individuals at hatching, three individuals at 1 dph, and three individuals at 3 dph were directly examined under transmission electron microscopy (TEM). Whole larval trunks with yolk sac were fixed immediately after fish were anesthetized (MS-222) for 3–12 hours at room temperature in 0.15 M cacodylate buffered with 2% glutaraldehyde and 4% paraformaldehyde at pH 7.3 with 2.5% sucrose to adjust for osmolarity to 350 mOsM according to Diaz et al. (2002). Samples were rinsed several times in the phosphate buffer and postfixed with 1% osmium tetraoxide in 0.15 M cacodylate buffer, dehydrated in a graded series of ethanol and propylene oxide, and embedded in Eponate 12 (Ted Pella Inc., Redding, CA). Electron microscopy studies were carried out on the yolk sac following fixation. The section (0.5 µm) obtained with a Leica EM UC6 (Leica Microsystems GmbH, Wetzlar, Germany) ultramicrotome were stained with 2% uranyl acetate and Reynold's lead citrate and examined with an FEI Tecnai G2 Spirit BioTwin (FEI, Hillsboro, OR) transmission electron microscope operated at 80 kV.

reported for sea bream, sea bass, pike perch, and turbot. No blood vessels in the yolk sac were observed in white bass larvae at hatching, 1 day posthatch (dph), and 3 dph. The *ysl* was adjacent to the perisyncytial space (Figure 6.2a). Poupard et al. (2000) and Kunz (2004) concluded that the absence of the blood circulatory system in the turbot yolk sac does not preclude the absorption and transfer of nutrients from the yolk cell to the developing embryo or larva. In their opinion, it is the result of the developmental scheme in teleosts, mentioned above, that the anterior lateral plate mesoderm gives rise to the heart tube formation. Therefore, the heart tube is in continuous association with the perisyncytial space at its venous end. In pelagic marine embryos, there is an enlargement of the perisyncytial space above the cephalic region into which the heart opens directly. This enables the transfer of the *ysl* lipoproteins to the embryonic circulation despite the absence of differentiated vitelline circulation. In contrast, there is an elaborated blood circulation system in the yolk sac of rainbow trout (*Oncorhynchus mykiss*) embryos, developed by differentiation of endothelium-lined vessels in the splanchnopleure of the coelomic mesoderm (Vernier 1969). In rainbow trout, the yolk sac is vascularized by one-fourth in embryos at 14 days postfertilization (dpf), with one afferent vein and two efferent yolk sac veins present. At 18 dpf, the vitelline circulation is 75% developed (Ignatieva 1991). At hatching, larvae of whitefish (*Coregonus alpinus*) exhibit a yolk circulation system taken over entirely by hepatic veins. A similar picture is found in the European perch embryo approaching hatching time (Kunz 2004).

The majority of nutrients from the endogenous reserves of the yolk sac must pass through the absorptive structure of the *ysl/ytl* in order to reach the embryo and larval tissues with *apoE* gene expression as the critical step in this process (Babin et al. 1997; Poupard et al. 2000). A few investigators have addressed the cytological basis of endogenous nutrition, particularly the utilization of yolk reserves in teleost fish embryos (Shimizu and Yamada 1980; Sire et al. 1994) and larvae/alevins (Vernier and Sire 1977a, 1977b; Sire and Vernier 1979; Walzer and Schönenberger 1979a, 1979b; Kjørsvik and Reiersen 1992; Mani-Ponset et al. 1994). Most studies thus far have dealt with the histochemical aspects of the yolk utilization sequence in teleost development (Vernier and Sire 1977a, 1977b; Walzer and Schönenberger 1979a, 1979b; Sire et al. 1994).

The *ysl* structure in brown trout (*Salmo trutta morpha fario*) immediately after hatching is composed of a cytoplasmic zone (*ycz*) placed above the vitellolysis zone (Walzer and Schönenberger 1979a, 1979b). The presence of cellular organelles in the cytoplasmic region of the *ysl* shows similarities among the alevins of trout, larvae of Atlantic halibut and pike perch, and with our studies on white bass (Figure 6.2; Walzer and Schönenberger 1979a, 1979b; Kjørsvik and Reiersen 1992; Mani-Ponset et al. 1994; Sire et al. 1994). The *ycz* is a very complex structure involved

Figure 6.2 TEM observation of the yolk syncytial layer (*ysl*) in white bass at hatching revealed several structures related to nutrient absorption: (a) 30,000×; numerous lipoprotein particles (VLDL, white arrowhead) in the adjacent extracellular perisyncytial space (*); (b) 49,000×; VLDL particles filling both rough endoplasmic reticulum (RER) and secretory vesicle (sv) filled with contents that are released by exocytosis (black arrowheads) into the perisyncytial space; (c) 49,000×; formation of small yolk vesicles cut out of the yolk platelets in the process of degradation occurring within the vitellolysis zone (vz), when large yolk platelets are divided into smaller secretory vesicles (white arrowhead); (d) 23,000×; mitochondria (m) in close apposition with RER cisterns; sv between m and RER; heterogeneous population of lipoprotein-filled sv with yolk nutrients in different stages of digestion (circle) inside the *ycz*; (e) 13,000×; the *ycz* composed of large nuclei of irregular shape, besides m and RER elements; (f) 13,000×; numerous microvilli (mv) protruded into the perisyncytial space. The yolk cytoplasmic zone is present above the vz. It contains all organelles and on the external surface is composed of mv. The lipoproteins were ~24 nm in diameter.

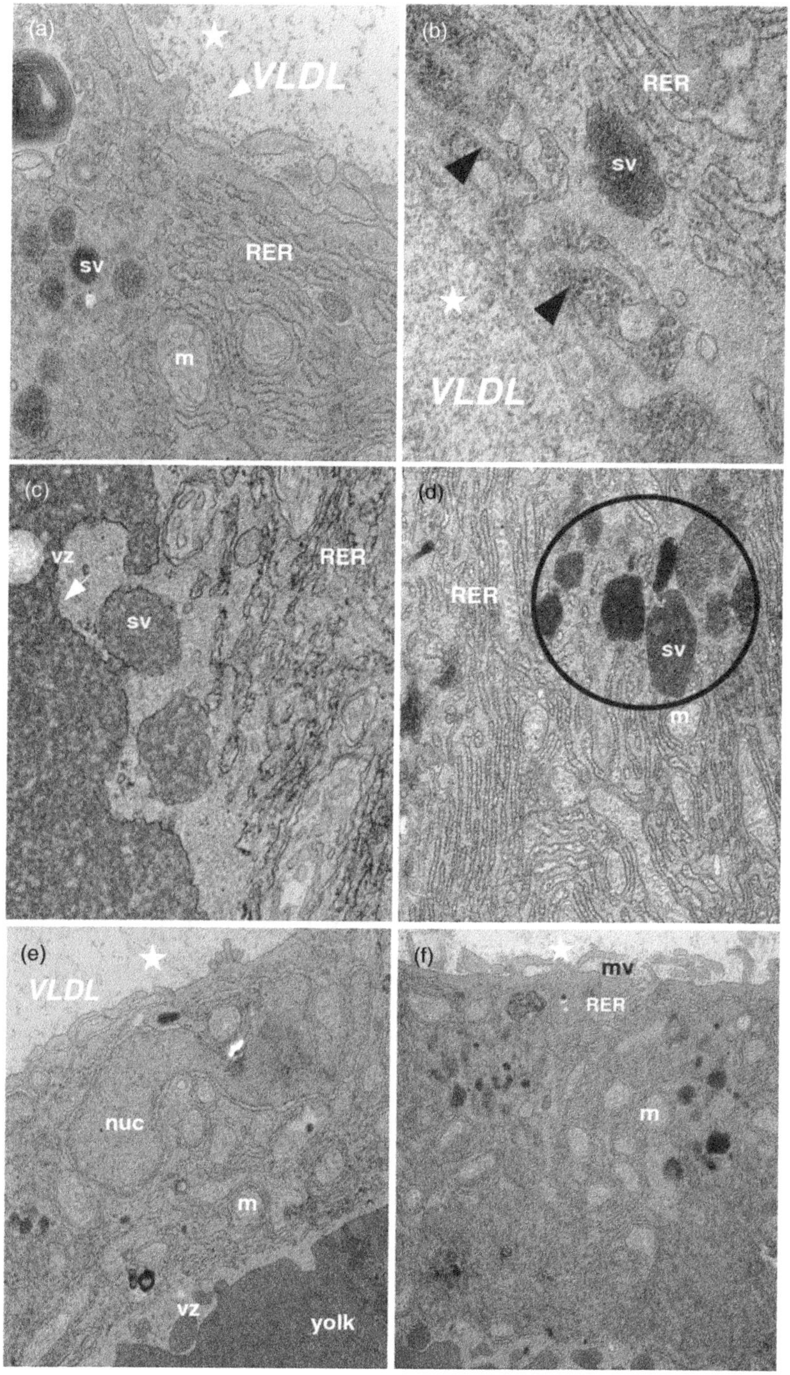
(a)
VLDL
sv
RER
m
(b)
RER
sv
VLDL
(c)
vz
RER
sv
(d)
RER
sv
m
(e)
VLDL
nuc
m
vz
yolk
(f)
mv
RER
m

in well-organized processes that include yolk hydrolysis, synthesis of lipoprotein particulates, and their transport to the vitelline circulation. The *ycz* contains smooth and rough ER, numerous mitochondria, and Golgi complexes (Figure 6.2). All these organelles, which extend in portions across the *ycz*, can form a stratified structure. However, some fish species do not seem to have a stratified "periblast," such as Atlantic cod, Atlantic halibut, sea bream, sea bass, walleye (*Sander vitreum*), and white bass, despite the presence of organelles (reviewed and summarized by Hoehne-Rejtan and Kjørsvik 2004).

6.3.3 Exocytosis in the *ysl* and *ytl*

Despite the fact that Walzer and Schönenberger (1979b) postulated the polarization and secretory function of the *ycz*, no exocytosis has been described in any published work on the development and ultrastructural organization of the *ysl* prior to or after hatching in teleost larvae (Vernier and Sire 1977a, 1977b; Long 1980; Shimizu and Yamada 1980; Sire et al. 1994). When describing European perch embryos, Krieger and Fleig (1999) suggested that exocytosis, endocytosis, and intracellular digestion of yolk proteins and lipid in the *ysl* and ICM are the pathways for yolk utilization. However, they did not provide experimental evidence for this, and they suggested that the morphological mechanism involved in nutrient mobilization from the yolk cell to the embryo in European perch is different from embryos of other teleosts. This hypothesis was based on the observation of special yolk morphology in the European perch, including the presence of the oil globule along with a nonvesiculated yolk mass (Krieger and Fleig 1999). These authors postulated other mechanisms of yolk utilization in fish species that possess yolk platelets in the yolk cell in contrast to the European perch, emphasizing that yolk processing during embryogenesis of perch takes place in the yolk compartment and that the function of the *ysl* is limited. Mani-Ponset et al. (1996), who studied sea bream, sea bass, and pike perch, suggested the presence of endocytosis vesicles that are involved in exchanges between the yolk cell and the perisyncytial space. Moreover, no pinocytotic or phagocytic vacuoles were observed by Ninhaus-Silveira et al. (2007) in the vicinity of the *ysl* microvilli in newly hatched larvae of the tropical fish curimba (*Prochilodus lineatus*). However, Poupard et al. (2000) showed the process of very low-density lipoprotein (VLDL) particle exocytosis into the perisyncytial space in the yolk of turbot larva. The newest studies on the "yolk sac juveniles" of silver arowana documented sequentially the process of exocytosis of yolk nutrients from the *ycz* into the microvillar interspace (Jaroszewska and Dabrowski 2009a). The process of lipoprotein particle exocytosis into the perisyncytial space was also confirmed in white bass larvae in our study (Figure 6.2a,b).

6.3.4 Variability of microvilli occurrence in fish yolk

Fishelson (1995) was the first author who hypothesized that microvilli described in the *ysl* of three species of cichlid (tilapias) larvae are transporters of nutrients across the monolayered capillary endothelium into the embryonic blood vessels. This author called it a "microvillar food transporting yolk sac syncytial surface." Mani-Ponset et al. (1994, 1996) postulated the presence of microvilli that protruded into the perisyncytial circulatory space in the yolk sac of sea bream, sea bass, and pike perch larvae. Our studies confirmed the presence of microvilli on the *ysl*/*ytl* surface in silver arowana "yolk sac juveniles" (Jaroszewska and Dabrowski 2009a) and white bass yolk sac larvae (Figure 6.2f). The report on turbot does not provide evidence of the microvillar structure of the plasma membrane of the yolk cell (Poupard et al. 2000).

In fish embryos, the microvilli were observed in the *ysl* of killifish (*Fundulus heteroclitus*; Trinkaus 1984b) and *Prochilodus lineatus* (Ninhaus-Silveira et al. 2007). Trinkaus (1984b) concluded that during killifish epiboly there is a gradual transition from long microvilli to much shorter ones on the *ysl* surface, but he provided no answer as to the purpose of this change.

6.3.5 Conclusions on the *ysl* function

Studies on zebrafish yolk sac larvae provide evidence of the *ysl* structure functioning in a manner very similar to higher vertebrates' (mouse) yolk endoderm (Ho et al. 1999). These authors alluded to the fact that the *ysl* in teleosts is the "functional equivalent" of mammalian visceral endoderm. In birds, the hypoblast of the embryo that becomes the origin of gut endodermis lies over the yolk and serves the major role in yolk absorption (Yoshizaki et al. 2004). When yolk reserves are exhausted, the *ysl/ytl* is resorbed and does not take part in the formation of the permanent fish body (Heming and Buddington 1988).

6.4 Yolk composition

Volume and density of the yolk in oviparous fishes, not necessarily the egg size, greatly affect ontogeny (Balon 1999). For example, eggs of the same diameter may have different amounts of yolk due to differences in perivitelline space, and this results in differences in yolk volume. Very little yolk of low density is commonly found in eggs of large size, whereas many smaller eggs have relatively large, high-density yolk (Balon 1986, 1999). Furthermore, large variation exists in the amount of yolk between various clutches of eggs or individual females (Balon 1986; Moodie et al. 1989).

The nutrient composition of fish eggs is species specific and their quality depends on maternal age, female weight, and diet (Sarasquete et al. 1995; Svensson et al. 2006). Balon (1999) concluded that yolk size and nutrient concentration together with cytoplasm volume may be more meaningful for characterizing reproductive investment. However, there are studies that also demonstrate correlations between the amount of yolk available at hatching and the original egg size (Dabrowski and Łuczyński 1984; Łuczyński et al. 1984). The absolute and relative amounts of dry yolk weight stores were higher in big rainbow trout alevins than in small ones at hatching; the alevins from small eggs had yolks that weighed 8.2 mg (or 72% of alevin weight), and those from big eggs had yolks that weighed 16.2 mg (Escaffre and Bergot 1984).

Sarasquete et al. (1995) concluded, based on results in gilthead sea bream, that the yolk reserves of fish contain proteins (rich in basic amino acids such as lysine and arginine, cystine and cysteine, and tyrosine and tryptophan), lipids, glycogen, and/or glycolipids. Simultaneously, the yolk is "equipped" with lysosomal enzymes and other enzymes related to protein, carbohydrate, and lipid mobilization. However, fish species differ in eggs and in yolk nutrient composition both qualitatively and quantitatively (see also Section 6.5).

Protein present in the egg, expressed as percentage of dry weight, ranges from 28.1 for red drum (*Sciaenops ocellata*) to 79.3 for winter flounder (Heming and Buddington 1988). The minimum and maximum values for lipids are 5.4 in common carp (*Cyprinus carpio*) and 52.0 in striped bass (*Morone saxatilis*), and for carbohydrates 0.4 and 5.3 in red drum and winter flounder, respectively (Heming and Buddington 1988). Demersal eggs of teleosts usually have lower water content (60–79%) compared with pelagic eggs (85–92%) (Craik and Harvey 1987; Kamler 2008).

6.4.1 Dry matter and water content in the eggs

The absolute amount of dry matter in fish egg varies between and within species (e.g., 7.0% in fertilized eggs of red drum and 46.3% for striped bass), but the percent water in the yolk and tissue is not constant throughout larval development (Dabrowski et al. 1984). In rainbow trout, yolk hydration increases significantly between hatching and swim-up stage, from 49% to 53–59% (Escaffre and Bergot 1984). The yolk volume decreases in Atlantic halibut larvae faster than the yolk protein content and results in an increase in protein concentration, and consequently, protein–to–dry weight ratio. This suggests a faster rate of yolk dehydration than of protein absorption. In turn, the decrease in the FAA pool is faster than the decrease in water (Rønnestad et al. 1993). As a result of water intake by yolk sac larvae, the percentage of dry matter in larvae of three marine species at the end of the endogenous phase of nutrition is lower than in the initial eggs (Kamler 2008; see also Section 6.6. and Table 6.1).

6.4.2 Proteins/amino acids

Pelagic marine fish eggs have very high water content (ca. 92%) and are characterized by low levels of phosphoproteins and high levels of FAA (Craik and Harvey 1984, 1987; Rønnestad et al. 1994). Rønnestad and Fyhn (1993) argued that 20–40% of the total amino acid pool used during embryonic and larval development of marine pelagic fish is used for catabolic purposes. The total amount of FAA is 92 nmol/mg egg in turbot, 122 nmol/mg egg in barfin flounder, 135 nmol/mg egg in Atlantic halibut, 190 nmol/mg egg in walleye pollock, and 200 nmol/mg egg in Atlantic cod (Fyhn 1990; Matsubara and Koya 1997; Ohkubo et al. 2006, 2008). Japanese eel eggs show a unique feature in their FAA concentration, with a value of 27 nmol/mg egg, which is lower than in other marine fish. On a percentage basis, it equates to only 22% of that reported in barfin flounder and 14% of that in walleye pollock (Ohkubo et al. 2008). FAA play a major role in osmoregulation in marine fish and are important osmolytes that contribute to egg buoyancy (Kamler 2008). In Japanese eel embryos, oil droplets ensure both energy supply and egg buoyancy (Ohkubo et al. 2008). Moreover, yolk amino acids (FAA and protein constituent) are the main substrates for energy metabolism and for protein synthesis in rapidly growing fish embryos and early yolk-feeding larvae (Rønnestad and Fyhn 1993; Rojas-García and Rønnestad 2003).

6.4.3 Lipids and fatty acids (FA)

Lipids are critical in embryonic and larval development. The lipid content of marine teleost pelagic eggs is 10–17% of the egg dry weight (Hilton et al. 2008; Iwamatsu et al. 2008). Lipids are present in eggs in two distinct forms: the lipoprotein yolk lipids, for example, phospholipids associated with the Lv; and the nonpolar lipids (TAG, wax esters, or cholesterol esters) (Poupard et al. 2000). Eggs containing low concentrations of total lipids (usually <15%) are relatively rich in polar lipids and their yolk sac larvae do not possess an oil globule (Finn et al. 1995b). Eggs rich in lipids (>15% of dry mass) contain oil globules composed mainly of the neutral lipids that have a high proportion of monounsaturated fatty acids (MUFA) (Desvilettes et al. 1997; Kamler 2008; Ohkubo et al. 2008). In some species, yolk lipids comprise only a small percentage of the yolk weight. Yolk lipids become more diverse in lipid composition than lipids in the oil globule. On average, yolk lipids are composed of 78.9% phospholipids, 6.5% TAG, 5.3% steryl and wax esters, and 2.8% sterols (striped bass; Eldridge et al. 1983).

Table 6.1 Chemical composition of fish eggs.

			Percentage of wet weight								
Species	Size: dry (d) or wet (w) weight	Dry matter (%)	Protein	Lipid	Glycogen	Ash	Lipid class: neutral lipids (NL); phospholipids (PL)	Docosahexaenoic acid (DHA)	Eicosapentaenoic acid (EPA)	Linoleic acid (LA)	Source
Red sea bream (*Chrysophrys major*)	38.5 μg (d)	12.73	7.67	3.82	0.22	0.91	–	–	–		Kimata (1983)
Atlantic salmon (*Salmo salar*; wild)	78.5 mg (w)	29.5	–	7.02	–	–	NL[a]	11.4	6.0	4.5	Cowey et al. (1985)
							PL[a]	25.5	9.7	1.4	
Atlantic salmon (wild)	–	–	–		–	–	NL	13.0	9.4	3.82	Pickova et al. (1999)
							PL	24.8	6.6	0.74	
Turbot (*Scophthalmus maximus*; wild)	–	–	–	13.2–13.8	–	–	NL	9.4	3.4	1.2	Silversand et al. (1996)
							PL	33.4	9.9	0.8	
Cod (*Gadus morhua*)	1.77 mg (w)	7.7	–	9.9	–	–	NL	19.0	12.0	1.0	Fraser et al. (1988)
							PL	33.0	15.0	traces	
White sturgeon (*Acipenser transmontanus*; wild)	–	–	–	13.5	–	–	NL	9.2	3.1	0.3	Czesny et al. (2000)
							PL	15.2	4.2	traces	
Walleye (*Sander vitreum*; wild, Lake Erie)	2.34–3.67 (w)	32.8–42.0	–	10.5–14.5	–	–	NL	10.7–12.8	6.1–6.8	3.10	Dabrowski et al. (unpublished data)
							PL	27.8–32.5	10.8–11.7	0.35	
Yellow perch (*Perca flavescens*; wild, Lake Erie)	–	–	–	4.2–5.0	–	–	NL	13.3	5.0	10.6	Rinchard et al. (2004)
							PL	30.4	10.1	1.8	

The information provided is based on data available in the literature and on our own previously unpublished data.
The compilation of biochemical data for drawing comparisons among fish species is frequently impossible because the data is expressed for individual organisms in order to demonstrate the utilization of nutrients during every step of ontogenesis (see Finn et al. 1995b; Hilton et al. 2008). Although more extensive analysis is required, it can be concluded that (1) there is no marked difference between lipid content of DHA and EPA (PUFA) between marine, freshwater, and migratory fishes; (2) PL contained considerably more PUFA than NL; (3) LA concentrations in PL are frequently below the level of detection in comparison with NL, which is perhaps indicative of selective deposition through vitellogenin synthesized in the liver; and (4) low-level LA concentrations are indicative of wild fish gonads in comparison with fish fed formulated feeds containing plant oils.
[a] Weight % of total identified fatty acids.

The oil globule lipids are composed of steryl and wax esters (90.4%) and TAG (9.6%). Lipids in yolk lipoproteins are composed primarily of polar lipids rich in (n-3) polyunsaturated fatty acids (PUFA) (Wiegand 1996b). Saturated fatty acids (SFA) and PUFA, eicosapentaenoic acid (EPA) C20:5(n-3), docosahexaenoic acid (DHA) C22:6(n-3), and arachidonic acid (ARA) C20:4(n-6) are initially found in the yolk, but they are preferentially incorporated into structural lipids in larval tissues. However, MUFA are preferentially utilized as an energy source in embryonic and larval fish metabolism (Wiegand 1996b; Kamler 2008). These preferences are influenced by low temperatures, presumably due to adaptive changes in membrane FA composition in larvae (Kamler 2008). This is particularly evident in salmonid embryos and alevins developing below 10°C where docosapentaenoic acid in phospholipids and DHA in triglycerides correlated with posthatch mortality of lake trout (*Salvelinus namaycush*) (Czesny et al. 2009). The total lipid concentration in Japanese eel eggs is higher (8.8 µg/0.6 mg egg; i.e., 1.5% of wet matter) than in fish eggs with no visible oil droplets such as that of barfin flounder (16.9 µg/2.9 mg egg; 0.6%) and walleye pollock (11.7 µg/1.6 mg egg; 0.7%). In pike, eggs contain 17.8% lipid dry weight. In this freshwater, piscivorous fish, polar lipids accounted for 40% of the total lipid pool, with PC and phosphatidylethanolamine (PE) as the dominant polar lipids, and phosphatidyloserine and phosphatidylnositol as the minor components (Desvilettes et al. 1997). Fish species that contain low amounts of total lipids in eggs utilize PC as the major lipid fuel during embryonic and larval development (Finn et al. 1995b). This rule also holds for marine species such as Atlantic cod, Atlantic herring (*Clupea harengus*), haddock (*Melanogrammus aeglefinus*), whiting, Atlantic halibut, plaice (*Pleuronectes platessa*), and Pacific halibut (*Hippoglossus stenolepis*). It was demonstrated that embryos of freshwater species, such as roach (*Rutilus rutilus*) and African catfish (*Clarias gariepinus*), also use phospholipids preferentially (summarized by Finn et al. 1995b).

6.4.4 Carbohydrates

Atlantic cod eggs at fertilization contain only 1% of carbohydrates, and this is similar to turbot. Glycogen concentration declines immediately after fertilization, suggesting that it is used as fuel during early embryogenesis. It has been hypothesized that the cytosolic fuel store must be able to maintain metabolism in the embryo until the *ysl* is sufficiently formed to access other energy sources contained in the yolk (Finn et al. 1995a, 1996).

6.5 Mechanism of yolk utilization

6.5.1 Structure, resistance to, and degradation of yolk platelets and granules

The *ysl/ytl* plays an important role in the process of protein transport from the yolk cells to the growing embryos and larvae. Krieger and Fleig (1999) suggested that final qualitative proteolysis occurs in the yolk platelets in the *ysl*. However, they also described "the existence of the yolk platelets in cells of ICM that imply that exocytosis, endocytosis, and later on the intracellular digestion of yolk proteins and lipids is the way to make potential nutrients in the yolk available to the developing embryo" (Krieger and Fleig 1999). Proteolysis of the yolk platelets occurs in the *ysl* because the yolk compartments themselves contain hydrolytic enzymes (Heming and Buddington 1988; Krieger and Fleig 1999). Vernier and Sire (1977b) described two types of yolk platelets in rainbow trout embryos with different enzyme contents. One form of the platelets has an enzyme that would allow nutrients to be released prior to the formation of the *ysl*. The second or common platelet type lacks this enzyme and is digested with the *ysl* structural enzymes (Heming and Buddington 1988). It is proposed that the region of the *ysl* with smooth ER and numerous mitochondria, and an abundance of glycogen granules, is responsible for carbohydrate and/or lipid metabolism. The second region is characterized by rough ER and Golgi complexes. It is associated with the synthesis and transport of proteinaceous substances (Heming and Buddington 1988). Yolk phosphoproteins are dephosphorylated by the action of calcium-dependent phosphatase to become soluble (Heming and Buddington 1988). Syncytial Golgi complexes are likely to supply acid hydrolases for the degradation of yolk platelets (Vernier and Sire 1977b; Hamlett et al. 1987).

YGs do not disintegrate their contents until specific embryonic developmental stages, even weeks after their formation. This is in contrast to the action of classical lysosomes that are capable of rapidly reducing almost any protein to FAA, sugars, and other small compounds. Yolk platelets are first found in the *ysl* at the morula stage, as described in European perch (Krieger and Fleig 1999). Two factors, pH and enzymatic latency, are responsible for the regulation of YG activation and yolk utilization (Fagotto 1995). YGs can modulate their pH; initial pH is neutral or slightly acidic, and during embryonic development acidification gradually occurs. This consequently triggers yolk degradation, as low (acidic) pH is required for cathepsin activity and proteolysis of the yolk proteins. One of the mechanisms required for enzyme activation in yolk compartments is H^+-adenosine triphosphatase (ATPase)-mediated acidification (Tingaud-Sequeira and Cerdà 2007). The induction of the maturation of the proteolytic enzymes that are active in a wide range of acidic pH (4–6.5) is the result of YG acidification at late developmental stages (Fagotto 1995).

It is likely that the resistance of YGs to degradation is similar to that in Vtg and this property is conserved during evolution, as Vtg is the predominant yolk protein that is present in most egg-laying animals (Fagotto 1995; Jorgensen 2008).

6.5.2 Utilization of yolk proteins and amino acids

Yolk proteins are derived from one form of Vtg, Vtg A, cleaved completely to generate a supply of amino acids. This pattern was identified by analyzing the expression of *ctlsa*, *ctlsb*, and *ctlsc* genes in tissues of zebrafish embryos and adults, as well as in killifish embryos. In killifish, procathepsin L (Ctsl) isoform is most likely involved in yolk protein hydrolysis (Tingaud-Sequeira and Cerdà 2007). The activity of aspartic protease–cathepsin D (Ctsd) appears to be involved in the degradation of yolk proteins in vertebrates in general, and zebrafish embryos and larvae in particular (Fagotto 1995; Tingaud-Sequeira and Cerdà 2007). As the proenzyme Ctsl, located in the *ysl*, is transferred to the yolk globules, detached from a central yolk mass, it is activated by Ctsd contained in yolk globules to cathepsin L, which hydrolyzes yolk proteins (Sire et al. 1994; Hartling and Kunkel 1999; Hiramatsu et al. 2002). Several enzymes in the yolk matrix are weakly positive, among them alkaline and acid phosphatase, ATPase, glucose-6-phosphatase, and trypsin. Sarasquete et al. (1993) also indicated that the cellular envelope covering the yolk sac shows high levels of activity of these enzymes, which may indicate their function in osmoregulation and defense (trypsin).

Despite the similar composition of the FAA pool found in newly fertilized eggs, Rønnestad et al. (1998) suggested that for many marine pelagic eggs, there is no rule describing protein synthesis rate from essential (EAA) or nonessential amino acids (NEAA) present in the yolk. As summarized by Kamler (2008), differential utilization of EAA and NEAA occurs in barfin flounder and Atlantic halibut in contrast to plaice, where no selective utilization of EAA and NEAA was observed. The mobilization of FAA can occur mainly before and just shortly after hatching as in turbot, in the embryo and early yolk sac larva as in lemon sole (*Microstomus kitt*), or in the yolk sac larva of species including Atlantic halibut (reviewed by Kamler 2008). In sea bass, approximately 60% (53 nmol/ind) of the FAA pool is exhausted before hatching (18°C), while neutral lipids derived from the oil globules appear to be the main energy substrate after hatching (Rønnestad et al. 1998). In walleye pollock, FAA were utilized mainly until hatching, around 17 dpf, whereas active utilization of Lv, derived from VtgB, phospholipids, and TAG (major lipid classes catabolized), occurred during the larval period, from 18 to 28 dpf at 5°C (Ohkubo et al. 2006). Yolk nutrient utilization in barfin flounder embryos (0–10 dpf at 8°C) and yolk sac larvae (days 11–21) was suggested by Ohkubo and Matsubara (2002) to have a triphasic sequence. It follows a distinct pattern: days 0–4 no FAA is utilized, and then the main depletion of FAA occurred after day 4 resulting in only 13% of the initial FAA level remaining by day 13. The Lv and phospholipids were mostly utilized during the 16–21 dpf period. In Atlantic cod, FAA comprised 75% of the metabolic fuel during embryonic development, followed by 13% of the polar lipids and 9% of TAG. After hatching, the fuels were used as follows: FAA, 2%; polar lipids, 20% (mainly PC); neutral lipids, 17%; and proteins, 31%. This resulted in the total use of 67% amino acids and 32% lipids (Finn et al. 1995a). In turbot, a species with pelagic eggs containing a single oil globule, the catabolic metabolism of embryos and endogenously feeding larvae was fueled by amino acids (50%, with similar amounts being supplied from both free and protein bound pools) and lipids (50%, mainly of neutral lipid origin). Also, FAA were the major metabolic fuel during embryonic development (84%) in turbot. Wax esters (33%) and TAG (25%) were the prominent fuels after hatching, and only 10% of FAA was used at the same time (Finn et al. 1996). During 12 days after hatching in Atlantic halibut yolk sac larvae, more than 70% of FAA pool is utilized from the yolk, with no significant changes in the yolk protein pool (Rønnestad et al. 1993). This suggests the sequential utilization of endogenous reserves in Atlantic halibut larvae, when FAA are the dominant energy substrate (relative catabolic contribution of 74% of oxygen consumed) during yolk absorption after hatching (Rønnestad et al. 1993). Later on, the additional amino acids are recruited from yolk proteins; 60% of total amino acids (free and protein amino acids) present at hatching are used as precursors for body protein synthesis, while the remaining 40% are used as the substrate in larval energy metabolism (Rønnestad et al. 1993).

Based on results obtained for many marine fishes, it was concluded that during the endogenous feeding stages FAA are mobilized earlier (embryogenesis and in newly hatched larvae) in the ontogeny than amino acids from the yolk protein pool (Rønnestad et al. 1995, 1998; reviewed by Kamler 2008). However, it is questionable how these pools, free and bound amino acids, can be distinguished without radiolabeling when protein synthesis and degradation proceed simultaneously at very high, and perhaps different, rates (Langer et al. 1993).

6.5.3 Lipid utilization

Substantial differences in the patterns of lipid class utilization by embryos and teleost larvae are reported (Wiegand 1996a; Desvilettes et al. 1997). Phospholipids, TAG, and wax esters alone, in sequence, or in combination may be consumed as energy sources at various stages of development. The existence of non-pancreatic lipase, supposedly yolk sac lipase, detected from eviscerated body in walleye pollock larvae was suggested to be related to the processing of phospholipids present in the yolk (Oozeki and Bailey 1995).

After hatching, lipids are the dominant fuel in many fishes, for example, turbot (about 60% lipids to 40% total FAA), sea bream (60–70% from neutral lipids of the oil globule), walleye pollock, African catfish, Senegal sole (*Solea senegalensis*), common dentex (*Dentex dentex*), and white sea bass (*Lates calcarifer*) (reviewed by Kamler 2008; Hilton et al. 2008). In yellowtail amberjack (*Seriola lalandi*) larvae, the wax and/or sterol esters (SE) and TAG are the most important energy sources between hatching and first feeding, being almost completely depleted at 5 and 3 dph, respectively (rearing temperature 20 ± 1°C; Hilton et al. 2008). In turn, in northern pike development, total lipids decreased by 41.3% during the embryonic phase (days 2–6, at 15.5°C), and then the lipid weights of the whole yolk sac larvae remained stable from hatching until day 11, with a tendency for lipid amounts to increase in larval bodies and to decrease in the yolk (Desvilettes et al. 1997). Based on the rate of lipid utilization it could be concluded that the period of yolk sac absorption (pike, 7–11 dpf; European sea bass until 100 hours postfertilization [hpf]) is characterized by low consumption of lipids, when proteins and FAA are used (Desvilettes et al. 1997; Rønnestad et al. 1998). A significant decrease in yolk lipids occurs in the swim-up stage of sea bass (between days 13 and 15, at 18°C); they are consumed as the energy source for movement, like in many other fish species (Desvilettes et al. 1997; Rønnestad et al. 1998).

In Japanese eel, there are two stages of nutrient utilization (Ohkubo et al. 2008). The first is when FAA utilization occurs between days 1 and 4 after fertilization and the second when TAG utilization occurs after day 2, and at the same time more than 30% of the FAA and Lv (380 kDa, ova-lipovitellin [oLV]) are utilized. Then TAG becomes the main energy provision from days 4 to 8, and phospholipids from days 2 to 10. Almost all the yolk nutrients in Japanese eel are absorbed by day 8 (100% of oLV; 77% of TAG; 32% of phospholipids), close to the onset of exogenous feeding at water temperature of 23°C (Ohkubo et al. 2008). Also, in gilthead sea bream, FAA appear to be a significant energy substrate during the embryonic stage (60–70%). However, FA from neutral lipids, of oil globule origin, become the main metabolic fuel after hatching (80–90% of energy) (Rønnestad et al. 1994).

The oil globule persists after yolk depletion for as long as 32 hours in rabbitfish (*Siganus guttatus*), 88 hours in barramundi (Bagarinao 1986), and several days in many species of freshwater fish larvae, beyond the "point of no-return" (Dabrowski 1989). In larvae of gilthead sea bream, 60% of the oil globule is still present when body proteins are being catabolized in the last part of the yolk sac stage (Rønnestad et al. 1994). The oil globule of European sea bass begins to be absorbed after hatching when yolk absorption is advanced, and occurs in parallel with catabolism of FA from yolk neutral lipids. At the onset of exogenous feeding, approximately 30% of the oil globule is still present in sea bass larvae (Rønnestad et al. 1998). In summary, absorption of the oil globule occurs principally after hatching and follows yolk absorption, mainly after consumption of the yolk, when larvae are free-swimming and are active. Oil globule exhaustion can be correlated with catabolism of FA, mostly from the neutral lipids (Poupard et al. 2000; Iwamatsu et al. 2008).

Similarly, in Atlantic halibut, a species whose eggs do not possess an oil globule, the larval body at hatching contained 11% lipids in larval body dry mass, while the yolk contained 93% of the remaining lipid pool. At the time of maximum larval body weight, the lipid retained amounted to 50.7% of the initial value (Rønnestad et al. 1995). During the period between hatching and the time of maximum larval size, the decline of the lipid classes in the yolk was shared equally, indicating a relatively minor change in the relative composition of lipids in the yolk (PC, 56 ± 5%; PE, 12 ± 1%; TAG, 12 ± 3%; SE, 7 ± 1%; cholesterol, 7 ± 1%). However, a large variation was described between incorporation of lipid classes in the larval body, with PC being the least conserved. Almost 100% of the cholesterol was conserved from hatch to the attainment of the maximum dry body weight after complete yolk absorption (Rønnestad et al. 1995), although synthesis of cholesterol at this stage cannot be refuted. The high catabolism of PC is thought to occur in the early stages of development in marine fish with a high content of PC in the eggs, when the catabolism of natural lipids, particularly TAG, seems to be correlated with the increased activity and metabolic rate of larvae (reviewed by Rønnestad et al. 1995; Rainuzzo et al. 1997).

Finn et al. (1995a, 1996) proposed that the sequence of catabolic substrate oxidation, when polar and neutral lipid utilization after hatching follows consumption of FAA during the embryonic stages, is in general similar in many cold-water fishes that spawn eggs with or without oil globules. As explained earlier, PC and/or TAG are the predominant lipid energy sources during larval development in fish with eggs containing less total lipids (no oil globule in the yolk), assuming that the oil globule supplies greater amounts of wax, SE, and/or TAG. However, it was pointed out by Finn et al. (1995b) that the utilization of PC is also important in salmon and turbot, and there are phospholipids found exclusively in marine species, sea bream and red drum, which possess an oil globule in the yolk sac. In larvae that have both the yolk and oil globule, TAG stored in the oil globule is the last to be utilized between hatching and first feeding (Bagarinao 1986). In turn, it is suggested that in eggs with no visible oil droplets, the catabolism of TAG during embryogenesis may occur earlier, that is, in the first moments of embryo development (Finn et al. 1995a, 1995b).

Lipids are present in yolk globules as polar and/or as neutral lipids. Among the species with an oil globule, there are marine (e.g., red drum; capelin, *Mallotus villosus*; sandeel, *Ammodytes lancea*; Atlantic menhaden, *Brevoortia tyrannus*; gilthead sea bream; and turbot) and freshwater species (e.g., rainbow trout; perch; burbot, *Lota lota*; and Atlantic salmon, *Salmo salar*) (reviewed by Finn et al. 1995b). In turbot larvae at hatching, two compartments with endogenous reserves, the yolk mass and the oil globule, are surrounded and separated from each other by the *ysl/ytl* (Poupard et al. 2000). ApoE is the essential molecule responsible for lipid metabolism in fish embryos and larvae because of its ability to communicate with lipoprotein receptors (Poupard et al. 2000). In turbot, for instance, the *apoE* gene is highly expressed in the *ysl* at sites of lipoprotein (VLDL particles) synthesis and secretion, in the ventral and posterior periphery of the yolk cell (Babin et al. 1997; Poupard et al. 2000). VLDL particles are found in the ER lumen and in secretory vesicles. They are then released into the extracellular perisyncytial space, and the intrasyncytial channels transfer the yolk- and oil-globule-derived lipids into the embryonic or larval tissue (Poupard et al. 2000; see also the results for white bass in Figure 6.2). As in many teleost species, the oil globule in turbot resorption occurs after hatching and principally at the time of the mixed, endoexotrophic nutrition. At the same time, a very high *apoE* expression is observed in the turbot *ysl* that is surrounding the oil globule. Poupard et al. (2000) also indicated that amino acids derived through exogenous nutrition are essential for apolipoprotein synthesis and that polar lipids are needed for lipoprotein synthesis (Poupard et al. 2000). As suggested by these authors, the pattern of apoE mRNA expression associated with the *ysl* lipoprotein synthesis suggested that *apoE* is a molecular marker for tracing endogenous lipid utilization during mixed feeding period in teleost fish early life.

6.5.4 Liver function in yolk utilization

The fish liver arises and remains in close proximity to the yolk. Because it is well-vascularized tissue, this organ plays an important role in the transfer of yolk nutrients to the gut and other organs (Morrison et al. 2001). The liver also regulates lipid metabolism by lipoprotein synthesis in association with the use of yolk reserves (Hoehne-Rejtan and Kjørsvik 2004).

Fishelson (1995) emphasized, although with no direct evidence, that the presence of YGs in the primordial liver cells of cichlid larvae can be interpreted as a similar process to that in avian embryos. In other species, like gilthead sea bream larvae (Guyot et al. 1995; Sarasquete et al. 1995) and pike perch larvae (Mani-Ponset et al. 1996), during the endogenous phase of nutrition, the liver accumulates mainly glycogen, until the endoexogenous feeding period starts (summarized by Hoehne-Rejtan and Kjørsvik 2004). No glycogen content was found in the early yolk sac stage in sharpsnout sea bream (*Diplodus puntazzo*), Atlantic cod, and Atlantic halibut hepatocytes (Figure 6.3; Hoehne-Rejtan and Kjørsvik

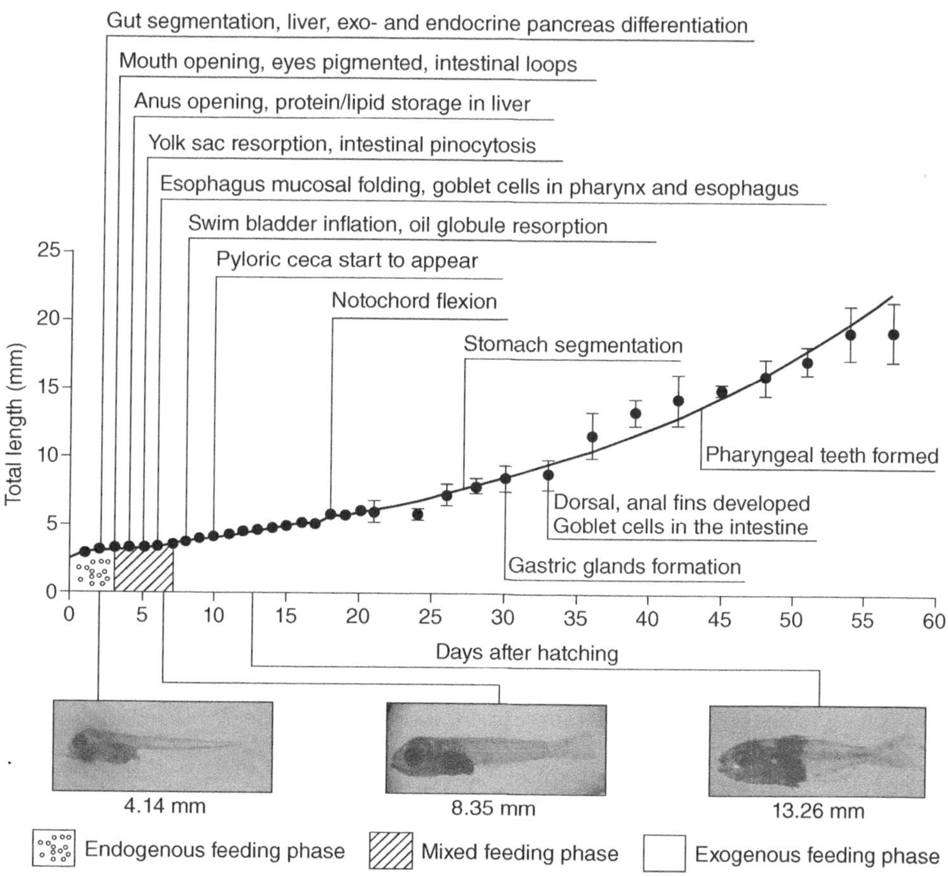

Figure 6.3 Growth (total length, mm) of sharpsnout sea bream larvae from hatching (0 dph) to 57 dph, and major events of digestive system differentiation that occur during development. Endogenous phase of nutrition: 0–3 dph; mixed feeding: 4–7 dph; exogenous phase of feeding: 8–57 dph. (Figure was redrawn based on figure 1 and table 1 from Micale et al. [2008].)

2004; Micale et al. 2008). As concluded by Hoehne-Rejtan and Kjørsvik (2004), the observed differences between fish species in hepatic glycogen content during the endogenous phase of nutrition could be related to the yolk lipid content and larval energy metabolism at the yolk sac stage.

6.5.5 Conclusions regarding processes involved in yolk utilization

Studies on liver development in sea bream larvae reveal the sequence of lipid appearances in the bloodstream. During the endotrophic period, yolk reserves are broken down and distributed as lipoproteins, which can be identified in the vitelline circulatory system. During the phase between mouth-opening and the ninth day, the absence or limited presence of lipoproteins in the blood is found in gilthead sea bream. However, the transfer of vitelline reserves in the form of VLDL is nearly or fully completed at this time, and lipids of food origin are not yet identified (Guyot et al. 1995). Signs of lipid digestion are found in the cells of the alimentary tract of sea bass larvae during the endogenous phase of nutrition, and this was discussed in reference to an ability of these larvae to synthesize and transport small lipoprotein particles before the start of exogenous feeding (García-Hernández et al. 2001).

In summary, energy substrates of eggs or the yolk are species specific. Energy dissipation based on lipids seems to be important for those species whose eggs contain oil globules, while amino acids play a more important role for those species whose eggs do not contain oil globules. The role of materials stored in the oil globule is not clear, but Iwamatsu et al. (2008) opined that the embryo does not require these nutrients, despite the fact that the oil droplets possess substances of high caloric value and vitamins A and E (Lubzens et al. 2003, 2010). It was also documented that starved larvae of striped bass retained the oil globule longer than fed larvae and the rate of oil absorption increased with an increase in food intake. Therefore, as concluded by Hilton et al. (2008) for yellowtail amberjack, lipid provisioning may be an important factor during the "critical period" for larval survival.

These findings may have important implications for the nutritional requirements at the onset of the first feeding (Rønnestad et al. 1994). Further investigations are recommended to clarify the specific role of oil droplets in maintaining nutritional balance, and consequently, in the survival and growth of larvae. Jorgensen (2008) has summarized the status of lipid/FA utilization in endogenous nutrition and concluded that there is still much to be discovered in terms of the molecular biology of yolk consumption. The mechanisms involved in controlling embryonic nutrition are still unknown. Questions remain such as how do embryos initiate yolk platelet breakdown and how is the cell signaling system activated to recognize the need for amino acids or other limiting nutrients. There are also questions regarding how embryo signaling systems communicate the rate of nutrient transfer into the blood to the gut or yolk sac. Additionally, there is a need to examine the differences in the nutrients deposited in the yolk (e.g., yolk platelets, lipid droplets, and glycogen) and how they are directed to provide buildings blocks for biomass gain and fuel for metabolism. It is assumed that the mechanisms of internal tissue amino acid sensing are evolutionarily conserved and that they control body weight gain in association with endocrine signaling. This hypothesis will provide new impetus for such studies.

6.6 Rate and efficiency of yolk absorption

6.6.1 Dynamics of yolk consumption

The preference and rate of yolk nutrients mobilized by developing larva differ as this process is species specific. The yolk composition and requirements for nutrients may change during the course of early development depending on many environmental factors. The dynamics of yolk consumption in most teleosts can be described by a sigmoid curve with three distinct phases: a slow absorption rate at the start (early embryonic development) and end (late larval stage) of yolk consumption, and the midphase (late embryonic stage) characterized by a relatively high and constant rate of absorption (Johns and Howell 1980; summarized by Rønnestad et al. 1993). Skjaerven et al. (2003) used a set of physiological (nitrogenous metabolites) and morphological (*ysl* differentiation) events during plaice embryonic development to characterize the step-by-step utilization of the yolk. These authors proposed four phases of yolk absorption, with the endotrophic prehatch phase divided into separate phases, first during the enhanced yolk consumption at a time between fertilization and gastrulation and second from the completion of the epiboly until hatching. This gradual transition in the modes of nutrient utilization is the result of increasing size of the larva, higher activity, and presence of an efficient circulatory system in the yolk sac (Kamler 2008). Yolk utilization begins with embryo development and is followed by increased oxygen consumption after the blastula stage is reached. To summarize, common features of yolk nutrient utilization are the following: FAA are utilized during the embryonic stages and the early yolk sac stage and are followed by FA from neutral lipids derived from the oil globule and amino acids derived from proteins as the main metabolic sources in the last part of the endogenous nutrient use (Rønnestad et al. 1998; Kamler 2008).

Protein is the major nutrient stored in the yolk. Protein in tissues of Atlantic halibut larvae when using the yolk between hatching and maximum body mass increases exponentially with age according to the equation Protein (μg/ind) = $42.67 e^{0.0677\tau}$ (τ, days post-hatch). In larval tissue, there is a linear increase in FAA content with age between hatching and final yolk sac resorption (FAA [nmol/ind] = $27.54 + 6.66\tau$). Protein content expressed per dry body mass consistently comprised about 54% (Rønnestad et al. 1993; Finn et al. 1995c). Protein synthesis rates in yolk sac larvae of African catfish are as high as 138% body protein weight/day (Conceição et al. 1997). In tissues of rainbow trout at the completion of yolk absorption, there was 22 times more FAA than in the tissues at the time of hatching (Zeitoun et al. 1977).

In embryos or yolk sac larvae, the concentration of ash is rarely determined, but available data suggests a low concentration of minerals in fish at hatching compared with the body at the completion of yolk absorption. It is apparently related to skeleton formation during ontogeny. Larvae with high mineral content emerge from eggs rich in minerals; however, extraoral mineral absorption, such as for instance of magnesium, is critical to larval survival (Van der Velden et al. 1991).

In principle, the carbon/nitrogen (C/N) ratio describes the ratio of fats to protein, with values near 3 for proteins and above 3 for lipids. The ratio was examined for the yolk sac turbot, Atlantic herring, and nase (*Chondrostoma nasus*), and it decreased with age (Korsgaard 1992). These results describe the mobilization pattern of yolk reserves and carbon-rich lipids that are preferentially catabolized to obtain energy, whereas proteins are saved for growth (Ostrowski and Divakaran 1991). In three species, rainbow trout, winter flounder, and red sea bream (*Pagrus major*), the concentration of RNA in the whole egg or larva/alevin (including the yolk) increased between early cleavage and the onset of exogenous feeding (Zeitoun et al. 1977; Buckley 1981; Seoka et al. 1997). This increase can be interpreted as an intense protein synthesis and cell proliferation (Kamler 2008).

As discussed by Raciborski (1987), the caloric value of the sea trout (*Salmo trutta morpha trutta*) yolk sac exceeds that of alevin body by about 1 kcal, in agreement with earlier studies on rainbow trout (Kamler and Kato 1983). The higher caloric value of the yolk sac as compared with the salmonid alevin body is the result of differences in the ratio of protein and lipids. It can be concluded that the yolk sac has higher caloric value per dry matter than the alevin body because the fish body contains less lipid and consists of many low-energy substances, including cartilage and bones forming the skeleton, and scales (Raciborski 1987). Moreover, it was observed that during sea trout development, the caloric value of the yolk sac did not change significantly, even when its weight decreased; therefore, there must be a proportional utilization of proteins and lipids (Raciborski 1987).

The duration of endogenous nutrition is related to larval size (reviewed by Kamler 2002). The relationship of the time from hatching to final yolk absorption ($\tau_{H\text{-}Re}$) on the body length of newly hatched larvae (TL, mm) in marine, freshwater, and anadromous fish (at 15°C) was described by the following equation: $\tau_{H\text{-}Re} = 4.76 + 1.3\,TL$, $n = 88$, $R^2 = 0.33$, $P < 0.0001$ (Miller et al. 1988).

The findings on yolk utilization rate were comprehensively reviewed recently by Kamler (2008), so this section will be brief and will summarize this work. The energy budget of an organism with an endogenous source of energy can be defined by the following equation: $C_Y = P + R + U$, where C_Y is the yolk energy consumed; P is the portion of energy invested in newly formed tissue; R is the energy expended in respiration, with no feces egested prior to the onset of external feeding; and U is the amount of energy excreted as nitrogen waste (Terjesen et al. 1997; Kamler 2008). In general, the yolk energy used (C_Y) and converted to tissue growth (P) can be determined from changes in the yolk and body tissue weights. The mass values multiplied by their respective energetic equivalents result in values of energy used or deposited in tissues (Johns et al. 1981).

6.6.2 Rate of yolk absorption

Yolk absorption rate is assumed to vary during development and is thought to be a function of the surface area of the absorptive layer (*ysl/ytl*) and the metabolic activity of this layer (Heming and Buddington 1988). The surface area available for yolk absorption is dependent on size; in teleosts, it is approximately equal to the area of the yolk sac (Skjaerven et al. 2003) and, to a lesser degree, the shape of the yolk mass (Heming and Buddington 1988). The activities of hydrolytic enzymes also contribute to the yolk resorption rate. As summarized by Kamler (2008), the combined effect of egg size (positive) and duration of embryonic development (negative) on yolk dry weight at hatching can be quantified by a polynomial equation, explaining 76% of the variance in yolk size. The faster absolute yolk absorption rate was observed in fish species producing larger eggs. Larvae of Atlantic cod, European sea bass, plaice, turbot, and winter flounder hatch from small eggs (1–2 mm) with a limited amount of yolk and have a short (2–6 days) yolk sac absorption period (reviewed by Gawlicka et al. 2000). Atlantic halibut, hatching from larger eggs (3–3.5 mm) with greater yolk reserves, have a longer endogenous phase of nutrition that lasts for 46–50 days (280–320 degree-days at 6°C) (Gawlicka et al. 2000). The effect of variation in the initial egg size on the yolk consumption rate within a pool of females of the same species was examined in rainbow trout. The results showed that the stage of complete yolk absorption in 50% of the trout alevins was reached earlier in small-egg alevins than in big-egg alevins and the difference was 2.5–3 days (Escaffre and Bergot 1984). The small-egg alevins and big-egg alevins were from different populations, with no intrafemale differences. The value for overall rate of yolk absorption was 70%, from fertilization to the time of maximum body weight, for both alevin classes of rainbow trout (Escaffre and Bergot 1984). However, the relative rate of yolk absorption (%/day) was lowest in the largest eggs (1.4%/day in chum salmon, *Oncorhynchus keta*) and highest in the smallest eggs (50.2%/day in bluegill sunfish, *Lepomis macrochirus*). Moreover, the mean relative yolk absorption rate (%/day) increases with time as development proceeds. The amount of energy consumed daily by an embryo is lower than in larvae between hatching and the larger swim-up stage, which are supplied with a more developed blood circulatory system and possess higher metabolic rates (activity). The absolute rate of yolk utilization decreases after the swim-up stage as a result of the limited amount of yolk remaining (Kamler 1992). In summary, the yolk consumption rate is assumed to be dependent on initial egg size, amount of remaining yolk, and environmental factors (Jaworski and Kamler 2002).

6.6.3 Influence of abiotic factors on the rate and efficiency of yolk absorption

Among the many extrinsic factors that influence the rate and efficiency of yolk absorption in fish larvae are temperature, light, oxygen concentration, and salinity. Species of viviparous fish or mouthbrooding fishes are able to manipulate egg incubation conditions. The variation within the acceptable environmental conditions will depend on those parameters experienced in the process of evolution of the species. In some species of fish, such as those in abyssal marine habitats, the environment is relatively constant, while in many freshwater and estuarine habitats, it may fluctuate in terms of temperature, salinity, pH, or ionic composition (Heming and Buddington 1988).

6.6.3.1 *Temperature*

Temperature affects larval size during yolk sac absorption because it plays a dominant role in the efficiency of yolk conversion into body tissues. For example, vendace (*Coregonus albula*) larvae caught in deep (2–6 m) cold waters had larger yolk than larvae from the near-surface (0–2 m; Karjalainen and Viljanen 1992). Łuczyński et al. (1984) also provided experimental evidence for this temperature effect on larval size and yolk sac volume in vendace at hatching. Typically, the yolk absorption rate increases with temperature within the optimal species-specific temperature range. Yolk utilization in nase, rainbow trout, chinook salmon (*Oncorhynchus tshawytscha*), spotted wolffish (*Anarhichas minor*), common pandora (*Pagellus erythrinus*), and mackerel (*Scomber scombrus*) (Heming 1982; Kamler and Kato 1983; Klimogainni et al. 2004; Mendiola et al. 2007) accelerated with higher water temperatures. Yolk utilization rate for Atlantic salmon and brown trout alevins incubated at temperatures between 4 and 12°C had a peak at 10°C. Yolk absorption efficiency was 81% at 10°C, whereas at 4°C the yolk was converted to brown trout alevin tissues at 57% (Johns et al. 1981; Heming 1982). Higher efficiency of yolk utilization for *Pagellus erythrinus* was observed at 18°C, with the highest rate of hatching and survival occurring at 18–21°C. Increasing growth rate in length was observed as temperature rose from 16 to 18°C and stabilized at 18–21°C (Klimogainni et al. 2004). In Nile tilapia, *Oreochromis niloticus*, the gross yolk utilization efficiency was between 55.4 and 61.7%; efficiency between hatching and maximum dry body weight was temperature dependent between 24 and 30°C and ranged from 55.4 to 61.7%. Rana (1990) noticed that at higher rearing temperatures, feeding commenced earlier. With increasing temperature from −1 to 7°C, the age at which 50% of Atlantic cod larvae absorbed their yolk decreased (Pepin et al. 1997). Summer flounder (*Paralichthys dentatus*) larvae, in contrast, showed no significant differences in time of total yolk sac absorption at experimental conditions of 5, 12, 16, and 21°C, as well as with temperature cycles 5–11, 11–16, and 16–21°C, but development rate was temperature dependent and increased with increasing temperature (Johns et al. 1981). This was not an exception, as these authors summarized, since there are several studies in which yolk–tissue conversion efficiency and larval size at yolk sac stage were found to be independent of temperature over the entire range of thermal tolerance, that is, *Bairdiella icista* and yellowtail flounder (*Limanda ferruginea*). The relatively small variation in utilization efficiency in the range of tolerated temperatures is due to the interaction between two processes: (1) the rate of increase in metabolic needs in the physiological temperature range, and (2) the shortening of development time with increase in temperatures. As a net result, the yolk utilization efficiency may be near constant assuming that both processes change at the same rate (i.e., measured as Q_{10} values) in the physiological temperature range (Johns and Howell 1980).

It is obvious that the rate of yolk absorption depends on species-specific tolerance to water temperatures. Practically, however, the highest yolk utilization efficiency occurs at the optimum temperature, the intermediate range, within the zone of tolerance. At the lower end of temperatures, yolk absorption increases, and further increases in temperature may be associated with a decrease in yolk utilization (Kamler 2008). Oil globule absorption appears to be more affected by an increase in temperature than yolk platelet utilization (reviewed by Heming and Buddington 1988).

6.6.3.2 Light

In salmonids, yolk absorption efficiency is not only maximized by dark conditions (Heming 1982) but is also affected by water exchange rates and substrate contour. Effects of light intensity on yolk absorption and development of the digestive tract have been examined in common dentex larvae, and yolk sac volume at the onset of external feeding was largest (17.6 μm^3) in total darkness and smaller (4.4 μm^3) under constant illumination of 450 lux for 24 hours a day. For oil globule volume, the values were 1.95 and 0.56 μm^3, respectively (Fírat et al. 2003). However, data obtained for Atlantic salmon were not as unambiguous (Kamler 2008).

6.6.3.3 Xenobiotic factors

The yolk absorption rate is also delayed under xenobiotic influences. Yolk absorption in Northern pike larvae was inhibited by 800 μg/L Al concentrations (in acidic water at pH 5.25 during 10 days) and in roach larvae (at pH 5.75). In Mozambique tilapia (*Oreochromis mossambicus*), yolk absorption was reduced during exposure to 0.2 mg/L Cd for 4 days. The same concentration of Cd or Cu or a mixture of both of these elements introduced during the embryonic stage of the common carp resulted in the reduction of the yolk absorption rate in comparison with carp larvae reared in pure water from the time of hatching (Kamler 2008). Oxygen saturation below 35% at both 5 and 10°C delayed yolk absorption in Atlantic salmon (Hamor and Garside 1977).

6.6.3.4 Salinity

Marine fish larvae have considerable species- and age-dependent variations in salinity tolerance ranges marked by reduced survival and developmental abnormalities at extreme salinities. Larvae of euryhaline fish species tolerate a wide range of salinities (summarized by Lein et al. 1997). However, in Atlantic halibut, 39‰ (parts per thousand [ppt]) reduced the survival of larvae transferred from 35‰ immediately after hatching or 49 days later (at first feeding). Although there were significant increases in yolk sac volume with increased salinity, larval dry weight was not affected (Lein et al. 1997).

6.6.3.5 Locomotory activity

Locomotory activity causes an increase in energy expenditure and results in the acceleration of yolk absorption rate as was shown for zebrafish yolk sac larvae and Atlantic salmon (reviewed by Kamler 2008). The reduction in yolk reserves slows down the yolk absorption rate in rainbow trout, nase, common carp, tench (*Tinca tinca*), and African catfish. Simultaneous transition to exogenous feeding, so-called mixed feeding, slowed the rate of yolk absorption in walleye pollock and sea trout. However, in other species, an excess of exogenous food increased the yolk absorption rate.

6.7 Nonyolk nutrient sources

In contrast to the endogenous nutrition phase, absorptive feeding supplies nutrients, such as minerals and amino acids, from the external environment or ovarian fluid and is common in oviparous as well as viviparous fishes (Balon 1986). Teleost embryos and larvae are able to take up dissolved organic matter, such as pyruvate, acetate, glycine, glucose, and albumin, from the surroundings, along with the endogenous yolk reserves (Heming and Buddington 1988). However, optimum surface area–to–volume ratio is a prerequisite for this type of nutrient acquisition. Organs that are involved in nutrient uptake include the body surface (transepidermal transport), finfold structures, appendages, and spines. Data are available on the acquisition of food molecules from the water via nonspecialized integument and via specialized external guts in oviparous fishes, for example, hypertrophied and trailing guts in the form of extreme rectal extensions, and hypertrophic finfolds (planktonic mycotrophid larva of the lanternfish, *Loweina rara*; viviparous embiotocid striped surfperch, *Embiotoca lateralis*) (Balon 1986; Finn 2007a). Information about absorptive feeding in oviparous fishes and the proportions of nutrients provided has been recently enhanced (black rockfish, *Sebastes melanops*; Berkeley et al. 2004). However, in oviparous species, exclusive parenteral nutrient acquisition cannot support fish growth.

6.7.1 Leptocephalous larvae: an extraordinary example of absorptive (parenteral) nutrition

Hulet (1978) provides an example of absorptive feeding in the leptocephalous larva of the elopomorph teleost, bandtooth conger (*Arisoma balaericum*). He argued that in this species the larvae have never been found with food in the gut and intestinal lumen. Pfeiler and Luna (1984) described metamorphosis in bonefish (*Albula* sp.) that has a leptocephalous larva with a transparent, compressed body that is composed of a gelatinous matrix covering a thin layer of myomers. Although the yolk appears to be exhausted in fish less than 1 cm in body length, bonefish leptocephali commonly reach 5–10 cm at metamorphosis. Also, food has not been found in the guts of these larvae. Pfeiler and Govoni (1993) concluded that low metabolic rates in *Albula* sp. leptocephalous larvae, the poorly differentiated gut, and high levels of FAA in the body, taken together, provide evidence for the epidermis and gut (water uptake) to be involved with taking dissolved organic material (DOM) from seawater as the major nutritional resource. On the contrary, Mochioka et al. (1993) were able to observe colored food (squid paste) intake and defecation in two species of leptocephalous eel larvae in captivity.

The hypothesis, therefore, as to the capability of utilizing dissolved organic matter via body surfaces in the case of species with leptocephalous larvae is contradicted by theories based on (1) oligotrophic, open ocean water with insufficient nutrients, (2) the presence of aloricate (naked) ciliates and tintinnids that after ingestion would not leave fecal material, (3) the presence of a set of fanglike teeth, suggesting these larvae may eat food, and (4) the fact that zooplanktonic particles have been found in the gut of leptocephalous larvae reared in aquaria (Sanderson and Kupferberg 1999). Moreover, leptocephali in the ocean were found to have ciliates, larvacean houses (Otake et al. 1993; Mochioka and Iwamizu 1996), and particulates including copepod fecal pellets (Otake et al. 1993; Mochioka and Iwamizu 1996) in their alimentary canals.

6.7.2 Uptake of FAA

Uptake of ^{14}C-labeled amino acids (alanine) from the ovarian fluid and/or external environment was observed in blenny fish, *Zoarces viviparus*, embryos and larvae (Korsgaard 1992). In yolk sac larvae of Atlantic halibut, the uptake of FAA from seawater was found to be positively correlated with FAA concentration in the range of 5–200 μM. The authors estimated that at the low concentration of 0.3 μM FAA present in seawater, only 0.6% of the metabolic needs of the fish larvae could be satisfied (Rønnestad et al. 1993). There is also evidence that at least some of the uptake of exogenous nutrients from natural seawater may result from direct drinking. Kamler (2008) concluded that parenteral contribution of soluble nutrients to the total energy needs of fish embryos and yolk sac larvae is negligible.

6.7.3 Water absorption

Kamler (2008) emphasized that the moisture in fish eggs is relatively low, but in comparison with reptiles and birds, fish larvae that use yolk increase environmental water absorption after hatching. For example, newly ovulated (spawned) common carp eggs contain ~69% water, while larvae at the end of yolk absorption contain ~85.5% water. Unfertilized eggs of nase contain 67.9% water, and higher moisture levels are found in larvae near the completion of yolk absorption (86.2, 86.4, 85.9, and 86.2% after incubation at 10, 13, 16, and 19°C, respectively) (Kamler 2008).

6.8 Mixed feeding stage

Balon (1986) introduced the term "mixed feeding," which refers to any combination of nutrient acquisition, endogenous, absorptive, or exogenous. At this stage, endogenous feeding is supplemented by nonyolk (parenteral, absorptive) nutrient uptake (Balon 1986). The duration of mixed feeding varies widely among fish species.

6.8.1 Duration of mixed feeding

Several marine species hatch in an advanced stage with respect to feeding ability, with open mouths and well-differentiated alimentary tracts, and are capable of feeding shortly after hatching. In Atlantic cod, haddock, Japanese sandeel (*Ammodytes personatus*), and turbot, the time from hatching to the beginning of exogenous feeding is shorter than from first feeding to completion of yolk and/or oil globule. The ratio of these periods for these four species is 1:6, 1:6, 1:3.5, and 1:2, respectively (reviewed by Kamler 1992). Larvae of whitefish (*Coregonus fera*) are able to ingest external food 4 days after hatching, long before the yolk nutrients are used up (7 dph), with the oil globule present for even 10 days longer (Loewe and Eckmann 1988).

However, yolk sac larvae of most freshwater species, for example, rainbow trout, brown trout, bluegill sunfish, and largemouth bass (*Micropterus salmoides*), are capable of mixed feeding during a short interval prior to complete yolk absorption, and the ratio of both periods is about 1.05 (i.e., Kjørsvik et al. 1991; Kamler 1992; Sanderson and Kupferberg 1999). In milkfish (*Chanos chanos*), barramundi (*Lates calcarifer*), and rabbitfish, the earliest food noticeable in the gut was at 78, 50, and 60 hours posthatch (hph), respectively, at rearing temperatures ranging from 26 to 30°C. That means that exogenous feeding begins 42 hours (milkfish), 70 hours (barramundi), and 12 hours (rabbitfish) before the completion of the yolk resources (Bagarinao 1986). Many examples relate egg size, larval size, and time since the commencement of exogenous feeding to complete yolk absorption in tropical marine fishes, such as those from the genera *Siganus*, *Mugil*, *Epinephelus*, and species such as *Anchoa mitchilli*, *Archosargus rhomboidalis*, and many others (reviewed and presented in an extended table by Bagarinao 1986).

6.8.2 Importance of mixed feeding

The duration and overlapping of feeding stages (endogenous, absorptive, and exogenous) in fish ontogeny are decisive in establishing evolutionary advantage for species with "mixed feeding" over the species that "transition" without (or with limited) backup from another phase. For instance, yolk sac larvae/juveniles may have considerable, highly nutritious resources in the yolk, whereas orally ingested food and acquisition of nutrients through digestive processes in the alimentary tract may be only a complementary source of nutrients. In other words, a highly nutritious yolk supplemented with "temporarily" imbalanced exogenous food does not hamper development of the juvenile, or its growth and survival. This phenomenon seems to be particularly relevant in the early ontogeny of cichlids. As Balon (1986) indicated, longer periods of mixed feeding result in a more secure nutrient supply and allow for feeding specialization in particular niches.

A mixed feeding phase affects increased growth, survival, and development by neutralizing any potential deficit in nutrient provision prior to completion of yolk absorption and allows a continuous dependence on endogenous yolk reserves during the "learning period" in transition to exclusively exogenous feeding (Sanderson and Kupferberg 1999). Thus, for practical purposes, food should be offered to larvae when they attain feeding capability (Kamler 1992). Larvae switch to an exogenous food supply, in the presence of the yolk reserves, when the alimentary canal is functional, although structural and functional development continues from the juvenile to the adult form (Govoni et al. 1986). There are several steps that are crucial in the development (metamorphosis) process of the digestive system in fish larvae (see Chapter 1). The attainment of pancreas secretion function constitutes one such crucial step, followed by the onset of brush border membrane enzymes in the intestine (Ma et al. 2005). In turn, after the initiation of exogenous feeding, there is regional gut differentiation, the appearance of the goblet cells, a valve indicating the beginning of the rectum, and development of mucosal folds. However, the differentiation (specialization) of the digestive system occurs at metamorphosis and continues during the juvenile period (Bisbal and Bengston 1995).

As Kamler (1992) concluded, the adjustment of larvae to exogenous food consists of two components, a behavioral and a physiological one, since the larvae must accept external food and then process it. Baltic herring (*Clupea harengus pallasi*) larvae offered food earlier did not accept it until 3 days after mouth-opening, and their growth began to increase in 4–5 days after first feeding. They prefer larger densities of prey (200 vs. 30 nauplii/dm^3; Pedersen et al. 1987).

Larvae of large yellow croaker (*Pseudosciaena crocea*) start to feed exogenously at 3 dph, when they have the buccopharynx open, a partially differentiated esophagus, and an intestine divided into anterior, middle, and posterior parts (Mai et al. 2005). A mixed feeding phase was observed for 3 days. As mentioned above, important morphological changes in the alimentary tract coincided with metamorphosis. Pyloric ceca differentiate as late as at 17 dph. The differentiation of the stomach into three distinct regions and gastric gland development was observed at 21 dph in large yellow croaker. At the same time, large longitudinal folds appeared in the middle and posterior intestines. These morphological features suggest that the digestive system of large yellow croaker attained characteristics of the juvenile at 40 dph (Mai et al. 2005).

Senegal sole larvae have a sufficiently developed digestive system for successful first feeding at 2 dph and the mixed phase of feeding is very short, as yolk reserves are exhausted by 3 dph. However, gastric glands in the stomach are distinguishable only at 27 dph (Ribeiro et al. 1999). In turn, when the mouth is opened in sharpsnout sea bream at 3 dph, the digestive tract possess buccopharynx, esophagus, developing stomach, and intestine; other organs of the digestive system are formed as well, including the pancreas, liver, and gallbladder, with both the bile and pancreatic duct opened into the anterior intestine. These larvae were able to ingest, digest, and absorb their first exogenous food (4 dph) before endogenous yolk reserves were completely exhausted (8 dph; Figure 6.3) (Micale et al. 2008). Within the first 3–4 days from hatching, at the onset of exogenous feeding, summer flounder and California halibut (*Paralichthys californicus*) complete morphological differentiation of the digestive tract, jaw suspension, and accessory organs for independent exogenous feeding. In both species at 3 dph, yolk reserves are completely depleted and the alimentary tract is differentiated into buccopharynx, esophagus, pre- and postvalvular intestine, and rectum (Bisbal and Bengston 1995; Gisbert et al. 2002, 2004). The completely developed stomach, with the main glandular region (fundic) and both nonglandular regions, the cardiac and pyloric regions, was not observed until 27–30 dph in California halibut, and at 31 dph in summer flounder (Bisbal and Bengston 1995; Gisbert et al. 2002, 2004). Gastric glands and pyloric ceca appear near metamorphosis, and for other flatfishes, these events occurs at the following time: 31 dph in turbot, later in yellowtail flounder at 36 dph, and in Japanese flounder (*Paralichthys olivaceus*) at 40–50 dph (Cousin and Laurencin 1987; Bisbal and Bengston 1995; Baglole et al. 1997). In sea bass larvae, mucosal folds and goblet cells in the esophagus appear between mouth-opening and completion of yolk absorption (García-Hernández et al. 2001); the stomach is completely developed at 55 dph. Early cecal development in sharpsnout sea bream is detected from 10 dph, and the first gastric glands appear at 30 dph (Figure 6.3; Micale et al. 2006; 2008).

These examples illustrate that the glandular stomach of fish larvae/alevins/“yolk sac juveniles” is developed much later than initial exogenous feeding (see also Chapter 1). Several authors have reported that the lack of gastric glands and, presumably, of a functional stomach in fish prevented weaning them onto a commercial diet during the early life stages. This can be associated with “a disrupted cascade” of acidic (stomach) and alkaline (pancreas) proteases characteristic of a mature digestive mechanism involved in protein hydrolysis (Dabrowski 1984; Segner et al. 1994; Baglole et al. 1997).

For practical purposes, the larval or alevin stages can be divided into three groups according to alimentary tract morphology and enzymes secreted in the gut. Salmonid alevins appear to have a functional stomach with developed gastric glands before changing from endogenous to external food. Therefore, during early ontogeny, salmonids and other fish with a functional stomach (cichlids) usually can easily adapt to a dry, compound diet. In the ontogeny of the cichlid digestive tract, the small stomach is visible before yolk sac absorption is completed. The fish take external food when the stomach appears as a sizeable blind pouch at the left side of the intestine. The most numerous fishes are probably those that have no functional stomach or gastric glands in the larval stage but later develop digestive regions of the alimentary tract. The third group includes those that remain stomachless throughout life (Dabrowski 1984; Jaroszewska et al. 2008). The differentiation of the stomach and pyloric ceca can be linked to the transition from larva to juvenile; their appearance takes place in fish during or after metamorphosis (García-Hernández et al. 2001). These authors observed granules of pancreatic zymogens, precursors of digestive enzymes, in Senegal sole larvae at 3 dph, 1 day after first feeding. In Japanese eel, the pancreas starts to synthesize digestive enzymes at 6 dph and acquires full function by the onset of exogenous feeding at 8 dph (Kurokawa et al. 2002). These studies are in agreement with many others on teleost larvae, for example, turbot, yellowfin tuna (*Thunnus albacores*), and striped bass, which showed that zymogene granules commonly appear in the pancreatic cells before or by the first feeding (reviewed by Kurokawa et al. 2002). It is suggested that this activity is not enhanced by food and that alkaline protease activity may play a compensatory function, replacing pepsin deficiency during the first stages of development in the absence of stomach (summarized by Moyano et al. 1996).

However, it is postulated that the presence of brush border peptidase activity in the larval intestine is not sufficient for digestion of complex proteins (Ribeiro et al. 1999). During the very long period of Atlantic halibut larval development, the exogenous phase of feeding begins at the age of 150–180 degree-days (at 6°C) when about 30–50% of the yolk sac is still present (Kjørsvik and Reiersen 1992; Gawlicka et al. 2000). The digestive enzyme activities reached their highest values much later; trypsin at 230 degree-days, lipase and alkaline phosphatase at 276 degree-days. Because the activity of alkaline phosphatase is associated with the capacity of extracellular nutrient absorption, high absorptive capacity of the intestine in Atlantic halibut yolk sac larvae is attained much later than when the first exogenous feeding takes place (Gawlicka et al. 2000).

Despite these facts, there is the proof of the functional absorption of food in the intestine of fish larvae after first feeding, for example, sea bass and summer flounder (Bisbal and Bengston 1995; García-Hernández et al. 2001). The presence of vacuolar supranuclear inclusions in enterocytes with neutral lipids in the anteromedian intestine along with vesicles containing protein in the posterior intestine corresponds to absorbed lipid droplets and endocytotic protein uptake, respectively. Luminal digestion of lipids leads to absorption by enterocytes of FA and monoglycerides. Inside the enterocytes, lipids are resynthesized and transported out of the cell as VLDL and chylomicrons (Sargent et al. 1989). Protein endocytosis and intracellular digestion in the posterior intestine represents a compensatory pathway for dietary protein utilization in teleost larvae, particularly in those with no developed gastric glands and slowly increasing pancreatic enzymatic system (Bisbal and Bengston 1995; García-Hernández et al. 2001). This mechanism of digestion is well known to occur in many other teleost larvae (Rombout et al. 1985).

Histochemical differences observed in the esophageal mucus of different teleosts indicate that goblet cells have more than a simple lubricating function—they also play a role in pregastric digestion since they contain neutral mucins (Murray et al. 1994). The presence of mucous cells in the esophagus in fish larvae is thought to be connected to the establishment of exogenous feeding. Goblet cells in the esophagus have been described in sharpsnout sea bream, sea bream, Senegal sole, sole *Solea solea*, yellowtail flounder, spotted sand bass, and sea bass larvae at the commencement of exogenous feeding (Figure 6.3; Boulhic and Gabaudan 1992; Sarasquete et al. 1995, 1996; Baglole et al. 1997; García-Hernández et al. 2001; Peña et al. 2003; Micale et al. 2008). These cells may increase in function as they become more numerous throughout larval development.

The association between yolk sac absorption and development of the digestive system has been investigated in a few teleost larvae. This contribution contains only part of this information. The differences that appear between and within species reflect variations in experimental conditions affecting larval development, and the results are of immense value to fish farming.

Acknowledgments

We thank Professor Elin Kjørsvik for the comments on the early draft of this chapter. We also thank the Ohio Division of Wildlife, Department of Natural Resources, Columbus, OH, for their support of the studies on the early life history of white bass (*Morone chrysops*).

Literature cited

Aoyama, J. 2009. Life history and evolution of migration in catadromous eels (genus *Anguilla*). *Aqua-BioScience Monographs (AMSM)* 2:1–42.

Babin, P.J., Thisse, C., Durliat, M., et al. 1997. Both apolipoprotein E and A-I genes are present in a nonmammalian vertebrate and are highly expressed during embryonic development. *Proceedings of the National Academy of Sciences of the United States of America* 94:8622–8627.

Bagarinao, T. 1986. Yolk resorption, onset of feeding and survival potential of larvae of three tropical marine fish species reared in the hatchery. *Marine Biology* 91:449–459.

Baglole, C.J., Murray, H.M., Goff, G.P., et al. 1997. Ontogeny of the digestive tract during larval development of yellowtail flounder: a light microscopic and mucus histochemical study. *Journal of Fish Biology* 51:120–134.

Balon, E. 1986. Types of feeding in the ontogeny of fishes and the life-history model. *Environmental Biology of Fishes* 16:11–24.

Balon, E. 1990. Epigenesis of an epigeneticist: the development of some alternative concepts on the early ontogeny and evolution of fishes. *Guelph Ichthyology Reviews* 1:1–48.

Balon, E. 1999. Alternative ways to become a juvenile or a definitive phenotype (and on some persisting linguistic offenses. *Environmental Biology of Fishes* 56:17–38.

Bemanian, V., Male, R., and Goksøyr, A. 2004. The aryl hydrocarbon receptor-mediated disruption of vitellogenin synthesis in the fish liver: cross-talk between AHR- and ERα-signaling pathways. *Comparative Hepatology* [Online]. Available at www.comparative-hepatology.com/content/3/1/2.

Berkeley, S.A., Chapman, C., and Sogard, S.M. 2004. Maternal age as a determinant of larval growth and survival in marine fish, *Sebastes melanops*. *Ecology* 85:1258–1264.

Betchaku, T., and Trinkaus, J.P. 1978. Contact relations, surface activity, and cortical microfilaments of marginal cells of the enveloping layer and of the yolk syncytial and yolk cytoplasmic layer of *Fundulus* before and during epiboly. *Journal of Experimental Zoology* 206:381–426.

Bisbal, G.A., and Bengtson, D.A. 1995. Development of the digestive tract in larval summer flounder. *Journal of Fish Biology* 47:227–291.

Boulhic, M., and Gabaudan, J. 1992. Histological study of the organogenesis of the digestive system and swimbladder of the Dover sole, *Solea solea*, Linnaeus 1758. *Aquaculture* 102:373–396.

Buckley, L.J. 1981. Changes in ribonucleic-acid deoxyribonucleic-acid, and protein content during ontogenesis in winter flounder, *Pseudopleuronectes americanus*, and effect of starvation. *Fishery Bulletin* 77:703–708.

Carnaveli, O., Carletta, R., Cambi, A., et al. 1999. Yolk formation and degradation during oocyte maturation in seabream *Sparus aurata*: involvement of two lysosymal proteinases. *Biology of Reproduction* 60:140–146.

Chen, S.R., and Kimelman, D. 2000. The role of the yolk syncytial layer in germ layer patterning in zebrafish. *Development* 127:4681–4689.

Conceição, L.E.C., Houlihan, D.F., and Verreth, J.A.J. 1997. Fast growth, protein turnover and costs of protein metabolism in yolk-sac larvae of African catfish (*Clarias gariepinus*). *Fish Physiology and Biochemistry* 16: 291–302.

Cooper, M.S., and Virta, V.C. 2007. Evolution of gastrulation in the ray-finned (Actinopterygian) fishes. *Journal of Experimental Zoology Molecular and Developmental Evolution* 308B:591–608.

Cousin, J., and Laurencin, F. 1987. Histological alterations observed in turbot, *Scophthalmus maximus* L., from days 15 to 40 after hatching. *Aquaculture* 67:218–220.

Cowey, C.B., Bell, J.G., Know, D., et al. 1985. Lipids and lipid antioxidant systems in developing eggs of salmon (*Salmo salar*). *Lipids* 20:567–572.

Craik, J.C.A., and Harvey, S.M. 1984. The magnitudes of three phosphorus-containing fractions in the blood plasma and mature eggs of fishes. *Comparative Biochemistry and Physiology* 78B:539–543.

Craik, J.C.A., and Harvey, S.M. 1987. The causes of buoyancy in eggs of marine teleosts. *Journal of the Marine Biological Association of the United Kingdom* 67:169–182.

Czesny, S., Dettmers, J.M., Rinchard, J., et al. 2009. Linking egg thiamine and fatty acid concentrations of Lake Michigan lake trout with early life stage mortality. *Journal of Aquatic Animal Health* 21:262–271.

Czesny, S., Dabrowski, K., Christensen, J.E., van Eenennaam, J., and Doroshov, S. 2000. Discrimination of wild and domestic origin of sturgeon ova based on lipids and fatty acid analysis. *Aquaculture* 189:145–153.

D'Amico, L.A., and Cooper, M.S. 2001. Morphogenetic domains in the yolk syncytial layer of axiating zebrafish embryos. *Developmental Dynamics* 222:611–624.

Dabrowski, K. 1984. The feeding of fish larvae: present "state of the art" and perspectives. *Reproduction Nutrition Development* 24:807–833.

Dabrowski, K. 1989. Formulation of a bioenergetic model for coregoninae early life history. *Transactions of the American Fisheries Society* 118:138–150.

Dabrowski, K., and Łuczyński, M. 1984. Utilization of body stores in the embryonated ova and larvae of two coregonid fishes. *Comparative Biochemistry and Physiology* 79A:329–334.

Dabrowski, K., Kaushik, S., and Luquet, P. 1984. Metabolic utilization of body stores during early life of whitefish (*Coregonus lavaretus* L.). *Journal of Fish Biology* 24:721–729.

Dabrowski, K., Ciereszko, R.E., Ciereszko, A., et al. 1996. Reproductive physiology of yellow perch (*Perca flavescens*): environmental and endocrinological cues. *Journal of Applied Ichthyology* 12:139–148.

Desvilettes, C., Bourdier, G., and Breton, J.C. 1997. Changes in lipid class and fatty acids composition during development in pike (*Esox lucius* L.) egg and larvae. *Fish Physiology and Biochemistry* 16:381–393.

Diaz, J.P., Mani-Ponset, L., Blasco, C., et al. 2002. Cytological detection of the main phase of lipid metabolism during early post-development in three teleost species: *Dicentrarchus labrax*, *Sparus aurata* and *Stizostedion lucioperca*. *Aquatic Living Resources* 15:169–178.

Ditty, J.G. 1989. Separating early larval sciaenids from the western north Atlantic: a review and comparison of larvae off Louisiana and the Atlantic Coast of the U.S. *Bulletin of Marine Science* 44(3):1083–1105.

Eldridge, M.B., Joseph, J.D., and Taberski, K.M. 1983. Lipid and fatty acids composition of the endogenous energy sources of striped bass (*Morone saxatilis*) eggs. *Lipids* 18: 510–513.

Escaffre, A.M., and Bergot, P. 1984. Utilization of the yolk in rainbow trout alevins (*Salmo gairdneri* Richardson): effect of egg size. *Reproduction Nutrition Development* 24:449–460.

Fagotto, F. 1995. Regulation of yolk degradation, or how to make sleepy lysosomes. *Journal of Cell Science* 108:3645–3647.

Fernández, J., Valladares, M., Fuentes, R., et al. 2006. Reorganization of cytoplasm in the zebrafish oocyte and egg during early steps of ooplasmic segregation. *Developmental Dynamics* 235:656–671.

Finn, R.N. 2007a. Vertebrate yolk complexes and the function implications of phosvitins and other subdomains in vitellogenesis. *Biology of Reproduction* 76:926–935.

Finn, R.N. 2007b. The maturational disassembly and differential proteolysis of paralogous vitellogenins in a marine pelagophil teleost: a conserved mechanism of oocyte hydration. *Biology of Reproduction* 76:936–948.

Finn, R.N., Fyhn, H.J., and Evjen, M.S. 1995a. Physiological energetic of developing embryos and yolk-sac larvae of Atlantic code (*Gadus morhua*). I. Respiration and nitrogen metabolism. *Marine Biology* 124:355–369.

Finn, R.N., Henderson, J.R., and Fyhn, H.J. 1995b. Physiological energetic of developing embryos and yolk-sac larvae of Atlantic code (*Gadus morhua*). II. Lipid metabolism and enthalpy balance. *Marine Biology* 124: 371–379.

Finn, R.N., Rønnestad, I., and Fyhn, H.J. 1995c. Respiration, nitrogen and energy metabolism of developing yolk-sac larvae of Atlantic halibut (*Hippoglossus hippoglossus* L.). *Comparative Biochemistry and Physiology* 111A:647–671.

Finn, R.N., Fyhn, H.J., Henderson, R.J., et al. 1996. The sequence of catabolic substrate oxidation and enthalpy balance of developing embryos and yolk-sac larvae of turbot (*Scophthalmus maximus* L.). *Comparative Biochemistry and Physiology* 115A:133–151.

Fírat, K., Saka, S., and Çoban, D. 2003. The effect of light intensity on early life development of common dentex *Dentex dentex* (L. 1758) larvae. *Aquaculture Research* 34:727–732.

Fishelson, L. 1995. Ontogenesis of cytological structures around the yolk sac during embryologic and early larval development of some cichlid fishes. *Journal of Fish Biology* 47:479–491.

Fraser, A.J., Gamble, J.C., and Sargent, J.R. 1988. Changes in lipid content, lipid class composition and fatty acid composition of developing eggs and unfed larvae of cod (*Gadus morhua*). *Marine Biology* 99:307–313.

Fyhn, H.J. 1990. Energy production in marine fish larvae with emphasis on free amino acids as a potential fuel. In: Mellinger, J. (ed.) *Animal Nutrition and Transport Processes. 1. Nutrition in Wild and Domestic Animals*. Karger, Basel, pp. 176–192.

García-Hernández, M.P., Lozano, M.T., Elbal, M.T., et al. 2001. Development of the digestive tract of sea bass (*Dicentrarchus labrax* L.). Light and electron microscopic studies. *Anatomy and Embryology* 204:39–57.

Gawlicka, A., Parent, B., Horn, M.H., et al. 2000. Activity of digestive enzymes in yolk-sac larvae of Atlantic halibut (*Hippoglossus hippoglossus*): indication of readiness for first feeding. *Aquaculture* 184:303–314.

Gisbert, E., Merino, G., Muguet, J.B., et al. 2002. Morphological development and allometric growth patterns in hatchery-reared California halibut larvae. *Journal of Fish Biology* 61:1217–1229.

Gisbert, E., Piedrahita, R.H., and Conklin, D.E. 2004. Ontogenetic development of the digestive system in California halibut (*Paralichthys californicus*) with note on feeding practices. *Aquaculture* 232:455–470.

Govoni, J.J., Boehlert, G.W., and Watanabe, Y. 1986. The physiology of digestion in fish larvae. *Environmental Biology of Fishes* 16:59–77.

Guyot, E., Diaz, P.J., and Connes, R. 1995. Ontogenesis of the liver in sea bream. *Journal of Fish Biology* 47:427–437.

Hamlett, W.C., Schwartz, F., and DiDio, L.J.A. 1987. Subcellular organization of the yolk syncytial-endoderm complex in the preimplantation yolk sac of the shark, *Rhizoprionodon terraenovae*. *Cell and Tissue Research* 247:275–285.

Hamor, T., and Garside, E.T. 1977. Size relations and yolk utilization in embryonated ova and alevins of Atlantic salmon *Salmo salar* L. in various combinations of temperature and dissolved oxygen. *Canadian Journal of Zoology* 55:1892–1898.

Hartling, R.C., and Kunkel, J.G. 1999. Development fate of the yolk protein lipovitellin in embryos and larvae of winter flounder, *Pleuronectes americanus*. *Journal of Experimental Zoology* 284:686–695.

Heming, T.A. 1982. Effects of temperature on utilization of yolk by chinook salmon (*Oncorhynchus tshawytscha*) eggs and alevins. *Canadian Journal of Fisheries and Aquatic Sciences* 39:184–190.

Heming, T.A., and Buddington, R.K. 1988. Yolk absorption in embryonic and larval fishes. In: Hoar, W.S., and Randall, D.J. (eds.) *Fish Physiology, Vol. XI. The Physiology and Developing Fish Part A Eggs and Larvae.* Academic Press, San Diego, CA, pp. 407–446.

Hilton, Z., Poortenaar, C.W., and Sewell, M.A. 2008. Lipid and protein utilization during early development of yellowtail kingfish (*Seriola lalandi*). *Marine Biology* 154:855–865.

Hiramatsu, N., Ichikawa, N., Fukada, H., et al. 2002. Identification and characterization of proteases involved in specific proteolysis of vitellogenin and yolk proteins in salmonids. *Journal of Experimental Zoology* 292:11–25.

Ho, C.Y., Houart, C., Wilson, S.W., et al. 1999. A role for the extraembryonic yolk syncytial layer in patterning the zebrafish embryo suggested by properties of the *hex* gene. *Current Biology* 9:1131–1134.

Hoehne-Rejtan, K., and Kjørsvik, E. 2004. Functional development of the liver and exocrine pancreas in teleost fish. *American Fisheries Society Symposium* 49:9–36.

Hulet, W.H. 1978. Structure and functional development of the eel leptocephalus *Ariosoma balearicum* (De La Roche, 1809). *Philosophical Transactions of the Royal Society of London. Series B, Biological Sciences* 282:107–138.

Hyodo, M., Aoki, A., Katsumata, M., et al. 1999. Yolk syncytial layer independent expression of *no tail* (*Brachyury*) or *goosecoid* genes in cultured explants from embryos of freshwater fish medaka. *Zoological Science* 16:453–457.

Ignatieva, G.M. 1991. The rainbow trout *Salmo gairdneri*. In: Dettlaff, T.A., and Vassetzky, S.G. (eds.) *Animal Species for Developmental Studies, Vol. 2. Vertebrates.* Consultants Bureau, New York, pp. 90–11.

Iwamatsu, T., Muramatsu, T., and Kobayashi, H. 2008. Oil droplets and yolk spheres during development of medaka embryos. *Ichthyological Research* 55:344–348.

Jaroszewska, M., and Dabrowski, K. 2009a. The nature of exocytosis in the yolk trophoblastic layer of silver arowana (*Osteoglossum bicirrhosum*) juvenile, the representative of ancient teleost fishes. *The Anatomical Record* 292: 1745–1755.

Jaroszewska, M., and Dabrowski, K. 2009b. Early ontogeny of Semionotiformes and Amiiformes (Neopterygii: Actinopterygii). In: Kunz, Y., Luer, C.A., and Kapoor, B.G. (eds.) *Development of Non-Teleost Fish.* Science Publishers, Enfield, NH, pp. 231–275.

Jaroszewska, M., Dabrowski, K., Wilczyńska, B., et al. 2008. Structure of the alimentary tract of the racer goby *Neogobius gymnotrachelus* (Kessler, 1857). *Journal of Fish Biology* 72:1773–1786.

Jaworski, A., and Kamler, E. 2002. Development of a bioenergetics model for fish embryos and larvae during the yolk feeding period. *Journal of Fish Biology* 60:785–809.

Johns, D.M., and Howell, W.H. 1980. Yolk utilization in summer flounder (*Paralichthys dentatus*) embryos and larvae reared at two temperatures. *Marine Ecology—Progress Series* 2:1–8.

Johns, D.M., Howell, W.H., and Klein-MacPhee, G. 1981. Yolk utilization and growth to yolk-sac absorption in summer flounder (*Paralichthys dentatus*) larvae at constant and cyclic temperature. *Marine Biology* 63:301–308.

Jorgensen, P. 2008. Yolk. *Current Biology* 18:R103–R104.

Kamler, E. 1992. *Early Life History of Fish. An Energetic Approach. Fish and Fisheries Series* 4. Chapman & Hall, London.

Kamler, E. 2002. Ontogeny of yolk-feeding fish: an ecological perspective. *Reviews in Fish Biology and Fisheries* 12:79–103.

Kamler, E. 2008. Resource allocation in yolk-feeding fish. *Reviews in Fish Biology and Fisheries* 18:143–200.

Kamler, E., and Kato, T. 1983. Efficiency of yolk utilization by *Salmo gairdneri* in relation to incubation temperature and egg size. *Polskie Archiwum Hydrobiologii* 30:371–306.

Karjalainen, J., and Viljanen, M. 1992. Size of vendace (*Coregonus albula*) and European whitefish (*C. lavaretus*) larvae sampled with different types of gear. *Polskie Archiwum Hydrobiologii (Polish Archives of Hydrobiology)* 39:371–380.

Kimata, M. 1983. Changes of chemical composition during early development in the red sea bream *Chrysophrys major* (Temminck et Schlegel) egg and larva. *Journal of the Faculty of Marine Science and Technology, Tokai University* 16:213–223.

Kjørsvik, E., and Reiersen, A.L. 1992. Histomorphology of the early yolk-sac larvae of the Atlantic halibut (*Hippoglossus hippoglossus* L.)—an indication of the timing of functionality. *Journal of Fish Biology* 41:1–19.

Kjørsvik, E., van der Meeren, T., Kryvi, H., et al. 1991. Early development of the digestive tract of cod larvae, *Gadus morhua* L., during start-feeding and starvation. *Journal of Fish Biology* 38:1–15.

Klimogainni, A., Koumoundouros, G., Kaspiris, P., et al. 2004. Effect of temperature on the egg and yolk-sac larval development of common pandora, *Pagellus erythrinus*. *Marine Biology* 145:1015–1022.

Korsgaard, B. 1992. Amino acid uptake and metabolism by embryos of the blenny *Zoarces viviparus*. *The Journal of Experimental Biology* 171:315–328.

Krieger, J., and Fleig, R. 1999. Yolk mobilization in perch, *Perca fluviatilis* L., embryos. *Fish Physiology and Biochemistry* 21:157–165.

Kunz, Y.W. 2004. *Development Biology of Teleost Fishes*. Springer, Dordrecht, the Netherlands.

Kurokawa, T., Suzuki, T., Ohta, H., et al. 2002. Expression of pancreatic enzyme genes during the early larval stages of Japanese eel *Anguilla japonica*. *Fisheries Science* 68:736–744.

Langer, H., Guillaume, J., Metailler, R., et al. 1993. Augmentation of protein synthesis and degradation by poor dietary amino acid balance in European sea bass (*Dicentrarchus labrax*). *The Journal of Nutrition* 123:1754–1761.

Lein, I., Tveite, S., Gjerde, B., et al. 1997. Effects of salinity on yolk sac larvae of Atlantic halibut (*Hippoglossus hippoglossus* L.). *Aquaculture* 156:291–303.

Li, Z., Korzh, V., and Gong, Z. 2007. Localized rpb4 expression in the yolk syncytial layer plays a role in yolk cell extension and early liver development. *BMC Developmental Biology*. Available at www.biomed central.com/1471-213X/7/117; doi:10.1186/1471-213X-7-117.

Liem, K.F. 1981. Larvae of air-breathing fishes as countercurrent flow devices in hypoxic environments. *Science* 211:1177–1179.

Loewe, H., and Eckmann, R. 1988. The ontogeny of the alimentary tract of coregonid larvae: normal development. *Journal of Fish Biology* 33:841–850.

Long, W.L. 1980. Proliferation, growth, and migration of nuclei in the yolk syncytium of *Salmo* and *Catostomus*. *Journal of Experimental Zoology* 214:333–343.

Lubzens, E., Lissauer, L., Levavi-Sivan, B., et al. 2003. Carotenoid and retinoid transport to fish oocytes and eggs: what is the role of retinol binding protein? *Molecular Aspects of Medicine* 24:441–457.

Lubzens, E., Young, G., Bobe, J., and Cerda, J. 2010. Oogenesis in teleosts: how fish eggs are formed. *General and Comparative Endocrinology* 165:367–389.

Łuczyński, M., Długosz, T., Szutkiewicz, B., et al. 1984. The influence of the incubation temperature on the body length and the yolk sac volume of *Coregonus albula* (L.) eleutheroembryos. *Acta Hydrochimica et Hydrobiologica* 12:615–628.

Ma, H., Cahu, C., Zambonino, J., et al. 2005. Activities of selected digestive enzymes during larval development of large yellow croaker (*Pseudosciaena crocea*). *Aquaculture* 245: 239–248.

Mai, K., Yu, H., Ma, H., et al. 2005. A histological study on the development of the digestive system of *Pseudosciaena crocea* larvae and juveniles. *Journal of Fish Biology* 67:1094–1106.

Mani-Ponset, L., Diaz, J.P., Schlumberger, O., et al. 1994. Development of yolk complex, liver and anterior intestine in pike-perch, *Stizostedion lucioperca* (Percidae), according to the first diet during rearing. *Aquatic Living Resources* 7:191–202.

Mani-Ponset, L., Guyot, E., Diaz, J.P., et al. 1996. Utilization of yolk reserves during post-embryonic development in three teleostean species: the sea bream *Sparus aurata*, the sea bass *Dicentrarchus labrax*, and the pike-perch *Stizostedion lucioperca*. *Marine Biology* 126:539–547.

Matsubara, T., and Koya, Y. 1997. Course of proteolytic cleavage in three classes of yolk proteins during oocyte maturation in barfin flounder *Verasper moseri*, a marine teleost spawning pelagic eggs. *The Journal of Experimental Zoology* 278:189–200.

Mendiola, D., Ibaibarriaga, L., and Alvarez, P. 2007. Thermal effect on growth and time to starvation during yolk-sac larval period of Atlantic mackerel *Scomber scombrus* L. *Journal of Fish Biology* 70:895–910.

Micale, V., Garaffo, M., Genovese, L., et al. 2006. The ontogeny of the alimentary tract during larval development in common pandora *Pagellus erythrinus* L. *Aquaculture* 251:354–365.

Micale, C., Di Giancamillo, A., Domeneghini, C., et al. 2008. Ontogeny of the digestive tract in sharpsnout sea bream *Diplodus puntazzo* (Cetti, 1777). *Histology and Histopathology* 23:1077–1091.

Miller, T.J., Crowder, L.B., Rice, J.A., et al. 1988. Larval size and recruitment mechanisms in fishes: toward a conceptual framework. *Canadian Journal of Fisheries and Aquatic Sciences* 45:1657–1670.

Mochioka, N., and Iwamizu, M. 1996. Diet of anguilloid larvae: leptocephali feed selectively on larvacean houses and fecal pellets. *Marine Biology* 125:447–452.

Mochioka, N., Iwamizu, M., and Kanda, T. 1993. Leptocephalus eel larvae will feed in aquaria. *Environmental Biology of Fishes* 36:381–384.

Moodie, G.E.E., Loadman, N.L., Wiegand, M.D., et al. 1989. Influence of eggs characteristics on survival, growth and feeding in larval walleye (*Stizostedion vitreum*). *Canadian Journal of Fisheries and Aquatic Sciences* 46:516–521.

Morrison, C.M., Miyake, T., and Wright, J. Jr. 2001. Histological study of the development of the embryo and early larva of *Oreochromis niloticus* (Pisces: Cichlidae). *Journal of Morphology* 247:172–195.

Moyano, F.J., Díaz, M., Alarcón, F.J., et al. 1996. Characterization of digestive enzyme activity during larval development of gilthead seabream (*Sparus aurata*). *Fish Physiology and Biochemistry* 15:121–130.

Murray, H.M., Wright, G.M., and Goff, G.P. 1994. A study of the posterior esophagus in winter flounder, *Pleuronectes americanus* and yellowtail flounder, *Pleuronectes ferruginea*: morphological evidence for pregastric digestion? *Canadian Journal of Zoology* 72: 1191–1198.

Nelson, J.S. 2006. *Fishes of the World*, 4th edition. John Wiley & Sons, Hoboken, NJ.

Ninhaus-Silveira, A., Foresti, F., de Azevedo, A., et al. 2007. Structural and ultrastructural characteristics of the yolk syncytial layer in *Prochilodus lineatus* (Valenciennes, 1836) (Teleostei; Prochilodontidae). *Zygote* 15:267–271.

Ohkubo, N., and Matsubara, T. 2002. Sequential utilization of free amino acids, yolk proteins and lipids in developing eggs and yolk-sac larvae of barfin flounder *Verasper moseri*. *Marine Biology* 140:187–196.

Ohkubo, N., Sawaguchi, S., Hamatsu, T., et al. 2006. Utilization of free amino acids, yolk proteins and lipids in developing eggs and yolk-sac larvae of walleye pollock *Theragra chalcogramma*. *Fisheries Science* 72:620–630.

Ohkubo, N., Sawaguchi, S., Nomura, K., et al. 2008. Utilization of free amino acids, yolk protein and lipids in developing eggs and yolk-sac larvae of Japanese eel *Anguilla japonica*. *Aquaculture* 282:130–137.

Oozeki, Y., and Bailey, K.M. 1995. Ontogenetic development of digestive enzyme activities in larval walleye pollock, *Theragra chalcogramma*. *Marine Biology* 122:177–186.

Ostaszewska, T., and Dabrowski, K. 2009. Early development of Acipenseriformes (Chondrostei, Actinopterygii). In: Kunz, Y. (ed.) *Development of Non-Teleost Fish*. Science Publishers, Enfield, NH, pp. 171–230.

Ostrowski, A.C., and Divakaran, S. 1991. Energy substrates for eggs and prefeeding larvae of the dolphin *Coryphaena hippurus*. *Marine Biology* 109:149–155.

Otake, T., Nogami, K., and Maruyama, K. 1993. Dissolved and particulate organic matter as possible food sources for eel leptacophali. *Marine Ecology Progress Series* 92:27–34.

Pedersen, B.H., Nilssen, E.M., and Hjelmeland, K. 1987. Variations in the content of trypsin and trypsinogen in larval herring (*Clupea harengus*) digesting copepod nauplii. *Marine Biology* 94:171–181.

Peña, R., Dumas, S., Villalejo-Fuerte, M., et al. 2003. Ontogenetic development of the digestive tract in reared spotted sand bass *Paralabrax maculatofasciatus* larvae. *Aquaculture* 219:633–644.

Pepin, P., Orr, D.C., and Anderson, J.T. 1997. Time to hatch and larval size in relation to temperature and egg size in Atlantic cod (*Gadus morhua*). *Canadian Journal of Fisheries and Aquatic Sciences* 54(Suppl. 1):2–10.

Pfeiler, E., and Govoni, J.J. 1993. Metabolic rates in early life history stages of elopomorph fishes. *Biological Bulletin* 185:277–283.

Pfeiler, E., and Luna, A. 1984. Changes in biochemical composition and energy utilization during metamorphosis of leptocephalus larvae of the bonefish (*Albula*). *Environmental Biology of Fishes* 10:243–251.

Pickova, J., Kiessling, A., Pettersson, A., et al. 1999. Fatty acid and carotenoid composition of eggs from two nonanadromous Atlantic salmon stocks of cultured and wild origin. *Fish Physiology and Biochemistry* 21:147–156.

Poupard, G., André, M., Durliat, M., et al. 2000. Apolipoprotein E gene expression correlates with endogenous lipid nutrition and yolk syncytial layer lipoprotein synthesis during fish development. *Cell and Tissue Research* 300: 251–261.

Raciborski, K. 1987. Energy and protein transformation in sea trout (*Salmo trutta* L.) larvae during transition from yolk to external food. *Polskie Archiwum Hydrobiologii* 34: 437–502.

Rainuzzo, J.R., Reitan, K.I., and Olsen, Y. 1997. The significance of lipids at early stages of marine fish: a review. *Aquaculture* 155: 103–115.

Rana, K.J. 1990. Influence of incubation temperature on *Oreochromis niloticus* (L.) eggs and fry II. Survival, growth and feeding of fry developing solely on their yolk reserves. *Aquaculture* 87:183–195.

Ribeiro, L., Sarasquete, C., and Dinis, M.T. 1999. Histological and histochemical development of the digestive system of *Solea senegalensis* (Kaup, 1858) larvae. *Aquaculture* 171: 293–308.

Rinchard, J., Ware, K., and Dabrowski, K. 2004. Evaluation of gamete quality in yellow perch collected from Lake Erie. Final report to Lake Erie Protection Fund, Toledo, OH.

Rojas-García, C., and Rønnestad, I. 2003. Assimilation of dietary free amino acids, peptides and protein in post-larval Atlantic halibut (*Hippoglossus hippoglossus*). *Marine Biology* 142:801–808.

Rombout, J.H.W., Lamers, C.H.J., Helfrich, M.H., Dekker, A., and Taverne-Thiele, J.J. 1985. Uptake and transport of intact macromolecules in the intestinal epithelium of carp (*Cyprinus carpio*) and the possible immunological implications. *Cell and Tissue Research* 239: 519–530.

Rønnestad, I., and Fyhn, H.J. 1993. Metabolic aspects of free amino acids in developing marine fish eggs and larvae. *Reviews in Fisheries Science* 1:239–259.

Rønnestad, I., Groot, E.P., and Fyhn, H.J. 1993. Compartmental distribution of free amino acids and protein in developing yolk-sac larvae of Atlantic halibut (*Hippoglossus hippoglossus*). *Marine Biology* 116:349–354.

Rønnestad, I., Koven, W.M., Tandler, A., et al. 1994. Energy metabolism during larval development of eggs and larvae of gilthead sea bream (*Sparus aurata*). *Marine Biology* 120:187–196.

Rønnestad, I., Finn, R.N., Lein, I., et al. 1995. Compartmental changes in the contents of total lipid, lipid classes and their associated fatty acids in developing yolk-sac larvae of Atlantic halibut, *Hippoglossus hippoglossus* (L.). *Aquaculture Nutrition* 1:119–130.

Rønnestad, I., Koven, W., Tandler, A., et al. 1998. Utilization of yolk fuels in developing eggs and larvae of European sea bass (*Dicentrarchus labrax*). *Aquaculture* 162:157–170.

Sakaguchi, T., Kikuchi, Y., Kuroiwa, A., et al. 2006. The yolk syncytial layer regulates myocardial migration by influencing extracellular matrix assembly in zebrafish. *Development* 133:4063–4072.

Sanderson, S.L., and Kupferberg, S.J. 1999. Development and evolution of aquatic larval feeding mechanism. In: Hall, B.K., and Wake, M.H. (eds.) *The Origin and Evolution of Larval Forms*. Academic Press, San Diego, CA, pp. 301–378.

Sarasquete, M.C., Polo, A., and de Canales, G. 1993. A histochemical and immunohistochemical study of digestive enzymes and hormones during the larval development of the sea bream, *Sparus aurata* L. *The Histochemical Journal* 25:430–437.

Sarasquete, M.C., Polo, A., and Yúfera, M. 1995. Histology and histochemistry of the development of the digestive system of larval gilthead seabream, *Sparus aurata* L. *Aquaculture* 130:79–92.

Sarasquete, C., González de Canales, M.L., Arellano, J.M., et al. 1996. Histological aspects of the yolk-sac and digestive tract of larvae of the Senegal sole, *Solea senegalensis* (Kaup, 1858). *Histology and Histopathology* 11: 881–888.

Sargent, J., Henderson, R.J., and Tocher, D.R. 1989. The lipids. In: Halver, J.E. (ed.) *Fish Nutrition*, 2nd edition. Academic Press, San Diego, CA, pp. 154–218.

Segner, H., Storch, V., Reinecke, M., et al. 1994. The development of functional digestive and metabolic organs in turbot *Scophthalmus maximus*. *Marine Biology* 119:471–486.

Seoka, M., Takii, K., Takaoka, O., et al. 1997. Biochemical phases in embryonic red sea bream development. *Fisheries Science* 63:122–127.

Shimizu, M., and Yamada, J. 1980. Ultrastructural aspects of yolk absorption in the vittelline syncytium of the embryonic rockfish, *Sebastes schlegeli*. *Japanese Journal of Ichthyology* 27:56–62.

Silversand, C., Norberg, B., and Haux, C. 1996. Fatty-acid composition of ovulated eggs from wild and cultured turbot (*Scophthalmus maximus*) in relation to yolk and oil globule lipids. *Marine Biology* 125:269–278.

Sire, M.F., and Vernier, J.M. 1979. Intestinal endogenous VLDL synthesis: ultrastructural study on a new model—trout embryo and fasted adults. *Biologie Cellulaire* 35(3):271–279.

Sire, M.F., Babin, P.J., and Vernier, J.M. 1994. Involvement of the lysosymal system in yolk protein deposit and degradation during vitellogenesis and embryonic development in trout. *Journal of Experimental Zoology* 269:69–83.

Skjaerven, K.H., Finn, R.N., Kryvi, H., et al. 2003. Yolk resorption in developing plaice (*Pleuronectus platessa*). In: Browman, H.I., and Skifttesvik, A.B. (eds.) *The Big Fish Bang Proceedings of the 26th Annual Larval Fish Conference*. Institute of Marine Research, Bergen, pp. 193–209.

Solnica-Krezel, L., and Driever, W. 1994. Microtubule arrays of the zebrafish yolk cell: organization and function during epiboly. *Development* 120:2443–2455.

Svensson, P.A., Pelabon, C., Blount, J.D., et al. 2006. Does female nuptial coloration reflect egg carotenoids and clutch quality in the two-spotted goby (*Gobiusculus flavescens*, Gobiidae)? *Functional Ecology* 20:689–698.

Terjesen, B.F., Verreth, J., and Fyhn, H.J. 1997. Urea and ammonia excretion by embryos and larvae of the African catfish *Clarias gariepinus* (Burchell 1822). *Fish Physiology and Biochemistry* 16:311–321.

Tingaud-Sequeira, A., and Cerdà, J. 2007. Phylogenetic relationship and gene expression pattern of three different cathepsin L (Ctsl) isoforms in zebrafish: Ctsl is the putative yolk processing enzyme. *Gene* 386:98–106.

Trinkaus, J.P. 1984a. Mechanism of *Fundulus* epiboly—a current view. *American Zoologist* 24:673–688.

Trinkaus, J.P. 1984b. *Cells into Organs*, 2nd edition. Prentice-Hall Inc., Englewood Cliffs, NJ, Yale University.

Trinkaus, J.P. 1993. The yolk syncytial layer of *Fundulus*: its origin and history and its significance for early embryogenesis. *The Journal of Experimental Zoology* 265:258–284.

Van der Velden, J.A., Spanings, F.A.T., Flik, G., et al. 1991. Early life stages of carp (*Cyprinus carpio* L.) depend on ambient magnesium for their development. *The Journal of Experimental Biology* 158:431–438.

Vernier, J.M. 1969. Table chronologique du développement embryonnaire de la truite arc-en-ciel, *Salmo gairdneri* Rich. *Annales des Embryologie et Morphologie* 2:495–520.

Vernier, J.M., and Sire, M.F. 1977a. Lipoprotéines de très basse densité et glycogène dans le syncytium vitellin, l'épithélium intestinal et le foie, aux stades précoces du dévelopment embryonnaire chez la truite arc-en-ciel. *Biologie Cellulaire* 29:45–53.

Vernier, J.M., and Sire, M.F. 1977b. Plaquettes vitellines et activité hydrolasique acide au cours du développement embryonnaire de la truite arc-en-ciel. Étude ulstrastructurale et biochimique. *Biologie Cellulaire* 29:99–112.

Walzer, C., and Schönenberger, N. 1979a. Ultrastructure and cytochemistry study of the yolk syncytial layer in the alevin of trout (*Salmo fario trutta* L.) after hatching. I. The vitellolysis zone. *Cell and Tissue Research* 196:59–73.

Walzer, C., and Schönenberger, N. 1979b. Ultrastructure and cytochemistry study of the yolk syncytial layer in the alevin of trout (*Salmo fario trutta* L. and *Salmo gairderi* R.) after hatching. II. The cytoplasmic zone. *Cell and Tissue Research* 196:75–93.

Webb, J.F. 1999. Larvae in fish development and evolution. In: Hall, B.K., and Wake, M.H. (eds.) *The Origin and Evolution of Larval Forms*. Academic Press, San Diego, CA, pp. 109–158.

Wiegand, M.D. 1996a. Utilization of yolk fatty acids by goldfish embryos and larvae. *Fish Physiology and Biochemistry* 15:21–27.

Wiegand, M.D. 1996b. Composition, accumulation and utilization of yolk lipids in teleost fish. *Reviews in Fish Biology and Fisheries* 6:259–286.

Yoshizaki, N., Soga, M., Ito, Y., et al. 2004. Two-step consumption of yolk granules during the development of quail embryos. *Development, Growth & Differentiation* 46:229–238.

Zeitoun, I.H., Ullrey, D.E., Bergen, W.G., et al. 1977. DNA, RNA, protein, and free amino-acids during ontogenesis of rainbow trout (*Salmo gairdneri*). *Journal of the Fisheries Research Board of Canada* 34:83–88.

Zhang, Y., Dabrowski, K., Hliwa, P., et al. 2006. Indispensable amino acid concentrations decrease in tissues of stomachless fish, common carp in response to free amino acid- or peptide-based diets. *Amino Acids* 31:165–172.

Chapter 7

Effects of larval nutrition on development

Elin Kjørsvik, Trina F. Galloway, Alicia Estevez, Øystein Sæle, and Mari Moren

7.1 Introduction

There are generally three different ways to become a fish juvenile (Pavlov 1999): by indirect development (also called altricial), intermediate development, and direct development (also called precocial). The transition from larval to juvenile form is characterized by a remodeling of all organ systems. Fish with direct development typically have a long embryonic period, and will have most of the juvenile characteristics when they start to eat. Fish with intermediate development, such as salmonids, may also have a long embryonic period and a relatively long period as yolk sac larvae (often called free embryo). Just as in fish with direct development, the yolk reserves are large, and the larva will have developed juvenile organ functionality when the yolk is absorbed. Fish with these types of development have no or a very short larval period and may be fed formulated diets from their first feed intake.

Our focus will be on fishes with indirect development. Indirect development is characterized by a larval stage adapted to a completely different environment and mode of life compared with their parents and the juvenile stage. Such fish larvae generally hatch from small eggs with little yolk reserves, and they are typically very small at the onset of first feeding. In order to increase their chances of survival, small teleost fish larvae spend most of their resources on the development of organs associated with feed intake and swimming movements (Osse et al. 1997). The larval stages are characterized by transient external characters, and the larva is so different from the juvenile that the entire organism needs a complete reprogramming to adapt from one life stage to the next. This reprogramming is often referred to as metamorphosis

Larval Fish Nutrition, First Edition. Edited by G. Joan Holt.

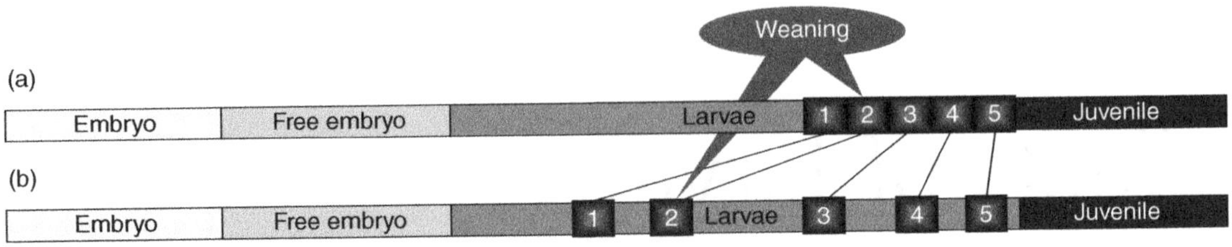

Figure 7.1 Heterochrony during the larval period. The first developmental periods in the life cycle of indirect developing fish: embryo (inside the egg); yolk sac larvae (also referred to as free embryo); larvae (exogenous feeding); and juvenile (small adult in form and function). 1–5 represents the presence of juvenile or adult morphological characters in a species with a distinct metamorphosis (a) versus a species with a gradual change (b). During the larval period, numerous organs change from a larval type to adult, for example, myosin in muscle (2), red blood cells (3), and maturation of the intestinal cells (4). Other systems appear for the first time, for example, the stomach (5). Most larvae are visual predators, but the timing of appearance of rods (1) varies between species.

and is the sum of changes from larval to juvenile characteristics. In some species the numerous changes are synchronized in time, and in others they are more spread out (Figure 7.1), which is referred to as heterochrony. However, this remodeling process is highly dependent on environmental and nutritional factors during the larval stage.

The larval stage is more than a way to become a juvenile; larvae also have anatomical adaptations to their habitat and food. For example, the body shape provides a high surface-to-volume ratio, allowing cutaneous respiration. The lack of scales and often sparse pigmentation render the pelagic larvae practically invisible. The gastrointestinal system is not fully developed, as the larvae, for example, lack a functional stomach. On the other hand, larvae in the wild have no need for a stomach since the copepod prey are rich in free amino acids. The larval stage is a vulnerable and highly specialized stage of life, evolved to survive in its natural habitat, and herein lie the challenges of culture: recognizing the particular nutritional and environmental needs of different species at a given stage of development.

7.2 Nutritional indicators of the digestive system

The capability of fish larvae to develop and grow depends not only on the quantity and quality of the food but also on the functional status of the digestive system (intestine, pancreas, and liver), and the interactions of the digestive and metabolic processes at different stages of development. The larval digestive organs are functional well before the first uptake of exogenous feed (Hoehne-Reitan and Kjørsvik 2004). The intestine is responsible for the digestion, absorption, and transport of nutrients, and its ontogenetic development is described in Chapter 1. The liver's main digestive functions are formation and secretion of bile. The liver also functions in the intermediate metabolism of proteins, carbohydrates, and lipids and in the detoxification of waste products and toxins. The exocrine pancreas is a main source of digestive enzymes for the intestinal hydrolysis of dietary nutrients before absorption by the gut epithelium cells.

The focus here will be on the suitability of nutritional indicators, such as digestive histology and enzyme activities, to evaluate functional larval development.

7.2.1 Nutritional biomarkers

7.2.1.1 *Histology*

Histology is commonly used in larval studies to describe nutritionally induced developmental changes and pathology of the liver, pancreas, and intestine (see, e.g., Segner and Juario 1986; Theilacker 1986; Theilacker and Watanabe 1989; Kjørsvik et al. 1991; Segner and Rösch 1998; MacQueen Leifson et al. 2003a, 2003b; Bransden et al. 2005; Morais et al. 2006; Gisbert et al. 2008). A few studies have also used three-dimensional stereological techniques (Howard and Reed 1998) to quantify characteristics of the liver in turbot larvae (Segner and Witt 1990; Segner et al. 1994; Fontagné et al. 1998) and developmental and dietary effects on gut and liver in cod larvae (Wold et al. 2008, 2009).

7.2.1.2 Digestive enzyme activity

Digestive enzyme activity is another commonly used indicator of nutritional condition. Small, pelagic fish larvae seem to possess the most important, if not all, pancreatic digestive enzymes before exogenous feeding starts. Of these, the bile salt–dependent lipase, a carboxylester lipase, is the most important digestive lipase in fish (Patton et al. 1975; Tocher and Sargent 1984; Lie and Lambertsen 1985; Gjellesvik 1991; Gjellesvik et al. 1992). Trypsin is another key enzyme, which activates other enzymes by cleavage and which is itself activated by an enterokinase from the intestinal enterocytes and by active trypsin (Hjelmeland et al. 1984; Brannon 1990; Rust 2002). The digestive process is supported by intestinal and gastric enzymes, which develop later during larval development and metamorphosis.

The absence of gastric digestion in fish larvae has led to the suggestion that intracellular digestion is important for first-feeding stomachless fish larvae (Govoni et al. 1986). High activity of intracellular digestive enzymes is observed early in larval ontogeny (Zambonino Infante et al. 1996; Cahu et al. 2003a; Mai et al. 2005), and the cytosolic enzyme activity progressively decreases with larval growth when activity of the brush border membrane enzymes increases (Zambonino Infante and Cahu 2001). The establishment of an efficient brush border membrane digestion thus represents a development toward the adult mode of digestion in enterocytes (Henning et al. 1994).

Isolation of several key genes related to larval feed ingestion, digestion, and assimilation processes has improved our knowledge of how genes may be differentially regulated in response to nutritional status (e.g., Murray et al. 2003; Morais et al. 2004, 2007; Perez-Casanova et al. 2004, 2006; Marza et al. 2005; Terova et al. 2007, 2008; Zambonino Infante and Cahu 2007; Kadereit et al. 2008). However, expression of these genes is currently not established as biological markers of larval nutritional condition, although there is a large potential in using these tools.

7.2.2 Liver

The liver (hepatic) energy stores respond rapidly to nutritional changes in fish larvae. Starvation and malnutrition in fish larvae can be observed in liver tissue as cellular changes and resorption or accumulation of glycogen and lipids. These hepatic changes seem more governed by nutritional factors than by developmental sequences.

However, the developmental patterns of glycogen storage in the liver seem very different between species (Hoehne-Reitan and Kjørsvik 2004). In larvae of gilthead sea bream and walleye, the liver accumulated glycogen during the endotrophic phase, accumulation decreased sharply during the mixed feeding stage, before being restored after feeding was established (Mani-Ponset et al. 1994; Guyot et al. 1995). Vernier and Sire (1986) also reported increasing glycogen storage in rainbow trout liver cells before hatching, and it increased steadily during development. In Atlantic cod (*Gadus morhua*), hepatic glycogen was only observed during the exponential growth period, well after yolk resorption (Kjørsvik et al. 1991; Wold et al. 2009).

Accumulation of lipid droplets and enlarged mitochondria in the hepatocytes has been observed when dietary composition is suboptimal (Bransden et al. 2005; Gisbert et al. 2005; Morais et al. 2006; Wold et al. 2009). A deficiency of (n-3) highly unsaturated fatty acids (HUFA) generally results in increased lipid content in the larval liver (Rösch and Segner 1990; Salhi et al. 1997, 1999). Higher levels of lipid accumulation in hepatocytes and enterocytes has also been observed in several species when larvae were fed increasing levels of dietary neutral lipids (NL) (Fontagné et al. 1998; Salhi et al. 1999; Gisbert et al. 2005).

The hepatonuclear size seems to be the most sensitive and dynamic parameter to monitor nutritional status in fish larvae (Figure 7.2), and it was recently found to be directly related to larval growth rates and not to larval size in cod (Wold et al. 2009). Fish larval hepatocyte and nucleus sizes generally decrease due to starvation and malnutrition (Appelbaum et al. 1986; Escaffre and Bergot 1986; Segner et al. 1988; Strüssmann and Takashima 1990; Fontagné et al. 1998; Diaz et al. 1998). Also, the rough endoplasmatic reticulum (RER) and Golgi apparatus became poorly developed in larvae fed suboptimal diets (Rösch and Segner 1990; Segner et al. 1993). These structures are involved in hepatic protein synthesis and secretion, and indicate a lower hepatic metabolic activity and protein synthesis than in more rapidly growing larvae fed optimal diets. Increasing hepatocyte nucleus size may be attributed to higher exchange rates between the nucleus and the cytoplasm, thereby denoting higher metabolic activity (Ghadially 1997).

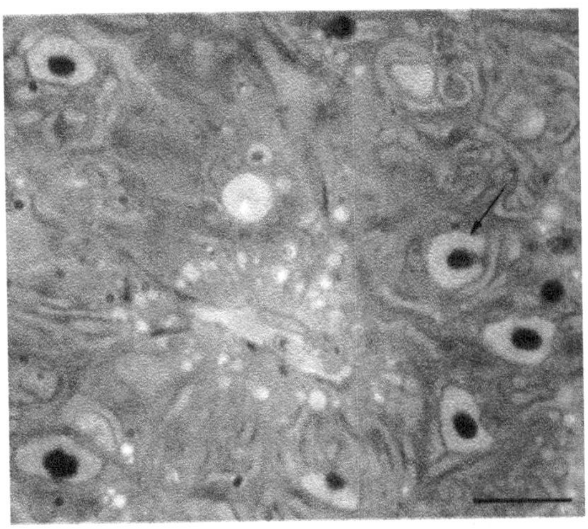

Figure 7.2 Cod larva liver, 39 days posthatch (dph), after weaning to formulated diets from 17 dph. The size and structure of the nuclei and mitochondria seem to be among the most sensitive parameters for larval nutritional status. Bar = 10 μm.

Swollen mitochondria in hepatocytes and enterocytes is described as a sign of starvation or malnutrition in fish larvae caused by autolytic bile emptied into the liver (Diaz et al. 1998; MacQueen Leifson et al. 2003a). Hepatocyte mitochondria in cod larvae fed marine phospholipids (PLs) were more electron dense and contained a denser intermembranous matrix than in larvae fed vegetal PL (Wold et al. 2009). Therefore, both the liver nuclei and mitochondria are found to be sensitive indicators of larval nutritional status.

7.2.3 Pancreatic digestive enzymes

In first-feeding larvae, total synthesis, content, and activity of digestive enzymes increase exponentially if they receive diets that support their growth and nutritional requirements (Izquierdo et al. 2000; Zambonino Infante and Cahu 2001; Hoehne-Reitan and Kjørsvik 2004).

During early developmental stages, the synthesis and secretion of pancreatic enzymes seems to be related to ingestion rates or amount of prey eaten (Hjelmeland et al. 1988; Pedersen and Andersen 1992; Zambonino Infante et al. 1996; Hoehne-Reitan et al. 2001). Secretion of pancreatic juice is regulated by hormonal stimulation, principally by secretin, cholecystokinin (CCK), and gastrin, and also by neuronal stimuli (Slack 1995; Chapter 9). CCK-positive cells have been observed in the marine pelagic halibut larvae during the first-feeding period (Kamisaka et al. 2001).

The mechanism behind pancreatic responses to food intake and diet composition in larval fish is not clear, but there seems to be an independent regulation of pancreatic enzymes (Péres et al. 1998; Hoehne-Reitan et al. 2003). The regulation of pancreatic secretion according to diet quality becomes functional during the first month of life in several species (Cahu and Zambonino Infante 1994, 1995; Pères et al. 1996; Péres et al. 1998; Ribeiro et al. 1999; Buchet et al. 2000; Izquierdo et al. 2001; Zambonino Infante and Cahu 2001).

The dietary composition is suggested to regulate enzyme synthesis on the transcriptional level in larval sea bass, whereas the food ration is proposed to regulate the synthesis on the translational level since only enzyme activities, and not mRNA levels, were affected by ration (Péres et al. 1998). The weaning process, suboptimal diets, and starvation result in a reduced content of pancreatic enzymes in fish larvae (Pedersen et al. 1987; Hoehne-Reitan and Kjørsvik 2004; Engrola et al. 2007; Zambonino Infante and Cahu 2007), which is probably due to a combination of differentially regulated enzyme synthesis and a starvation-induced intracellular degradation of pancreatic tissue.

7.2.4 Intestinal enzymes and gut maturation

Digestive capacity depends not only on digestive enzyme activity but also on the functional development of the gut, and the gut maturation process is known to be affected by dietary factors (Zambonino Infante and Cahu 1999, 2007).

Enterocyte morphology in fish larvae is influenced by developmental stage and dietary composition, and it also reflects the larval nutritional condition. In cod larvae, enterocytes are small and contain few and poorly developed mitochondria during the earliest feeding, and the number of enterocytes and their energetic capacity increases concurrently with the onset of exponential growth (Figure 7.3; from Wold 2007). Pathological effects are swollen mitochondria (MacQueen Leifson et al. 2003a, 2003b) and reduced epithelium height due to starvation and suboptimal feeding (Oozeki et al. 1989; Theilacker and Watanabe 1989; Kjørsvik et al. 1991; Segner

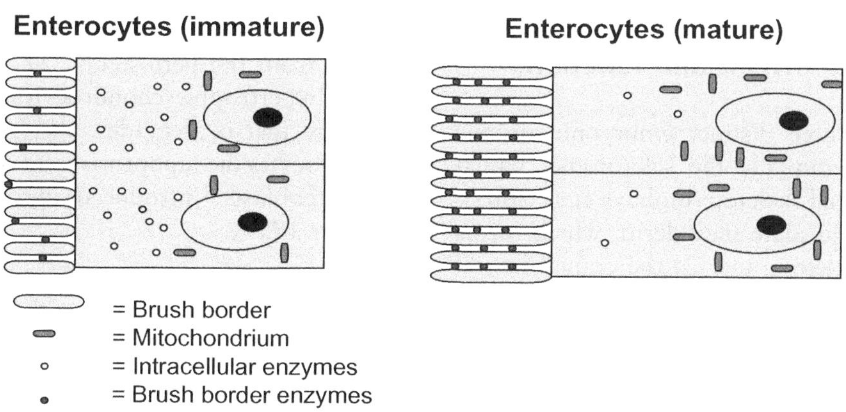

Figure 7.3 Intestinal cell (enterocyte) maturation during larval development. Assays of the intestinal enzyme activities seem to be good indicators of larval developmental status, and thus, they are good criteria for the evaluation of the dietary responses during early ontogeny (from Wold 2007).

et al. 1993; Gisbert and Doroshov 2003; Gisbert et al. 2008).

Marine fish larvae seem to have high requirements for PLs (Izquierdo et al. 2001). Larvae fed diets with high levels of NLs or with low levels of PLs generally have an accumulation of lipid vacuoles (steatosis) in the basal zone of the enterocytes, which indicates a reduced intestinal transport capacity of triacylglycerides (TAG). The accumulation of lipid droplets may be explained by limitations in lipoprotein synthesis and chylomicrone assembly due to a deficiency of PLs since supplementation of formulated diets with PLs alleviates this problem (Fontagné et al. 1998). If the essential n-3 fatty acids are present in the dietary PLs, they are more efficiently utilized and result in a more rapid larval gut maturation (Cahu et al. 2003a; Gisbert et al. 2005; Wold et al. 2007). PLs have an important function in nutrient transport mechanisms from gut epithelial cells to the blood (Izquierdo et al. 2001) and as components of all cell membranes (Sargent et al. 2002). A widely accepted hypothesis now is that fish larvae have limited capacity of PL biosynthesis, and that dietary PLs therefore are necessary for the formation and transport of lipoproteins (Tocher et al. 2008).

7.3 Skeletal development

7.3.1 Cell origins and functions

There are three distinct embryonic origins of the skeleton: (1) the sclerotome, which forms the axial skeleton (Inohaya et al. 2007); (2) the lateral plate mesoderm, which forms the limb skeleton; and (3) the cranial neural crest (NC), which forms the branchial arches and the craniofacial bones and cartilage (Gilbert 1997).

Furthermore, there are two basic types of bones based on their development: Chondral, or substitute bone, develops from a cartilaginous "model" or template with more or less the same shape as the future bone. Centers of ossification occur inside the cartilage (endochondral ossification) or in the perichondrum surrounding the cartilage (perichondral ossification). Eventually, the cartilage will be completely replaced by bone.

Dermal bone originates in the ossification zones of the dermal connective tissue. It is derived directly from mesenchyme in the deeper layers of the dermis (intramembranous ossification). Intramembranous ossification appears to be more ancient than chondral ossification (Morriss-Kay 2001). The frontal bones that play a key role in eye migration in flatfishes (Sæle et al. 2006a, 2006b) are of dermal origin. The derivation of the frontals is not very clear, as they differentiate later than both cartilage and dermal bones of the viscerocranium (Morriss-Kay 2001). Different avian studies conclude differently; Noden (1988) describes the dermal skull roof to be of mesodermal origin, whereas later research traces its origin to the NC (Couly et al. 1993).

In chondral ossification, the mesenchymal cells will form a cartilage model for the future bone. The formation of cartilage takes place in three stages: proliferation of the mesenchyme; condensation of the precartilaginous mesenchyme; and finally, the differentiation of the chondrocytes. When the cartilage is formed, cells in the central part of the cartilage become larger and start secreting a matrix more susceptible to invasion by blood vessels from the periosteum. These cells are called hypertrophic chondrocytes. When the cartilage matrix degrades, the hypertrophic chondrocytes die (apoptosis) and are replaced by osteoblasts introduced by the blood (Gilbert 1997).

7.3.2 Fish skeletons and typical characteristics of fish bone and cartilage

The majority of teleost species have bone without osteocytes, termed acellular bone

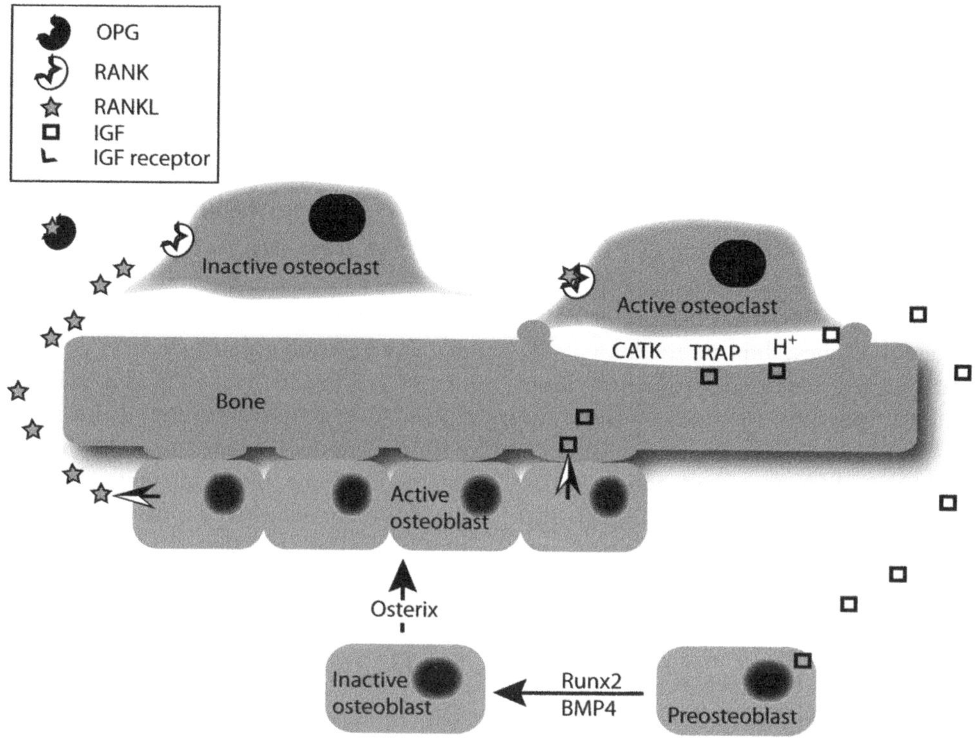

Figure 7.4 Schematic overview of some nutrients' effect on bone metabolism. Osteoblasts produce RANKL upon stimulation from PGE2 and vitamin D. RANKL will then bind to RANK on osteoclasts, stimulating maturation and activation as well as inhibition of apoptosis. Production of the proteolytic enzymes TRAP and CATK are enhanced, and degradation of bone occurs, releasing IGF entrapped in the bone matrix. CATK is also upregulated by retinoic acid (vitamin A). IGF initiates the proliferation and maturation of osteoblasts. Preosteoblasts require upregulation of BMP4 and Runx2/Cbfa1 and later osterix to become mature and active. Runx2 and osterix are upregulated by PGE2 and BMP4 is influenced by retinoic acid. Active osteoblasts deposit bone with IGF. The amount of IGF deposited may increase with higher levels of dietary n-3 fatty acids. Osteoblasts may also produce a RANK inhibitor, named OPG, which bind RANK and therefore block the binding of RANK to RANKL, reducing osteoclast activity. While vitamin D is shown to upregulate both OPG and RANKL, PGE2 downregulates OPG production.

(Meunier and Huysseune 1992; Figure 7.4). In addition to the absence of osteocytes, acellular bone has smaller calcified crystals and a greater amount of organic substance, presumably collagen (Moss 1961). There is no mix of cellular and acellular bone in one skeleton (Meunier and Huysseune 1992). In general, cellular bone is found in less advanced groups of teleosts and acellular bone is found in advanced groups.

Compact mesenchymal cells form a periosteum around the osteoblasts and their calcified matrix (hydroxyapatite salts) in both acellular and cellular bone. Cells on the inside of the periosteum become osteoblasts that deposit bone matrix parallel with the central calcified matrix, in this way building the bone layer by layer.

Remodeling of bone by osteoclastic activity is not only common in teleost fish, it is also necessary for growth (Witten and Villwock 1997a, 1997b; Witten et al. 2000, 2001). Tartrate-resistant acid phosphatase (TRAP) and cathepsin K (CATK) are the main lytic enzymes that resolve minerals in bone. The osteoclast forms a sealed compartment between itself and the bone surface, and the enzymes are activated by the excretion of protons to the compartment (Figure 7.4). Both mono- and multinucleated osteoclasts in teleosts have been shown to function in this same way (Persson et al. 1995, 1998, 1999; Witten et al. 1999, 2000, 2001).

Most knowledge about ossification is based on mammals, but increasing amounts of information on these processes in fish are emerging. In mouse Pax1 is essential for the development of the vertebral body and Pax9 for the neural arches, but in medaka both Pax1 and Pax9 are needed for neural arch and vertebral body development (Mise et al. 2008).

Osteoclast proliferation and activation is controlled by osteoprotegerin (OPG), the receptor activator for nuclear factor κB (RANK) and its ligand (RANKL) system. RANKL is the osteoclast differentiation factor expressed by osteoblasts (Figure 7.4), stromal cells (Khosla 2001; Boyle et al. 2003), and chondroblasts (Komuro et al. 2001). Retinoic acid (RA) directly upregulates osteoclast activity, partly regulating bone formation and metabolism (Harada et al. 1995), but the mechanism is unknown (Rohde and Deluca 2003). This system is not yet described in fish.

7.3.3 Eye migration in flatfish

Eye migration in flatfish is the most radical asymmetric development in any vertebrate. The complex remodeling of tissues is restricted to the area around the ethmoid plate and the eyes (Sæle et al. 2006b). Other osseous asymmetries such as those in the jaw are not unifying in flatfishes in the same way as eye migration (Gibb 1997). The complexity of eye migration is due to the large array of different tissues involved and comprises their intricate interactions. Most striking is the interaction between cartilaginous elements, perichondral and dermal ossification (Sæle et al. 2006a). The first sign of asymmetric growth is seen in the dorsal parts of the ethmoid plate and takes place around day 20–25 post first feeding in halibut. There is a correlation between the position of the eyes and the osteoclastic modeling of the surrounding dermal bones (Sæle et al. 2006a).

7.3.4 Nutrients known to affect skeletal development

Several nutritional factors have been suggested to play a role in skeletal development in fish (reviewed by Lall and Lewis-McCrea 2007), but few studies have been conducted with marine fish larvae to reveal the underlying mechanisms. How nutrients affect skeletal development is not well understood (reviewed by Cahu et al. 2003b).

7.3.4.1 Lipid-soluble vitamins (A, D, and K)

The fat-soluble vitamins A, D, and K are known to affect bone growth and development in fish as well as in terrestrial animals. Vitamin A (VA) has received most attention from larval researchers, probably because several studies have confirmed the teratogeneity of excess VA (reviewed by Cahu et al. 2003b). Immersing Japanese flounder larvae in relatively high concentrations of RA or retinol results in a high number of skeletal deformities (Takeuchi et al. 1998; Haga et al. 2002a, 2002b). Villeneuve et al. (2005a) fed European sea bass larvae increasing levels of VA and observed a linear correlation between VA level and malformation rate. They linked this to alterations in expression patterns of the family of VA nuclear receptors (RXRα, RARα, and RARγ). Copepods and unenriched commercial live feed contain insignificant levels of VA (Moren et al. 2005; Hamre et al. 2008), but both copepods and *Artemia* have high levels of carotenoids, which are precursors of VA (Moren et al. 2002, 2004). Rotifers, on the other hand, contain little or no carotenoids (Hamre et al. 2008). Possibly VA-deficient live prey can be avoided by enrichment with VA or its precursors. More interesting than the known teratogenic effect of excess VA is that the dietary content of other nutrients can affect the expression of the VA-activated nuclear receptors. Villeneuve et al. (2005b) altered the expression of RARα and RXRα in European sea bass larvae by changing the PL level and composition in the diet. Mazurais et al. (2008) fed European sea bass larvae diets with different amounts of vitamin mixture (both water and fat soluble) and found that elevated expression of peroxisome proliferator–activated receptor γ (PPARγ) coincided with low bone morphogenic protein 4 (BMP4) expression in larvae fed low levels of the vitamin mixture. The nuclear receptor PPARγ is highly expressed in adipocytes, while BMP4 is a bone morphogenic protein that is produced in osteoblasts. The larvae fed low levels of vitamins also exhibited a higher degree of deformities and a lower ratio of bone to cartilage compared with those fed higher levels, suggesting that high levels of PPARγ expression may have converted some osteoblasts into adipocytes and that this loss of osteoblasts caused the skeletal deformities. Which component or combination of vitamins that caused this is yet to be shown.

Vitamin D, or calcitriol, is a major osteoclast stimulator, but does so by stimulating osteoblasts, which in turn activates osteoclasts through the RANKL/OPG system mentioned above (reviewed by Witten and Huysseune 2009). This activation increases the release of calcium and phosphate from mineralized tissues in many, but not in all, investigated species (reviewed by Lock et al. 2009). Little or nothing is known about the effects of vitamin D in fish larvae, but it is most likely to give similar effects as in older fish.

Vitamin K is required for the function of calcium-specific binding proteins, often called Gla proteins due to the presence of γ-carboxyglutamate, or Gla, residues. Gla is essential for the production of osteocalcin, which in turn is required for mineralization to occur (Gavaia et al. 2006). Little data exist on marine fish larvae, although vitamin K deficiency has resulted in vertebral deformities in haddock juveniles (Roy and Lall 2007) and in killifish (*Fundulus heteroclitus* L.) (Udagawa 2001).

7.3.4.2 Lipids and fatty acids

The dietary lipid quantity, lipid class, fatty acid composition (in particular the ratios between arachidonic acid [ARA] and eicosapentaenoic acid [EPA] and docosahexaenoic acid [DHA]), and lipid peroxidation have proven to be important in relation to skeletal deformities in marine fish larvae (Cahu et al. 2003b; Villeneuve et al. 2005b; Lall and Lewis-McCrea 2007). By increasing the amount of PLs in the diet of European sea bass larvae from 3 to 12%, Cahu et al. (2003b) drastically reduced the incidence of malformations. On the other hand, high inclusion of EPA and DHA in the PL fraction (12%) produced a higher number of skeletal deformities in sea bass (Villeneuve et al. 2005b). Incorporation of DHA and EPA in the diet PL fraction rather than in the NL fraction (1 and 3% DHA/EPA) enhanced the vertebral ossification process in cod (Kjørsvik et al. 2009). Several aspects of the cellular mechanisms here are yet to be unraveled. Copeman et al. (2002) improved eye migration in yellowtail flounder with a diet low in ARA and high in DHA + EPA. The runt-related transcription factors (Runx) family are, together with bone morphogenetic proteins (BMPs), essential regulators in osteoblast and chondroblast differentiation (Fisher and Pendergast 1997; Flores et al. 2004). These are again regulated by the prostaglandin PGE2, a derivative of ARA (Zhang et al. 2002; Nakagawa et al. 2006). Liu et al. (2006) discovered that the cyclooxygenase COX enzymes (converting ARA to PGE2) of brook trout had a much higher affinity toward ARA compared with the COX enzymes of human and bovine origin. An unbalanced diet (the level of ARA can often be too high compared with EPA and DHA in enriched live prey) may give too high PGE2 levels. In rats, the bone formation decreased with increasing PGE2 levels (Watkins et al. 2000).

7.3.4.3 Minerals

Minerals are an important part of bone structures and are necessary in metabolic processes and signal transductions. Calcium (Ca), phosphorus (P), boron (B), zinc (Zn), copper (Cu), silicon (Si), vanadium (V), selenium (Se), manganese (Mn), and fluorine (F) are known to affect either bone formation or mineralization in terrestrial animals (reviewed by Beattie and Avenell 1992). A comparison of the mineral contents of enriched rotifers and *Artemia* with that of copepods revealed that the diets used in commercial rearing may have significantly lower levels of P, Zn, Cu, Se, and Mn, as well as iodine (I) (Hamre et al. 2007, 2008). Atlantic cod larvae fed copepods contain higher levels of Cu, Se, Mn, and I compared with cod larvae fed enriched rotifers (Hamre, unpublished data). Roy et al. (2002) showed that haddock juveniles fed a P-deficient diet (0.42% P in diet dry weight) had increased osteoclast activity and numbers, as well as a lower number of osteoblasts, compared with fish fed a diet with sufficient levels of P (1.02% P in diet dry weight). This effect may be valid for the larval stages of marine fish species.

7.3.4.4 Other dietary factors involved in skeletal deformations

There are a few studies that correlate certain essential amino acids (arginine and lysine) to the synthesis of insulin-like growth factors (IGFs) and collagen in osteoblasts in terrestrial animals/cell cultures (Chevalley et al. 1998; Fini et al. 2001). The main focus in studies with marine fish larvae has been the bioavailability of dietary proteins and lately the involvement of CCK (Koven et al. 2002; Kvåle et al. 2002; Cahu et al. 2003b, 2004). Apparently, if the larvae receive insufficient amounts of proteins or an unbalanced amino acid profile, bone deformities appear. Whether this is due to impaired protein synthesis in general or whether it is bone specific is unknown.

Some persistent organic pollutants (POPs) are known to be strong endocrine disruptors and to be fatal for developing embryos, where most organs can be altered, including the skeleton. One example is dioxin 2,3,7,8-tetrachlorodibenzo-*p*-dioxin (TCDD), which affects bone tissue, and it has been shown that osteopontin is downregulated in osteoblast cultures when TCDD is introduced (Wejheden et al. 2006). There are, however, few studies on the contents of POPs in larval diets. But by adding TCDD to *Artemia*, which in turn were fed to medaka, Kim and Cooper (1998) observed major deformities in craniofacial structures as well as other lesions such as multifocal hemorrhages, edema, and lack of swim bladder inflation. Hence, there are potentially other dietary factors aside from nutrients that may play a major role in larval development.

7.4 Swimming musculature

7.4.1 Normal development and growth

In newly hatched larvae of, for example, the common carp (*Cyprinus carpio*), the axial skeletal musculature constitutes approximately 20% of the larval body volume (Alami-Durante 1990), increasing to 60% of the body volume in juveniles (Osse and van den Boogaart 1995). The axial musculature is thus the most rapidly growing tissue in larval fish.

Different skeletal muscle cell (fiber) types are segregated to a much greater degree in

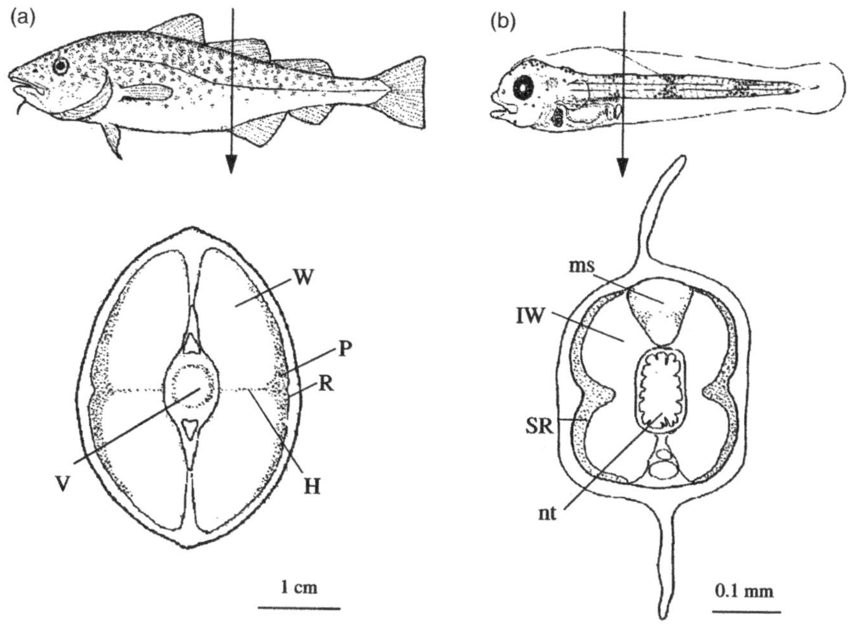

Figure 7.5 Schematic drawings showing the position of different muscle fiber types in the tail of (a) an adult cod (40–50 cm) and (b) a cod larva at the onset of first feeding (approximately 4.5 mm standard length). W = white fibers, P = pink or intermediate fibers, R = red fibers, H = horizontal septum, V = vertebra, IW = inner white muscle, SR = superficial red muscle layer, ms = medulla spinalis, nt = notochord (from Galloway 1999).

fishes than in terrestrial vertebrates (Bone 1978). In adult fishes, a lateral wedge of red muscle fibers at the level of the horizontal septum is separated by a zone of intermediate fibers from the white bulk of the muscle mass (more than 90%) (Figure 7.5a). The red fibers have slow contraction speeds, high mitochondrial and myoglobin content, a rich blood supply, and an aerobic metabolism, and they function during sustained swimming movements (Bone 1978). The white fibers have larger diameters than the red fibers, low mitochondrial and myoglobin content, a poor blood supply, and an anaerobic metabolism. These fibers are used during short bursts of high swimming speeds and rapidly become exhausted (Bone 1978). Pink or intermediate muscle fibers operate at intermediate swimming speeds (Scapolo and Rowlerson 1987) and are located between the red and white fibers.

Newly hatched fish larvae have one layer of superficial red fibers covering several layers of inner white fibers (Figure 7.5b). Both fiber types are aerobic at hatching and have different myosin isoforms from the adult muscle fiber types (Hinterleitner et al. 1987). During larval development, the distribution pattern of adult muscle fiber types, expression of adult myosin isoforms, and specialization of metabolism in the different fiber types occur gradually (Hinterleitner et al. 1987).

Muscle tissue grows by hyperplasia (formation of new fibers) and hypertrophy (increase in size of existing fibers). During the embryonic stage, the first myotubes are formed by fusion of several myoblasts and are therefore multinucleated (Nag and Nursall 1972). The myotubes differentiate into muscle fibers during or soon after formation, and produce contractile proteins, which eventually constitute the bulk of the fiber volume. Multiplication by simple mitotic division is therefore virtually impossible. New muscle fiber production (myogenesis) in larvae, juveniles, or adults arises from mitotic multiplication of myogenic progenitor cells (MPCs) that originate in the embryo (Koumans and Akster 1995). MPCs are also the source of additional nuclei for hypertrophic growth of existing muscle fibers. Fish that grow to a large final size (typical for many aquaculture species) are special in that they grow by hyperplasia during a large part of their adult life (Weatherley et al. 1988), whereas in terrestrial vertebrates the number of skeletal muscle fibers is fixed at or shortly after hatching or birth (Goldspink 1974).

Most of our knowledge about the regulation of myogenesis originates from an extensive literature on mammals and other terrestrial vertebrates, which, given the conservative features of vertebrate genomes, probably is a good starting point for unraveling the corresponding mechanisms in fishes (Johnston 2006). However, one should be aware of the significant differences between fishes and terrestrial vertebrates—the fate of muscle cells is specified relatively much earlier in fishes; adaxial cells (precursors of red muscle cells) are present only in fish embryos; hyperplasia occurs long after hatching in fish that reach a large final size; fishes are poikilothermic; and in nature fishes are often subject to marked seasonal fluctuations in food supply, with conditions for growth varying throughout the year (Currie and Ingham 2001; Rowlerson and Veggetti 2001; Johnston 2006).

Myogenesis in fishes that reach a large final size occurs in three distinct phases: embryonic (the first myotubes are formed in the embryo), stratified (new myotubes are formed in dorsal and ventral growth zones during the late embryonic and early larval stages), and mosaic hyperplasia (new myotubes are formed between existing muscle fibers and give rise to a mosaic of fiber diameters during the late larval and juvenile and adult stages) (Rowlerson and Veggetti 2001). Figure 7.6 (adapted from Johnston 2006) summarizes

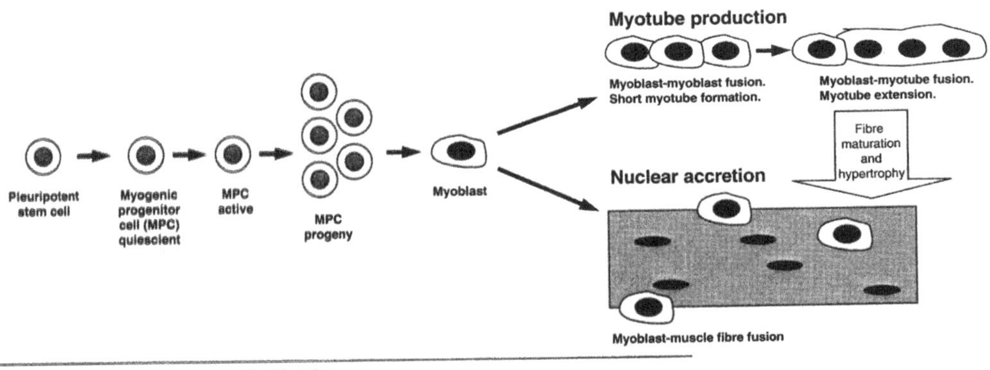

Event	Specification	Activation	Proliferation Cell cycle exit	Differentiation	Migration	Fusion
MRF	MyoD myf-5			Myogenin MRF-4 MEF-2 gene family members		
External signals	Sonic hedgehog fgfs	HGF/SF Myostatin	Peptide growth factors Myostatin			IGF-I IGF-II
Markers		c-met Pax-7 Sox-8 Fox-K.1	PCNA p21			Marker genes

Figure 7.6 Schematic description of the main events of myogenesis in teleost skeletal muscle. Explanation of the figure is given in the text (adapted from Johnston 2006).

our current knowledge of the main events during myogenesis in teleost skeletal muscle.

Embryonic myogenesis starts during segmentation of the embryo and is under the control of signals arising from structures neighboring the somites (Currie and Ingham 2001). Hedgehog proteins, particularly sonic hedgehog (Shh), represent the major signaling pathway in zebrafish myogenesis. Hedgehog genes are expressed in axial structures (floor plate and notochord) during myogenesis and are responsible for the maintenance of the early myogenic factors MyoD and myf-5 in adaxial cells (Devoto et al. 1996; Blagden et al. 1997; Du et al. 1997; Coutelle et al. 2001). The presence of MyoD in adaxial cells of the presomitic mesoderm in many fish species shows that these cells are destined to become muscle cells very early in organogenesis (zebrafish, Weinberg et al. 1996; trout, Delalande and Rescan 1999; Xie et al. 2001; herring, Temple et al. 2001; gilthead sea bream, Tan and Du 2002; Atlantic cod, Hall et al. 2003; common carp, Cole et al. 2004; flounder, Zhang et al. 2006; Atlantic halibut, Galloway et al. 2006). MyoD and myf-5 induce the adaxial cells to become MPCs, which are committed to a myogenic lineage (Rudnicki and Jaenisch 1995). The MPCs are then activated by hepatocyte growth factor/scatter factor and committed to final differentiation (Johnston 2006). A proliferation phase increases the number of active MPCs. MPC proliferation is controlled by positive (e.g., peptide growth factors) and negative (e.g., myostatin) signaling pathways (Buckingham 2001). Following proliferation, the MPC progeny initiate the differentiation program involving the expression of the myogenic regulating factors (MRFs) myogenin and MRF4 and MEF2 gene family members.

The adaxial MPCs fuse to form short myotubes in a myoblast–myoblast fusion event and migrate to the surface of the somite, where they mature into superficial slow red fibers (Devoto et al. 1996). The medial to lateral migration of the slow red fibers may provide a signal for the remaining myogenic cells to form myotubes that mature into the bulk of inner fast white muscle fibers (Devoto et al. 1996; Henry and Amacher 2004). White muscle differentiation is also thought to be regulated by fibroblast growth factors (fgfs), but it is unclear how this occurs (Ochi and Westerfield 2007). Fusion of myoblasts and differentiation into multinucleated myotubes is activated by myogenin (Rescan 2001). Once formed, the myotubes mature into muscle fibers by initiating the production of muscle-specific contractile proteins, such as myosin heavy chain (MyHC) and myosin light chain (MyLC) (reviewed in Goldspink et al. 2001). The process of myogenesis during the stratified and mosaic hyperplasia phases is very similar to that described for the embryonic phase. During the mosaic hyperplasia phase, MPC progeny can also fuse with mature muscle fibers (myoblast–muscle fiber fusion) in order to add nuclei to growing muscle fibers and thereby maintain the volume of cytoplasm controlled by each nucleus within optimal limits. The regulation of fiber size is thought to be controlled by signaling pathways involving IGFI and IGFII.

7.4.2 Nutritional effects

It is well known that different diets for larval fishes can create very different somatic growth rates between fish batches of the same genetic background (Claramunt and Wahl 2000), but few studies have focused on how muscle growth mechanisms in fish larvae are affected by nutrition. The onset of stratified hyperplastic muscle growth follows shortly after the transition to exogenous feeding in several species (Rowlerson and Veggetti 2001). In experiments with larval African catfish (Akster et al. 1995), common carp (Alami-Durante et al. 1997), Atlantic cod (Galloway et al. 1999), pike perch (Ostaszewska et al. 2008), and pacu (Leitão et al. 2009), the diets which gave the best somatic growth rates also gave the highest rates of muscle fiber recruitment, indicating that the myoblast proliferation phase may be prolonged in well-fed larvae. Suboptimal larval diets resulted in reduced rates of somatic and hyperplastic muscle growth, and sometimes even muscular atrophy and slower maturation (Leitão et al. 2009). However, at a given length, the larvae generally had similar numbers and sizes of white muscle fibers. Indeed, there is generally a positive correlation between length and ontogenetic state within a fish species (Laurence 1979; Fuiman and Higgs 1997).

Most of our knowledge on the effect of nutrients on the regulation of somitogenesis and myogenesis arises from terrestrial vertebrates. However, some information is available for fishes. Several observations suggest that RA plays an important role during somitogenesis and myogenesis in fishes. First, RA and its precursors have been detected in fish eggs from several species (zebrafish, Costaridis et al. 1996; chinook salmon, Li et al. 2005). In addition, retinaldehyde dehydrogenase 2 (raldh2), the main RA synthesizing enzyme, and BCox, an enzyme converting provitamin A to retinal, are present in the early somites of zebrafish embryos (Begemann et al. 2001; Grandel et al. 2002; Lampert et al. 2003; Hamade et al. 2006). Also, RA receptors RAR and RXR are expressed in the somites of developing zebrafish (Joore et al. 1994; Jones et al. 1995; Kawakami et al. 2005). Finally, *MyoD* and *myogenin* expression have been shown to increase in somites of RA-treated zebrafish embryos, thereby promoting muscle formation and differentiation (Hamade et al. 2006). Deficiencies in RA or its precursors lead to defects in somite formation and downregulation of *myogenin* expression in quail embryos (Maden et al. 2000).

Myoblast differentiation and the regulation of muscle fiber mass (myotube maturation and hypertrophy) are thought to be controlled by signaling pathways involving insulin-like growth factors (IGFI and IGFII). IGFII transcription required for myoblast differentiation in mice is controlled by amino acid sufficiency and mTOR, which is a serine/threonine protein kinase that regulates cell growth, proliferation, motility, survival, protein synthesis, and transcription (Erbay et al. 2003). This indicates that the mTOR–IGF axis is a molecular link between nutritional levels and skeletal muscle development. It is likely that the amino acid composition of fish larval diets regulates muscle development and growth in many ways since most of the regulating factors (transcription factors, growth factors, etc.) as well as the contractile units are proteins, but little is known about how specific amino acids and protein regulate muscle growth in fish larvae.

Mammalian studies have shown that MPCs are able to transdifferentiate toward that of adipocytes, especially under the influence of long-chain fatty acids (Grimaldi et al. 1997), but it is not known how this works in fish.

Since fish larvae are subject to fluctuations in food supply, it is natural to speculate that nutrient availability will influence muscle fiber recruitment and differentiation. Indeed, skeletal myoblast differentiation was inhibited in mice exposed to restricted glucose availability (starvation) (Fulco et al. 2008). However, no such studies exist for fish larvae.

7.4.3 Other factors affecting muscle development and growth

Fish muscle development, growth, and functionality show great plasticity to changing environmental conditions such as temperature, salinity, oxygen concentration, pH, light availability, and water flow. Particularly, temperature and dissolved oxygen concentrations determine the rate of myogenesis, the composition of subcellular organelles, the patterns of gene expression, and the number and size of muscle fibers in fishes. The plasticity of fish myogenesis to environmental changes was recently reviewed by Johnston (2006).

7.5 Pigmentation

7.5.1 Normal development

Pigmentation patterns in fish are very diverse, also for the larval stage. Pigmentation patterns are formed during embryogenesis and have a camouflage and predator avoidance role. The larval pigmentation patterns are replaced by the adult type at metamorphosis, after which it is more variable and has functions ranging from camouflage to courtship and warning coloration. Pigments are contained within specialized cells in the skin known as chromatophores due to their ability to translocate intracellular pigment organelles in response to varied stimuli, lightening or darkening the overall appearance of the fish. Fish color changes have been studied since ancient times and are grouped into two categories, physiological and morphological color changes. The former are involved in quick changes caused by the immediate aggregation or dispersion of pigmentary organelles within chromatophores in response to environmental signals and/or stress. The latter, long-term changes resulting from the gradual decrease or increase in the number and/or size of chromatophores, are, for example, found after background adaptation. The mechanisms regulating these changes (reviewed by Fujii 1993a, 1993b, 2000) are hormonally (melanocyte-stimulating hormone [MSH] and melanin-concentrating hormone [MCH]) and neurally (sympathetic innervation) controlled. All fish chromatophores, except for those of the pigmented retinal epithelium (which are derived from the outer neuroectodermal sheath of the eyecup), are derived from a transient embryonic tissue called the NC (Thomas and Erickson, 2008). The chromatophores are formed by a population of pleuripotent cells that migrate along defined pathways to specific sites of the embryo and give rise to a wide range of cell types (neurons and glias, endocrine cells, enteric ganglia cells, craniofacial cartilage and bone, etc.), depending on the pathway selected (Figure 7.7). Melanoblasts (chromatophore precursors) use a ventral–lateral pathway in fish, whereas neurogenic cells migrate ventrally along the neural tube, the differences in migratory behavior being determined by the expression of signaling molecules and receptors (reviewed by Harris and Erickson 2007; Kelsh and Parichy 2008).

In fish, six different types of chromatophores are recognized (Table 7.1), and only in the case of leukophores and cyanophores has the embryonic origin not been determined. Thus, a prerequisite for formation of pigment cells is the formation of the NC. The cells migrate from the NC, with a pleuripotent character being specified shortly after migration. According to Harris and Erickson (2007), a lineage switch occurs, by which NC cells are specified first as neuro/gliogenic cells, and later, some convert to the melanogenic fate. This event is controlled by two competing signaling molecules, the Wnts and BMPs, which specify melanocytes and neuronal and glial cells, respectively. Although most NC cells are specified before migration, a few remain pleuripotent NC stem cells into adulthood. In mammals, the specification of NC cells into melanoblasts (downstream of Wnt signaling) is regulated by several transcription factors, including *Mitf*, *Pax3*, and *Sox10*; the first one regulating melanogenesis and the last two regulating melanoblast specification. *Mitf* is not only crucial for survival and specification of melanoblast but also affects melanophore growth and survival and regulates the expression of the genes required for melanogenesis, including tyrosinase and tyrosinase-related proteins 1 and 2 (TRP1, TRP2). Thus, *Mitf* expression is induced in NC cells at the time when melanoblasts are to be specified, and *Mitf* expression in turn upregulates the other genes that confer melanophore characteristics to the cell. *Mitf* and *Sox10* have also been found in zebrafish, with similar functions to those found in mammals (Elsworthy et al. 2003). Once assigned, chromatophores migrate and differentiate. In the zebrafish,

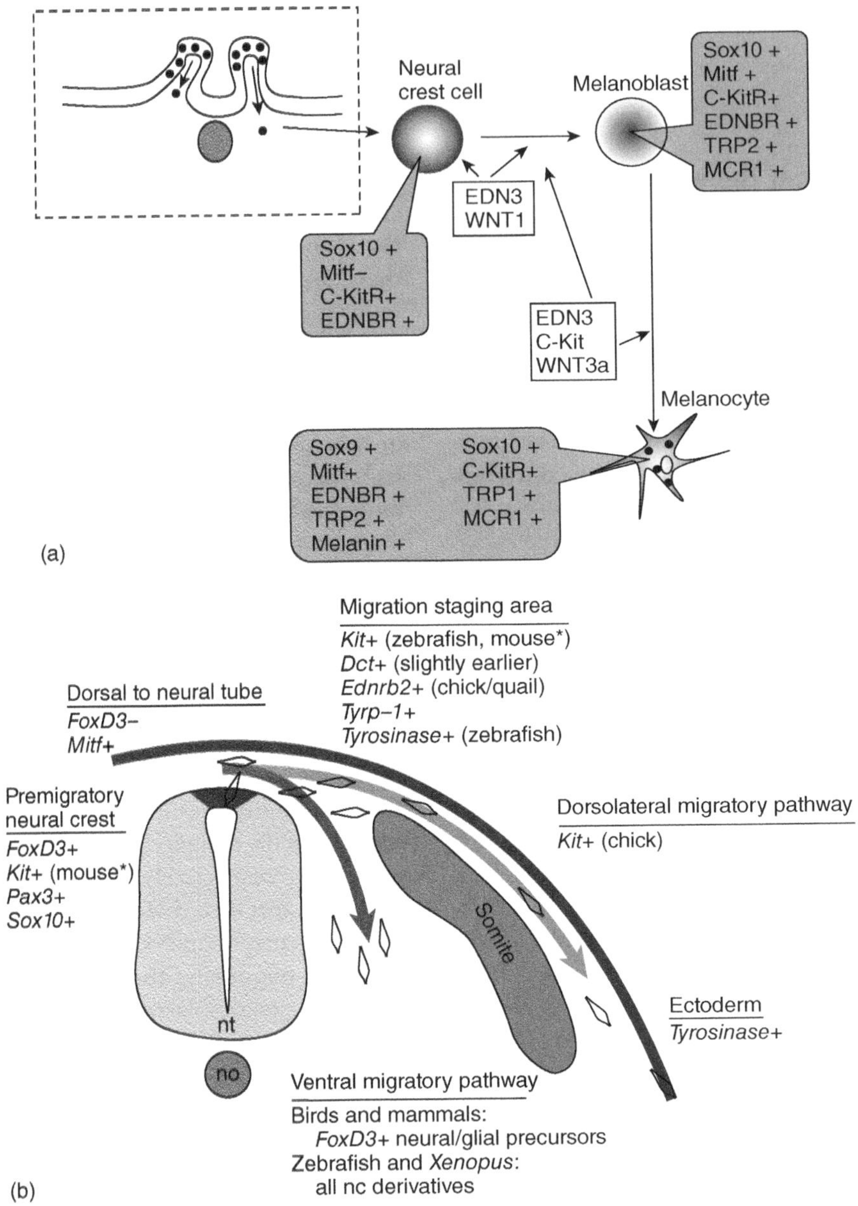

Figure 7.7 (a) Premigratory neural crest (NC) cells in the dorsal neural tube express *FoxD3*, *Sox10*, and *Pax3*. *Sox10* and *Pax3* continue to be expressed during the delamination and migration of melanoblast. *FoxD3* is downregulated in melanoblasts when they exit the neural tube. *Mitf* expression continues throughout melanoblast migration. As melanoblasts migrate, other genes are upregulated at a different time. In birds and mammals, neuroblasts and glioblasts migrate in the ventral migratory pathway; in frogs and fish, melanophores are observed in both migratory pathways (from Thomas and Erickson 2008). (b) The expression of early markers (*Mitf*, *kit*, and *Sox10*) is followed by downregulation of *Sox10* and an increase in *Sox9*, indicating differentiation processes. *Sox10* and *Mitf* act synergistically on the TRP2 promotor. EDN3 and WNT1 and WNT3 promote cell differentiation, resulting in mature chromatophores expressing the melanosomal genes tyrosinase and TRP1 and melanocortin receptor 1 (MCR1), which mediate the activity of MSH, triggering melanin synthesis (from Hamre et al. 2007).

Table 7.1 Fish chromatophores.

Chromatophore	Color	Pigment organelle	Pigment
Melanophore	Black or brown	Melanosome	Melanin
Xantohpore	Yellow-orange	Pterinosome + carotenoid vesicles	Pteridines + carotenoids
Erythrophore	Red	Pterinosome + carotenoid vesicles	Pteridines + carotenoids
Iridiophore	Shiny silver or blue	Reflecting platelet	Guanine crystals
Leukophore	Milky cream or white	Reflecting platelet	Guanine crystals
Cyanophore	Electric blue	Cyanosome	?

From Kelsh and Parichy (2008).

sparse/kita function and *touchtone/trmp7* gene are known to be critical for migration of melanophores and survival (Kelsh et al. 2000; Kelsh and Parichy 2008). *Kit* (*sparse* in fish) is essential for proper melanoblast differentiation.

Pigment patterns change markedly during the fish life cycle, in coordination with changes occurring in other traits. In the case of zebrafish, the pigment pattern at metamorphosis is completely remodeled from larval type to adult type including migration of melanophores, new melanophore formation, larval melanophore death, and inclusion of xantophores and interstripe iridiophores (Kelsh and Parichy 2008). Similarly, in flatfish, larval symmetrical pigmention patterns with big-sized melanophores change at metamorphosis, in parallel with skull remodeling, eye migration, and 90° rotation of the body, into an asymmetric adult coloration. Adult small-sized melanophores appear in clusters around larval pigment cells, becoming densely distributed over the ocular side of the body (Seikai et al. 1987, 1993), whereas larval chromatophores disappear on the ocular and blind sides; the blind side being covered only with iridiophores. To form the adult pigmention pattern, not only do the appropriate chromatophores have to develop, but they have to be correctly organized through interactions among chromatophores and at precise times and locations of chromatophore differentiation.

7.5.2 Nutrients involved in pigmentation

One of the main problems in cultured marine fish, especially in the case of flatfish, is the presence of abnormally pigmented individuals with low market value. In mammals, albinism is characterized by the skin, hair, and eyes containing melanocytes that do not produce melanin, and it is associated with mutations in several genes (including tyrosinase and TRP1 and TRP2) that are required for melanin synthesis. Pigmentation in fish is nutritionally controlled, with nutrients probably arresting the development of pigment cells at a specific stage in larval development and/or avoiding melanin synthesis (Guo et al. 2003).

Among the nutrients found to influence pigmentation success in flatfish is VA, which is required for the development and maintenance of the epithelial cells, mucous membranes, and skin. High VA levels in live prey or in the rearing water enhances pigmentation, stimulating the development of adult-type chromatophores and inducing undesired cranial and spinal deformities (Dedi et al. 1997; Haga et al. 2002a). In some cases, true ambicoloration occurs when the blind side of the flatfish body is covered with all the chromatophore types with the same distribution as on the ocular side (Haga et al. 2005). Retinoids have also been described as hypopigmenting agents in mammals and are used (all-*trans* RA) in topical application to reduce melasma. However, retionid action is mostly on keratinocytes, pigment transfer, epidermal turnover, and induction of desquamation (Nair et al. 1993), phenomena not occurring in fish.

Fatty acids are required for membrane permeability and function and for neural and visual development. Dietary content of EPA (20:5n-3) and DHA (22:6n-3) and the DHA : EPA ratio have been found to correlate with normal pigmentation in turbot (Reitan et al. 1994), although no such effect could be found in a later study with turbot (Estévez et al. 1999) and halibut. On the other hand, ARA (20:4n-6) used for live prey enrichment reduced pigmentation success in turbot and halibut (McEvoy et al. 1998; Estévez et al. 1999), yellowtail flounder (Copeman et al. 2002), Japanese flounder (Estévez et al. 2001), and Senegalese sole (Villalta et al. 2005a), with EPA addition slightly improving pigmentation. Hence, the EPA : ARA ratio seems to be more important than the DHA : EPA ratio in improving the pigmentation pattern in flatfish. Recent experiments with halibut using *Artemia* enriched with monoacylglycerol-rich oils have indicated a certain importance of DHA levels for halibut pigmentation (Hamre and Harboe 2008) not found for other flatfish species (Villalta et al. 2005b). The reasons why ARA induces pseudoalbinism are still unknown, although nutritional stress has been suggested by Sargent et al. (1999), considering the importance of ARA as an eicosanoid (prostaglandin) precursor. Villalta et al. (2008) found higher PGE and PGF production in fish fed high dietary ARA, although no significant differences could be found between albino and pigmented fish. In humans, Ando et al. (2006) found that linoleic acid (but not arachidonic acid) decreased tyrosinase activity inside the melanocytes and reduced the synthesis of melanin.

A different problem to pseudoalbinism in flatfish is the so-called ambicoloration that Norman (1934) classified in three categories: "staining," "spotting," and "true ambicoloration." Ambicoloration has been described in wild and cultured flatfish (Burton 1988) and has been considered to be induced by light and temperature during larval life (Dartner 1986), by demersal behavior in some Pleuronectidae (Norman 1934), by crowding during intensive culture (Takahashi 1994), and by problems with eye migration. In the case of cultured fish, the use of a sandy substrate helps avoid this problem (Stickney and White 1975).

In a recent review on pigmentation and eye migration in Atlantic halibut larvae, Hamre et al. (2007) tried to correlate all the factors involved in flatfish metamorphosis (thyroid hormones), eye migration, and pigmentation (VA and fatty acids). All of them are ligands to specific nuclear receptors (TRs, RARs, RXRs, and PPARs) that modulate gene expression for some of the genes needed to produce a juvenile fish from a larva or a melanophore from an NC cell. The establishment of body and organ axis and the mechanisms of asymmetric gene expression over the left–right axis might also be involved in both eye migration and pigmentation.

7.6 Conclusions and recommendations

The transition phase between endo- and exotrophic nutrition in fish larvae is a critical phase during their development since it is often also accompanied by low enzymatic activity, increased mortality, and reduced growth. The functional larval development seems to follow the larval growth pattern and food availability, but various nutrients such as the essential n-3 fatty acids and VA may affect mechanisms regulating developmental sequences and may induce abnormalities. The development and maturation of regulatory mechanisms of synthesis and secretory responses seem to start during the transition phase, and larvae seem especially vulnerable to malnutrition at this earliest stage of feeding.

The gut and the liver are key organs to study when digestive system capacity and larval nutritional condition is to be evaluated in fish larvae. The liver seems more directly affected by dietary composition than the gut and the pancreas, and the size and structure of the nuclei and mitochondria seem to be among the most sensitive parameters for larval nutritional status. In addition, assays of intestinal enzyme activities seem to be good indicators of larval developmental status, and thus, they are good criteria for the evaluation of the dietary responses during early ontogeny.

Several studies show clear correlations between certain nutrients and the development of deformities as discussed above, so there are no doubts that diet quality can determine skeletal quality, but skeletal development is influenced by physical forces as well. Factors such as current/water flow, temperature, and handling may inflict damage to the skeleton as well. However, high-quality feed results in stronger skeletal structures and may help prevent or minimize the effects of these physical factors. Finally, when more details are available on how the different nutrients affect skeletal development, in terms of critical phases and signal pathways between and within different bone cells, one may be able to design a more optimal diet.

There is a surprising lack of literature on the effects of nutrition on mechanisms regulating muscle growth in fish larvae, considering that MPC recruitment in embryonic and larval life may have a profound effect on the future growth potential of the larvae and juveniles and on the final size of the adult fish. Future research on this topic will benefit from comparative studies of several fish species.

The relative numbers of slow and fast growers between and within batches of fish larvae is a matter little studied. However, this is of major significance for marine fish hatcheries. Fast growers may be preferred to increase productivity, but slow growers may be more resilient to stress and disease and have better flesh quality at harvest. Growth dispersion may result in cannibalism in some species and will increase operational costs of grading procedures. Differential growth in a given larval batch may depend on genetic factors and environmental and nutritional conditions. The differences in growth mechanisms between slow and fast growers remain to be studied.

Considering all the factors involved in fish pigmentation, more research is needed to determine the relationship between dietary nutrients, melanin synthesis (tyrosinase activity), melanophore differentiation, and sidedness definition in marine fish.

Molecular approaches offer valuable and sensitive techniques that will certainly contribute more in the future to find controlling factors and developmental markers, especially when combined with established histological, immunochemical, and biochemical techniques.

Literature cited

Akster, H.A., Verreth, J.A.J., Spierts, I.L.Y., Berbner, T., Schmidbauer, M., and Osse, J.M.W. 1995. Muscle growth and swimming in larvae of *Clarias gariepinus* (Burchell). *ICES Marine Science Symposia* 201:45–50.

Alami-Durante, H. 1990. Growth of organs and tissues in carp (*Cyprinus carpio* L.) larvae. *Growth, Development, and Aging: GDA* 54(3):109–116.

Alami-Durante, H., Fauconneau, B., Rouel, M., Escaffre, A.M., and Bergot, P. 1997. Growth and multiplication of white skeletal muscle fibres in carp larvae in relation to somatic growth rate. *Journal of Fish Biology* 50:1285–1302.

Ando, H., Wen, Z.M., and Kim, H.Y. 2006. Intracellular composition of fatty acids affects the processing and function of tyrosinase through the ubiquitin-proteasome pathway. *The Biochemical Journal* 394:43–50.

Appelbaum, S., Segner, H., and Storch, V. 1986. Electron microscopic study on the influence of different nutritional conditions on liver and white trunk muscles of whitefish (*Coregonus lavaretus*) larvae. *Zoologischer Anzeiger* 217:54–64.

Beattie, J.H., and Avenell, A. 1992. Trace element nutrition and bone metabolism. *Nutrition Research Reviews* 5:167–188.

Begemann, G., Schilling, T.F., Rauch, G.J., Geisler, R., and Ingham, P.W. 2001. The zebrafish neckless mutation reveals a requirement for raldh2 in mesodermal signals that pattern the hindbrain. *Development* 128:3081–3094.

Blagden, C.S., Currie, P.D., Ingham, P.W., and Hughes, S.M. 1997. Notochord induction of zebrafish slow muscle mediated by Sonic hedgehog. *Genes & Development* 11:2163–2175.

Bone, Q. 1978. Locomotor muscle. In: Hoar, W.S., and Randall, D.J. (eds.) *Fish Physiology*, Vol. VII. Academic Press, New York, pp. 361–424.

Boyle, W.J., Simonet, W.S., and Lacey, D.L. 2003. Osteoclast differentiation and activation. *Nature* 423:337–342.

Brannon, P.M. 1990. Adaptation of the exocrine pancreas to diet. *Annual Review of Nutrition* 10:85–105.

Bransden, M.P., Cobcroft, J.M., Battaglene, S.C., Morehead, D.T., Dunstan, G.A., Nichols, P.D., and Kolkovski, S. 2005. Dietary 22:6n-3 alters gut and liver structure and behaviour in larval striped trumpeter (*Latris lineata*). *Aquaculture* 248:275–285.

Buchet, V., Zambonino Infante, J.L., and Cahu, C.L. 2000. Effect of lipid level in a compound diet on the development of red drum (*Sciaenops ocellatus*) larvae. *Aquaculture* 184:339–347.

Buckingham, M. 2001. Skeletal muscle formation in vertebrates. *Current Opinion in Genetics & Development* 11:440–448.

Burton, D. 1988. Melanophore comparisons in different forms of ambicoloration in the flatfish *Pseudopleuronectes americanus* and *Reinhardtius hippoglossoides*. *Journal of Zoology (London)* 214:353–360.

Cahu, C.L., and Zambonino Infante, J.L. 1994. Early weaning of sea bass (*Dicentrarchus labrax*) larvae with a compound diet: effect on digestive enzymes. *Comparative Biochemistry and Physiology* 109A:213–222.

Cahu, C.L., and Zambonino Infante, J.L. 1995. Effect on the molecular form of dietary nitrogen supply in sea bass larvae: response to pancreatic enzymes and intestinal peptidases. *Fish Physiology and Biochemistry* 14:209–214.

Cahu, C.L., Zambonino Infante, J.L., and Barbosa, V. 2003a. Effect of dietary phospholipid level and phospholipid:neutral lipid value on the development of sea bass (*Dicentrarchus labrax*) larvae fed a compound diet. *The British Journal of Nutrition* 90:21–28.

Cahu, C., Infante, J.Z., and Takeuchi, T. 2003b. Nutritional components affecting skeletal development in fish larvae. *Aquaculture* 227:245–258.

Cahu, C., Ronnestad, I., Grangier, V., and Infante, J.L.Z. 2004. Expression and activities of pancreatic enzymes in developing sea bass larvae (*Dicentrarchus labrax*) in relation to intact and hydrolyzed dietary protein; involvement of cholecystokinin. *Aquaculture* 238:295–308.

Chevalley, T., Rizzoli, R., Manen, D., Caverzasio, J., and Bonjour, J.P. 1998. Arginine increases insulin-like growth factor-I production and collagen synthesis in osteoblast-like cells. *Bone* 23:103–109.

Claramunt, R.M., and Wahl, D.H. 2000. The effects of abiotic and biotic factors in determining larval fish growth rates: a comparison across species and reservoirs. *Transactions of the American Fisheries Society* 129(3):835–851.

Cole, N.J., Hall, T.E., Martin, C.I., Chapman, M.A., Kobiyama, A., Yoshiaki, N., Watabe, S., and Johnston, I.A. 2004. Temperature and the expression of myogenic regulatory factors (MRFs) and myosin heavy chain isoforms during embryogenesis in the common carp *Cyprinus carpio* L. *The Journal of Experimental Biology* 207:4239–4248.

Copeman, L.A., Parrish, C.C., Brown, J.A., and Harel, M. 2002. Effects of docosahexaenoic, eicosapentaenoic and arachidonic acids on the early growth, survival, lipid composition and pigmentation of yellowtail flounder (*Limanda ferruginea*): a live food enrichment experiment. *Aquaculture* 210:285–304.

Costaridis, P., Horton, C., Zeitlinger, J., Holder, N., and Maden, M. 1996. Endogenous retinoids in the zebrafish embryo and adult. *Developmental Dynamics: An Official Publication of the American Association of Anatomists* 205:41–51.

Couly, G.F., Coltey, P.M., and Ledouarin, N.M. 1993. The triple origin of skull in higher vertebrates—a study in quail-chick chimeras. *Development* 117:409–429.

Coutelle, O., Blagden, C.S., Hampson, R., Halai, C., Rigby, P.W., and Hughes, S.M. 2001. Hedgehog signalling is required for maintenance of myf5 and myoD expression and timely terminal differentiation in zebrafish adaxial myogenesis. *Developmental Biology* 236:136–150.

Currie, P.D., and Ingham, P.W. 2001. Induction and patterning of embryonic skeletal muscle cells in the zebrafish. In: Johnston, I.A. (ed.) *Fish Physiology*, Vol. XVIII. Academic Press, San Diego, CA, pp. 1–17.

Dartner, J.V. Jr. 1986. Observations on anomalous conditions in some flatfishes (Pisces: Pleuronectiformes) with a new record of partial albinism. *Environmental Biology of Fishes* 17:141–152.

Dedi, J., Takeuchi, T., Seikai, T., Watanabe, T., and Hosoya, K. 1997. Hypervitaminosis A during vertebral morphogenesis in larval Japanese flounder. *Fisheries Science: FS* 63:466–473.

Delalande, J.M., and Rescan, P.Y. 1999. Differential expression and interpretation of positional MyoD genes in developing and adult myotomal musculature of the trout (*Oncorhynchus mykiss*). *Development Genes and Evolution* 209:432–437.

Devoto, S.H., Melancon, E., Eisen, J.S., and Westerfield, M. 1996. Identification of separate slow and fast muscle precursor cells *in vivo*, prior to somite formation. *Development* 122:3371–3380.

Diaz, J.P., Mani-Ponset, L., Guyot, E., and Connes, R. 1998. Hepatic cholestasis during the post-embryonic development of fish larvae. *The Journal of Experimental Zoology* 280:277–287.

Du, S.J., Devoto, S.H., Westerfield, M., and Moon, R.T. 1997. Positive and negative regulation of muscle cell identity by members of the hedgehog and TGF-beta gene families. *Journal of Cell Biology* 139:145–156.

Elsworthy, S., Lister, J.A., Carney, T.J., Raible, D.W., and Kelsh, R.N. 2003. Transcriptional regulation of mitfa accounts for the sox10 requirement in zebrafish melanophore development. *Development* 130:2809–2818.

Engrola, S., Conceicao, L.E.C., Dias, L., Pereira, R., Ribeiro, L., and Dinis, M.T. 2007. Improving weaning strategies for Senegalese sole: effects of body weight and digestive capacity. *Aquaculture Research* 38:696–707.

Erbay, E., Park, I.-H., Nuzzi, P.D., Schoenherr, C.J., and Chen, J. 2003. IGF-II transcription in skeletal myogenesis is controlled by mTOR and nutrients. *Journal of Cell Biology* 163(5): 931–936.

Escaffre, A.M., and Bergot, P. 1986. Morphologie quantitative du foie des alevins de truite arc-en-ciel (*Salmo gairdnerii*) issus de gros ou de petits oeufs: incidence de la date premier repas. *Archiv für Hydrobiologie* 107:331–348.

Estévez, A., McEvoy, L.A., Bell, J.G., and Sargent, J.R. 1999. Growth, survival, lipid composition and pigmentation of turbot (*Scophthalmus maximus*) larvae fed live prey enriched in arachidonic and eicosapentaenoic acids. *Aquaculture* 180:321–343.

Estévez, A., Kaneko, T., Seikai, T., Dores, R., Tagawa, M., and Tanaka, M. 2001. Ontogeny of ACTH and MSH cells in Japanese flounder (*Paralichthys olivaceus*) in relation to albinism. *Aquaculture* 202:131–143.

Fini, M., Torricelli, P., Giavaresi, G., Carpi, A., Nicolini, A., and Giardino, R. 2001. Effect of L-lysine and L-arginine on primary osteoblast cultures from normal and osteopenic rats. *Biomedicine & Pharmacotherapy* 55: 213–220.

Fisher, N.M., and Pendergast, D.R. 1997. Reduced muscle function in patients with osteoarthritis. *Scandinavian Journal of Rehabilitation Medicine* 29:213–221.

Flores, M.V., Tsang, V.W.K., Hu, W.J., Kalev-Zylinska, M., Postlethwait, J., Crosier, P., Crosier, K., and Fisher, S. 2004. Duplicate zebrafish runx2 orthologues are expressed in developing skeletal elements. *Gene Expression Patterns* 4:573–581.

Fontagné, S., Geurden, I., Escaffre, A.M., and Bergot, P. 1998. Histological changes induced by dietary phospholipids in intestine and liver of common carp *Cyprinus carpio* L. larvae. *Aquaculture* 161:213–223.

Fuiman, L.A., and Higgs, D.M. 1997. Ontogeny, growth and the recruitment process. In: Chambers, R.C., and Trippel, E.A. (eds.) *Early Life History and Recruitment in Fish Populations*. Chapman and Hall, London, pp. 225–249.

Fujii, R. 1993a. Cytophysiology of fish chromatophores. *International Review of Cytology* 143:19–255.

Fujii, R. 1993b. Coloration and chromatophores. In: Evans, D.H. (ed.) *The Physiology of Fishes*. CRC Press, Boca Raton, FL, pp. 535–562.

Fujii, R. 2000. The regulation of motile activity in fish chromatophores. *Pigment Cell Research* 13:300–319.

Fulco, M., Cen, Y., Zhao, P., Hoffman, E.P., McBurney, M.W., Sauve, A.A., and Sartorelli, V. 2008. Glucose restriction inhibits skeletal myoblast differentiation by activating SIRT1 through AMPK-mediated regulation of Nampt. *Developmental Cell* 14:661–673.

Galloway, T.F. 1999. Muscle growth and development in early life stages of the Atlantic cod (*Gadus morhua* L.) and halibut (*Hippoglossus hippoglossus* L.). Dr.Scient. thesis, Norwegian University of Science and Technology, Trondheim, Norway.

Galloway, T.F., Kjørsvik, E., and Kryvi, H. 1999. Muscle growth and development in Atlantic cod larvae (*Gadus morhua* L.) related to different somatic growth rates. *The Journal of Experimental Biology* 202:2111–2120.

Galloway, T.F., Bardal, T., Kvam, S.N., Dahle, S.W., Nesse, G., Randøl, M., Kjørsvik, E., and Andersen, Ø. 2006. Somite formation and expression of MyoD, myogenin and myosin in Atlantic halibut (*Hippoglossus hippoglossus* L.) embryos incubated at different temperatures: transient asymmetric expression of MyoD. *The Journal of Experimental Biology* 209:2432–2441.

Gavaia, P.J., Simes, D.C., Ortiz-Delgado, J.B., Viegas, C.S.B., Pinto, J.P., Kelsh, R.N., Sarasquete, M.C., and Cancela, M.L. 2006. Osteocalcin and matrix Gla protein in zebrafish (*Danio rerio*) and Senegal sole (*Solea senegalensis*): comparative gene and protein expression during larval development through adulthood. *Gene Expression Patterns* 6:637–652.

Ghadially, F.N. 1997. *Ultrastructural Pathology of the Cell and Matrix*, Vol. 1, 4th edition. Butterworth-Heinemann, Woburn, MA.

Gibb, A.C. 1997. Do flatfish feed like other fishes? A comparative study of percomorph prey-capture kinematics. *The Journal of Experimental Biology* 200:2841–2859.

Gilbert, S.F. 1997. Early vertebrate development: mesoderm and endoderm. In: Gilbert, S.F. (ed.) *Developmental Biology*, 5th edition. Sinauer Associates Inc., Sunderland, MA, pp. 341–388.

Gisbert, E., and Doroshov, S.I. 2003. Histology of the developing digestive system and the effect on food deprivation in larval green sturgeon (*Acipenser medirostris*). *Aquatic Living Resources* 16:77–89.

Gisbert, E., Villeneuve, L., Zambonino Infante, J.L., Quazuguel, P., and Cahu, C.L. 2005. Dietary phospholipids are more efficient than neutral lipids for long-chain polyunsaturated fatty acid supply in European sea bass *Dicentrarchus labrax* larval development. *Lipids* 40(6):1–11.

Gisbert, E., Ortiz-Delgado, J.B., and Sarasquete, C. 2008. Nutritional cellular biomarkers in early life stages of fish. *Histology and Histopathology* 23:1525–1539.

Gjellesvik, D.R. 1991. Enzymatic lipid digestion in teleosts: bile salt–dependent lipase as major lipolytic enzyme in cod pancreas. PhD thesis, University of Bergen, Bergen, Norway.

Gjellesvik, D.R., Lombardo, D., and Walthers, B.T. 1992. Pancreatic bile salt dependent lipase from cod (*Gadus morhua*): purification and properties. *Biochimica et Biophysica Acta* 1124:123–134.

Goldspink, G. 1974. Development of muscle. In: Goldspink, G. (ed.) *Differentiation and Growth of Cells in Vertebrate Tissues*. Chapman and Hall, London, pp. 69–99.

Goldspink, G., Wilkes, D., and Ennion, S. 2001. Myosin expression during ontogeny, post-hatching growth and adaptation. In: Johnston, I.A. (ed.) *Fish Physiology*, Vol. XVIII. Academic Press, San Diego, CA, pp. 43–72.

Govoni, J.J., Boehlert, G.W., and Watanabe, Y. 1986. The physiology of digestion in fish larvae. *Environmental Biology of Fishes* 16:59–77.

Grandel, H., Lun, K., Rauch, G.J., Rhinn, M., Piotrowski, T., Houart, C., Sordino, P., Kuchler, A.M., Schulte-Merker, S., Geisler, R., Holder, N., Wilson, S.W., and Brand, M. 2002. Retinoic acid signalling in the zebrafish embryo is necessary during pre-segmentation stages to pattern the anterior–posterior axis of the CNS and to induce a pectoral fin bud. *Development* 129:2851–2865.

Grimaldi, P.A., Teboul, L., Inadera, H., Gaillard, D., and Amri, E.Z. 1997. Trans-differentiation of myoblasts to adipoblasts: triggering effects of fatty acids and thiazolidinediones. *Prostaglandins, Leukotrienes, and Essential Fatty Acids* 57:71–75.

Guo, H., Huang, B., Zhang, S., and Qi, F. 2003. Biochemical and histochemical activities of tyrosinase in the skins of normal and albino turbot *Scophthalmus maximus*. *Fish Physiology and Biochemistry* 29:67–76.

Guyot, E., Diaz, J.P., and Connes, R. 1995. Organogenesis of the liver in sea bream. *Journal of Fish Biology* 47:427–437.

Haga, Y., Suzuki, T., and Takeuchi, T. 2002a. Retinoic acid isomers produce malformations in postembryonic development of the Japanese flounder, *Paralichthys olivaceus*. *Zoological Science* 19:1105–1112.

Haga, Y., Takeuchi, T., and Seikai, T. 2002b. Influence of all-trans retinoic acid on pigmentation and skeletal formation in larval Japanese flounder. *Fisheries Science: FS* 68:560–570.

Haga, Y., Nataami, K., and Takeuchi, T. 2005. Process of true ambicoloration in larval and juvenile Japanese flounder *Paralichthys olivaceus*: an ultrastructural study. *Nippon Suisan Gakkaishi* 71:782–790.

Hall, T.H., Cole, N.J., and Johnston, I.A. 2003. Temperature and expression of seven muscle-specific protein genes during embryogenesis in the Atlantic cod *Gadus morhua* L. *The Journal of Experimental Biology* 206:3187–3200.

Hamade, A., Deries, M., Begemann, G., Bally-Cuif, L., Genêt, C., Sabatier, F., Bonnieu, A., and Cousin, X. 2006. Retinoic acid activates myogenesis *in vivo* through Fgf8 signalling. *Developmental Biology* 289:127–140.

Hamre, K., and Harboe, T. 2008. Critical levels of essential fatty acids for normal pigmentation in Atlantic halibut (*Hippoglossus hippoglossus* L.) larvae. *Aquaculture* 277:101–108.

Hamre, K., Holen, E., and Moren, M. 2007. Pigmentation and eye migration in Atlantic halibut (*Hippoglossus hippoglossus* L.) larvae: new findings and hypotheses. *Aquaculture Nutrition* 13:65–80.

Hamre, K., Srivastava, A., Ronnestad, I., Mangor-Jensen, A., and Stoss, J. 2008. Several micronutrients in the rotifer *Brachionus* sp may not fulfil the nutritional requirements of marine fish larvae. *Aquaculture Nutrition* 14:51–60.

Harada, H., Miki, R., Masushige, S., and Kato, S. 1995. Gene-expression of retinoic acid receptors, retinoid-X receptors, and cellular retinol-binding protein-I in bone and its regulation by vitamin-A. *Endocrinology* 136:5329–5335.

Harris, M.L., and Erickson, C.A. 2007. Lineage specification in neural crest cell pathfinding. *Developmental Dynamics* 236:1–19.

Henning, S.J., Rubin, D.C., and Shulman, R.J. 1994. Ontogeny of the intestinal mucosa. In: Johnson, L.R. (ed.) *Physiology of the Gastrointestinal Tract*, 3rd edition. Raven Press, New York, pp. 571–610.

Henry, C.A., and Amacher, S.L. 2004. Zebrafish slow muscle cell migration induces a wave of fast muscle morphogenesis. *Developmental Cell* 7:917–923.

Hinterleitner, S., Platzer, U., and Wieser, W. 1987. Development of the activities of oxidative, glycolytic and muscle enzymes during early larval life in three families of freshwater fish. *Journal of Fish Biology* 30(3):315–326.

Hjelmeland, K., Huse, I., Jørgensen, T., Molvik, G., and Raa, J. 1984. Trypsin and trypsinogen as indices of growth and survival potential of cod (*Gadus morhua* L.) larvae. In: Dahl, E., Danielssen, D.S., Moksness, E., and Solemdal, P. (eds.) *The Propagation of Cod Gadus morhua L.* Flødevigens Rapportserie, Arendal, Norway, pp. 189–202.

Hjelmeland, K., Pedersen, B.H., and Nilssen, E.M. 1988. Trypsin content in intestines of herring larvae, *Clupea harengus*, ingesting inert polystyrene spheres or live crustacean prey. *Marine Biology* 98:331–335.

Hoehne-Reitan, K., and Kjørsvik, E. 2004. Functional development of the exocrine pancreas and liver in teleost fish. In: Govoni, J.J. (ed.) *The Development of Form and Function in Fishes, and the Question of Larval Adaptation*, Symposium 40. American Fisheries Society Symposium, Bethesda, MD, pp. 9–36.

Hoehne-Reitan, K., Kjørsvik, E., and Reitan, K.I. 2001. Bile salt–dependent lipase in larval turbot, as influenced by density and lipid content of fed prey. *Journal of Fish Biology* 58:746–754.

Hoehne-Reitan, K., Kjørsvik, E., and Reitan, K.I. 2003. Lipolytic activities in developing turbot larvae as influenced by the diet. *Aquaculture International* 11:477–489.

Howard, C.V., and Reed, M.G. 1998. *Unbiased Stereology—Three-Dimensional Measurement in Microscopy*. Bios Scientific Publishers, Springer-Verlag, New York.

Inohaya, K., Takano, Y., and Kudo, A. 2007. The teleost intervertebral region acts as a growth center of the centrum: *in vivo* visualization of osteoblasts and their progenitors in transgenic fish. *Developmental Dynamics: An Official Publication of the American Association of Anatomists* 236:3031–3046.

Izquierdo, M.S., Socorro, J., Arantzamendi, L., and Hernàndez-Cruz, C.M. 2000. Recent advances in lipid nutrition in fish larvae. *Fish Physiology and Biochemistry* 22:97–107.

Izquierdo, M.S., Tandler, A., Salhi, M., and Kolkovski, S. 2001. Influence of dietary polar lipids quantity and quality on ingestion and assimilation of labelled fatty acids by larvae gilthead seabream. *Aquaculture Nutrition* 7:153–160.

Johnston, I.A. 2006. *Environment* and plasticity of myogenesis in teleost fish. *Journal of Experimental Biology* 209:2249–2264.

Jones, B.B., Ohno, C.K., Allenby, G., Boffa, M.B., Levin, A.A., Grippo, J.F., and Petkovich, M. 1995. New retinoid X receptor subtypes in zebra fish (*Danio rerio*) differentially modulate transcription and do not bind 9-cis retinoic acid. *Molecular and Cellular Biology* 15:5226–5234.

Joore, J., van der Lans, G.B., Lanser, P.H., Vervaart, J.M., Zivkovic, D., Speksnijder, J.E., and Kruijer, W. 1994. Effects of retinoic acid on the expression of retinoic acid receptors during zebrafish embryogenesis. *Mechanisms of Development* 46:137–150.

Kadereit, B., Kumar, P., Wang, W.J., Miranda, D., Snapp, E.L., Severina, N., Torregroza, I., Evans, T., and Silver, D.L. 2008. Evolutionarily conserved gene family important for fat storage. *Proceedings of the National Academy of Sciences of the United States of America* 105:94–99.

Kamisaka, Y., Totland, G.K., Tagawa, M., Kurokawa, T., Suzuki, T., Tanaka, M., and Rønnestad, I. 2001. Ontogeny of cholecystokinin-immunoreactive cells in the digestive tract of Atlantic halibut, *Hippoglossus hippoglossus*, larvae. *General and Comparative Endocrinology* 123:31–37.

Kawakami, Y., Raya, A., Raya, R.M., Rodriguez-Esteban, C., and Belmonte, J.C. 2005. Retinoic acid signalling links left–right asymmetric patterning and bilaterally symmetric somitogenesis in the zebrafish embryo. *Nature* 435:165–171.

Kelsh, R.N., and Parichy, D. 2008. Chapter 1.2. Pigmentation. In: Finn, R.N., and Kapoor, B.G. (eds.) *Fish Larval Physiology*. Science Publishers, Enfield, NH.

Kelsh, R.N., Schmid, B., and Eisen, J.S. 2000. Genetic analysis of melanophore development in zebrafish embryos. *Developmental Biology* 225:277–293.

Khosla, S. 2001. Mini review: the OPG/RANKL/RANK system. *Endocrinology* 142:5050–5055.

Kim, Y., and Cooper, K.R. 1998. Interactions of 2.3.7.8-tetrachlorodibenzo-P-dioxin (TCDD) and 3,3′4,4′5-pentachlorobiphenyl (PCB 126) for producing lethal and sublethal effects in the Japanese medaka embryos and larvae. *Chemosphere* 36:409–418.

Kjørsvik, E., Van der Meeren, T., Kryvi, H., Arnfinnson, J., and Kvenseth, A.M. 1991. Early development of the digestive tract of cod larvae, *Gadus morhua* L., during start-feeding and starvation. *Journal of Fish Biology* 38:1–15.

Kjørsvik, E., Olsen, C., Wold, P-A., Hoehne-Reitan, K., Cahu, C.L., Rainuzzo, J., Olsen, A.I., Øie, G., and Olsen, Y. 2009. Comparison of dietary phospholipids and neutral lipids on skeletal development and fatty acid composition in Atlantic cod (*Gadus morhua*). *Aquaculture* 294:246–255.

Komuro, H., Olee, T., Kuhn, K., Quach, J., Brinson, D.C., Shikhman, A., Valbracht, J., Creighton-Achermann, L., and Lotz, M. 2001. The osteoprotegerin/receptor activator of nuclear factor kappa B/receptor activator of nuclear factor kappa B ligand system in cartilage. *Arthritis and Rheumatism* 44:2768–2776.

Koumans, J.T.M., and Akster, H.A. 1995. Myogenic cells in development and growth of fish. *Comparative Biochemistry and Physiology* 110A:3–20.

Koven, W., Rojas-Garcia, C.R., Finn, R.N., Tandler, A., and Ronnestad, I. 2002. Stimulatory effect of ingested protein and/or free amino acids on the secretion of the gastro-endocrine hormone cholecystokinin and on tryptic activity, in early-feeding herring larvae, *Clupea harengus*. *Marine Biology* 140:1241–1247.

Kvåle, A., Harboe, T., Espe, M., Naess, T., and Hamre, K. 2002. Effect of predigested protein on growth and survival of Atlantic halibut larvae (*Hippoglossus hippoglossus* L.). *Aquaculture Research* 33:311–321.

Lall, S.P., and Lewis-McCrea, L.M. 2007. Role of nutrients in skeletal metabolism and pathology in fish—an overview. *Aquaculture* 267:3–19.

Lampert, J.M., Holzschuh, J., Hessel, S., Driever, W., Vogt, K., and von Lintig, J. 2003. Provitamin A conversion to retinal via the beta,beta-carotene-15,15-oxygenase (bcox) is essential for pattern formation and differentiation during zebrafish embryogenesis. *Development* 130:2173– 2186.

Laurence, G.C. 1979. Larval length–weight relations for seven species of northwest Atlantic fishes reared in the laboratory. *Fishery Bulletin* 76:890–895.

Leitão, N.J., Dal Pai-Silva, M., Almeida, F.L.A., and Portella, M.C. 2009. Influence of initial feeding on muscle growth and expression of myogenic regulating factors in pacu *Piaractus mesopotamicus* larvae. In: Hendry, C.I., van Stappen, G., Wille, M., and Sorgeloos, P. (eds.) *Larvi '09—Fish & Shellfish Larviculture Symposium*, Special Publication No. 38. European Aquaculture Society, Oostende, Belgium, pp. 226–227.

Li, H.X., Tyndale, S.T., Heath, D.D., and Letcher, R.J. 2005. Determination of carotenoids and all-trans-retinol in fish eggs by liquid chromatography–electrospray ionization–tandem mass spectrometry. *Journal of Chromatography B* 816(1–2):49–56.

Lie, Ø., and Lambertsen, G. 1985. Digestive lipolytic enzymes in cod (*Gadus morhua*): fatty acid specificity. *Comparative Biochemistry and Physiology* 80B:447–450.

Liu, W., Cao, D.Z., Oh, S.F., Serhan, C.N., and Kulmacz, R.J. 2006. Divergent cyclooxygenase responses to fatty acid structure and peroxide level in fish and mammalian prostaglandin H synthases. *FASEB Journal* 20:1097–1108.

Lock, E.-J., Waagbø, R., Bonga, S.W., and Flik, G. 2009. The significance of Vitamin D for fish: a review. *Aquaculture Nutrition* 16:100–116.

MacQueen Leifson, R., Homme, J.M., Jøstensen, J.P., Lie, Ø., Myklebust, R., and Strøm, T. 2003a. Phospholipids in formulated start-feeds: effect on turbot (*Scophthalmus maximus*, L.) larval growth and mitochondrial alterations in enterocytes. *Aquaculture Nutrition* 9:43–54.

MacQueen Leifson, R., Homme, J.M., Lie, Ø., Myklebust, R., and Strøm, T. 2003b. Three different lipid sources in formulated start-feeds for turbot (*Scophthalmus maximus*, L.) larvae. Effect on growth and mitochondrial alteration in enterocytes. *Aquaculture Nutrition* 9: 33–42.

Maden, M., Graham, A., Zile, M., and Gale, E. 2000. Abnormalities of somite development in the absence of retinoic acid. *The International Journal of Developmental Biology* 44:151–159.

Mai, K., Yu, H., Ma, Q., Gisbert, E., Zambonino Infante, J.L., and Cahu, C.L. 2005. A histological study on development of the digestive system of *Pseudosciaena crocea* larvae and juveniles. *Journal of Fish Biology* 67: 1094–1106.

Mani-Ponset, L., Diaz, J.P., Schlumberger, O., and Connes, R. 1994. Development of yolk complex, liver and anterior intestine in pikeperch larvae, *Stizostedion luciperca* (Percidae), according to the first diet during larval rearing. *Aquatic Living Resources* 7:191–202.

Marza, E., Barthe, C., Andre, M., Villeneuve, L., Helou, C., and Babin, P.J. 2005. Developmental expression and nutritional regulation of a zebrafish gene homologous to mammalian microsomal triglyceride transfer protein large subunit. *Developmental Dynamics* 232:506–518.

Mazurais, D., Darias, M.J., Gouillou-Coustans, M.F., Le Gall, M.M., Huelvan, C., Desbruyeres, E., Quazuguel, P., Cahu, C., and Zambonino-Infante, J.L. 2008. Dietary vitamin mix levels influence the ossification process in European sea bass (*Dicentrarchus labrax*) larvae. *American Journal of Physiology. Regulatory, Integrative and Comparative Physiology* 294:R520–R527.

McEvoy, L.A., Estévez, A., Bell, J.G., Shields, R.J., Gara, B., and Sargent, J.R. 1998. Influence of dietary levels of eicosapentaenoic and arachidonic acid on the pigmentation success of turbot (*Scophthalmus maximus*) and halibut (*Hippoglossus hippoglossus*). *Bulletin of the Aquaculture Association of Canada* 98:17–20.

Meunier, F.J., and Huysseune, A. 1992. The concept of bone tissue in osteichthyes. *Netherlands Journal of Zoology* 42:445–458.

Mise, T., Iijima, M., Inohaya, K., Kudo, A., and Wada, H. 2008. Function of Pax1 and Pax9 in the sclerotome of medaka fish. *Genesis* 46:185–192.

Morais, S., Cahu, C., Zambonino-Infante, J.L., Robin, J., Rønnestad, I., Dinis, M.T., and Conceição, L.E.C. 2004. Dietary TAG source and level affect performance and lipase expression in larval sea bass (*Dicentrarchus labrax*). *Lipids* 39:449–458.

Morais, S., Caballero, M.J., Conceição, L.E.C., Izquierdo, M.S., and Dinis, M.T. 2006. Dietary neutral lipid level and source in Senegalese sole (*Solea senegalensis*) larvae: effect on growth, lipid metabolism and digestive capacity. *Comparative Biochemistry and Physiology. Part B* 144:57–69.

Morais, S., Knoll-Gellida, A., André, M., Barthe, C., and Babin, P.J. 2007. Conserved expression of alternative splicing variants of peroxisomal acyl-CoA oxidase 1 in vertebrates and developmental and nutritional regulation in fish. *Physiological Genomics* 28:239–252.

Moren, M., Naess, T., and Hamre, K. 2002. Conversion of beta-carotene, canthaxanthin and astaxanthin to vitamin A in Atlantic halibut (*Hippoglossus hippoglossus* L.) juveniles. *Fish Physiology and Biochemistry* 27:71–80.

Moren, M., Opstad, I., and Hamre, K. 2004. A comparison of retinol, retinal and retinyl ester concentrations in larvae of Atlantic halibut (*Hippoglossus hippoglossus* L.) fed *Artemia* or zooplankton. *Aquaculture Nutrition* 10:253–259.

Moren, M., Gundersen, T.E., and Hamre, K. 2005. Quantitative and qualitative analysis of retinoids in *Artemia* and copepods by HPLC and diode array detection. *Aquaculture* 246:359–365.

Morriss-Kay, G.M. 2001. Derivation of the mammalian skull vault. *Journal of Anatomy* 199:143–151.

Moss, M.L. 1961. Studies of the acellular bone of teleost fish. *Acta Anatomica* 46:343–462.

Murray, H.M., Gallant, J.W., Perez-Casanova, J.C., Johnson, S.C., and Douglas, S.E. 2003. Ontogeny of lipase expression in winter flounder. *Journal of Fish Biology* 62:816–833.

Nag, A.C., and Nursall, J.R. 1972. Histogenesis of white and red muscle fibres of trunk muscles of a fish *Salmo gairdneri*. *Cytobios* 6:226–247.

Nair, X., Parah, P., Suhr, L., and Tramposch, K.M. 1993. Combination of 4-hydroxyanisole and all trans retinoic acid produces synergistic skin depigmentation in swine. *The Journal of Investigative Dermatology* 101:145–149.

Nakagawa, K., Ohta, Y., Imai, Y., Nomura, C., and Takaoka, K. 2006. Prostaglandin E2 EP4 agonist (ONO-4819) accelerates BMP-induced osteoblastic differentiation. *Bone* 38:S16–S17.

Noden, D.M. 1988. Interactions and fates of avian craniofacial mesenchyme. *Development* 103:121–140.

Norman, J.R. 1934. *A Systematic Monograph of the Flatfishes (Heterosomata). I. Psettodidae, Bothidae, Pleuronectidae.* British Museum (Natural History), London.

Ochi, H., and Westerfield, M. 2007. Signaling networks that regulate muscle development: lessons from zebrafish. *Development, Growth and Differentiation* 49:1–11.

Oozeki, Y., Ishii, T., and Hirano, R. 1989. Histological study of the effects of starvation on reared and wild-caught larval stone flounder *Kareius bicoloratus*. *Marine Biology* 100:269–276.

Osse, J.W.M., and Boogaart, J.G.M. 1995. Fish larvae, development, allometric growth, and the aquatic environment. *ICES Marine Science Symposia* 201:21–34.

Osse, J.W.M., Boogaart, J.G.M., van Snik, G.M.J., and van der Sluys, L. 1997. Priorities during early growth of fish larvae. *Aquaculture* 155:249–258.

Ostaszewska, T., Dabrowski, K., Wegner, A., and Krawiec, M. 2008. The effects of feeding on muscle growth dynamics and the proliferation of myogenic cells during pike perch development (*Sander lucioperca*). *Journal of the World Aquaculture Society* 39(2):184–195.

Patton, J.S., Nevenzel, J.C., and Benson, A.A. 1975. Specificity of digestive lipases in hydrolysis of wax esters and triglycerides studied in anchovy and other selected fish. *Lipids* 10:575–583.

Pavlov, D.A. 1999. Features of transition from larva to juvenile in fishes with different types of early ontogeny. *Environmental Biology of Fishes* 56(1–2):41–52.

Pedersen, B.H., and Andersen, K.P. 1992. Induction of trypsinogen secretion in herring larvae (*Clupea harengus*). *Marine Biology* 112:559–565.

Pedersen, B.H., Nilssen, E.M., and Hjelmeland, K. 1987. Variations in the content of trypsin and trypsinogen in larval herring (*Clupea harengus*) digesting copepod nauplii. *Marine Biology* 94:171–181.

Pères, A., Cahu, C.L., Zambonino Infante, J.L., Le Gall, M.M., and Quazuguel, P. 1996. Amylase and trypsin responses to intake of dietary carbohydrate and protein depend on the developmental stage in sea bass (*Dicentrarchus labrax*) larvae. *Fish Physiology and Biochemistry* 15:237–242.

Péres, J.A., Zambonino Infante, J.L., and Cahu, C. 1998. Dietary regulation of activities and mRNA levels of trypsin and amylase in sea bass (*Dicentrarchus labrax*) larvae. *Fish Physiology and Biochemistry* 19:145–152.

Perez-Casanova, J.C., Murray, H.M., Gallant, J.W., Ross, N.W., Douglas, S.E., and Johnson, S.C. 2004. Bile salt–activated lipase expression during larval development in the haddock (*Melanogrammus aeglefinus*). *Aquaculture* 235:601–617.

Perez-Casanova, J.C., Murray, H.M., Gallant, J.W., Ross, N.W., Douglas, S.E., and Johnson, S.C. 2006. Development of digestive capacity in larvae of haddock (*Melanogrammus aeglefinus*) and Atlantic cod (*Gadus morhua*). *Aquaculture* 251:377–401.

Persson, P., Takagi, Y., and Bjornsson, B.T. 1995. Tartrate-resistant acid-phosphatase as a masker for scale resorption in rainbow-trout, *Oncorhynchus mykiss*: effects of estradiol-17-beta treatment and refeeding. *Fish Physiology and Biochemistry* 14:329–339.

Persson, P., Sundell, K., Bjornsson, B.T., and Lundqvist, H. 1998. Calcium metabolism and osmoregulation during sexual maturation of river running Atlantic salmon. *Journal of Fish Biology* 52:334–349.

Persson, P., Bjornsson, B.T., and Takagi, Y. 1999. Characterization of morphology and physiological actions of scale osteoclasts in the rainbow trout. *Journal of Fish Biology* 54:669–684.

Reitan, K.I., Rainuzzo, J.R., and Olsen, Y. 1994. Influence of lipid composition of live feed on growth, survival and pigmentation of turbot larvae. *Aquaculture International* 2:33–48.

Rescan, P.-Y. 2001. Regulation and functions of myogenic regulatory factors in lower vertebrates. *Comparative Biochemistry and Physiology* 130B:1–12.

Ribeiro, L., Zambonino Infante, J.L., Cahu, C., and Dinis, M.T. 1999. Development of digestive enzymes in larvae of *Solea senegalensis*, Kaup 1858. *Aquaculture* 179:465–473.

Rohde, C.M., and Deluca, H. 2003. Bone resorption activity of all-trans retinoic acid is independent of vitamin D in rats. *Journal of Nutrition* 133:777–783.

Rösch, R., and Segner, H. 1990. Development of dry food for larvae of *Coregonus lavaretus* L. I. Growth, food digestion and fat absorption. *Aquaculture* 91:101–115.

Rowlerson, A., and Veggetti, A. 2001. Cellular mechanisms of postembryonic muscle growth in aquaculture species. In: Johnston, I.A. (ed.) *Fish Physiology*, Vol. XVIII. Academic Press, San Diego, CA, pp. 103–140.

Roy, P.K., and Lall, S.P. 2007. Vitamin K deficiency inhibits mineralization and enhances deformity in vertebrae of haddock (*Melanogrammus aeglefinus* L.). *Comparative Biochemistry and Physiology. Part B, Biochemistry & Molecular Biology* 148:174–183.

Roy, P.K., Witten, P.E., Hall, B.K., and Lall, S.P. 2002. Effects of dietary phosphorus on bone growth and mineralisation of vertebrae in haddock (*Melanogrammus aeglefinus* L.). *Fish Physiology and Biochemistry* 27:35–48.

Rudnicki, M.A., and Jaenisch, R. 1995. The MyoD family of transcription factors and skeletal myogenesis. *BioEssays* 17:203–209.

Rust, M.B. 2002. Nutritional physiology. In: Halver, J.E., and Hardy, R.W. (eds.) *Fish Nutrition*, 3rd edition. Academic Press, San Diego, CA, pp. 367–452.

Sæle, Ø., Silva, N., and Pittman, K. 2006a. Post-embryonic remodelling of neurocranial elements: a comparative study of normal versus abnormal eye migration in a flatfish, the Atlantic halibut. *Journal of Anatomy* 209:31–41.

Sæle, Ø., Smáradóttir, H., and Pittman, K. 2006b. The twisted story of eye migration in flatfish. *Journal of Morphology* 267:730–738.

Salhi, M., Izquierdo, M.S., Hernandez-Cruz, C.M., Socorro, J., and Fernandez-Palacios, H. 1997. The improved incorporation of polyunsaturated fatty acids and changes in liver structure in larval gilthead seabream fed on microdiets. *Journal of Fish Biology* 51:869–879.

Salhi, M., Hernández-Cruz, C.M., Bessonart, M., Izequierdo, M.S., and Fernández-Palacios, H. 1999. Effect of different dietary polar lipid levels and different n-3 HUFA content in polar lipids on gut and liver histological structure of gilthead seabream (*Sparus aurata*) larvae. *Aquaculture* 179:253–263.

Sargent, J.R., McEvoy, L.A., Estévez, R.J., and Tocher, D.R. 1999. Lipid nutrition in marine fish during early development: current status and future directions. *Aquaculture* 179:217–229.

Sargent, J.R., Tocher, D.R., and Bell, J.G. 2002. The lipids. In: Halver, J.E., and Hardy, R.W. (eds.) *Fish Nutrition*. Academic Press, San Diego, CA, pp. 181–257.

Scapolo, P.A., and Rowlerson, A. 1987. Pink lateral muscle in the carp (*Cyprinus carpio* L.): histochemical properties and myosin composition. *Experimentia* 43:384–386.

Segner, H., and Juario, J.V. 1986. Histological observations on the rearing of milkfish, *Chanos chanos*, fry using different diets. *Journal of Applied Ichthyology* 2:162–173.

Segner, H., and Rösch, R. 1998. Ontogeny of digestive and metabolic functions in *Coregonus lavaretus*. *Archiv für Hydrobiologie, Special Issues. Advances in Limnology* 50:1–14.

Segner, H., and Witt, U. 1990. Weaning experiments with turbot (*Scophthalmus maximus*): electron microscopy of liver. *Marine Biology* 105:353–361.

Segner, H., Rösch, R., Schmidt, H., and von Poppinghausen, K.I. 1988. Studies on the suitability of commercial dry diets for rearing of *Coregonus lavaretus* from Lake Constance. *Aquatic Living Resources* 1:231–238.

Segner, H., Rösch, R., Verreth, J., and Witt, U. 1993. Larval nutritional physiology: studies with *Clarias gariepinus*, *Coregonus lavaretus* and *Scophthalmus maximus*. *Journal of the World Aquaculture Society* 24:121–134.

Segner, H., Storch, V., Reinecke, M., Kloas, W., and Hanke, W. 1994. The development of the functional and metabolic organs in turbot, *Scophthalmus maximus*. *Marine Biology* 119:471–486.

Seikai, T., Matsumoto, J., Shimozaki, M., Oikawa, A., and Akiyama, T. 1987. An association of melanophores appearing at metamorphosis as vehicles of asymmetric skin color formation with pigment anomalies developed under hatchery conditions in the Japanese flounder, *Paralichthys olivaceus*. *Pigment Cell Research* 1:143–151.

Seikai, T., Hirose, E., and Matsumoto, J. 1993. Dual appearances of pigment cells from *in vitro* cultured embryonic cells of Japanese flounder: an implication for a differentiation-associated clock. *Pigment Cell Research* 6:423–431.

Slack, J.M.W. 1995. Developmental biology of the pancreas. *Development* 121:1569–1580.

Stickney, R.R., and White, D.B. 1975. Ambicoloration in tank cultured flounder, *Paralichthys dentatus*. *Transactions of the American Fisheries Society* 104:158–160.

Strüssmann, C.A., and Takashima, F. 1990. Hepatocyte nuclear size and nutritional condition of larval pejerry, *Odontesthes bonariensis* (Cuvier et Valenciennes). *Journal of Fish Biology* 36:59–65.

Takahashi, Y. 1994. Influence of stocking density and food at late phase of larval period on hypermelanosis on the blind body side in juvenile Japanese flounder. *Nippon Suisan Gakkaishi* 60:593–598.

Takeuchi, T., Dedi, J., Haga, Y., Seikai, T., and Watanabe, T. 1998. Effect of vitamin A compounds on bone deformity in larval Japanese flounder (*Paralichthys olivaceus*). *Aquaculture* 169:155–165.

Tan, X., and Du, S.J. 2002. Differential expression of two MyoD genes in fast and slow muscles of gilthead seabream (*Sparus aurata*). *Development Genes and Evolution* 212:207–217.

Temple, G.K., Cole, N.J., and Johnston, I.A. 2001. Embryonic temperature and the relative timing of muscle-specific genes during development in herring (*Clupea harengus* L.). *Journal of Experimental Biology* 204:3629–3637.

Terova, G., Rimoldi, S., Chini, V., Gornati, R., Bernardini, G., and Saroglia, M. 2007. Cloning and expression analysis of insulin-like growth factor I and II in liver and muscle of sea bass (*Dicentrarchus labrax*, L.) during long-term fasting and refeeding. *Journal of Fish Biology* 70(Supplement B):219–233.

Terova, G., Rimoldi, S., Bernardini, G., Gornati, R., and Saroglia, M. 2008. Sea bass ghrelin: molecular cloning and mRNA quantification during fasting and refeeding. *General and Comparative Endocrinology* 155/2:341–351.

Theilacker, G.H. 1986. Starvation-induced mortality of young sea-caught jack mackerel, *Trachurus symmetricus*, determined with histological and morphological methods. *Fishery Bulletin* 84:1–17.

Theilacker, G.H., and Watanabe, Y. 1989. Midgut cell height defines nutritional status of laboratory-raised larval northern anchovy, *Engraulis mordax*. *Fishery Bulletin* 87: 457–469.

Thomas, A.J., and Erickson, C.A. 2008. The making of a melanocyte: the specification of melanoblasts from the neural crest. *Pigment Cell Melanoma Research* 21:598–610.

Tocher, D.R., and Sargent, J.R. 1984. Studies on triacylglycerol, wax esters and sterol ester hydrolases in intestinal caeca of rainbow trout (*Salmo gairdneri*) fed diets rich in triacylglycerols and wax esters. *Comparative Biochemistry and Physiology* 77B:561–571.

Tocher, D.R., Bendiksen, E.Å., Campbell, P.J., and Bell, J.G. 2008. The role of phospholipids in nutrition and metabolism of teleost fish. *Aquaculture* 280:21–34.

Udagawa, M. 2001. The effect of dietary vitamin K (phylloquinone and menadione) levels on the vertebral formation in mummichog *Fundulus heteroclitus*. *Fisheries Science* 67:104–109.

Vernier, J.-M., and Sire, M.-F. 1986. Is the Golgi apparatus the obligatory final step for lipoprotein secretion by intestinal cells? *Tissue Cell* 18:447–460.

Villalta, M., Estévez, A., and Bransden, M.P. 2005a. Arachidonic acid enriched live prey induces albinism in Senegalese sole (*Solea senegalensis*) larvae. *Aquaculture* 245:193–209.

Villalta, M., Estévez, A., Bransden, M.P., and Bell, J.G. 2005b. The effect of graded concentrations of dietary DHA on growth, survival and fatty acid profile of Senegal sole (*Solea senegalensis*) larvae during the *Artemia* feeding period. *Aquaculture* 249:353–365.

Villalta, M., Estévez, A., Bransden, M.P., and Bell, J.G. 2008. Arachidonic acid, arachidonic/eicosapentaenoic acid ratio, stearidonic acid and eicosanoids are involved in dietary-induced albinism in Senegal sole (*Solea senegalensis*). *Aquaculture Nutrition* 14:120–128.

Villeneuve, L., Gisbert, E., Delliou, H.L., Cahu, C.L., and Zambonino-Infante, J.L. 2005a. Dietary levels of all-trans retinol affect retinoid nuclear receptor expression and skeletal development in European sea bass larvae. *British Journal of Nutrition* 93:791–801.

Villeneuve, L., Gisbert, E., Zambonino-Infante, J.L., Quazuguel, P., and Cahu, C.L. 2005b. Effect of nature of dietary lipids on European sea bass morphogenesis: implication of retinoid receptors. *British Journal of Nutrition* 94:877–884.

Watkins, B.A., Li, Y., Allen, K.G.D., Hoffmann, W.E., and Seifert, M.F. 2000. Dietary ratio of (n-6)/(n-3) polyunsaturated fatty acids alters the fatty acid composition of bone compartments and biomarkers of bone formation in rats. *Journal of Nutrition* 130:2274–2284.

Weatherley, A.H., Gill, H.S., and Lobo, A.F. 1988. Recruitment and maximal diameter of axial muscle fibres in teleosts and their relationship to somatic growth and ultimate size. *Journal of Fish Biology* 33:851–859.

Weinberg, E.S., Allende, M.L., Kelly, C.S., Abdelhamid, A., Murakami, T., Andermann, P., Doerre, G., Grunwald, D., and Riggleman, B. 1996. Developmental regulation of zebrafish MyoD in wild-type, no tail and spadetail embryos. *Development* 122:271–280.

Wejheden, C., Brunnberg, S., Hanberg, A., and Lind, P.M. 2006. Osteopontin: a rapid and sensitive response to dioxin exposure in the osteoblastic cell line UMR-106. *Biochemical and Biophysical Research Communications* 341:116–120.

Witten, P.E., and Huysseune, A. 2009. A comparative view on mechanisms and functions of skeletal remodelling in teleost fish, with special emphasis on osteoclasts and their function. *Biological Reviews* 84:315–346.

Witten, P.E., and Villwock, W. 1997a. Bone resorption by mononucleated cells during skeletal development in fish with acellular bone. *Journal of Bone and Mineral Research: The Official Journal of the American Society for Bone and Mineral Research* 12:F252–F252.

Witten, P.E., and Villwock, W. 1997b. Growth requires bone resorption at particular skeletal elements in a teleost fish with acellular bone (*Oreochromis niloticus*, Teleostei: Cichlidae). *Journal of Applied Ichthyology* 13:149–158.

Witten, P.E., Holliday, L.S., Delling, G., and Hall, B.K. 1999. Immunohistochemical identification of a vacuolar proton pump (V-ATPase) in bone-resorbing cells of an advanced teleost species, *Oreochromis niloticus*. *Journal of Fish Biology* 55:1258–1272.

Witten, P.E., Villwock, W., Peters, N., and Hall, B.K. 2000. Bone resorption and bone remodelling in juvenile carp, *Cyprinus carpio* L. *Journal of Applied Ichthyology* 16:254–261.

Witten, P.E., Hansen, A., and Hall, B.K. 2001. Features of mono- and multinucleated bone resorbing cells of the zebrafish *Danio rerio* and their contribution to skeletal development, remodeling, and growth. *Journal of Morphology* 250:197–207.

Wold, P-A. 2007. Functional development and response to dietary treatment in larval Atlantic cod (*Gadus morhua* L.): focus on formulated diets and early weaning. PhD thesis, Department of Biology, Norwegian University of Science and Technology (NTNU).

Wold, P-A., Hoehne-Reitan, K., Cahu, C.L., Zambonino Infante, J.L., Rainuzzo, J., and Kjørsvik, E. 2007. Phospholipids vs. neutral lipids: effects on digestive enzymes in Atlantic cod (*Gadus morhua* L.) larvae. *Aquaculture* 272:502–513.

Wold, P-A., Hoehne-Reitan, K., Rainuzzo, J., and Kjørsvik, E. 2008. Allometric growth and functional development of the gut in developing cod (*Gadus morhua* L.) larvae. *Journal of Fish Biology* 72:1637–1658.

Wold, P-A., Hoehne-Reitan, K., Cahu, C.L., Zambonino-Infante, J., Rainuzzo, J., and Kjørsvik, E. 2009. Phospholipids vs. neutral lipids: effects on intestine, liver and pancreas histology in Atlantic cod (*Gadus morhua* L.) larvae. *Aquaculture Nutrition* 15:73–84.

Xie, S.Q., Mason, P.S., Wilkes, D., Goldspink, G., Faconneau, B., and Stickland, N.C. 2001. Lower environmental temperature delays and prolongs myogenic regulatory factor expression and muscle differentiation in rainbow trout (*Oncorhynchus mykiss*) embryos. *Differentiation* 68:106–114.

Zambonino Infante, J.L., and Cahu, C.L. 1997. Partial substitution of di- and tripeptides for native proteins in sea bass diet improves *Dicentrarchus labrax* larval development. *Journal of Nutrition* 127(4):608–614.

Zambonino Infante, J.L., and Cahu, C.L. 1999. High dietary lipid levels enhance digestive tract maturation and improve *Dicentrarchus labrax* larval development. *Journal of Nutrition* 129:1195–1200.

Zambonino Infante, J.L., and Cahu, C.L. 2001. Ontogeny of the gastrointestinal tract of marine fish larvae. *Comparative Biochemistry and Physiology C* 130:477–487.

Zambonino Infante, J.L., and Cahu, C.L. 2007. Dietary modulation of some digestive enzymes and metabolic processes in developing marine fish: applications to diet formulation. *Aquaculture* 268:98–105.

Zambonino Infante, J.L., Cahu, C.L., Peres, A., Quazuguel, P., and Le Gall, M.M. 1996. Sea bass (*Dicentrarchus labrax*) larvae fed different *Artemia* rations: growth, pancreas enzymatic response and development of digestive functions. *Aquaculture* 139:129–138.

Zhang, X., Schwarz, E.M., Young, D.A., Puzas, J.E., Rosier, R.N., and O'Keefe, R.J. 2002. Cyclooxygenase-2 regulates mesenchymal cell differentiation into the osteoblast lineage and is critically involved in bone repair. *Journal of Clinical Investigation* 109:1405–1415.

Zhang, Y., Tan, X., Zhang, P.-J., and Xu, Y. 2006. Characterization of muscle-regulatory gene, MyoD, from flounder (*Paralichthys olivaceus*) and analysis of its expression patterns during embryogenesis. *Marine Biotechnology* 8: 139–148.

Chapter 8

Bioenergetics of growth in commercially important developing fishes

Ione Hunt von Herbing and Meredith Turnbough

8.1 Background and overview of problems

For hundreds of years, fish from the ocean have provided the world with much of its protein. In the 1990s, the collapse of the commercial fishery in New England and the listing of Atlantic salmon as an endangered species created an acute shortage of fish and a critical need for alternative sources of protein. More recently, our growing awareness of climate change and its effect on fisheries and aquaculture has added further concern as to the sustainability of commercial fisheries and the development of sustainable aquaculture.

In 2004, about 75% (105.6 million tonnes) of the estimated world fish production was used for direct human consumption, and the remaining 25% (34.8 million tonnes) was processed into feeds, mostly fish meal and oil (FAO 2009), and 7.3 million tonnes was discarded. Fish for nonfood purposes comes mostly from natural stocks of small pelagics, and some 90% of such catches were processed into fish meal, with the remaining 10% being utilized directly for aquafeed or as feed for domestic livestock. Future projections predict increased demand for fishery products, and it is evident that aquaculture will play a crucial role in satisfying that demand. The large quantity of fish entering the international fish trade will continue to keep fish prices relatively high, and this may compromise access of the poor to adequate fish protein.

Aquaculture of cold-water fishes is an alternative source of protein, and its share of the market has been steadily increasing as culture technologies of important cold-water species, such as Atlantic cod and halibut, improve. However, conversion of fish meal into high-quality protein is still inefficient,

Larval Fish Nutrition, First Edition. Edited by G. Joan Holt.

and new types of meal, as well new culture methods, are being developed to improve fillet yield. For cold-water fish culture at high latitudes, temperature is one of the most important factors affecting growth and development at all life history stages. Ninety-seven percent of all marine fishes (temperate and tropical) produce planktonic larvae, which are temperature sensitive and have high growth rates (20–50%/day), as well as the highest cumulative mortality rates (>95%) of their entire life history. After metamorphosis from the larval to the juvenile stage, fish are more temperature tolerant, and growth and mortality rates decline dramatically. We have little understanding of the physiological processes that facilitate changes in temperature sensitivity and/or responses to other environmental variables such as oxygen concentration and pH throughout the life history of fishes. This is particularly important now that increases in global temperatures of more than 2°C have been predicted by 2050 (Moss et al. 2008). How these increases in temperature will affect fish populations is largely unknown, although some changes in distributions of fish populations have already been noted (Brander 2007). More important perhaps is the unpredictability of how global temperature increases will affect the most sensitive of fish stages (embryonic and larval), as we still have only a rudimentary understanding of the effects of temperature on the metabolism of growth. Yet, robust bioenergetic models for fish early life history stages will likely help predict the subsequent survival or demise of adult populations under changing conditions.

8.2 Suboptimal growth and development: limiting factors for productivity

The most significant problem that limits fish productivity is suboptimal growth and development in marine fish species, specifically in cold-water species such as Atlantic cod, *Gadus morhua*. In all fishes, growth rates decrease from high values (20–50%/day) at the larval stage to lower values (1–5%/day) characteristic of juveniles and adults (Wieser 1995; Hunt von Herbing 2006). In adult fish, with lower growth rates, metabolic costs of growth account for about 40% of the energy content of newly synthesized tissue and represent a large amount of energy available (Pedersen 1997). How larvae fuel their high growth rates and whether they have higher energetic costs, or are more energetically efficient, than adults is presently unknown. In Atlantic cod, a commercially important cold-water marine fish currently under intensive culture in the US market, growth rates can be as high as 30%/day in the larval stages under optimal conditions, but decrease to less than 1%/day in the adult stages (Hunt von Herbing 2006).

A central axiom of nutrition physiology states that in animals, the rate of growth is directly proportional to the growth-related increment of the rate of oxygen consumption above maintenance (Wieser 1994). This axiom is true for juvenile and adult fish but may not hold above a critical level in fast-growing larvae, and growth and metabolism become uncoupled (Rombough 1994; Wieser 1995). At high growth rates, some species of fish larvae can experience "cost-free growth" (Pedersen 1997). We do not understand the physiological basis for "cost-free growth," but two hypotheses have been suggested. The first hypothesis states that as growth rate increases, the growth efficiency also increases by decreasing the cost of synthesis of a unit protein. In the second hypothesis, energy supporting metabolic functions such as maintenance and/or locomotory activity is suppressed at high growth rates, and energy is directed to growth (Pannevis and Houlihan 1992; Wieser 1995; Wieser and Kauffman 1998). The former strategy may involve an increase in the efficiency of protein deposition, analogous to the reduction in the ATP cost of protein synthesis found in trout hepatocytes at different temperatures (Pannevis and Houlihan 1992; Wieser 1994). As global temperatures rise past those that are considered to be optimal for development and growth of fish embryos and larvae, two possibilities arise that would affect larval growth rates and bioenergetics: (1) the cost of maintenance metabolism may increase due to higher metabolic demands placed on the tissues and organ systems to develop and grow faster (metabolic stress response), thus reducing the energy available for other energy-requiring processes such as protein deposition; and (2) increases in temperature may decrease the efficiency of protein deposition (turnover) and increase the ATP cost of protein synthesis, resulting in higher metabolic load per unit protein formed. Ultimately, this may be reflected in less efficient growth and protein production in adults.

8.3 Bioenergetics: mechanisms of growth and development

Bioenergetics provides a framework for the study of interrelationships between feeding rates and growth rates of fish exposed to different environmental conditions, and can also provide insights into the fundamental mechanisms of these relationships based on the study of the partitioning of energetic resources within the organism (Jobling 1993). To increase productivity is to increase our understanding of mechanisms of growth, development, and metabolism in fish. The development and use of bioenergetic models can aid in reaching this objective, and perhaps even more importantly, they can be used as predictive models to understand the impact of temperature/environmental change on developing fishes.

The past 40 years have seen a continued increase in research efforts directed to the study of fish growth and bioenergetics (reviewed in Jobling 1993). In general, bioenergetics can be understood as the way in which energy obtained from food is distributed among several metabolic compartments. This can also be expressed in the following general equation:

$$I = (R_s + R_a + R_f) + (G_s + G_g) + (E_f + E_u + E_s), \tag{8.1}$$

where I is the ingested energy from food; R_s is the standard metabolism; R_a is the active metabolism (maximal aerobic metabolism); R_f is the feeding metabolism (also called specific dynamic action [SDA] or food-induced thermogenesis [FIT]); G_s is somatic tissue growth; G_g is reproductive tissue growth; and E_f, E_u, and E_s are energy lost to feces, urea, and ammonia, respectively. With respect to developing fishes, and particularly larval fishes, in which the reproductive organs are not developed, energy is not directed to reproductive activities and can therefore be preferentially directed to growth, probably accounting for the high growth rates during this stage.

Historically, studies of bioenergetics have focused on adults because larvae and juvenile fishes are more difficult to work with, as they are fragile and easily killed. Metabolic measurements in adult fish have been carried out at a variety of environmental conditions, by monitoring the difference in oxygen concentration between water flowing into and out of an animal chamber (respirometry), taking into account the flow rate through the chamber and the weight of the animal. This method allows the derivation of power performance curves between respiration (R) and swimming speed (U):

$$R_u = R_a - R_s = aU^b - R_s. \tag{8.2}$$

Here R_a equals the active metabolic rate; R_s equals the standard metabolic rate; and U equals the swimming speed and a and b are constants. Power performance curves have been successfully derived for small juvenile

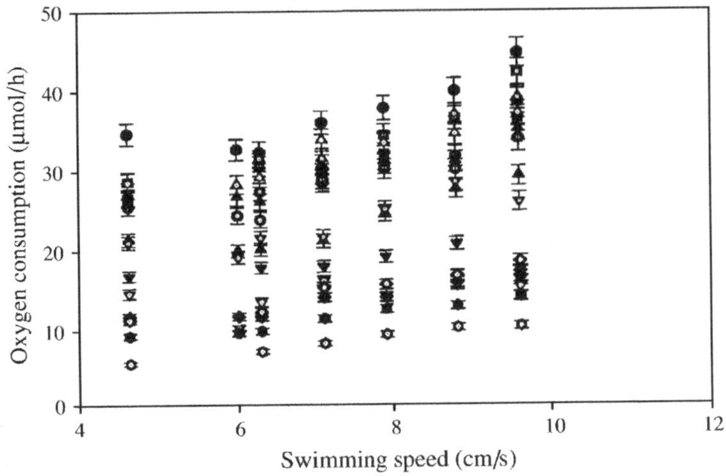

Figure 8.1 Power performance curves of individual juvenile *Gadus morhua* in relation to wet mass (☐ = 0.53, ✹ = 0.55, ◇ = 0.73, ◆ = 0.81, ▽ = 0.83, ▽ = 1.14, ▼ = 1.18, △ = 1.2, △ = 1.42, ▲ = 1.53, ▲ = 1.88, ◘ = 1.9, ■ = 2.07, ◉ = 2.07, ○ = 2.37, ● = 2.89 g) (from Hansen and Hunt von Herbing 2009).

cod that range in mass from 0.53 to 2.89 g at 10°C (Hansen and Hunt von Herbing 2009; Figure 8.1).

Relationships between metabolic rate and mass can differ among developmental stages and activity levels (Wieser 1995; Finn et al. 2002; Hunt von Herbing and White 2002; Hansen and Hunt von Herbing 2009). Generally, adult metabolic rates scale with a metabolic mass exponent (b) between 0.69 and 0.80. In larvae, metabolic mass exponents vary across species and activity levels (Table 8.1), but in many studies metabolic rates are referred to as *routine* rates as activity levels were not controlled when metabolic rates were measured. Only two studies (Kauffman 1990; Hunt von Herbing and Boutilier 1996) have controlled for activity, and results from these studies provide further evidence to show that metabolic mass exponents vary with activity level (Hunt von Herbing 2006).

Temperature does not affect the metabolic mass exponent (Finn et al. 2002). But studies that considered the effects of temperatures on swimming performance revealed that the metabolic costs of swimming of larvae in warmer water was greater than that of larvae in colder water (Fuiman and Batty 1997; Hunt von Herbing 2002). This unexpected result may be related to the way in which temperature differentially affects the structural and metabolic properties of swimming muscles in larvae versus juveniles/adults. This is supported by results from the effects of rearing temperature on fiber structure, distribution of fiber types, and protein polymorphism in swimming muscles of fish larvae (Calvo and Johnston 1992; Vieira and Johnston 1992; Usher et al. 1994).

For larvae, traditional swim-tunnel trials are not appropriate, as larval fish are incapable of sustained activity. Recent technological advances in the fields of microcalorimetry and microrespirometry, however, have made development of partial energy budgets possible for larvae (Hunt von Herbing et al. 2001). The total energy budget includes both aerobic and anaerobic metabolic components. As the anaerobic component represents only a small proportion of the energy budget of feeding in larval fishes (Finn et al. 1995), we will focus on the aerobic energy budget (Equation 8.2) in this chapter.

Table 8.1 Metabolic scaling relationships in larvae, juvenile, and adult fishes at three activity metabolic levels.

	b	Activity level	References
Juveniles and adults			
Micropterus salmoides	0.65	R_r	Beamish (1970)
Oncorhynchus nerca	0.9–1.0	R_a	Brett and Glass (1973)
Salmo gairdneri	1.11–1.14	R_a	Wieser (1995)
Micropterus salmoides	0.92	R_a	Beamish (1970)
Gadus morhua	0.79	R_a	Hunt von Herbing and White (2002)
Larvae			
Rutilus rutilus, Chalcalburnus chaloides	0.87, 0.87	R_s	Kauffman (1990)
Rutilus rutilus, Chalcalburnus chaloides	0.77, 0.77	R_r	Kauffman (1990)
Rutilus rutilus	1.0	R_r	Wieser and Medgyesy (1990)
Oncorhynchus tshawytscha	1.0	R_r	Rombough (1988, 1994)
Hippoglossus hippoglossus	1.0	R_r	Finn (1994)
Scomber scombrus	1.0	R_r	Giguere et al. (1988)
Rutilus rutilus, Chalcalburnus chaloides	0.84, 0.87	R_a	Kauffman (1990)
Gadus morhua	0.85	R_r	Hunt von Herbing and Boutilier (1996)

R_s = standard, R_r = routine, R_a = active, b = the metabolic mass exponent in $R = aM^b$, M = mass, a = mass coefficient.
From Hunt von Herbing (2006).

A simple bioenergetic model separates total aerobic metabolism into three compartments:

$$R_{aer} = R_s + R_f + R_a, \quad (8.3)$$

where R_{aer} is the total aerobic metabolism; R_s is the standard metabolic rate; R_f is the heat increment of feeding (HIF) or SDA (Jobling 1981); and R_a is the active metabolic rate. The standard metabolic rate is the oxygen uptake (mgO_2/h) of an unfed fish at rest. The active metabolic rate is the oxygen uptake of a fish swimming at its maximum aerobic sustainable velocity, and the difference between R_a and R_s is the aerobic scope or "scope for activity" (Fry 1947). The aerobic scope has been studied in the context of aerobic efficiency: For a given amount of calorie intake, those animals that expend energy at a low proportion of aerobic scope maximize their surplus energy, which can be allocated to growth (Priede 1985). Thus, differences in aerobic scope between individuals may translate into differences in growth rate and reflect differences in the cost of growth. Changes in global temperatures are likely to reduce aerobic scope for growth as standard metabolism increases with temperature stress, thus requiring better feeds or different culture methods to improve protein deposition efficiency and fish productivity in an aquaculture setting.

8.4 Energetics of growth through development

Energetic expenditures associated with the ingestion, digestion, and assimilation of food have been termed SDA, HIF, and/or FIT, and are thought to reflect the metabolic cost of growth. In this chapter, SDA does not include digestion (here considered as both ingestion and digestibility) or assimilation, even though they are considered important processes affecting growth rate and efficiencies in juvenile and adult fishes. In adult fish, ingestion and digestibility have been determined by using an inert marker such as chromium oxide in the diet. For larvae, there has been no suitable marker, and ingestion and digestibility have not been quantified, although several projects are ongoing to develop a technique using fluorescent microspheres as inert markers (see also Chapter 3). This technique is in its early stages of development, but if successful, it could be used to investigate the effects of diet composition and digestibility on SDA and growth in future studies.

In fast-growing fish larvae, a greater proportion of energy must be allocated to processing food. This allows for a much larger scope for growth and rapid growth rate (Goolish and Adelman 1988). Scaling studies in the largemouth bass, *Micropterus salmoides*, showed that the SDA of a satiated fish exceeded the active metabolic rate in fish weighing less than 100g, but was less than 50% of active metabolism in larger fish (Tandler and Beamish 1980). In both small (0.5–3.0g) and larger (<80g) cod juveniles, SDA comprised 80–100% of the aerobic scope (Soofiani and Hawkins 1982; Hunt von Herbing and White 2002). This suggests that at peak SDA there is little or no energy left for other activities and that juvenile cod may convert large amounts of ingested food directly to growth. Furthermore, the SDA response should be relatively larger in fast-growing larval fish, and SDA may better reflect maximum aerobic metabolic rates than locomotory energy demands (Soofiani and Hawkins 1982; Soofiani and Priede 1985; Hunt von Herbing and White 2002). Studies of the relationship between SDA and growth are few in marine fish larvae, but one study indicates a lower cost of growth and high growth efficiency in cod larvae compared with juveniles (Kiorboe et al. 1987).

In a recent study, microcalorimetry was used to measure total metabolic heat output (R_t) and SDA (R_f) in larval cod (McCollum et al. 2006; Figure 8.2). In this case, R_t estimated both aerobic and anaerobic metabolic rates. For a cod larva fed to satiation on roti-

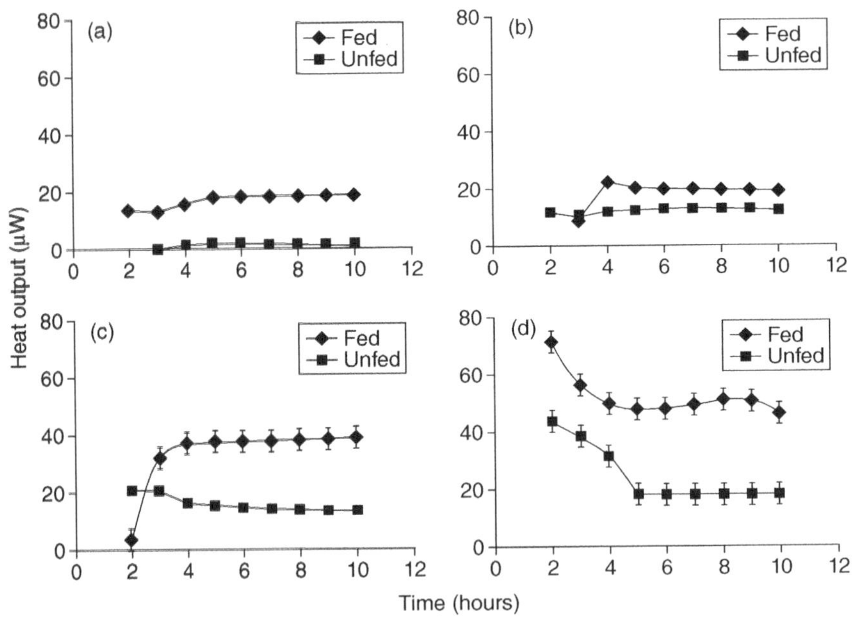

Figure 8.2 Comparison of heat output between fed and unfed larval cod at 10°C for (a) 4dph, (b) 13dph, (c) 21dph, and (d) 28dph. Each point represents mean (±standard error) heat outputs per hour for fish reared at 10°C. This figure is a representation of larval heat output from 4dph to 4 weeks posthatch (from McCollum et al. 2006).

fers, metabolic rates were first detected as a small increase (0.75 times that of standard rates) in heat output at 13 days posthatch (7 mm). The magnitude of SDA continued to increase, with larval size reaching metabolic rates that were four to five times that of standard metabolic rates at 28 days posthatch (11 mm) (McCollum et al. 2006).

Results for feeding metabolism are rare for fish larvae, and McCollum et al. (2006) suggest that high levels of SDA may provide energy for the high growth rates observed in fish larvae. After feeding, SDA results in a typical pattern, which is a rapid rise in oxygen uptake shortly after ingestion of a meal to a peak at two to three times the prefed level within a few hours and then a slow decline to prefed levels (Jobling 1981). The increase in metabolism can temporarily reduce the aerobic scope for activity and may limit further feeding and locomotion, thus curtailing growth. A fish with high SDA will have reduced energy for feeding and locomotion and lower growth, whereas a fish with low SDA may have greater foraging endurance and higher growth. Conversely, Jobling (1985) suggested that SDA may be considered an index of metabolic efficiency, with higher SDA resulting from increased growth rates and lower SDA reflecting reduced growth rates. These conflicting hypotheses have not yet been resolved for fish larvae, although some preliminary data from Chabot (2009) suggest that, at least for cod juveniles, higher levels of SDA correlate with high growth rates.

In recent studies of juvenile marine fishes, 20–40% of postprandial oxygen consumption was found to be entirely due to protein synthesis (Houlihan et al. 1988; Houlihan 1991). In one landmark study, amino acids were infused directly into the bloodstream, and the increase in metabolic rate was similar to that of SDA after feeding (Brown and Cameron 1991). Protein synthesis in larvae has rarely been measured, but one study on nace (*Chondrostoma nasus* L.) showed an increase of 5% in the efficiency in the retention of synthesized protein, with a 5% increase in specific growth rate (SGR). Rates of retention of synthesized protein were high (45–50%) and close to those recorded for juvenile fish (Houlihan 1991). For larval nace, increases in growth rate and protein synthesis rate (k_s) suggest that there is an increase in protein turnover rate, without a significant increase in protein degradation rate. Cost of protein synthesis, however, increased with increasing rates of growth to levels that were three times higher than minimum theoretical costs. To test the hypothesis that the cost of protein synthesis decreases with increasing rates of growth, rates of protein synthesis must be recorded in conjunction with SDA in fish at different growth rates. This has yet to be done for species important to aquaculture, but would aid in our understanding of the factors that control growth.

8.4.1 Relationships between temperature and growth

Hundreds of studies over the past 40 years have investigated the effects of temperature on growth in fishes, but few have studied the effects of temperature on feeding fishes. In fact, Jobling (1996) states that studies on the effects of temperature on feeding fish are scarce and "sorely needed." There are presently no studies on the energetics of feeding in marine fish larvae. At high latitudes, fish can experience changes in temperature that fluctuate 10 or 15°C over a season and they can undergo a 5–10°C change by vertically migrating through the water column. Temperature, therefore, is one of the most important factors affecting feeding, metabolic rates, growth, and ultimately, survival. In general, extremes of temperature tend to decrease aerobic scope (Brett 1964; Elliot 1976). Fish should therefore display a preference for the temperature that maximizes aerobic scope. The magnitude of SDA tends to decrease with increasing temperature (Brett and Groves 1979; Jobling 1981). This may result in faster growth rates at higher temperatures. Since smaller size classes of fish experience exponentially higher mortality, faster growth often translates to lower mortality rates. Thus, the temperature at which SDA is minimized (i.e., energy conversion is maximized) may be the optimal temperature for growth and survival of fish (Tupper 1994).

Generally, the temperature at which conversion efficiency is maximized is slightly higher than the optimum temperature for growth (Jobling 1993). Some adjustment of this relationship may be required because there is evidence that there are thermal optima for growth and ingestion rates in some species of fishes that may change with size, age, and developmental stage (Jobling et al. 1994). For example, temperature may have a more profound effect on larvae than on older stages as Q_{10}s for larvae range from 1.5 to 4.9 compared with juveniles and adults, which have Q_{10}s of 2.0 (Finn et al. 2002). This is supported by studies that show that temperature and size are the greatest determinant of growth in Norwegian coastal and Arctic cod larvae (Otterlei et al. 1999; Otterlei 2000). Higher Q_{10} values in larvae suggest that they are more stenothermal than juveniles and adults, but studies are still needed to determine how temperature sensitivity affects growth and feeding in fish at different stages of development. These are even more critical at the current level of global warming as changes in ocean temperatures are likely to impact larval growth, survival, and distribution.

8.5 Growth, bioenergetics, and stress: why fish do not grow at maximal rates

Growth is a critical time in the life of all organisms, but most animals do not grow at their maximal rate (Case 1978; Arendt 1997; Mangel and Stamps 2001). In poultry and cattle, growth (defined here as increases in mass) is determinate (i.e., it stops after maturity); but in 90% of all fishes, growth is indeterminate (i.e., growth never stops). Fishes have their highest rates of growth (20–50%/day) in their larval stages, after which growth rate decreases in the juvenile (5–10%/day) and adult stages (1–3%/day). One of the reasons why animals may not attain their highest growth rates is because the increase in metabolic activity needed to fuel rapid growth may cause damage to the organism. Examples of possible damage include studies that have reported positive correlations between growth rate and oxidative stress in the form of reactive oxygen species (ROS) (Merry 1995; Rollo 2002; Brown-Borg and Rakoczy 2003). The positive covariation between growth and oxidative stress may also be related to the link between growth and longevity (Jennings et al. 1999; Miller et al. 2000; Bartke et al. 2001; Rollo 2002). Further, in early life history stages, very rapid growth may prevent the correct development of key structural components such as skeletal muscles, as well as complex systems such as the nervous and immune systems (Fisher et al. 2006; Pihlaja et al. 2006). The mechanisms that result in the relationship between oxidative stress, muscle development and integrity, and metabolic costs of growth remain very poorly understood especially in fishes, a majority of which are indeterminate growers.

Stress is of central concern in aquaculture as the various stressors that accompany intensive fish husbandry can predispose the fish to compromised growth and health, and promote disease. For purposes of clarity, stress will be defined as the response of the cell, or organism, to any demand placed on it such that it causes an extension of a physiological state beyond the normal resting state (Barton 1997). Stress in fish is caused by physical disturbances encountered in aquaculture, such as handling and transport, and evokes a variety of responses that may be adaptive or maladaptive (Barton and Iwama 1991). There are two major types of stress; acute or short-term stress (e.g., handling, point exposure to high concentrations of toxins or changes in environmental variables, hypoxia), and chronic or long term (oxidative damage accumulated over time as a function of fast growth, long-term exposure to environmental toxins). The animals' response to stress may be broadly categorized into those that are either adaptive, which allows homeostatic recovery, or maladaptive and compromise performance and growth.

The overall effect of stress may be considered as a change in biological condition beyond the normal resting state that challenges homeostasis and thus presents a threat to the fish's health. These stress-induced changes are grouped as primary; secondary, which includes metabolic, hematological, hydromineral, and structural; and animal responses. Many of these responses can be used as quantitative indicators of stress, although investigators need to be aware of the various "nonstress" factors that can also influence these conditions. A major focus of current research is on the response of the hypothalamic–pituitary–interrenal axis and the resultant elevation of circulating corticosteroids. Stress, through the action of corticosteroids, may (1) reduce immunocompetence by influencing lymphocyte numbers and antibody-production capacity; and (2) affect reproduction by altering levels and patterns of reproductive hormones that influence maturation.

8.5.1 Compensatory growth (CG)

In aquaculture, one of the goals is to increase feed conversion efficiency (FCE) of commercially important species. An approach that has been recently adopted is that of inducing CG mechanisms. CG is described as a phase of accelerated growth, often resulting when favorable conditions are restored after a period of growth depression (Ali et al. 2003). Costs of CG with respect to oxidative damage can result from two nonexclusive mechanisms. The first is that CG may impair the antioxidant defense system, making organisms more susceptible to oxidants. The second is that CG may itself cause oxidative stress and greater ROS production due to increased levels of oxygen consumption and metabolism (Loft et al. 1994; Rollo 2002). Support for the first mechanism is provided by Blount et al. (2003), showing that after a period of low-quality nutrition, zebra finches showed lower lipophilic antioxidant levels, suggesting that they assimilated less lipophilic antioxidants from the diet. Crescenzo et al. (2006) showed CG in fat, but not in mass, after a period of semistarvation in rats, and a higher superoxide dismutase (SOD) activity (reflective of oxidative stress) in previously semistarved rats after they were fed ad libitum again. Alonso-Alvarez et al. (2007) demonstrated higher susceptibility to oxidative damage after CG in zebra finches when challenged with an oxidant, but could not separate out both mechanisms as they assessed combined effects of the exogenous (oxidant) and endogenous (due to CG) sources of ROS.

One of the costs of CG is that rapid growth leads to an accumulation of oxidative damage due to the formation of ROS (Mangel and Munch 2005) at the cellular level that is expressed at the level of the organism (De Block and Stoks 2008). Oxidative damage/stress has been defined as the disturbance in the balance between the production of ROS, or free radicals, and antioxidant defenses, which may lead to tissue injury. Also, a free radical can be defined as any chemical species that contains unpaired electrons in their outer orbit and thus can react virtually with all cell components. Although ROS are crucial to normal biological processes, they are potentially dangerous and are commonly referred to as pro-oxidants. Stress may also alter metabolic scope in fish and affect growth, partly as a result of either the catabolic or gluconeogenic effect of corticosteroids. Further, it may alter the metabolic cost of growth and bioenergetic budget in rapidly growing larval stages, compromising the amount of energy allocated to growth.

8.5.2 Mechanisms of ROS production

Molecular oxygen (O_2) is required by animals for the oxidation of ingested food and the generation of energy. During this process, O_2 undergoes tetravalent reduction to water. However, partial reduction of O_2 results in the formation of ROS, including both radical species, such as the superoxide anion radical ($O_2 \cdot -$) and hydroxyl radical (OH·), and non-radical species such as H_2O_2. ROS are continually produced as toxic by-products of normal metabolism from several endogenous processes, and approximately 1–3% of oxygen consumed in animal systems is converted to ROS, with OH· being the most potent (Livingstone 2003). The production of the different ROS can be interrelated, yielding the highly damaging OH·. Thus, $O_2 \cdot -$ can be converted to H_2O_2 via dismutation:

$$2H + + 2O_2 \rightarrow H_2O_2$$

and O_2 and H_2O_2 can react together to yield OH· via a metal-catalyzed Haber-Weiss reaction:

$$O_2 + H_2O_2 \rightarrow OH + OH - O_2.$$

These ROS can in turn give rise to other ROS such as peroxyl and alkoxyl radicals (ROO· and RO·) through reaction with other biological molecules. Thus, an initial pro-oxidant event can give rise to a spreading web of ROS production within an animal.

In a normal healthy cell, antioxidant defenses include free radical scavengers (e.g., vitamin C, reduced glutathione [GSH], carotenoids, lipid-soluble molecules [vitamins A and E]) and specific antioxidant enzymes (SOD, catalase, and glutathione peroxidase [GPX]). A balance is thought to exist between pro-oxidant production and antioxidant defense, although low levels of oxidative damage, specifically to key biological molecules such as lipid, protein, and DNA, are always present. Increases in ROS production can overcome antioxidant defenses, resulting in increased oxidative damage to macromolecules, alterations in critical cellular processes, and systemic oxidative stress. Exposure to increased ROS production can also lead to induction of antioxidant enzymes through specific gene expression and increased transcription.

Any process that leads to increased ROS production, either directly or indirectly via organic radical formation or other mechanisms, can potentially result in enhanced oxidative stress and biological damage. The same general scenario of ROS production, antioxidant defense, and oxidative damage as seen for mammals is indicated for aquatic organisms, although much less is known, particularly in *in vivo* events and in the relationship of oxidative damage, growth, and development.

8.5.3 Gene expression and bioenergetics of growth under stress

Mutations that affect growth factor pathways (GH/IGF1/insulin) also influence the expression of antioxidant defenses (Rollo 2002). Studies on cell senescence also suggest a link between growth rate and oxidative stress. Oxidative stress accelerates the rate of telomere shortening (von Zglinicki 2002), which is associated with cell death (Harley et al. 1990) and mortality (Wong and Collins 2003), and the rate of telomere shortening is accelerated during a period of somatic growth in vertebrates (Jennings et al. 2000; Brümmendorf et al. 2002; Hall et al. 2004).

Stressors such as overcrowding, changes in pH, variations in salinity, temperature changes, hypoxia, and even hyperoxia have been evaluated in relation to their individual or additive effects on the growth and health of cultured fish (Guerriero et al. 2002). SGR are lower in Atlantic cod exposed to hypoxia (Olsvik et al. 2006); long-term stress causes immunosuppression in Atlantic salmon (Fast et al. 2008). These results seem straightforward and intuitive; it is expected that an animal under stress will not grow as quickly or resist disease as well as an unstressed animal.

The question of *if* various stressors affect growth has largely been answered. The question of *why* these factors affect growth is simplistic in a global way, in that if an organism is expending extra energy to maintain homeostasis despite the effects of a stressor, then that organism has less energy to devote to growth. The question of *how* really answers the question of *why*, and is key to understanding the mechanisms by which growth is delayed or derailed by stress. For example, overcrowding of Atlantic cod for 1 hour elicits a dramatic upregulation in the transcription of catalase (as well as other antioxidant enzymes). At 2 hours post crowding, the levels of catalase mRNA transcripts are twice what they were before crowding. This level of catalase mRNA is maintained for 24 hours after the crowding event (Caipang et al. 2008). Every nucleotide and amino acid devoted to the doubled production of catalase represents RNA and protein molecules that were diverted from the growth of the organism. Only through a more detailed knowledge of the actual physiochemical events involved in the stress response will there be a clear path to designing methods, and particularly broodstocks that are better able to cope with stressors. The possibility of developing broodstock that are resistant to particular stressors is hinted at in another example. In this case, the median SGR of Atlantic cod exposed to hyperoxia was unchanged from that of the SGR of Atlantic cod exposed to normoxia, but the standard deviation of the group of fish in hyperoxia was 1.2 ± 0.5%/day compared with 0.8 ± 0.2%/day for the group in normoxia (Olsvik et al. 2006). Although more studies would certainly be required, these results suggest that some individual cod can cope with hyperoxia (and likely increased ROS) while maintaining a higher growth rate than others. With advances in technology, more genetic sequence data has become available; and indeed, the list of organisms whose whole genome has been sequenced seems to be growing at a geometric rate. These tools should allow for greater functional discrimination between individual fish in relation to their suitability for aquaculture, at least for the suitability of the juveniles or adults. While there is mounting research dealing with stress in juvenile and adult fish, there is a dearth of such information regarding larval fish. This most critical of stages, in terms of survival, is also the least studied and understood in terms of what might represent a selective advantage to the larval phenotype.

8.6 Concluding remarks

Much of stress research can be referred to as "stamp collecting" involving various stress responses, from qualitative to quantitative responses, in an increasing number of species (Iwama et al. 2006). While it may be reasonable to state that the number of significant insights into the mechanisms underlying the stress response in fishes have been few over the last 20 years, the integration of bioenergetics with stress studies is likely to further our understanding of the response. How stress affects energetic partitioning is a new approach and one likely to provide further understanding of the link between metabolism and growth under different environmental conditions and in different aquaculture settings. New genomic and proteomic capabilities will allow us to examine traditional approaches to stress response in new ways and will provide a more comprehensive understanding of the energetics of stress.

Other advances in instrumentation and techniques such as microcalorimetry and advanced kinetic enzyme analyses will further our understanding of whole-animal energetic partitioning. The increasing number of organisms that are being sequenced, together with advanced microarray techniques, will integrate our understanding of the stress response and how it affects other functions such as growth, from the molecular to the organismal level. The interface between bioenergetics and stress physiology will provide much needed information about growth regulation and its costs, and will lead to ways to optimize growth in the aquaculture industry, while minimizing stress-related damage.

Literature cited

Ali, M., Nicieza, A., and Wootton, R.J. 2003. Compensatory growth in fishes: a response to growth depression. *Fish and Fisheries* 4:147–190.

Alonso-Alvarez, C., Bertrand, S., Faivre, B., and Sorci, G. 2007. Increased susceptibility to oxidative damage as a cost of accelerated somatic growth in zebra finches. *Functional Ecology* 21:873–879.

Arendt, J.D. 1997. Adaptive intrinsic growth rates: an integration across taxa. *The Quarterly Review of Biology* 72:149–177.

Bartke, A., Coschigano, K., Kopchick, J., Chandrashekar, V., Mattison, J., Kinney, B., and Hauck, S. 2001. Genes that prolong life: relationships of growth hormone and growth to aging and life span. *The Journals of Gerontology. Series A, Biological Sciences and Medical Sciences* 56:B340–B349.

Barton, B.A. 1997. *Stress in Finfish: Past, Present, and Future—A Historical Perspective.* Cambridge University Press, Cambridge.

Barton, B.A., and Iwama, G.K. 1991. Physiological changes in fish from stress in aquaculture with emphasis on the response and effects of corticosteroids. *Annual Review of Fish Diseases* 1:3–26.

Beamish, F.W.H. 1970. Oxygen consumption of largemouth bass, *Micropterus salmoides*, in relation to swimming speed and temperature. *Canadian Journal of Zoology* 45:1221–1228.

Blount, J.D., Metcalfe, N.B., Arnold, K.E., Surai, P.F., Devevey, G.L., and Monaghan, P. 2003. Neonatal nutrition, adult antioxidant defences and sexual attractiveness in the zebra finch. *Proceedings. Biological Sciences* 270:1691–1696.

Brander, K.M. 2007. Global fish production and climate change. *Proceedings of the National Academy of Sciences of the United States of America* 104:19709–19714.

Brett, J.R. 1964. The respiratory metabolism and swimming performance of young sockeye salmon. *Journal of the Canadian Fisheries Research Board* 21:1183–1226.

Brett, J.R., and Glass, N.R. 1973. Metabolic rates and critical swimming speeds of sockeye salmon (*Oncorhynchus nerka*) in relation to size and temperature. *Journal of the Fisheries Research Board of Canada* 30:379–387.

Brett, J.R., and Groves, T.D.D. 1979. Physiological energetics. In: Hoar, W.S., Randall, D.J., and Brett, J.R. (eds.) *Fish Physiology, Vol. 8, Bioenergetics and Growth.* Academic Press, New York, p. 786.

Brown, C.R., and Cameron, J.N. 1991. The induction of specific dynamic action in channel catfish by infusion of essential amino acids. *Physiological zoology* 64:276–297.

Brown-Borg, H.M., and Rakoczy, S.G. 2003. Growth hormone administration to long-living dwarf mice alters multiple components of the antioxidative defense system. *Mechanisms of Ageing and Development* 124:1013–1024.

Brümmendorf, T.H., Mak, J., Sabo, K.M., Baerlocher, G.M., Dietz, K., Abkowitz, J.L., and Lansdorp, P.M. 2002. Longitudinal studies of telomere length in feline blood cells: implications for hematopoietic stem cell turnover *in vivo*. *Experimental Hematology* 30:1147–1152.

Caipang, C.M.A., Brinchmann, M.F., and Kiron, V. 2008. Short-term overcrowding of Atlantic cod, *Gadus morhua*: effects on serum-mediated antibacterial activity and transcription of glucose transport and antioxidant defense related genes. *Comparative Biochemistry and Physiology. Part A, Molecular & Integrative Physiology* 151:560–565.

Calvo, J., and Johnston, I.A. 1992. Influence of rearing temperature on the distribution of muscle fibre types in the turbot *Scophthalmus maximus* at metamorphosis. *Journal of Experimental Marine Biology and Ecology* 161:45–55.

Case, T.J. 1978. On the evolution and adaptive significance of postnatal growth rates in the terrestrial vertebrates. *The Quarterly Review of Biology* 53:243–282.

Chabot, D. 2009. Comparison of a commercial and three experimental diets for their effects on the SDA characteristics of Atlantic cod, *Gadus morhua*. Society for Experimental Biology, Glasgow, Scotland.

Crescenzo, R., Lionetti, L., Mollica, M.P., Ferraro, M., D'Andrea, E., Mainieri, D., Dulloo, A.G., Liverini, G., and Iossa, S. 2006. Altered skeletal muscle subsarcolemmal mitochondrial compartment during catch-up fat after caloric restriction. *Diabetes* 55:2286–2293.

De Block, M., and Stoks, R. 2008. Compensatory growth and oxidative stress in a damselfly. *Proceedings. Biological Sciences* 275:781–785.

Elliot, J.M. 1976. Energy losses in the waste products of brown trout (*Salmo trutta* L.). *The Journal of Animal Ecology* 45:561–580.

FAO. 2009. National Aquaculture Sector Overview. Mexico. National Aquaculture Sector Overview fact sheets. (Text by Montero Rodríguez, M.) In: FAO Fisheries and Aquaculture Department [online]. Rome. Updated February 1, 2005. Available at www.fao.org/fishery/countrysector/naso_mexico/en.

Fast, M.D., Hosoya, S., Johnson, S.C., and Alfonso, L.O.B. 2008. Cortisol response and immune-related effects of Atlantic salmon (*Salmo salar* Linnaeus) subjected to short- and long-term stress. *Fish & Shellfish Immunology* 24:194–204.

Finn, R.N. 1994. Physiological energetics of developing marine fish embryos and larvae. Dr. thesis, University of Bergen, Bergen, Norway.

Finn, R.N., Fyhn, H.J., and Evjen, M.S. 1995. Physiological energetics of developing embryos and yolk-sac larvae of Atlantic cod (*Gadus morhua*). I. Respiration and nitrogen metabolism. *Marine Biology* 124:355–369.

Finn, R.N., Ronnestad, I., van der Meeren, T., and Fhyn, H.J. 2002. Fuel and scaling during the early life stages of Atlantic cod (*Gadus morhua*). *Marine Ecology Progress Series* 243:217–234.

Fisher, M.O., Nager, R.G., and Monaghan, P. 2006. Compensatory growth impairs adult cognitive performance. *PLoS Biology* 4:e251.

Fry, F.E.J. 1947. Effects of the environment on animal activity. *Publications of the Ontario Fisheries Research Laboratory* 68:1–62.

Fuiman, L., and Batty, R. 1997. What a drag it is getting cold: partitioning the physical and physiological effects of temperature on fish swimming. *The Journal of Experimental Biology* 200:1745–1755.

Giguere, L.A., Cote, B., and St. Pierre, J.F. 1998. Metabolic rates scale isometrically in larval fishes. *Marine Ecology—Progress Series* 50:13–19.

Goolish, E.M., and Adelman, I.R. 1988. Tissue-specific allometry of aerobic respiratory enzyme in a large and small species of cyprinid (Teleostei). *Canadian Journal of Zoology* 66:2199–2208.

Guerriero, G., Di Finizio, A., and Ciarcia, G. 2002. Stress-induced changes of plasma antioxidants in aquacultured sea bass, *Dicentrarchus labrax*. *Comparative Biochemistry and Physiology. Part A, Molecular & Integrative Physiology* 132:205–211.

Hall, M.E., Nasir, L., Daunt, F., Gault, E.A., Croxall, J.P., Wanless, S., and Monaghan, P. 2004. Telomere loss in relation to age and early environment in long-lived birds. *Proceedings. Biological Sciences* 271:1571–1576.

Hansen, S.L., and Hunt von Herbing, I. 2009. Aerobic scope for activity in age 0 year Atlantic cod *Gadus morhua*. *Journal of Fish Biology* 74:1355–1370.

Harley, C.B., Futcher, A.B., and Greider, C.W. 1990. Telomeres shorten during ageing of human fibroblasts. *Nature* 345:458–460.

Houlihan, D.F. 1991. Protein turnover in ectotherms and its relationships to energetics. In: Giles, R. (ed.) *Advances in Comparative and Environmental Physiology*. Vol. 7, Springer-Verlag, Berlin, pp. 1–43.

Houlihan, D.F., Hall, S.J., Gray, C., and Noble, B.S. 1988. Growth rates and protein turnover in Atlantic cod, *Gadus morhua*. *Canadian Journal of Fisheries and Aquatic Sciences* 45:961–964.

Hunt von Herbing, I. 2002. Effects of temperature on larval fish swimming performance: the importance of physics to physiology. *Journal of Fish Biology* 61:865–876.

Hunt von Herbing, I. 2006. The physiological basis for metabolic scaling in animals: a developing perspective. In: Warburton, S.J., Burggren, W.W., Pelster, B., Reiber, C.L., and Spicer, J. (eds.) *Comparative Developmental Physiology: Contributions, Tools, and Trends.* Oxford University Press, New York, pp. 83–98.

Hunt von Herbing, I., and Boutilier, R.G. 1996. Activity and metabolism of larval Atlantic cod (*Gadus morhua*) from Scotian Shelf and Newfoundland source populations. *Marine Biology* 124:607–617.

Hunt von Herbing, I., and White, L. 2002. The effects of body mass and feeding on metabolic rate in small juvenile Atlantic cod. *Journal of Fish Biology* 61:945–958.

Hunt von Herbing, I., Gallager, S.M., and Halteman, W. 2001. Metabolic costs of pursuit and attack in early larval Atlantic cod. *Marine Ecology Progress Series* 216:201–212.

Iwama, G.K., Afonso, L.O.B., and Vijayan, M.M. 2006. Stress in fish. In: Evans, D.H., and Claiborne, J.B. (eds.) *The Physiology of Fishes.* CRC Press, Boca Raton, FL, pp. 319–342.

Jennings, B.J., Ozanne, S.E., Dorling, M.W., and Hales, C.N. 1999. Early growth determines longevity in male rats and may be related to telomere shortening in the kidney. *FEBS Letters* 448:4–8.

Jennings, B.J., Ozanne, S.E., and Hales, C.N. 2000. Nutrition, oxidative damage, telomere shortening, and cellular senescence: individual or connected agents of aging? *Molecular Genetics and Metabolism* 71:32–42.

Jobling, M. 1981. The influences of feeding on the metabolic rate of fishes: a short review. *Journal of Fish Biology* 18:385–400.

Jobling, M. 1985. Growth. In: Tytler, P., and Calow, P. (eds.) *Fish Energetics: New Perspectives.* Croom Helm, London, pp. 213–230.

Jobling, M. 1993. Bioenergetics: feed intake and energy partitioning. In: Rankin, J.C., and Jensen, F.B. (eds.) *Fish Ecophysiology.* Chapman and Hall, London, pp. 1–44.

Jobling, M. 1996. Temperature and growth: modulation of growth rate via temperature. In: Wood, C.M., and McDonald, D.G. (eds.) *Global Warning: Implication for Freshwater and Marine Fish, Seminar Series Society for Experimental Biology*, Vol. 61. Cambridge University Press, Cambridge, pp. 225–253.

Jobling, M., Meløy, O.H., Santos, J., and Christiansen, B. 1994. The compensatory growth response of the Atlantic cod: effects of nutritional history. *Aquaculture International* 2:75–90.

Kauffman, R. 1990. Respiratory cost of swimming in larval and juvenile cyprinids. *The Journal of Experimental Biology* 150:343–366.

Kiorboe, T., Munk, P., and Richardson, K. 1987. Respiration and growth of larval herring (*Clupea harengus*): relation between specific dynamic action and growth efficiency. *Marine Ecology Progress Series* 40:1–10.

Livingstone, D.R. 2003. Oxidative stress in aquatic organisms in relation to pollution and aquaculture. *Revue de Medecine Veterinaire* 154:427–430.

Loft, S., Astrup, A., Buemann, B., and Poulsen, H.E. 1994. Oxidative DNA damage correlates with oxygen consumption in humans. *The FASEB Journal* 8:534–537.

Mangel, M., and Munch, S.B. 2005. A life-history perspective on short- and long-term consequences of compensatory growth. *The American Naturalist* 166:E155–E176.

Mangel, M., and Stamps, J.A. 2001. Tradeoffs between growth and mortality and the maintenance of individual variation in growth. *Evolutionary Ecology Research* 3:583–593.

McCollum, A., Geubtner, J., and Hunt von Herbing, I. 2006. Metabolic cost of feeding in Atlantic Cod (*Gadus morhua*) larvae using microcalorimetry. *ICES Journal of Marine Science* 63:335–339.

Merry, B.J. 1995. Effect of dietary restriction on aging: an update. *Reviews in Clinical Gerontology* 5:247–258.

Miller, R.A., Chrisp, C., and Atchley, W. 2000. Differential longevity in mouse stocks selected for early life growth trajectory. *The Journals of Gerontology. Series A, Biological Sciences and Medical Sciences* 55:B455–B461.

Moss, R., Babiker, M., Brinkman, S., Calvo, E., Carter, T., Edmonds, J., Elgizouli, I., Emori, S., Erda, L., Hibbard, K., Jones, R., Kainuma, M., Kelleher, J., Lamarque, J.F., Manning, M., Matthews, B., Meehl, G., Meyer, L., Mitchell, J., Nakicenovic, N., O'Neill, B., Pichs, T., Riahi, K., Rose, S., Runci, P., Stouffer, R., van

Vuuren, D., Weyant, J., Wilbanks, T., van Ypersele, J.P., and Zurek, M. 2008. Towards new scenarios for analysis of emissions, climate change, impacts, and response strategies. In: *IPCC Expert Meeting Report*. Intergovernmental Panel on Climate Change, Geneva.

Olsvik, P.A., Kristensen, T., Waagbo, R., Tollefsen, K.E., Rosseland, B.O., and Toften, H. 2006. Effects of hypo- and hyperoxia on transcription levels of five stress genes and the glutathione system in liver of Atlantic cod *Gadus morhua*. *The Journal of Experimental Biology* 209:2893–2901.

Otterlei, E. 2000. Temperature- and size-dependent growth of larval and early juvenile Atlantic Cod (*Gadus morhua* L.). PhD thesis, University of Bergen, Bergen, Norway.

Otterlei, E., Nyhammer, G., Folkvord, A., and Stefansson, S.O. 1999. Temperature- and size-dependent growth of larval and early juvenile Atlantic cod (*Gadus morhua*): a comparative study of Norwegian coastal cod and northeast Arctic cod. *Canadian Journal of Fisheries and Aquatic Sciences* 56:2099–2111.

Pannevis, M.C., and Houlihan, D.F. 1992. The energetic cost of protein synthesis in isolated hepatocytes of rainbow trout (*Oncorhynchus mykiss*). *Journal of Comparative Physiology B: Biochemical, Systemic, and Environmental Physiology* 162:393–400.

Pedersen, B.H. 1997. The cost of growth in young fish larvae: a review of new hypotheses. *Aquaculture* 155:259–269.

Pihlaja, M., Siitari, H., and Alatalo, R.V. 2006. Maternal antibodies in a wild altricial bird: effects on offspring immunity, growth and survival. *The Journal of Animal Ecology* 75:1154–1164.

Priede, I.G. 1985. Metabolic scope in fishes. In: Tytler, P., and Calow, P. (eds.) *Fish Energetics: New Perspectives*. Croom Helm, London, pp. 33–64.

Rollo, C.D. 2002. Growth negatively impacts the life span of mammals. *Evolution & Development* 4:55–61.

Rombough, P.J. 1988. Respiratory gas exchange, aerobic metabolism, and effects of hypoxia during early life. In: Hoar, W.S., and Randall, D.J. (eds.) *Fish Physiology*, Vol. XI, Part A. Academic Press, San Diego, CA, pp. 59–161.

Rombough, P.J. 1994. Energy partitioning during fish development: additive or compensatory allocation of energy to support growth. *Functional Ecology* 8:178–186.

Soofiani, N.M., and Hawkins, A.D. 1982. Energetic costs at different levels of feeding in juvenile cod *Gadus morhua* L. *Journal of Fish Biology* 21:577–592.

Soofiani, N.M., and Priede, I.G. 1985. Aerobic metabolic scope and swimming performance in juvenile cod, *Gadus morhua* L. *Journal of Fish Biology* 26:121–138.

Tandler, A., and Beamish, F.W. 1980. Specific dynamic action and diet in largemouth bass, *Micropterus salmoides* (Lacepede). *The Journal of Nutrition* 110:750–764.

Tupper, M. 1994. Settlement and post-settlement processes in the population regulation of a temperate reef fish: the role of energy. PhD thesis, Dalhousie University, Halifax.

Usher, M.L., Stickland, N.C., and Thorpe, J.E. 1994. Muscle development in Atlantic salmon (*Salmo salar*) embryos and the effect of temperature on muscle cellularity. *Journal of Fish Biology* 44:953–964.

Vieira, V.I.A., and Johnston, I.A. 1992. Influence of temperature on muscle-fibre development in larvae of the herring *Clupea harengus*. *Marine Biology* 112:333–341.

Wieser, W. 1994. Cost of growth in cells and organisms: general rules and comparative aspects. *Biological Reviews* 69:1–33.

Wieser, W. 1995. Energetics of fish larvae, the smallest vertebrates. *Acta Physiologica Scandinavica* 154:279–290.

Wieser, W., and Kauffman, R. 1998. A note on the interactions between temperature, viscosity, body size and swimming energetics in fish larvae. *The Journal of Experimental Biology* 201:1369–1372.

Wieser, W., and Medgyesy, N. 1990. Aerobic maximum for growth in the larvae and juveniles of cyprinid fish, *Rutilus rutilus* (L.): implications for energy budgeting in small poikilotherms. *Functional Ecology* 4:233–242.

Wong, J.M.Y., and Collins, K. 2003. Telomere maintenance and disease. *Lancet* 362:983–988.

von Zglinicki, T. 2002. Oxidative stress shortens telomeres. *Trends in Biochemical Sciences* 27:339–344.

Chapter 9

Regulation of digestive processes

Ken Webb and Ivar Rønnestad

9.1 Introduction

Digestion is fundamental to all animals, and it is sometimes considered a straightforward process involving complex food being reduced to simpler forms that can be absorbed into the body. While this is true in the very broadest of senses, it fails to consider the myriad of regulatory processes by which animals extract the maximum value from each unit of food. Most of what we know about the regulation of digestive processes comes from work with terrestrial food animals or from the study of disease states in humans. These studies have allowed for rapid advancement in our understanding of digestive processes in fish, but it is important that we remember that fish have had hundreds of millions of years to adapt to their environment since the amphibians first moved onto land. Fish are known to share many of the same digestive controls as other monogastric vertebrates, and yet we must be careful in assuming that systems that are well defined in mammals or avians function in the same way, or even exist at all, in fish. Fish are not just a more "primitive" vertebrate; they are also extremely successful animals that must regularly deal with extremes in pressure, salinity, temperature, and oxygen content among others that terrestrial vertebrates do not have to face. These differences often set fish apart due to the relatively large phylogenetic distance between fish and mammals and the physiological adaptations necessary to deal with a marine existence. Nowhere is this more clearly demonstrated than in the altricial larvae of marine fish.

In many ways, altricial larvae can be considered as free-swimming embryos that must undergo dramatic transformations before metamorphosing into the juvenile stage. The first part of this transformative process is

Larval Fish Nutrition, First Edition. Edited by G. Joan Holt.

fueled by endogenous nutrients provided by the mother, but the majority of the ontogenetic process requires that larvae capture and process exogenous nutrients even though the digestive system may not be fully formed. We have an adequate understanding of how the tissues and organs of the digestive system develop in the larvae of most commercially important fish, but we have only a cursory understanding of the systems that regulate digestive function in adult fish, much less their larvae. Like most other fundamental processes in animals, there are a host of interwoven systems of signals, responses, and modulators that are all active in ensuring the proper digestion and absorption of food in fish. In order to understand how these processes work in larvae, we must first verify that they exist even in the adult. Once this is established, it then becomes possible to establish if the system works similarly to the analogous system in terrestrial vertebrates and when it becomes functional in developing larvae. Currently, much of our understanding is still based on work done in mammals and, in some cases, on work done in juvenile and adult fish. In this chapter, we will discuss some of the major regulatory systems and signals in the digestive process of vertebrates and highlight what is currently known in larval fish.

9.2 Digestion in fish

Digestion in fish is much like that in other vertebrates. Food enters the digestive tract, where it is processed both mechanically by body structures and chemically by a series of acidic and/or alkaline enzymes. After being reduced by these processes, it is then absorbed into the body, and waste is evacuated. This series of processes is tightly regulated and can be grouped into three phases: the cephalic phase, the gastric phase, and the intestinal phase. Under normal conditions, the motility of the gut is regulated through feedback loops to allow sufficient digestion and absorption to occur in each of these phases before the food bolus is passed to the next phase. Under certain instances, however, the regulatory systems fail and food is passed either too slowly (hypomotility of the gut) or too quickly (hypermotility) to allow for proper digestion. One of the more notable cases in fish larvae is when the larvae of some marine fish species such as red drum (*Sciaenops ocellatus*) exhibit insufficient digestion of live prey. Larvae have been observed to overconsume prey items such as *Artemia* nauplii to the point that as more and more nauplii are ingested, it forces previously ingested nauplii through the digestive tract so that live nauplii are seen to pass from the anus. This does not appear to be a disease state and it may occur because the feedback mechanisms by which gut fullness promotes satiety have not yet become active in the larvae.

9.2.1 Cephalic phase

The cephalic phase is the initial digestive regulatory phase that is initiated by visual, olfactory, or gustatory sensations, and it serves as a forewarning to the digestive tract (also known as a feed-forward system) to prepare for the imminent arrival of a food bolus. In terrestrial animals, this is commonly seen in animals such as dogs, which often exhibit a salivary response to the sight or smell of food. This process is stimulated by the release of neural transmitters such as acetylcholine and norepinephrine (Hjelmeland et al. 1988) and leads to the activation of reflexes mediated by the sympathetic and parasympathetic nervous systems. These reflexes have several functions; among them is an increase in esophageal mucus secretion. This serves to ease the passage of food and acts as a protective layer against damage or infection (Reifel and Travill 1977; Scocco et al. 1998). This lubricant role is especially important in fish since they do not possess salivary glands (Scocco et al. 1998). Also, during the cephalic phase, there is an increased secretion of insulin, glucagon, and catecholamines from the pancreas (LeBlanc 2000), although the reaction in fish may be somewhat different from that in other vertebrates due to differences in hormones such as glucagon.

The exact initiator of the cephalic phase is not well understood, but the optic and olfactory centers are major components. In humans and dogs, when food is tasted and/or swallowed but not allowed to reach the stomach, the cephalic responses are even greater than if the food item is ingested normally (LeBlanc 2000). In Atlantic herring larvae (*Clupea harengus*) that have been fed polystyrene spheres, which have no gustatory or digestible substrates, there is a small but detectable increase in intestinal trypsin response, suggesting that stomach distension alone is capable of initiating secretion (Hjelmeland et al. 1988). On the other hand, Hjelmeland et al. (1988) found that secretion was much greater when the larvae were fed live prey, suggesting that some other mechanism is active as well. Kolkovski et al. (1997a, 1997b) found that in larval gilthead sea bream (*Sparus aurata*), visual and chemical cues from *Artemia* nauplii were sufficient to increase ingestion rates of microparticulate diets and suggested that the observed increases in the hormone bombesin (BBS) were responsible. Certainly, the use of feed attractants such as protein hydrolysates, phospholipids, polyamines, and live algae have been shown to be beneficial to the use of microparticulate diets as larval feeds (Browman 2005; Kvåle et al. 2006; Rocha et al. 2008) and to stimulate intestinal maturation (Péres et al. 1997); but there is little information on the exact nature of the neural and biochemical responses of fish during the cephalic phase.

9.2.2 Gastric phase

The gastric phase is the second regulatory phase and is initiated by the distension and acidification caused by the arrival of food in the stomach. In larval marine fish, the gastric phase is typically somewhat reduced in relevance due to the absence of a functional stomach for part or all of the larval period. In juvenile and adult fish, the gastric phase is primarily focused on the digestion of the food bolus, although there are a number of reactions that also serve to prime the intestine for the arrival of the food bolus as it passes out of the stomach. Larval fish may only possess a stomach anlage, but this second role of the gastric phase may already be functional as early as first feeding. Certainly, some of the digestive hormones involved in the gastric phase have been shown to be present in larval fish, but their activity and function remains to be demonstrated at these early ages (Figure 9.1).

The initiation of the gastric phase occurs when distension of the stomach stimulates the myenteric reflexes to release acetylcholine, producing a cascade of responses within the stomach itself. The most obvious response during the gastric phase is the production of gastric acid (hydrochloric acid—HCl) from

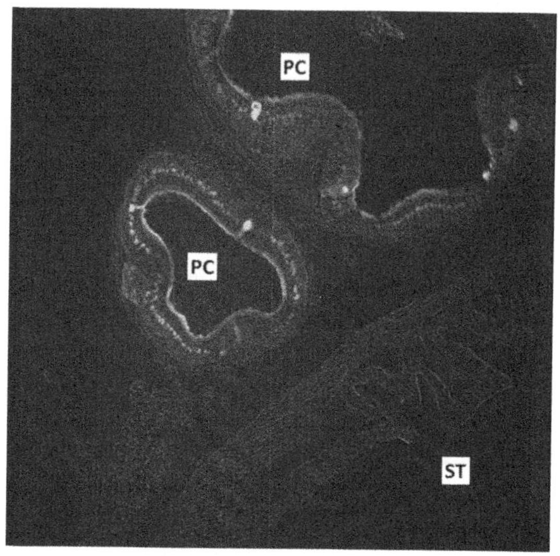

Figure 9.1 Cholecystokinin (CCK) immunoreactive cells in the gut of larval red drum (*Sciaenops ocellatus*); 100× image of CCK immunoreactivity in the pyloric ceca of a 26-dph red drum larva. PC = pyloric ceca, ST = stomach (from Webb et al. 2010).

the parietal cells. This is stimulated by the action of the hormone gastrin and serves both to activate the protease pepsin and to directly hydrolyze the food bolus. Other notable substances produced during the gastric phase are histamine and somatostatin.

In the larvae of a typical freshwater fish with a short larval period such as the African catfish (*Clarias gariepinus*), gastrin-producing cells are seen in the anterior intestine at 4 days posthatch (dph) as are cells that are immunoreactive to gastrin antibodies (Verreth et al. 1992), suggesting that the gastric phase is fully functional at this point. In a typical marine larva such as that of Japanese flounder (*Paralichthys olivaceus*), however, mRNA for the gastrin preprohormone is not detectable until the larva enters metamorphosis at 35 dph (Kurokawa et al. 2003), coinciding with the development of functional gastric glands and an acidic pH in the stomach (Rønnestad et al. 2000a). As discussed later, gastrin plays an integral role in the gastric phase, and the absence of gastrin mRNA until approximately the time of the initiation of gastric activity is suggestive that there is no substantial gastric phase in these larvae.

9.2.3 Intestinal phase

The intestinal phase is the final regulatory phase that begins with the passage of chyme from the stomach into the intestine. This is perhaps the most important phase in larval marine fish as all digestion and absorption must occur during this phase until the stomach becomes functional. Even in juvenile and adult fish, the intestinal phase is of key importance since a substantial portion of digestion as well as all absorption occurs in the intestine (Clements and Raubenheimer 2005). While the intestine has been primed by signals received during the cephalic and gastric phases, the initiator of the intestinal phase is the passage of acidic chyme through the pyloric valve from the stomach. This stimulates the release of bicarbonate rich fluids from the pancreatic ducts and the release of bile through the bile ducts. Bile is a greenish yellow fluid secreted by the liver and stored in the gallbladder. It contains a number of key elements, including bicarbonate and bile acids (normally as salts) that neutralize the chyme and emulsify lipids through the formation of micelles. This process also delivers pancreatic digestive enzymes such as trypsin and bile salt–dependent lipase into the intestinal lumen.

Due to the importance of the intestinal phase in larval fish, the structures and peptides necessary for this phase become active either slightly before or just at the onset of exogenous feeding. This can be seen in larvae of red drum (*Sciaenops ocellatus*), where there is a dramatic increase in trypsin and chymotrypsin activity at 3 dph before the larvae have been fed (Lazo 1999; Lazo et al. 2000b; Applebaum et al. 2001). At the same time, the first cells containing the hormone cholecystokinin (CCK) can be detected using immunohistochemistry (IHC), though they are not seen in all fish until a few days later (Webb et al. 2010).

9.3 Regulatory systems

Digestion in vertebrates is controlled through a complex series of intrinsic and extrinsic regulatory systems. These systems function to synchronize the flow of food through the digestive system, the release of digestive juices and enzymes, the absorption of nutrients, and the evacuation of the gut. Intrinsic digestive regulatory systems are principally paracrine hormonal systems such as somatostatin, while extrinsic systems include factors such as the enteric nervous system (ENS) and exocrine hormone systems.

9.3.1 ENS

The ENS is a collection of neurons in the gastrointestinal (GI) tract that represents one of the three major parts of the autonomic nervous system defined by Langley (1903). The ENS can be thought of as a displaced portion of the central nervous system (Goyal and Hirano 1996), and it provides neural control of the digestive process through interaction of sympathetic and parasympathetic neurons arranged into two plexuses or sets of ganglia known as the myenteric (Auerbach's) plexus and the submucosal (Meissner's) plexus (Goyal and Hirano 1996; Burns and Thapar 2006; Furness 2006) as well as through the action of a wide variety of neuropeptides (Bjenning and Holmgren 1988). The ENS is considered the "brain of the gut" and is well studied in many vertebrate species (see recent reviews by Newgreen and Young 2002; Burns and Thapar 2006; Furness 2006), including the zebrafish (*Danio rerio*) (Holmberg et al. 2008; Olsson et al. 2008).

The ENS regulates gut motility, controls blood flow, coordinates secretion and absorption, and modulates immune and endocrine functions reflexively (Burns and Thapar 2006). It is composed of approximately 10^8 neurons in humans, nearly as many as the spinal cord (Lundgren et al. 1989; Goyal and Hirano 1996), and is capable of functioning by reflex without relying on direct commands from the brain or spinal cord (Kunze and Furness 1999; Furness 2006). This independence of function allows for *in vitro* studies on GI function as demonstrated by Rønnestad et al. (2000b) in juvenile Atlantic halibut (*Hippoglossus hippoglossus*). Both the myenteric and submucosal plexuses in fish have similar structures as that in mammals, with only minor variations such as a lower number of ganglia in the two plexuses (Bjenning and Holmgren 1988) and an irregular representation of neurotransmitters (Li and Furness 1993).

The role of the ENS in gut motility is fairly well understood. Electrical slow waves spread from the interstitial cells of Cajal and stimulate the contractile activity of the gut (Holmberg et al. 2008). The slow propagation of these waves and the resulting contractions form peristaltic waves in an anterograde (oral to anal) direction. These peristaltic waves are generated spontaneously (i.e., not in response to drugs, food, or mechanical expansion) by the action of receptors with tyrosine kinase activity (KIT receptors) (Goyal and Hirano 1996; Holmberg et al. 2004) and are generally accepted to have a housekeeping function to maintain gut health. In addition to the spontaneous waves, the ENS also regulates feeding-induced peristalsis and thereby determines the rate at which food passes.

In fish larvae, rhythmic contractile waves can be seen even before the onset of exogenous feeding in the zebrafish (Holmberg et al. 2004, 2008). Through these waves, the food bolus can be moved along the digestive tract so that digestion, absorption, and evacuation can take place in a controlled fashion. These waves can be stimulated with acetylcholine and inhibited with atropine, suggesting cholinergic gut innervations (Holmberg et al. 2004, 2008). In addition to the anterograde peristaltic waves, retrograde (anal to oral) waves have been observed in fish such as Atlantic halibut (*Hippoglossus hippoglossus*) (Rønnestad et al. 2000b). It is believed that rather than moving food through the gut, this retrograde motion serves to mix the digestive secretions into the food bolus as well as to push food into the pyloric ceca (Rønnestad et al. 2000b).

9.3.2 GI regulatory molecules

GI regulatory molecules include paracrine and endocrine hormones, neuropeptides, and cytokines. These substances often have a specific site of action where they produce a directed response such as the stimulation of gastric acid secretion by gastrin as they simultaneously produce an indirect response such as the inhibition of feeding behavior in neonatal chicks (Furuse et al. 1999). These molecules function in a tightly interwoven web of feedback and feed-forward loops that are incompletely understood in well-studied terrestrial vertebrates such as humans. In larval fish, study of these molecules and their interrelationships is in its infancy, and we are often forced to make assumptions based on what is known in other animals.

9.3.2.1 BBS

BBS is one of a number of anorexigenic hormones, which has been shown to reduce feed intake (Gibbs et al. 1979; Bjenning and Holmgren 1988; Lee et al. 1994; Furuse et al. 2007). Primarily found in the digestive tract, BBS is also found at high concentrations in localized areas of the brain (Lee et al. 1994). Named for the frog species from the skin of which it was originally extracted, BBS is a tetradecapeptide that is characterized by a Gly-His-Leu-Met-NH_2 at the C-terminus (Gonzalez et al. 2008). It is closely related to two other mammalian peptides, the gastrin-releasing peptide (GRP) and neuromedian B (NMB). These three peptides together represent the three subfamilies that make up the group commonly referred to as the BBS/GRP-like family (Xu and Volkoff 2009). While there are structural differences in BBS and GRP, they share many of the same functions and their roles are often examined together.

Early work in rats by Gibbs et al. (1979) showed that intraperitoneal injections of BBS produced selective satiety where food consumption was reduced without affecting the consumption of water. Perhaps most interestingly, the authors reported that rats began feeding as eagerly as nontreated animals, they just became satiated sooner, suggesting that BBS does not reduce the desire for food, only affecting the amount necessary to reach satiety. In work with goldfish (*Carassius auratus*), Himick and Peter (1994) found that BBS reduced the appetite by stimulating the secretion of the hormone ghrelin, suggesting that this may be the method by which BBS acts on satiety. The authors also found that BBS may increase growth hormone, suggesting a link between BBS and growth. In addition to effects on satiety, BBS is known to have other functions including cardiac/vascular effects, the stimulation of other digestive hormones such as CCK, gastrin, and pancreatic polypeptide (PP), and stimulating the release of pancreatic enzymes (Bjenning and Holmgren 1988; Liehr et al. 1993). BBS has also been shown to stimulate the peristalsis of the gut by directly exciting cholinergic and noncholinergic excitatory neurons in the intestinal wall, although this effect was much weaker than that produced by CCK (Barthó et al. 1982).

BBS/GRP has been found in a wide variety of fish ranging from elasmobranchs to teleosts (Bjenning and Holmgren 1988; Volkoff et al. 2005). Distribution of BBS immunoreactive cells in the gut tissues and nerves of fish has been shown to be species specific and not sensitive to short-term starvation (Burkhardt-Holm and Holmgren 1989). Work with GRP in embryonic and larval Atlantic cod (*Gadus morhua*) found that GRP mRNA is weakly present as early as the blastula stage, but becomes elevated around the time the eggs hatch, indicating that *de novo* production begins at this point (Xu and Volkoff 2009). In larvae, BBS was found to increase three times more in gilthead sea bream (*Sparus auratus*) that were fed *Artemia* nauplii compared with those fed formulated dry diets (Kolkovski et al. 1997b), providing some insight into how live prey stimulates the digestive system in a way that inert particles do not (Koven et al. 2001).

9.3.2.2 Gastrin

Gastrin was first recognized in 1905 as a hormone for its ability to induce gastric acid secretion, although in the following years, there was much debate about its actual structural identity (Dockray et al. 2001). In 1964, the true identity of the hormone now known as gastrin was identified from the antral mucosa of hogs (Gregory and Tracy 1964). Since that time, gastrin has been extensively studied and we now know it has numerous functions in addition to this role. The precursor to gastrin (preprogastrin) is a peptide of 101 (human), 104 (rat), or 114 (Japanese flounder, *Paralichthys olivaceus*) amino acid residues long (Kurokawa et al. 2003; Dockray et al. 2005), which is cleaved after translation to yield progastrin. Progastrin and the various COOH-terminal Gly-gastrins are referred to as the "nonclassical gastrins" since we now recognize that these various cleavage products have additional activities besides the traditional role of stimulating gastric acid secretion assigned to the "classical gastrins" (G34-Gly and G17-Gly) (Dockray et al. 2005). Some of these roles include the proliferation and differentiation of intestinal cells (Koh et al. 1999), differentiation and maturation of parietal cells (Dockray et al. 2001; Paterson et al. 2004), and control of histamine production in the enterochromaffin-like cells (Dockray et al. 2005).

Gastrin is well studied in fish, though most of this work has focused on its role in gastric acid secretion and tissue differentiation. In larval Japanese flounder, it was found that gastrin began to be expressed in the intestine synchronously with the differentiation of the stomach, supporting the concept of gastrin as a regulator of gastric acid secretion (Kurokawa et al. 2003). In turbot (*Scophthalmus maximus*) larvae, gastrin immunoreactive cells were also first seen coinciding with the development of the stomach (Reinecke et al. 1996), while in European sea bass (*Dicentrarchus labrax*), gastrin immunoreactive cells were seen in the intestine prior to the development of the stomach and in the stomach after it had differentiated (García Hernández et al. 1994). In the spotted river puffer (*Tetraodon nigroviridis*), gastrin was found in the intestine despite the fact that spotted river puffers do not possess a stomach (Kurokawa et al. 2003). This illustrates the multiple roles gastrin plays in the ontogeny of digestion in larval fish.

9.3.2.3 Somatostatin

Somatostatins are a family of cyclopeptides that are also known as somatotropin release-inhibiting factors (SIRFs) and are produced by a variety of cell types, including both endocrine and GI cells (Weckbecker et al. 2003). SIRFs have a large number of inhibitory effects, which include the ability to inhibit the secretion of digestive hormones such as CCK or gastrin as well as the ability to reduce the secretion of gastric acid and pancreatic enzymes (Cahu and Infante 1995a; Weckbecker et al. 2003).

SIRFs are widely studied in fish and have been shown to produce a wide variety of effects. They strongly inhibit the secretion of growth hormone in a number of species in response to stress factors (Boeuf and Le Bail 1999). In juvenile red drum (*Sciaenops ocellatus*), stress induced by background colors and altered illumination produced increased somatolactin levels, which may explain reductions in growth following stressful periods during culture of this species (Zhu and Thomas 1995, Boeuf and Le Bail 1999).

9.3.2.4 CCK

CCK is a peptide that has both a primary gut form and a carboxy-terminal octapeptide form that is found both in the central nervous system (CNS) and the gut (de Pedro and Björnsson 2001; Chandra and Liddle 2007). CCK shares an amidated C-terminal pentapeptide sequence (Gly-Trp-Met-Asp-Phe-NH_2) with gastrin, the amphibian hormone caerulin, and the ancestral hormone cionin (Bjenning and Holmgren 1988). In addition to sharing similar structures, CCK and gastrin are bound by the same CCK/gastrin receptor and share many of the same functions. Like gastrin, CCK has received a great deal of attention due to its role as a regulator of pancreatic enzyme secretion, gut motility, and gallbladder contractions (Grider 1994; Fink et al. 1998). CCK is also an anorexigenic hormone, which functions to reduce food intake (Konturek et al. 2004; Little et al. 2005), and has been widely investigated in mammalian systems for this reason. CCK in fish is known to have a strong role in the regulation of GI function and provides evidence of a regulatory loop for pancreatic enzyme secretion (Einarsson et al. 1997; De Pedro and Björnsson 2001; Koven et al. 2002; Rønnestad 2002; Rønnestad et al. 2003).

Regulation of pancreatic secretion in fish is poorly understood and is generally based on what is known in mammalian systems, primarily those of rats and humans. Making it somewhat more complex, research has shown that control systems may even be different between rats and humans. In the rat, *in vitro* work with pancreatic acini has shown that CCK works through CCK-A receptors directly on pancreatic acini to stimulate enzymatic secretion rather than through a neural, cholinergic-dependent pathway (Owyang and Logsdon 2004). This is different from that in human cells where research has shown that human pancreatic acini do not express CCK-A receptors, only CCK-B receptors (Nishimori et al. 1999; Morisset et al. 2003; Owyang and Logsdon 2004), and do not respond directly to CCK (Ji et al. 2001). The neural component of CCK-induced pancreatic stimulation appears universal, however, as *in vivo* work has shown that physiological doses of CCK are capable of producing a cholinergic neural-mediated pancreatic response in dogs (Konturek et al. 1972), humans (Valenzuela et al. 1983), and rats (Soudah et al. 1992). Regardless of the method of action, CCK is considered one of the most important, if not the most important, regulator of pancreatic enzyme secretion (Olsson and Holmgren 2001).

In fish, CCK-induced pancreatic enzyme secretion has been studied in a number of species. In red drum, CCK-immunoreactive (CCK-IR) cells are seen as early as first feeding and become increasingly common as the fish age (Webb et al. 2010; Table 9.1). These CCK-IR cells are concentrated in the pyloric ceca and in the adjacent anterior intestine where secretions of pancreatic enzymes occur, although in the larvae of some species this concentration does not occur (Rønnestad et al. 2007). In Atlantic herring, cells containing CCK mRNA are seen as early as hatching and their numbers increase dramatically by 8 dph (Kamisaka et al. 2005). Tube feeding experiments with herring have shown that CCK and trypsin are both elevated in larvae after tube feeding with liposomes containing

Table 9.1 Distribution of CCK-IR cells in red drum larvae by gut section and age.

	Age (days posthatch)							
	3	4	5	6	9	12	18	26
Number of larvae with CCK-IR cells	1/10	2/5	4/5	5/5	5/5	5/5	5/5	5/5
Foregut/stomach	–	–	–	–	–	–	–	–
Anterior midgut/intestine	+	+	+	+	+	++	++	++
Pyloric ceca	NP	NP	NP	NP	NP	NP	++	+++
Posterior midgut/intestine	–	–	+	+	+	+	+	+
Hindgut/rectum	–	–	–	–	–	–	–	–

Presence of five or less cells/section is indicated by (+), between 5 and 10 cells by (++), and more than 10 cells/section by (+++). Absence of CCK-IR cells is indicated by (–), and (NP) indicates that the gut section is not present at that age.
From Webb et al. (2010).

free amino acids, bovine serum albumin, or a 1:1 mixture of the two but not when fed liposomes containing physiological saline (Koven et al. 2002). The work by Koven et al. (2002) demonstrates that CCK likely functions to stimulate pancreatic secretion in fish the same as it does in mammals. Similar responses have also been seen in larvae of Atlantic halibut (Rojas-Garcia and Rønnestad 2002), European sea bass (Cahu et al. 2004), and red drum (Webb 2008) among others.

Certain foods have been shown to have a stronger stimulatory effect on CCK secretion than others (Liddle 1995), and this may play a significant role in the inability of marine fish larvae to utilize inert diets at first feeding. Products of proteolysis and lipolysis stimulate the CCK (endocrine I) cell to release CCK, which then acts humorally on pancreatic acinar cells to release pancreatic enzymes (Klein et al. 1999; Liddle 2000). In larval fish, which do not yet possess a functional stomach, these products may not be produced in sufficient amounts to properly stimulate pancreatic enzyme secretion. Work with Atlantic salmon (*Salmo salar*) has shown that exogenous CCK produces a dose-dependent release of pancreatic proteases into the intestine (Einarsson et al. 1997), supporting this hypothesis. When larval fish such as red drum (*Sciaenops ocellatus*) are cofed live prey or algae along with inert diets, growth and survival are enhanced along with pancreatic enzyme activities (Lazo 1999; Lazo et al. 2000a). Within 30 minutes after a meal of live prey, larval red drum have significantly increased CCK and trypsin levels compared with their unfed counterparts (Webb 2008). While it is possible that simple gut distension is responsible for these increases, work with larvae of Atlantic herring (*Clupea harengus*) has shown that distension of the gut with an empty carrier solution did not produce this response, while introduction of proteins and amino acids showed results similar to that seen in red drum (Koven et al. 2002). Endogenous CCK secretion has been stimulated in fish by bovine serum albumin (Koven et al. 2002), potato starch (Cahu et al. 2004), or free methionine (Koven et al. 2001). CCK therefore represents a likely candidate for investigation into the mechanism by which digestive function may be controlled in larval fishes.

9.3.2.5 Secretin/vasoactive intestinal polypeptide (VIP)

Secretin is a 27-amino-acid peptide synthesized as a prohormone in the S cells found in the intestinal membrane. Structurally, the mature peptide is similar to glucagon-like peptide (GLP) and VIP. The principal function of secretin is to stimulate the pancreas to produce bicarbonate-rich secretions in response to acidification of the intestine by chyme passing from the stomach. Secretin is released in response to acidic chyme arriving in the anterior intestine and acts on the duct cells to release bicarbonate from the pancreas in addition to bile from the liver and is modulated by a secretin-releasing peptide. Secretin and the secretin-releasing peptide operate through a feedback loop to regulate the release of pancreatic bicarbonate into the anterior intestine (Song et al. 1999). This feedback system can be either positive or negative and is attributable to species differences (Song et al. 1999).

VIP is a 28-amino-acid peptide first identified by its activity as a vasodilatory hormone (Bjenning and Holmgren 1988; Olsson and Holmgren 2001), hence the name. VIP is a member of the VIP family along with secretin and GLP and is highly conserved among mammals and teleosts (Olsson and Holmgren 2001). VIP mediates some of the secretory effect of serotonin (Wapnir and Teichberg 2002), thereby inhibiting gallbladder contraction (Aldman and Holmgren 1992). VIP in rainbow trout (*Oncorhynchus mykiss*) is particularly found in the circular layer of the intestine associated with nerve fibers of the myenteric and submucosal plexuses

(Holmgren et al. 1982). VIP has also been shown to be important in the osmoregulation of postembryonic fish (Varsamos et al. 2005).

9.3.2.6 Pancreatic polypeptides (PP, PYY)

First discovered as a by-product of insulin purification from chicken pancreas (Kimmel et al. 1975), PP functions as a regulator of pancreatic and GI function. Pancreatic peptide YY (PYY, sometimes referred to as intestinal peptide YY) was first isolated from porcine pancreas and has functions similar to PP, in addition to regulating several other digestive processes (Larhammar 1996). PP and PYY are members of the neuropeptide Y (NPY) family of homologous peptides and are thought to have evolved from a common ancestor through a series of gene duplication events (Larhammar 1996; Conlon 2002). Though both PP and PYY are short, only 36 amino acid residues long, and share a common ancestry and roles in restricting the secretion of pancreatic enzymes (Guan et al. 1991; Kurokawa et al. 2004), the amino acid sequences of the two can vary greatly among species (Conlon 2002). Despite the similarity in name, PYY is structurally more similar to NPY than to PP (Conlon 2002).

PP has only been identified in tetrapods (Larhammar 1996; Conlon 2002) and has an amino acid sequence that is much more variable than that of NPY or PYY (Larhammar 1996). It is secreted in the pancreas by the PP (or F) cells (Slack 1995), which are almost totally confined to exocrine pancreatic tissues (Kurokawa et al. 2000; Hennig et al. 2002), as well as in the distal GI tract (Mendieta-Zerón et al. 2008). Postprandial increases in PP concentration are rapid, occurring within 10 minutes after ingestion of a meal (Hornnes et al. 1980). While there is evidence that PP has direct inhibitory effects on the secretion of pancreatic digestive enzymes and gallbladder contraction, this has not been demonstrated (Hennig et al. 2002). On the other hand, Okumura et al. (1995) found that injection of doses of PP into the dorsal motor nucleus of rats inhibited pancreatic flow and protein output, suggesting that PP works by modulating the pancreatic tone through the ENS rather than through direct action. These injections were sufficient to inhibit pancreatic secretions in response to both a normal meal and exogenous CCK-8 in a specific and dose-dependent manner (Okumura et al. 1995).

PP has been reported in numerous fish species including Korean aucha perch (*Coreoperca herzi*) (Lee et al. 2004), Japanese flounder (*Paralichthys olivaceus*) (Kurokawa et al. 2000), and European sea bass (*Dicentrarchus labrax*) (García Hernández et al. 1994). Distribution of PP cells in larvae is found in the exocrine pancreas as expected but PP-immunoreactive (PP-IR) cells are typically not seen until late in the larval stage (García Hernández et al. 1994; Kurokawa et al. 2000). PP-IR cells are also found in the stomach and intestine of European sea bass larvae (García Hernández et al. 1994), though not in Japanese flounder (Kurokawa et al. 2000). Kurokawa et al. (2000) suggest that the presence of PP-IR cells seen in European sea bass larvae may be due to the use of antibodies raised against PP in nontarget species and the structural similarity between PP and PYY causing false positives.

Like PP, PYY regulates GI function in response to a meal. Unlike PP, there are two primary forms of PYY that have been shown to be active in mammals. The two forms ($PYY_{1\text{-}36}$ and $PYY_{3\text{-}36}$) differ only in the presence or absence of the first two amino acid residues, but $PYY_{1\text{-}36}$ has been shown to bind four different PYY receptors, while $PYY_{3\text{-}36}$ binds more specifically to one type of PYY receptor (Mendieta-Zerón et al. 2008). In addition to these two forms, fish of the superorder Acanthomorpha have another form known as peptide Y (PY) whose ortholog is referred to as PYYb in zebrafish (Fredriksson et al. 2006) and sometimes in other fish as well (Murashita et al. 2009).

PYY is secreted by endocrine L cells in the GI tract in response to intraluminal nutrients and digestive products (Grudell and Camilleri 2007). Interestingly, the secretion of PYY has also been shown to be positively correlated with caloric content of the food bolus (Mendieta-Zerón et al. 2008), suggesting its role in sensing caloric "fullness" as opposed to volumetric fullness. In rats, release of PYY has been shown to be mediated by neural reflex even before the arrival of a food bolus in the intestine (Grudell and Camilleri 2007). Secreted mostly in the distal intestine, PYY works to delay gastric emptying and modulates intestinal motility (Tatemoto 1982). PYY is also considered to be the major hormone regulating absorption in the small intestine of mammals through the action of a specific type of protein kinase C (PKC_{ε}) (Wapnir and Teichberg 2002), though this role has not been demonstrated in fish.

PYY/PY have been identified in numerous fish including Atlantic salmon (*Salmo salar*) (Murashita et al. 2009), Atlantic cod (*Gadus morhua*) (Jensen and Conlon 1992), yellowtail (*Seriola quinqueradiata*) (Murashita et al. 2006), and Japanese eel (*Anguilla japonica*) (Kurokawa et al. 2004). In European sea bass, cells reacting to the PYY antisera (PYY-IR) were first seen in the intestine of larvae during phase II (9–15 dph), and by phase IV (55–60 dph) were found in the stomach as well (García Hernández et al. 1994). In larvae of Japanese flounder, PY mRNA was found in the intestinal epithelia as early as 3 dph and in the endocrine pancreas at 30 dph but PYY was not found at any age (Kurokawa and Suzuki 2002). In Japanese eel, PYY mRNA was found in the intestinal epithelia when the fish began feeding at 8 dph (Kurokawa et al. 2004). Since PYY/PY is found around the time exogenous feeding begins in the larvae of the Japanese eel and Japanese flounder, it suggests synchronicity with the development of the exocrine pancreas (Kurokawa and Suzuki 2002). Why European sea bass does not also exhibit this same type of synchronicity is unknown, but the difference may be related to the use of antibodies for detection in sea bass, which may not have shown detectable signals as early as for the other two fish.

9.3.2.7 Motilin/ghrelin

Motilin is a 22-amino-acid peptide first identified for its ability to stimulate gastric activity and produce strong, recurring contractions (Itoh et al. 1978). Motilin is released during periods of fasting and is believed to be involved in regulating interdigestive pancreatic secretions through its role in regulating the migrating myoelectric complex (Konturek et al. 2003; Sanger 2008). Ghrelin is similar in structure to motilin, and the receptors for the two peptides share 52% homology, leading to the original name for ghrelin as the motilin-related peptide (Sanger 2008). Recently, however, major differences between the two have been established, most notably the stimulation of appetite by ghrelin (Sanger 2008) and the ability of ghrelin to serve as a secretagogue for growth hormone (Xu and Volkoff 2009). It is the stimulatory nature of ghrelin on appetite that has made it of great interest recently since it is the only known orexigenic GI peptide to date (Murashita et al. 2009).

Since ghrelin is produced in adult fish primarily in the stomach, it is not clear what role it may have in the development of larval fish before the development of a functional stomach. In larvae of Atlantic halibut, ghrelin mRNA expression was detected at the time of hatching (Manning et al. 2008) even though the stomach in this species does not become functional until approximately 73 dph (Luizi et al. 1999; Sæle et al. 2004). In Atlantic cod, this was even more dramatic as ghrelin mRNA was first detected at the cleavage stage of cod embryos (Xu and Volkoff 2009). This suggests that ghrelin may be active in a developmental role even before the onset of exogenous feeding, but the digestive role of ghrelin in stomachless fish larvae is not known.

9.3.2.8 GLP

GLP is produced by endocrine L cells in response to a meal and inhibits short-term feed intake, gastric emptying, and intestinal motility in mammals (Mojsov 2000). This has been established in catfish (*Ictalurus punctatus*) as well (Silverstein et al. 2001), but there are substantial differences between the role of GLP in mammals and in teleosts. In mammals, GLP modulates the secretion of insulin by the pancreas in response to plasma glucose levels (Claus et al. 2007). As blood glucose increases, GLP is secreted from the liver and increases the release of insulin, which causes body tissues to take up glucose from the bloodstream and store it as glycogen. In fish, however, GLP is more similar to glucagon in that it has the opposite effect and appears to function more like glucagon than like insulin (Mojsov 2000; Volkoff et al. 2005). GLP in fish leads to increased gluconeogenesis and glycogenolysis and is considered one of the most potent direct metabolic hormones in fish (Mojsov 2000).

Studies of GLP in larval fish is relatively limited, but studies with larvae of European sea bass have shown that immunoreactivity to bovine/porcine glucagon antisera could first be detected between 25 and 46 dph. This is in contrast to work with larval lamprey (*Geotria australis* and *Mordacia mordax*) in which GLP/glucagon immunoreactivity was not seen until the fish had completed metamorphosis to the juvenile stage (Youson and Potter 1993). The significance of this difference in GLP immunoreactivity is unclear but may be related to phylogenetic differences between the Cephalaspidomorphi and the Osteichthyes (Youson and Al-Mahrouki 1999).

9.4 Future research

The study of the regulation of digestive processes in larval fish is a wide-open field with significant opportunities for future researchers. Our understanding of these processes in terrestrial vertebrates continues to grow and this helps us understand better how these processes might work in fish. Furthermore, the use of zebrafish as a model species has opened a whole new range of tools for understanding the ontogeny of control systems in fish.

Literature cited

Aldman, G., and Holmgren, S. 1992. Vip inhibits cck-induced gallbladder contraction involving a beta-adrenoceptor mediated pathway in the rainbow-trout, *Oncorhynchus mykiss, in vivo. General and Comparative Endocrinology* 88:287–291.

Applebaum, S.L., Perez, R., Lazo, J.P., et al. 2001. Characterization of chymotrypsin activity during early ontogeny of larval red drum (*Sciaenops ocellatus*). *Fish Physiology and Biochemistry* 25:291–300.

Barthó, L., Holzer, P., Donnerer, J., et al. 1982. Effects of substance P, cholecystokinin octapeptide, bombesin, and neurotensin on the peristaltic reflex of the guinea-pig ileum in the absence and in the presence of atropine. *Naunyn-Schmiedeberg's Archives of Pharmacology* 321:321–328.

Bjenning, C., and Holmgren, S. 1988. Neuropeptides in the fish gut. *Histochemistry and Cell Biology* 88:155–163.

Boeuf, G., and Le Bail, P.-Y. 1999. Does light have an influence on fish growth? *Aquaculture* 177:129–152.

Browman, H. 2005. Applications of sensory biology in marine ecology and aquaculture. *Marine Ecology Progress Series* 287:266–269.

Burkhardt-Holm, P., and Holmgren, S. 1989. A comparative study of neuropeptides in the intestine of two stomachless teleosts (*Poecilia reticulata, Leuciscus idus melanotus*) under conditions of feeding and starvation. *Cell and Tissue Research* 255:245–254.

Burns, A.J., and Thapar, N. 2006. Advances in ontogeny of the enteric nervous system. *Neurogastroenterology and Motility* 18:876–887.

Cahu, C., Rønnestad, I., Grangier, V., et al. 2004. Expression and activities of pancreatic enzymes in developing sea bass larvae (*Dicentrarchus labrax*) in relation to intact and hydrolyzed dietary protein; involvement of cholecystokinin. *Aquaculture* 238:295–308.

Cahu, C.L., and Infante, J.L.Z. 1995a. Effect of the molecular form of dietary nitrogen supply in sea bass larvae: response of pancreatic enzymes and intestinal peptidases. *Fish Physiology and Biochemistry* 14:209–214.

Cahu, C.L., and Infante, J.L.Z. 1995b. Maturation of the pancreatic and intestinal digestive functions in sea bass (*Dicentrarchus labrax*): effect of weaning with different protein sources. *Fish Physiology and Biochemistry* 14:431–437.

Calzada, A., Medina, A., and Canales, M.L.G. 1998. Fine structure of the intestine development in cultured sea bream larvae. *Journal of Fish Biology* 53:340–365.

Chandra, R., and Liddle, R.A. 2007. Cholecystokinin. *Current Opinion in Endocrinology, Diabetes, and Obesity* 14:63–67.

Claus, T.H., Pan, C.Q., Buxton, J.M., et al. 2007. Dual-acting peptide with prolonged glucagon-like peptide-1 receptor agonist and glucagon receptor antagonist activity for the treatment of type 2 diabetes. *The Journal of Endocrinology* 192:371–380.

Clements, K.D., and Raubenheimer, D. 2005. Feeding and nutrition. In: Evans, D.H., and Claiborne, J.B. (eds.) *The Physiology of Fishes*, 3rd edition. CRC Press, New York, pp. 47–82.

Conlon, M.J. 2002. The origin and evolution of peptide YY (PYY) and pancreatic polypeptide (PP). *Peptides* 23:269–278.

Curnow, J., King, J., Bosmans, J., et al. 2006. The effect of reduced *Artemia* and rotifer use facilitated by a new microdiet in the rearing of barramundi *Lates calcarifer* (Bloch) larvae. *Aquaculture* 257:204–213.

Dockray, G., Dimaline, R., and Varro, A. 2005. Gastrin: old hormone, new functions. *Pflügers Archiv European Journal of Physiology* 449:344–355.

Dockray, G.J., Varro, A., Dimaline, R., et al. 2001. The gastrins: their production and biological activities. *Annual Review of Physiology* 63:119–139.

Einarsson, S., Davies, P.S., and Talbot, C. 1997. Effect of exogenous cholecystokinin on the discharge of the gallbladder and the secretion of trypsin and chymotrypsin from the pancreas of the Atlantic salmon, *Salmo salar* L. *Comparative Biochemistry and Physiology. Toxicology & Pharmacology: CBP* 117:63–67.

Faulk, C.K., Benninghoff, A.D., and Holt, G.J. 2007. Ontogeny of the gastrointestinal tract and selected digestive enzymes in cobia *Rachycentron canadum* (L.). *Journal of Fish Biology* 70:567–583.

Fink, H., Rex, A., Voits, M., et al. 1998. Major biological actions of CCK—a critical evaluation of research findings. *Experimental Brain Research* 123:77–83.

Fontagne, S., Robin, J., Corraze, G., et al. 2000. Growth and survival of European sea bass (*Dicentrarchus labrax*) larvae fed from first feeding on compound diets containing medium-chain triacylglycerols. *Aquaculture* 190:261–271.

Fredriksson, R., Sjödin, P., Larson, E.T., et al. 2006. Cloning and characterization of a zebrafish Y2 receptor. *Regulatory Peptides* 133:32–40.

Furness, J.B. 2006. *The Enteric Nervous System*. Blackwell Publishing, Malden, MA.

Furuse, M., Ao, R., Bungo, T., et al. 1999. Central gastrin inhibits feeding behavior and food passage in neonatal chicks. *Life Sciences* 65:305–311.

Furuse, M., Yamane, H., Tomonaga, S., et al. 2007. Neuropeptidergic regulation of food intake in the neonatal chick: a review. *The Journal of Poultry Science* 44:349–356.

García Hernández, M.P., Lozano, M.T., and Agulleiro, B. 1994. Ontogeny of some endocrine cells of the digestive tract in sea bass (*Dicentrarchus labrax*): an immunocytochemical study. *Cell and Tissue Research* 277:373–383.

Gibbs, J., Fauser, D.J., Rowe, E.A., et al. 1979. Bombesin suppresses feeding in rats. *Nature* 282:208–210.

Giffard-Mena, I., Charmantier, G., Grousset, E., et al. 2006. Digestive tract ontogeny of *Dicentrarchus labrax*: implication in osmoregulation. *Development, Growth and Differentiation* 48:139–151.

Gonzalez, N., Moody, T.W., Igarashi, H., et al. 2008. Bombesin-related peptides and their receptors: recent advances in their role in physiology and disease states. *Current Opinion in Endocrinology, Diabetes, and Obesity* 15:58–64. doi:10.1097/MED.0b013e3282f3709b.

Goyal, R.K., and Hirano, I. 1996. The enteric nervous system. *New England Journal of Medicine* 334:1106–1115.

Gregory, R.A., and Tracy, H.J. 1964. The constitution and properties of two gastrins extracted from hog antral mucosa. *Gut* 5:103–117.

Grider, J.R. 1994. Role of cholecystokinin in the regulation of gastrointestinal motility. *The Journal of Nutrition* 124:1334S–1339S.

Grudell, A.B.M., and Camilleri, M. 2007. The role of peptide YY in integrative gut physiology and potential role in obesity. *Current Opinion in Endocrinology, Diabetes, and Obesity* 14:52–57.

Guan, D., Maouyo, D., Taylor, I.L., et al. 1991. Peptide-YY, a new partner in the negative feedback control of pancreatic secretion. *Endocrinology* 128:911–916.

Hennig, R., Kekis, P.B., Friess, H., et al. 2002. Pancreatic polypeptide in pancreatitis. *Peptides* 23:331–338.

Himick, B.A., and Peter, R.E. 1994. Bombesin acts to suppress feeding behavior and alter serum growth hormone in goldfish. *Physiology & Behavior* 55:65–72.

Hjelmeland, K., Pedersen, B.H., and Nilssen, E.M. 1988. Trypsin content in intestines of herring larvae, *Clupea harengus*, ingesting inert polystyrene spheres or live crustacea prey. *Marine Biology* 98:331–335.

Holmberg, A., Schwerte, T., Pelster, B., et al. 2004. Ontogeny of the gut motility control system in zebrafish *Danio rerio* embryos and larvae. *The Journal of Experimental Biology* 207: 4085–4094.

Holmberg, A., Holmgren, S., and Olsson, C. 2008. Enteric control. In: Finn, R.N., and Kapoor, B.G. (eds.) *Fish Larval Physiology*. Science Publishers, Enfield, NH, pp. 553–572.

Holmgren, S., Vaillant, C., and Dimaline, R. 1982. VIP-, substance P-, gastrin/CCK-, bombesin-, somatostatin- and glucagon-like immunoreactivities in the gut of the rainbow trout, *Salmo gairdneri*. *Cell and Tissue Research* 223: 141–153.

Hornnes, P.J., Kuhl, C., Holst, J.J., et al. 1980. Simultaneous recording of the gastro-entero-pancreatic hormonal peptide response to food in man. *Metabolism* 29:777–779.

Itoh, Z., Takeuchi, S., Aizawa, I., et al. 1978. Changes in plasma motilin concentration and gastrointestinal contractile activity in conscious dogs. *Digestive Diseases and Sciences* 23:929–935.

Jensen, J., and Conlon, J.M. 1992. Characterization of peptides related to neuropeptide tyrosine and peptide tyrosine-tyrosine from the brain and gastrointestinal tract of teleost fish. *European Journal of Biochemistry* 210:405–410.

Ji, B., Yan, B.I., Simeone, D., et al. 2001. Human pancreatic acinar cells lack functional responses to cholecystokinin and gastrin. *Gastroenterology* 121:1380–1390.

Kamisaka, Y., Drivenes, O., Kurokawa, T., et al. 2005. Cholecystokinin mRNA in Atlantic herring, *Clupea harengus*—molecular cloning, characterization, and distribution in the digestive tract during the early life stages. *Peptides* 26:385–393.

Kimmel, J.R., Hayden, L.J., and Pollock, H.G. 1975. Isolation and characterization of a new pancreatic polypeptide hormone. *Journal of Biological Chemistry* 250:9369–9376.

Klein, S., Cohn, S.M., and Alpers, D.H. 1999. The alimentary tract in nutrition: a tutorial. In: Shils, M.E., Olson, J.A., Shike, M., et al. (eds.) *Modern Nutrition in Health and Disease*, 9th edition. Lippincott Williams & Wilkins, New York, pp. 605–625.

Koh, T.J., Dockray, G.J., Varro, A., et al. 1999. Overexpression of glycine-extended gastrin in transgenic mice results in increased colonic proliferation. *Journal of Clinical Investigation* 103:1119–1126.

Kolkovski, S. 2001. Digestive enzymes in fish larvae and juveniles—implications and applications to formulated diets. *Aquaculture* 200:181–201.

Kolkovski, S., Arieli, A., and Tandler, A. 1997a. Visual and chemical cues stimulate microdiet ingestion in sea bream larvae. *Aquaculture International* 5:527–536.

Kolkovski, S., Koven, W., and Tandler, A. 1997b. The mode of action of *Artemia* in enhancing utilization of microdiet by gilthead seabream *Sparus aurata* larvae. *Aquaculture* 155:193–205.

Konturek, S., Tasler, J., and Obtułowicz, W. 1972. Effect of atropine on pancreatic responses to endogenous and exogenous cholecystokinin. *Digestive Diseases and Sciences* 17:911–917.

Konturek, S.J., Zabielski, R., Konturek, J.W., et al. 2003. Neuroendocrinology of the pancreas;

role of brain–gut axis in pancreatic secretion. *European Journal of Pharmacology* 481:1–14.

Konturek, S.J., Konturek, J.W., Pawlink, T., et al. 2004. Brain–gut axis and its role in the control of food intake. *Journal of Physiology and Pharmacology* 55:137–154.

Koven, W., Kolkovski, S., Hadas, E., et al. 2001. Advances in the development of microdiets for gilthead seabream, *Sparus aurata*: a review. *Aquaculture* 194:107–121.

Koven, W., Rojas-Garcia, C.R., Finn, R.N., et al. 2002. Stimulatory effect of ingested protein and/or free amino acids on the secretion of the gastro-endocrine hormone cholecystokinin and on tryptic activity, in early-feeding herring larvae, *Clupea harengus*. *Marine Biology* 140:1241–1247.

Kunze, W.A.A., and Furness, J.B. 1999. The enteric nervous system and regulation of intestinal motility. *Annual Review of Physiology* 61:117–142.

Kurokawa, T., and Suzuki, T. 2002. Development of neuropeptide Y-related peptides in the digestive organs during the larval stage of Japanese flounder, *Paralichthys olivaceus*. *General and Comparative Endocrinology* 126:30–38.

Kurokawa, T., Suzuki, T., and Andoh, T. 2000. Development of cholecystokinin and pancreatic polypeptide endocrine systems during the larval stage of Japanese flounder, *Paralichthys olivaceus*. *General and Comparative Endocrinology* 120:8–16.

Kurokawa, T., Suzuki, T., and Hashimoto, H. 2003. Identification of gastrin and multiple cholecystokinin genes in teleost. *Peptides* 24:227–235.

Kurokawa, T., Iinuma, N., Unuma, T., et al. 2004. Development of endocrine system regulating exocrine pancreas and estimation of feeding and digestive ability in Japanese eel larvae. *Aquaculture* 234:513–525.

Kvåle, A., Yúfera, M., Nygård, E., et al. 2006. Leaching properties of three different microparticulate diets and preference of the diets in cod (*Gadus morhua* L.) larvae. *Aquaculture* 251:402–415.

Langley, J.N. 1903. The autonomic nervous system. *Brain* 26:1–26.

Larhammar, D. 1996. Evolution of neuropeptide Y, peptide YY and pancreatic polypeptide. *Regulatory Peptides* 62:1–11.

Lazo, J.P. 1999. Development of the digestive system in red drum (*Sciaenops ocellatus*) larvae. PhD dissertation, Department of Marine Science, The University of Texas at Austin, Austin, TX.

Lazo, J.P., Dinis, M.T., Holt, G.J., et al. 2000a. Co-feeding microparticulate diets with algae: toward eliminating the need of zooplankton at first feeding in larval red drum (*Sciaenops ocellatus*). *Aquaculture* 188:339–351.

Lazo, J.P., Holt, G.J., and Arnold, C.R. 2000b. Ontogeny of pancreatic enzymes in larval red drum *Sciaenops ocellatus*. *Aquaculture Nutrition* 6:183–192.

LeBlanc, J. 2000. Nutritional implications of cephalic phase thermogenic responses. *Appetite* 34:214–216.

Lee, J.H., Ku, S.K., Park, K.D., et al. 2004. Immunohistochemical study of the gastrointestinal endocrine cells in the Korean aucha perch. *Journal of Fish Biology* 65:170–181.

Lee, M.C., Schiffman, S.S., and Pappas, T.N. 1994. Role of neuropeptides in the regulation of feeding behavior: a review of cholecystokinin, bombesin, neuropeptide Y, and galanin. *Neuroscience & Biobehavioral Reviews* 18:313–323.

Li, Z.S., and Furness, J.B. 1993. Nitric oxide synthase in the enteric nervous system of the rainbow trout, *Salmo gairdneri*. *Archives of Histology and Cytology* 56:185–193.

Liddle, R.A. 1995. Regulation of cholecystokinin secretion by intraluminal releasing factors. *American Journal of Physiology—Gastrointestinal Liver Physiology* 269:G319–G327.

Liddle, R.A. 2000. Regulation of cholecystokinin secretion in humans. *Journal of Gastroenterology* 35:181–187.

Liehr, R.-M., Reidelberger, R.D., Varga, G., et al. 1993. Mechanism of bombesin-induced pancreatic secretion in unanesthetized rats. *Peptides* 14:717–723.

Little, T.J., Horowitz, M., and Feinle-Bisset, C. 2005. Role of cholecystokinin in appetite control and body weight regulation. *Obesity Reviews* 6:297–306.

Luizi, F.S., Gara, B., Shields, R.J., et al. 1999. Further description of the development of the digestive organs in Atlantic halibut (*Hippoglossus hippoglossus*) larvae, with notes on differential absorption of copepod and *Artemia* prey. *Aquaculture* 176:101–116.

Lundgren, O., Svanvik, J., and Jivegård, L. 1989. Enteric nervous system. *Digestive Diseases and Sciences* 34:264–283.

Manning, A.J., Murray, H.M., Gallant, J.W., et al. 2008. Ontogenetic and tissue-specific expression of preproghrelin in the Atlantic halibut, *Hippoglossus hippoglossus* L. *The Journal of Endocrinology* 196:181–192.

Mendieta-Zerón, H., López, M., and Diéguez, C. 2008. Gastrointestinal peptides controlling body weight homeostasis. *General and Comparative Endocrinology* 155:481–495.

Mojsov, S. 2000. Glucagon-like peptide-1 (GLP-1) and the control of glucose metabolism in mammals and teleost fish. *American Zoologist* 40:246–258.

Morisset, J., Julien, S., and Laine, J. 2003. Localization of cholecystokinin receptor subtypes in the endocrine pancreas. *Journal of Histochemistry and Cytochemistry* 51:1501–1513.

Moyano, F.J., Díaz, M., Alarcón, F.J., et al. 1996. Characterization of digestive enzyme activity during larval development of gilthead seabream (*Sparus aurata*). *Fish Physiology and Biochemistry* 15:121–130.

Murashita, K., Fukada, H., Hosokawa, H., et al. 2006. Cholecystokinin and peptide Y in yellowtail (*Seriola quinqueradiata*): molecular cloning, real-time quantitative RT-PCR, and response to feeding and fasting. *General and Comparative Endocrinology* 145:287–297.

Murashita, K., Kurokawa, T., Nilsen, T.O., et al. 2009. Ghrelin, cholecystokinin, and peptide YY in Atlantic salmon (*Salmo salar*): molecular cloning and tissue expression. *General and Comparative Endocrinology* 160:223–235.

Newgreen, D., and Young, H.M. 2002. Enteric nervous system: development and developmental disturbances—part 1. *Pediatric and Developmental Pathology* 5:224–247.

Nishimori, I., Kamakura, M., Fujikawa-Adachi, K., et al. 1999. Cholecystokinin A and B receptor mRNA expression in human pancreas. *Pancreas* 19:109–113.

Okumura, T., Pappas, T.N., and Taylor, I.L. 1995. Pancreatic polypeptide microinjection into the dorsal motor nucleus inhibits pancreatic secretion in rats. *Gastroenterology* 108:1517–1525.

Olsson, C., and Holmgren, S. 2001. The control of gut motility. *Comparative Biochemistry and Physiology. Part A, Molecular & Integrative Physiology* 128:479–501.

Olsson, C., Holmberg, A., and Holmgren, S. 2008. Development of enteric and vagal innervation of the zebrafish (*Danio rerio*) gut. *The Journal of Comparative Neurology* 508:756–770.

Owyang, C., and Logsdon, C.D. 2004. New insights into neurohormonal regulation of pancreatic secretion. *Gastroenterology* 127: 957–969.

Paterson, A.C., Lockhart, S.M., Baker, J., et al. 2004. Identity and regulation of stored and secreted progastrin-derived peptides in sheep. *Endocrinology* 145:5129–5140.

de Pedro, N., and Björnsson, B.T. 2001. Regulation of food intake by neuropeptides and hormones. In: Houlihan, D., Boujard, T., and Jobling, M. (eds.) *Food Intake in Fish*. Blackwell Publishing, Oxford, UK, p. 440.

Péres, A., Cahu, C.L., and Infante, J.L.Z. 1997. Dietary spermine supplementation induces intestinal maturation in sea bass (*Dicentrarchus labrax*) larvae. *Fish Physiology and Biochemistry* 16:479–485.

Reifel, C.W., and Travill, A.A. 1977. Structure and carbohydrate histochemistry of the esophagus in ten teleostean species. *Journal of Morphology* 152:303–313.

Reinecke, M., Müller, C., and Segner, H. 1996. An immunohistochemical analysis of the ontogeny, distribution and coexistence of 12 regulatory peptides and serotonin in endocrine cells and nerve fibers of the digestive tract of the turbot, *Scophthalmus maximus* (Teleostei). *Anatomy and Embryology* 195:87–101.

Rocha, R.J., Ribeiro, L., Costa, R., et al. 2008. Does the presence of microalgae influence fish larvae prey capture? *Aquaculture Research* 39:362–369.

Rojas-Garcia, C.R., and Rønnestad, I. 2002. Cholecystokinin and tryptic activity in the gut and body of developing Atlantic halibut larvae: evidence for participation in the regulation of protein digestion. *Journal of Fish Biology* 61:973–986.

Rønnestad, I. 2002. Control and efficiency of digestive function of marine fish larvae. *Avances en Nutrición Acuícola VI. Memorias del VI Simposium Internacional de Nutrición Acuícola* 3:152–165.

Rønnestad, I., Dominguez, R.P., and Tanaka, M. 2000a. Ontogeny of digestive tract functionality in Japanese flounder, *Paralichthys olivaceus* studied by *in vivo* microinjection: pH and assimilation of free amino acids. *Fish Physiology and Biochemistry* 22:225–235.

Rønnestad, I., Rojas-Garcia, C.R., and Skadal, J. 2000b. Retrograde peristalsis: a possible mechanism for filling the pyloric caeca. *Journal of Fish Biology* 56:216–218.

Rønnestad, I., Tonheim, S.K., Fyhn, H.J., et al. 2003. The supply of amino acids during early feeding stages of marine fish larvae: a review of recent findings. *Aquaculture* 227:147–164.

Rønnestad, I., Kamisaka, Y., Conceição, L.E.C., et al. 2007. Digestive physiology of marine fish larvae: hormonal control and processing capacity for proteins, peptides and amino acids. *Aquaculture* 268:82–97.

Sæle, Ø., Solbakken, J.S., Watanabe, K., et al. 2004. Staging of Atlantic halibut (*Hippoglossus hippoglossus* L.) from first feeding through metamorphosis, including cranial ossification independent of eye migration. *Aquaculture* 239:445–465.

Sanger, G.J. 2008. Motilin, ghrelin and related neuropeptides as targets for the treatment of GI diseases. *Drug Discovery Today* 13:234–239.

Scocco, P., Accili, D., Menghi, G., et al. 1998. Unusual glycoconjugates in the oesophagus of a tilapine polyhybrid. *Journal of Fish Biology* 53:39–48.

Silverstein, J.T., Bondareva, V.M., Leonard, J.B.K., et al. 2001. Neuropeptide regulation of feeding in catfish, *Ictalurus punctatus*: a role for glucagon-like peptide-1 (GLP-1). *Comparative Biochemistry and Physiology. Part B, Biochemistry & Molecular Biology* 129:623–631.

Slack, J. 1995. Developmental biology of the pancreas. *Development* 121:1569–1580.

Song, Y., Li, P., Lee, K.Y., et al. 1999. Canine pancreatic juice stimulates the release of secretin and pancreatic secretion in the dog. *American Journal of Physiology. Gastrointestinal and Liver Physiology* 277:G731–G735.

Soudah, H.C., Lu, Y., Hasler, W.L., et al. 1992. Cholecystokinin at physiological levels evokes pancreatic enzyme secretion via a cholinergic pathway. *American Journal of Physiology. Gastrointestinal and Liver Physiology* 263:G102–G107.

Tatemoto, K. 1982. Isolation and characterization of peptide YY (PYY), a candidate gut hormone that inhibits pancreatic exocrine secretion. *Proceedings of the National Academy of Sciences of the United States of America* 79:2514–2518.

Valenzuela, J.E., Lamers, C.B., Modlin, I.M., et al. 1983. Cholinergic component in the human pancreatic secretory response to intraintestinal oleate. *Gut* 24:807–811.

Varsamos, S., Nebel, C., and Charmantier, G. 2005. Ontogeny of osmoregulation in postembryonic fish: a review. *Comparative Biochemistry and Physiology. Part A, Molecular & Integrative Physiology* 141:401–429.

Verreth, J.A.J., Torreele, E., Spazier, E., et al. 1992. The development of a functional digestive system in the African catfish *Clarias gariepinus* (Burchell). *Journal of the World Aquaculture Society* 23:286–298.

Volkoff, H., Canosa, L.F., Unniappan, S., et al. 2005. Neuropeptides and the control of food intake in fish. *General and Comparative Endocrinology* 142:3–19.

Wapnir, R.A., and Teichberg, S. 2002. Regulation mechanisms of intestinal secretion: implications in nutrient absorption. *Journal of Nutritional Biochemistry* 13:190–199.

Webb, K.A. 2008. Cholecystokinin and the ontogeny of digestion in the red drum (*Sciaenops ocellatus*). PhD dissertation, Marine Science Institute, The University of Texas at Austin, Port Aransas, TX.

Webb, K.A., Khan, I.A., Nunez, B.S., et al. 2010. Cholecystokinin: molecular cloning and immunohistochemical localization in the gastrointestinal tract of larval red drum, *Sciaenops ocellatus* (L.). *General and Comparative Endocrinology* 166:152–159.

Weckbecker, G., Lewis, I., Albert, R., et al. 2003. Opportunities in somatostatin research: biological, chemical and therapeutic aspects. *Nature Reviews. Drug Discovery* 2:999–1017.

Xu, M., and Volkoff, H. 2009. Molecular characterization of ghrelin and gastrin-releasing peptide in Atlantic cod (*Gadus morhua*): cloning, localization, developmental profile and role in food intake regulation. *General and Comparative Endocrinology* 160:250–258.

Youson, J.H., and Al-Mahrouki, A.A. 1999. Ontogenetic and phylogenetic development of the endocrine pancreas (Islet organ) in fishes. *General and Comparative Endocrinology* 116:303–335.

Youson, J.H., and Potter, I.C. 1993. An immunohistochemical study of enteropancreatic endocrine cells in larvae and juveniles of the southern-hemisphere lampreys *Geotria australis* and *Mordacia mordax*. *General and Comparative Endocrinology* 92:151–167.

Zhu, Y., and Thomas, P. 1995. Red drum somatolactin: development of a homologous radioimmunoassay and plasma levels after exposure to stressors or various backgrounds. *General and Comparative Endocrinology* 99:275–288.

Section 3

Feeds and Feeding

Chapter 10

Feeding behavior in larval fish

Manuel Yúfera

10.1 Introduction

After egg hatching, fish larvae go through important changes to reach the juvenile stage, the most evident being a dramatic biomass increase. From a feeding point of view, the larval stage ends when all organs and structures related to food acquisition are completely developed and functional, and the definitive feeding habits have been achieved (Koumoundouros et al. 1999; Yúfera et al. 2004; Russo et al. 2007). Feeding success in fish larvae is critical for obtaining the nutrients and the energy necessary for healthy growth and development that allows them to survive to the end of the larval period. Irrespective of the juveniles' and adults' feeding modes, the larval fish are planktivorous. The ontogeny of feeding-related systems in fish is highly variable. The different species may be precocial or altricial, according to the degree of development attained at the time of hatching and opening of the mouth (Balon 1986). Larvae of precocial species hatch in an advanced stage of development. Larvae of altricial species hatch with undeveloped organs and structures and exhibit a more complicated development pattern. In these species, the yolk sac continues as the sole source of nutrients for the developing embryo immediately after hatching. After the start of exogenous feeding, the larvae continue the transformation to juveniles. During this period of strong anatomical and physiological changes, the feeding process undergoes a continuous adaptation to optimize the new body capacities for efficient food intake and nutrient assimilation.

Feeding behavior is the result of an interaction of complex processes that include several pre- and postconsumptive steps: searching, detection, attack, capture, ingestion, digestion,

Larval Fish Nutrition, First Edition. Edited by G. Joan Holt.

and evacuation. Each of them has a specific pattern that changes throughout development. The feeding strategy is related to the specific characteristics of each species, and it is adapted for successful exploitation of a given trophic niche. Availability of suitable prey is one of the most determinant biotic factors, but feeding mode and amount of food intake are also influenced by prevailing environmental conditions.

Absence of predators and competitors, abundant food, and low variety of prey are the primary differences when fish larvae are transferred from their natural habitats to rearing tanks. Although food is abundant, only a few zooplankton species are used as larval prey in aquaculture (Bengtson 2003; Conceição et al. 2010). On the other hand, there is a significant research effort in developing microdiets for advanced weaning and for the complete replacement of live prey (Baskerville-Bridges and Kling 2000; Fontagné et al. 2000; Langdon 2003; Yúfera et al. 2005). A good understanding of how environmental factors influence development and behavioral adaptations in aquaculture conditions are necessary for successfully rearing the target species. This review presents an overview of the feeding behavior and foraging strategies during the larval stage of marine teleosts, with special emphasis on those aspects determining the mode of food intake that are relevant in advancing feeding protocols with live prey and designing formulated microdiets for larval rearing.

10.2 Searching for and detecting food

Searching for prey and detecting them depend on the appropriate functioning of some organs and tissues that become progressively available throughout development, in parallel with the structures and organs related to feeding function (O'Connell 1981; Fukuhara 1985; Osse and van den Boogaart 1999). In fact, at the start of feeding and the transition from endogenous to exogenous nutrient sources, it is necessary that all structures and organs associated with food uptake and processing are, in some degree, ready and functional (Osse and van den Boogaart 1999; Yúfera and Darías 2007). From hatching, larvae are progressively aware of different external stimuli that indicate the presence of potential food items. Movement capacity develops in parallel, and consequently, so does the possibility for exploring the medium and moving toward those potential prey. Searching depends basically on swimming capacity, while detection depends largely on sensory organs.

10.2.1 Role of sensory organs

In fish, as in all vertebrates, sensory organs give direct information from the immediate surroundings and therefore play a relevant role in defining the subsequent responses. In the first stages, the behavior pattern tends primarily to favor survival by acquiring suitable food and avoiding being eaten by larger animals. The role of sensory organs is crucial in feeding behavior and escape from predators (Blaxter 1986).

Food detection occurs by means of visual, chemical, and mechanical stimuli. Olfaction allows detection of distant stimuli, sight allows the identification of objects at medium and relatively short distance, while touch and gustation need very close or direct contact with the source of the stimulus. Most marine fish hatch with immature anatomical features. Nevertheless, the sensory organs develop quickly during the first days after hatching. Details of the morphogenesis and functional ontogeny of eyes, chemoreceptors (responsible for smell and taste), and neuromasts (responsible for detection of movements) in fish larvae have been the subject of various recent reviews (see, for instance, Evans and Browman 2004; Døving and

Kasumyan 2008; Loew and Wahl 2008; Pankhurst 2008).

10.2.1.1 Vision

Sight allows the larvae to perceive objects that are relatively close, within the range of half to several body lengths. The morphogenesis of the retina is highly variable in teleosts (Evans and Browman 2004). Altricial marine teleosts hatch with undeveloped eyes, although the pigment in the retina appears in a few hours or days and functional pigmented eyes with the optic nerve connected to the optic tectum are present by the time of first feeding in many marine fish. This early retina has only one type of photoreceptor, formed predominantly by green-sensitive single cones (Helvik et al. 2001; Evans and Browman 2004) that allow relatively good acuity under bright light. Double and mosaic cone structures and rod photoreceptors appear later at metamorphosis and enable vision at low light intensity as well as motion sensitivity (Kawamura et al. 1984; Higgs and Fuiman 1996; Shand et al. 2002; Evans and Browman 2004).

Although with notable interspecific variability in the visual pigment types, this general pattern has been observed in species belonging to different families such as flounder *Paralichthys olivaceus* (Kawamura et al. 1989), spotted sand bass *Paralabrax maculatofasciatus* (Peña et al. 2004), halibut *Hippoglossus hippoglossus* (Kvenseth et al. 1996; Helvik et al. 2001), red sea bream *Pagrus major* (Kawamura et al. 1984); snapper *Pagrus auratus* (Pankhurst and Eagar 1996), red porgy *Pagrus pagrus* (Roo et al. 1999), and bluefin tuna *Thunnus orientalis* (Kawamura et al. 2003). Increase in lens size and the development of the retina with larval growth enhance the visual field and acuity; therefore, older larvae have increased reactive distance and perception angle (Arnold and Holford 1990; Hunt von Herbing and Gallager 2000; Cobcroft and Pankhurst 2006). Most fish are visual feeders during the first larval stages and require a minimum light intensity threshold to develop and grow normally (Boeuf and Le Bail 1999). Higher feeding intensity under illuminated conditions has been observed in larvae of many species, as in cod *Gadus morhua* (Puvanendran and Brown 1998), greenback flounder *Rhombosolea tapirina* (Cox and Pankhurst 2000), haddock *Melanogrammus aeglefinus* (Downing and Litvak 2001), or *Pagrus auratus* (Fielder et al. 2002). Nevertheless, vision is not always necessary for feeding. Some species are able to growth at very low light intensities (Chesney 1989) or in total absence of light. The African catfish, *Clarias gariepinus*, a freshwater species, can be reared in the dark from first feeding (Appelbaum and Kamler 2000) and is able to eat actively in the dark from mouth-opening, probably helped by neuromasts and taste buds for detecting and recognizing the food (Mukai et al. 2008). Similarly, striped bass *Morone saxatilis* larvae are able to feed independent of light as an adaptation to turbid estuarine waters (MacIntosh and Duston 2007). First-feeding larvae of *Melanogrammus aeglefinus* can also eat in the dark, although they eat better at moderate light intensities (Downing and Litvak 2001).

10.2.1.2 Olfaction

Olfaction allows for more remote detection of a stimulus. The olfactory organ appears early during embryonic development. Olfactory placodes and pits are already present at the onset of feeding (O'Connell 1981; Noakes and Godin 1988; Tanaka et al. 1991; Boglione et al. 2003; Kawamura et al. 2003) and develop further to complete the deepening of the pits, the formation of nares, the folding of olfactory epithelium, and the formation of olfactory lamellae by the late larval stage. Functionality and the role of olfactory sense in prey detection and identification from first feeding as well as in older larvae have been probed in herring *Clupea harengus* (Dempsey 1978), *Pagrus major* (Iwai 1980; Tanaka et al. 1991), gilthead sea bream *Sparus aurata* (Kolkovski et al. 1997), turbot *Psetta maxima* (Knutsen 1992; Kasumyan et al. 1998), and common sole *Solea solea* (Knutsen 1992) using chemical stimuli that evoke specific behavior. In later stages, when the olfactory organs are more developed, olfaction is involved in detection of nursery areas (James et al. 2008; Lara 2008). Chemical signals have a relevant role in many species in searching for the appropriate habitat for settling after pelagic larval life (Adverlund and Takemura 2006; Lara 2008).

10.2.1.3 Taste

The gustatory system is the other chemoreceptor. In many marine species, intra- and extraoral taste buds develop or proliferate some days or weeks after first feeding (Iwai 1980; Cobcroft and Pankhurst 2006; Sánchez-Amaya et al. 2007), allowing better prey discrimination and the possibility for rejecting prey (Kentouri and Divanach 1982; Boglione et al. 2003). However, some teleosts with demersal eggs develop taste buds at very early stages, and they are present at first feeding as in the willow shiner *Gnathopogon elongates caulescens* (Mukai 2006) and *Clarias gariepinus* (Mukai et al. 2008).

10.2.1.4 Neuromasts

Mechanical stimuli such as touching or water movements are detected by neuromasts and the lateral line system. In larval fish, some few free neuromasts are already present at hatching and progressively proliferate during growth and development (Higgs and Fuiman 1996; Kawamura et al. 2003). Cephalic and trunk canalized neuromasts within lateral lines appear later in larval development, conferring a better perception of water acceleration. Evidence of their involvement in prey detection has been shown in the larvae of mottled sculpin *Cottus bairdi* (Jones and Janssen 1992) and of *Gnathopogon elongates caulescens* (Mukai 2006) by blocking the neuromasts with streptomycin or by immobilizing zooplanktonic prey to prevent vibrations.

The progressive development and completion of all these sensory organs increase the capacity for detection and recognition of potential prey. All sensory organs are involved to different degrees during the development of prey detection. At first feeding with limited locomotor capacity, the detection of prey is restricted to a small area right in front of the head, and is consequently highly dependent on encounter opportunity. Fish larvae smell the prey metabolites and, when their swimming capacity improves, can steer their way toward the zooplankton patches. Specific prey items are primarily detected by vision helped by neuromasts, and subsequently an attack is attempted. Once taste buds proliferate, larvae may make the decision to swallow or regurgitate prey according to its palatability and previous experiences. In later stages of development, vision and olfactory organs are linked with localization and migration toward different areas and adaptation to new trophic niches.

10.3 Locomotor capacity and searching strategies

Both predation efficiency and escape ability are closely related to swimming capacity. The ability of larval fish to move within the aqueous medium is affected by some anatomical features that are changing during the larval stage, such as the musculature anatomy and structure, the presence of the primordial finfold and the appearance of the caudal fin and median and lateral fins, swim bladder inflation, and the body size and shape (Fukuhara 1985; Blaxter 1986; Webb and Weihs 1986). On the other hand, the swimming capacity is also dependent on the hydrodynamic environmental conditions characterized by the Reynolds number, a parameter indicating the ratio of inertial to viscous forces in fluids, that is based on larval length and swimming speed (Webb and Weihs 1986). Low Reynolds numbers indicate that viscous forces are dominant. Larval fish experience low and intermediate values (Webb and Weihs 1986; Batty and Blaxter 1992), particularly during yolk sac stage and the first days of feeding when the total length (TL) of marine fish larvae ranges between approximately 2 and 4 mm. Under such conditions, frictional forces clearly dominate. Before notochord flexion, swimming speed and endurance are limited. In fact, those species with very low larval length at first feeding have short swimming times and need to increase larval length quickly (Hunt von Herbing and Gallager 2000). There is a relatively less viscous hydrodynamic environment with increased TL, with a consequent decrease in the energetic cost of swimming. After notochord flexion, usually occurring at 5–6-mm TL, and caudal fin development, larvae swim in an inertial environment and swimming endurance increases markedly. Consequently, larval fish acquire better swimming performance with growth (Webb and Weihs 1986; Fuiman et al. 1999), although each species improves this swimming competency at a specific rate (Clark et al. 2005). In later larval stages, larvae are able to swim faster than the ambient current speed and cover long distances (Clark et al. 2005), allowing dispersal and recruitment. But Koumoundouros et al. (2009) working in *Sparus aurata* found that in late metamorphosis, this sparid exhibits a decrease in relative sustainable swimming speed as a consequence of muscle anatomy and changes in body size and shape as an adaptation to demersal habitat requiring more maneuverability.

Basically, fish larvae exhibit alternating periods of swimming and inactivity (Browman and O'Brien 1992; Fuiman et al. 1999; Hunt von Herbing and Gallager 2000). Swimming speed, pause duration and frequency, burst duration, reactive distance, perception angles, and duration of predation cycle are measurable aspects that define the changes in behavior during searching and attack throughout development (Hunt von Herbing and Gallager 2000; Chesney 2008). This pause–travel foraging behavior has been described in many marine fish species. Larvae swim by intense short bursts followed by gliding and scanning for prey during pauses, as described in *Gadus morhua* (Hunt von Herbing and Gallager 2000). As larvae grow, swimming speed and movement distance increase, pause duration tends to decrease, and the absolute reactive distance and perception horizontal and vertical angles become greater (Blaxter 1986; Hunt von Herbing and Gallager 2000; Chesney 2008; Figure 10.1). Thus, attack and feeding success increase in progressively larger larvae. Three different searching strategies, ambush, swimming and pause, and cruiser have been described at the later stages of the larval period (Browman and O'Brien 1992). At first feeding, even the smallest larvae have some primordial hunting abilities, but the efficacy increases with development and growth, changing from passive feeding highly dependent on encounter opportunity to an active prey searching capability.

10.4 Capture and ingestion

Capture success relies not only on development stage and concomitant hunting capacity but also on the availability and accessibility of prey. Once a prey is perceived, the foraging has three possible results: unsuccessful attacks, aborted attacks, and successful attacks (Hunt von Herbing and Gallager 2000). After mouth-opening, fish larvae need to learn hunting and have to do it quickly. Not all larvae of the same cohort start feeding at the same time. The percentage of actively feeding larvae at the onset of feeding exhibits a pattern of increase that is characteristic for

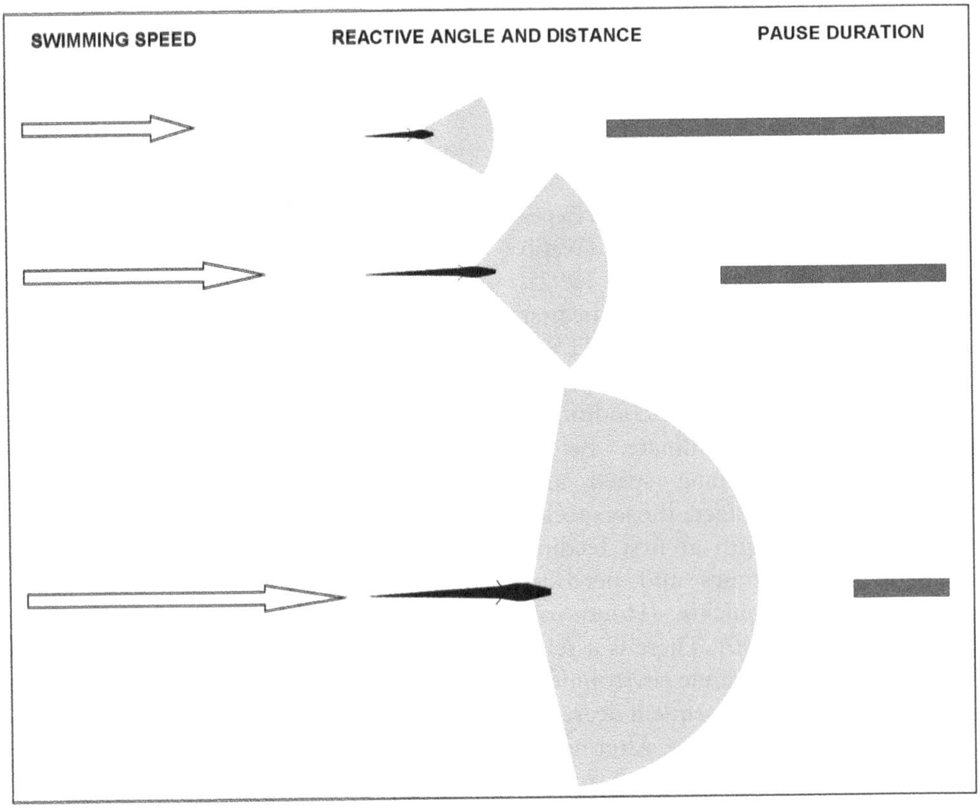

Figure 10.1 Schematic diagram showing the changes in swimming speed (arrows), reactive distance and angle of perception, and pause duration (bars) in relation to fish larvae age and total length.

each species. The observed differences in feeding ability among species are mainly due to anatomical characteristics and aggressiveness of the larvae. In some species, the totality of the population is able to start feeding in a few hours, while others take several days (Kiørboe et al. 1985; Polo et al. 1992; Næss et al. 1995; Doi et al. 1997; Olsen et al. 2000; Parra and Yúfera 2000; Dou et al. 2002; Carton 2005; Nhu et al. 2009; Russo et al. 2009; Figure 10.2). In the latter case, high prey availability and accessibility seem to be crucial for successfully initiating feeding (Kiørboe et al. 1985; Polo et al. 1992). Prey size and swimming ability are primary factors determining the efficacy with which the prey is caught. In fact, the ability to start feeding after mouth-opening is typically affected by prey size (Figure 10.3). During the very early stages with low swimming capacity, encounter opportunity depends on prey density. As larvae grow in length with age, attack and feeding success increase (Hunt von Herbing and Gallager 2000; Monk et al. 2006), although capture success increases mainly during the first days of feeding (Monk et al. 2006).

10.4.1 Prey selection

The detection of a potential prey does not always elicit a strike response. Selection of food by fish larvae depends on prey accessibility and organoleptic preferences. The accessibility of a given prey is determined by

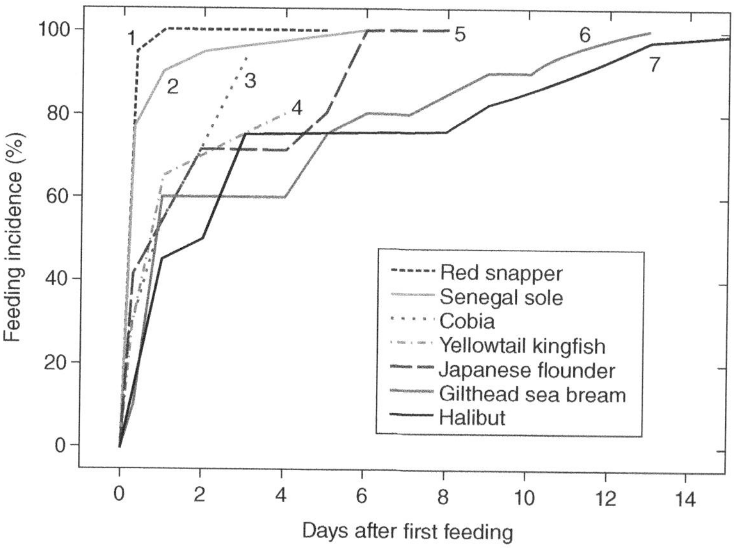

Figure 10.2 Feeding incidence (percentage of larvae with some content within the gut) at the onset of exogenous feeding in different fish species. Curves elaborated from data reported in the literature: (1) *Lutjanus argentimaculatus*, 31°C (Doi et al. 1997); (2) *Solea senegalensis*, 19.5°C (Cañavate et al. 2006; Yúfera, unpublished data); (3) *Rachycendron canadum*, 28°C (Nhu et al. 2009); (4) *Seriola lalandi*, 20°C (Carton 2005); (5) *Paralichthys olivaceus*, 17.1°C (Dou et al. 2002); (6) *Sparus aurata*, 19.5°C (Parra and Yúfera 2000); and (7) *Hippoglossus hippoglossus*, 11–12°C (Næss et al. 1995; Olsen et al. 2000).

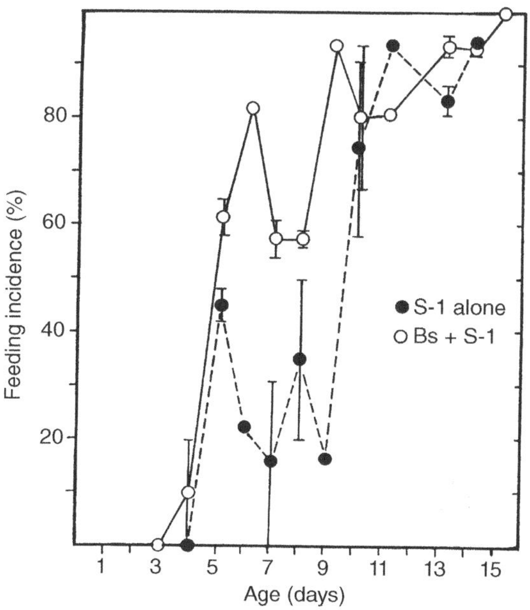

Figure 10.3 Feeding incidence at the onset of feeding in gilthead sea bream larvae fed on rotifers with different body sizes, *Brachionus plicatilis* s.s. (S-1 strain, body length 140–276 μm; filled circles and dashed lines) and *Brachionus rotundiformis* (Bs strain, body length 92–176 μm; open circles and continuous lines) (from Polo et al. 1992, with permission).

its size, shape, and ability to escape from predators. Prey size is the primary characteristic defining choice in early larvae. Mouth gape limits the dimensions of prey that can be ingested. Jaw length and mouth gape increase almost proportionally to larval length. This increase is particularly rapid during the first few days after mouth-opening as may be observed in the measurements reported for different species (Shirota 1970; Sumida and Moser 1980; Fernández-Díaz et al. 1994; Doi et al. 1997), indicating that improving feeding capacity is a developmental priority in early larvae. The proportional increase of mouth size with larval length is not permanent and is slower in older larvae after reaching an inflexion point (Rowlands et al. 2006). Consequently, larvae tend to select larger prey as growth progresses as has been observed in the field (Munk 1992; Østergaard et al. 2005) and in laboratory experiments with different species, for example, Western Atlantic sea bream *Archosagus rhomboidalis* (Stepien 1976), striped mullet *Mugil cephalus* (Oozeki

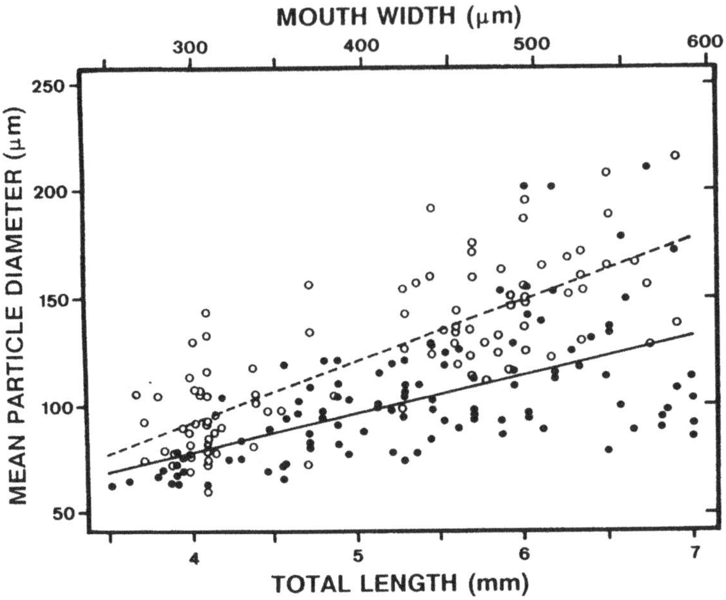

Figure 10.4 Mean diameter of ingested particles in relation to total length in gilthead sea bream larvae fed on hard microcapsules (filled circles and continuous line) and soft microcapsules (open circles and dashed line) (from Fernández-Díaz et al. 1994, with permission).

et al. 1992), *Sparus aurata* (Polo et al. 1992; Fernández-Díaz et al. 1994), halibut *Hippoglossus hippoglossus* (Olsen et al. 2000), and dusky grouper *Epinephelus marginatus* (Russo et al. 2009). Nevertheless, Hamasaki et al. (2009) did not find such increasing size preference in greater amberjack *Seriola dumerili* larvae fed on rotifers with different lorica sizes. This increasing pattern is not equal for all species, and the preferred sizes or prey species may vary in order to reduce prey niche overlapping among species (Østergaard et al. 2005; Simonsen et al. 2006).

Although fish larvae are able to ingest prey with similar size to mouth gape, they tend to ingest smaller prey. Prey/gape ratio determined in different species usually ranges between 25 and 60% (Shirota 1970; Fernández-Díaz et al. 1994; Busch 1996; Cunha and Planas 1999; Østergaard et al. 2005). The size of prey that can be ingested is also determined by the diameter of the esophagus (Busch 1996; Yúfera and Darías 2007), mainly early when the pressure of the yolk sac constrains the opening of the lumen of the esophagus (Busch 1996). Fernández-Díaz et al. (1994), working with *Sparus aurata* larvae fed on inert microdiets, showed that larvae fed on soft particles were able to select larger sizes than when fed hard particles (Figure 10.4). This fact reveals the importance of the physical characteristics of food items for larvae that ingest by directly swallowing the entire prey during the grand part of the larval stage, and the prey has to accommodate to the diameter of the esophagus. With the development of buccal teeth and jaw musculature, larval fish change capture mode from simple swallowing to biting and swallowing. After this developmental event, organism size and shape is not such a constraint to larvae in choosing the appropriate prey.

Factors other than size are also involved in the food preferences of larval fish. Some metabolites from planktonic organisms (amino acids and nucleotides) have been identified as olfactory and taste stimuli for feeding (Kolkovski et al. 1997). The ability to discriminate chemical clues by olfaction and gustation is present to some degree at first feeding and improves with development. Kentouri and Divanach (1982) observed in different sparids (*Diplodus sargus*, *Diplodus vulgaris*, and *Sparus aurata*) how early larvae eat a variety of prey items but regurgitate annelid larvae and the ciliate *Fabrea salina*, organisms that were then omitted. Fernández-Díaz et al. (1994) found that *Sparus aurata* larvae prefer live prey over inert microdiets, both within the same size range. In fact, the poor ingestion rates obtained in some cases with microdiets may improve by adding small amounts of live prey or exudates of live organisms. Kolkovski et al. (1997) identified arginine, alanine and glycine, and sodium salt of betaine from metabolites excreted by *Artemia* as strong feeding stimulants. Interestingly, a well-ingested microencapsulated diet (Yúfera et al. 1995) released higher amounts of three of these compounds when immersed in water (Yúfera et al. 2002).

In spite of the theoretical prey size spectrum increase in parallel with larval size, the preferred prey is able to supply appropriate biomass and energy. Swimming and hunting are costly activities (O'Brien et al. 1989; Hunt von Herbing et al. 2001), and the optimization of effort for best efficiency is a requisite to maintain a reasonable growth rate. Therefore, growing fish larvae tend to choose those prey that result in a favorable ratio between energy gain and energy spent in the capture of such prey (Hunt von Herbing et al. 2001; Puvanendran et al. 2004).

A particular case in prey preference is cannibalism behavior. This is a phenomenon affecting mainly older larvae and juveniles of piscivorous species (Katavić et al. 1979; Dou et al. 2000; Sabate et al. 2009) but is also observable in early larval stages. Freshwater species of the *Brycon* genus exhibit strong cannibalism behavior from first feeding that may have an impact, with up to 98% of the population cannibalized in the first week (Baras et al. 2000). Cannibalism has been described in many very different cultured species (e.g., in sea bass *Dicentrarchus labrax* after 50 dph, Katavić et al. 1979; in bluefin porgy *Sparidentex hasta* after the first month [17–20-mm TL], Teng et al. 1999; in common dentex *Dentex dentex* beyond 10 dph [>10-mm TL], Koumoundouros et al. 1999; in *Paralichthys olivaceus*, after larvae settlement, Dou et al. 2000; in *Clarias gariepinus*, Mukai et al. 2008; in barramundi *Lates calcarifer*, Parazo et al. 1991; and in seven-band grouper *Ephinephelus septemfasciatus* beyond 52 dph, Sabate et al. 2009). Some of the particular conditions and factors inducing cannibalism are high larval density, low prey availability or episodic prey scarcity, and most important, larval size variation within the sibling cohort (Katavić et al. 1979; Hetch and Pienaar 1993; Dou et al. 2000). Consequently, during larval rearing, low larval density, abundant prey supply, and size sorting are scheduled to prevent or reduce cannibalism behavior.

10.4.2 Ingestion patterns

Many factors influence the quantity and the manner in which food biomass is taken in by larval fish. Some of the factors depend on species-specific characteristics of the larvae, while others depend on environmental conditions, including food availability and accessibility. Obtaining reliable estimates of the number of prey items and mass ingested by larval fish per unit time is a difficult task strongly influenced by experimental conditions and methodology (Dunbrack et al. 2009). Different experimental approaches have been used such as gut fullness plus evacuation rates, gut filling rates, and decline of prey number in water, combined with prey staining, autofluorescence, or ^{14}C labeling (Stepien 1976; Houde and Schekter 1980; Tandler and Mason 1984; Kelly et al. 2000; Parra and Yúfera 2000; Wuenschel and Werner 2004; Getchis and Bengtson 2006). Independent of the method of determination, food consumption is variable among individuals as well as over the 24-hour cycle (Shoji et al. 2001). In addition, the feeding capacity is changing quickly, mainly after the onset of feeding, during which feeding incidence and intensity are sharply increasing (Parra and Yúfera 2000; Carton 2005; Monk et al. 2006; Yoseda et al. 2008). Another aspect to take into account is that feeding responses determined under simplified laboratory conditions should be compared with caution with field determination (MacKenzie et al. 1990; Bochdansky et al. 2008). Principal factors influencing feed intake by larval fish are prey density, temperature, and illumination, as well as larval size and developmental status. Prey density is a defining factor in feeding success. Effects of increasing prey concentrations on foraging behavior have been tested in different species, showing an increase in the ingestion rate according to a functional response (the consumption rate keeps practically constant at prey densities above a satiation threshold) as the most common response to increasing prey density (Houde and Schekter 1980; MacKenzie et al. 1990; Dou et al. 2000; Parra and Yúfera 2000; Lika and Papandroulakis 2005; Shaw et al. 2006; Bochdansky et al. 2008). Nevertheless, each species has different abilities at low prey density (Houde and Schekter 1980). Although densities between 5 and 15 individuals/mL are usual in intensive larval rearing, 1 prey/mL seems to be enough for achieving close to maximum feeding rates (Iizawa 1984; Parra and Yúfera 2000; Lika and Papandroulakis 2005; Shaw et al. 2006). In general, low prey concentration affects ingestion—decreasing specific ration and increasing time of evacuation—assimilation, and growth efficiency (Boehlert and Yoklavich 1984; Theilacker 1987). Evacuation of undigested, or even still living, rotifers is commonly observed when high prey density is supplied to fish larvae. On the other hand, Iizawa (1984) found that *Dicentrarchus labrax* larvae only evacuate the gut contents when the food is completely digested, independent of prey density.

Although the effect of temperature has been studied in several species in terms of development and growth performance, including some feeding- and energetics-related aspects (e.g., Laurence 1977; Yúfera et al. 1993a), the direct effect on ingestion rate has been scarcely examined. Changes in temperature affect all larval metabolic rates, and it is expected to influence also the ingestion rates according to an optimum temperature range. Thus, daily specific ingestion of larval spotted sea trout *Cynoscion nebulosus* increased with increasing temperature in the range 24–32°C (Wuenschel and Werner 2004), in part due to the parallel increase in swimming activity (Hunter 1981; Batty and Blaxter 1992). Houde and Zastrow (1993) compiled information on the dynamics and energetics of many species from different habitats and families during larval stages. The estimated daily specific ingestion increased with temperature in the range 5–26°C. Average values on a weight basis during the overall larval period ranged from 0.15/day to 1.20/day approximately. On the other hand, Dou et al. (2000), working with larvae of *Paralichthys olivaceus* in the range 12–26°C, found a Gaussian pattern, with maximum ingestion at 19°C.

In reared fish larvae, food consumption has been usually examined in short experiments of a few hours, yielding an hourly feeding rate, or alternatively estimated from prey number decrease over 24-hour cycles, yielding a daily food ration. Nevertheless, feeding activity may change throughout the circadian period. Photoperiod and illumination level are determinant factors for most species, and like other vertebrates the feeding rhythm in fish is conditioned by the light–dark cycle. In most species, daily consumption, and growth, is related to the duration of the light period as well as to light intensity and quality as reported, for instance, for *Gadus morhua* (Puvanendran and Brown 2002), *Rhombosolea tapirina* (Cox and Pankhurst 2000), leopard coral grouper *Plectropomus leopardus* (Yoseda et al. 2008), and *Dicentrarchus labrax* (Villamizar et al. 2009). In addition to light conditions, the foraging activity of fish larvae is not always constant during the light period. In the field, the highest feeding activity usually has been observed during sunrise and sunset under crepuscular light (McLaren et al. 1997; Reiss et al. 2004) or just during sunset (e.g., in *Melanogrammus aeglefinus*, MacKenzie et al. 1999; and Japanese Spanish mackerel *Scomberomorus niphonius*, Shoji et al. 2001). In contrast, red drum and spotted sea trout larvae appeared to feed continuously throughout the 12-hour light period (Holt and Holt 2000). In larvae reared under light and dark conditions, ingestion may be higher at the beginning of the light period. Interestingly, a circadian rhythm pattern under conditions of continuous illumination was observed in larvae of *Dicentrarchus labrax* (Ronzani Cerqueira and Chatain 1991) and *Plectropomus leopardus* (Yoseda et al. 2008).

Flatfish larvae experience remarkable anatomical and behavioral transformation during development, changing from pelagic to benthic preferences. Some flatfish species continue feeding actively during eye migration, while others do not eat (Lagardère et al. 1999; Parra and Yúfera 2001; Geffen et al. 2007). Changes in life habits of flatfish larvae may induce notable changes in feeding behavior, mainly in those species with nocturnal habits. Ma et al. (2006) reported that in tongue sole *Cyconoglossus semilaevis*, feeding rhythm increases with larval age and changes after metamorphosis. During the pelagic phase, the larvae fed preferentially during daytime, while after eye migration, postlarvae prefer feeding during nighttime. Similarly, Cañavate et al. (2006) found an increase in feeding activity of Senegal sole *Solea senegalensis* in dark conditions when larvae were approaching the eye migration phase.

As development and growth advance, the larvae ingest progressively more food, eating either more or larger prey (Houde and Schekter 1980; Bailey et al. 1995; Parra and Yúfera 2001). In farmed species under regular feeding conditions, the sequence of increasing prey size generally offered to larvae includes small and large rotifers first, followed by *Artemia* nauplii and metanauplii. Daily rotifer consumption is roughly between 50 and 1,000 rotifers per larva during the first 2 weeks. For instance, in *Sparus aurata*, ingestion increased from 4 to 50 rotifers/larva/h from 6 to 15 dph (Parra and Yúfera 2000); in striped trumpeter *Latris lineata* it increased from 9 to 18 rotifers/larva/h from 8 to 15 dph (Bransden et al. 2005); in summer flounder *Paralichthys dentatus* it increased from 62 to 301 rotifers/larva/day from 6 to 13 dph (Bengtson et al. 1999); and in *Rhombosolea tapirina* it increased from 10 to 22 rotifers/larva/h from 12 and 18 dph under the best assayed conditions (Shaw et al. 2006). In terms of specific ingestion, fish larvae may ingest above their own weight daily during the first days of feeding (Figure 10.5; Houde and Schekter 1983; Baras et al. 2000; Parra and Yúfera 2000, 2001; Papandroulakis et al. 2002; Wuenschel and Werner 2004; Yúfera and Darías 2007). The specific ingestion tends to decrease with larval growth and the achievement of more efficient digestion and assimilation (Laurence 1977; Theilacker 1987; Day et al. 1996; Baras et al. 2000; Parra and Yúfera 2000; Wuenschel and Werner 2004; Cañavate et al. 2006; Getchis and Bengtson 2006; Ma et al. 2006). In addition, gut passage time also tends to increase with larval growth (Yamashita and Bailey 1989; Wuenschel and Werner 2004). Nevertheless, a continuous increase in specific

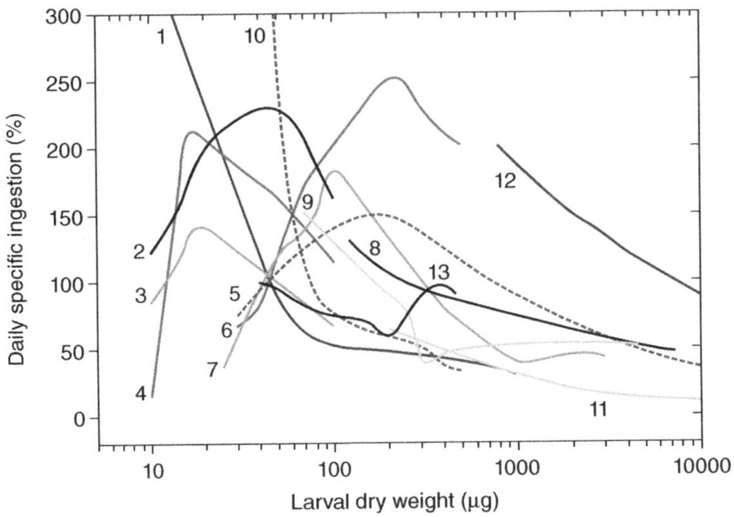

Figure 10.5 Some examples of daily specific ingestion (mg or calories ingested per mg or calorie of larva) of different fish species during the larval stage. Curves elaborated from data reported in the literature: (1) *Pseudopleuronectes americanus* (Laurence 1977); (2) *Anchoa mitchilli*, 26°C (Houde and Schekter 1983); (3) *Archosargus rhomboidalis*, 26°C, (Houde and Schekter 1983); (4) *Achirus lineatus*, 28°C (Houde and Schekter 1983); (5) *Cynoscion nebulosus*, 28°C (Wuenschel and Werner 2004); (6) *Sparus aurata*, 19.5°C (Parra and Yúfera 2001); (7) *Solea senegalensis*, 19.5°C (Parra and Yúfera 2001); (8) *Solea solea*, 15°C (Day et al. 1996); (9) *Clarias gariepinus*, 28°C (Conceição et al. 1998); (10) *Diplodus sargus*, 19.5°C (Yúfera and Darías 2007); (11) *Cynoglossus semilaevis*, 23°C (Ma et al. 2006); (12) *Brycon moorei*, 27°C (Baras et al. 2000); (13) *Pagrus pagrus*, 17°C (Papandroulakis et al. 2004). When necessary, reported wet weight values have been converted to dry weight on the basis of 20%.

ingestion during the first weeks has also been observed in *Dicentrarchus labrax* (Barahona-Fernandes and Conan 1981), *Anchoa mitchilli*, *Anchoa lamprotaenia* (Chitty 1981), and *Sparus aurata* (Tandler and Mason 1984; Parra and Yúfera 2000).

10.5 Effect of feeding and nutrition limitation on foraging ability

Food limitation in quantity and quality has evident effects on many biological processes of developing larvae and also on the foraging capacity. The most obvious consequence of food restriction is observed at the onset of feeding. Time to start feeding from the opening of the mouth to the moment of irreversible starvation (defined as the time after which at least 50% of the larvae are unable to begin feeding even when offered food) depends mainly on temperature and larval length (Yúfera and Darías 2007). Nevertheless, some anatomical structures may be affected before reaching this point (e.g., Theilacker 1987; Yin and Blaxter 1987; Yúfera et al. 1993b; Dou et al. 2002; Gisbert et al. 2004), influencing swimming and predation capacity (Skiftesvik 1992). Once feeding starts, suboptimal feeding induces developmental delays, with a decline in the growth rate and consequent effect on TL and swimming capacity. Very low prey density may drastically reduce prey encounters. Therefore, swimming time increases at low prey density (Munk and Kiørboe 1985) in an effort to maintain the ingestion threshold, but in total absence of food the initial increase is followed by a strong decrease to preserve energy. On the other hand, nutritional deficiencies, particularly in essential fatty acids, clearly affect foraging behavior. Bell et al. (1995) reported reduced vision at low light intensities in *Clupea harengus* juveniles under dietary deficiency of docosahexanoic acid. O'Brien-MacDonald et al. (2006) found that the swimming activity and attacks on prey by *Gadus morhua* larvae were higher when fed on rotifers enriched with high lipid levels. These authors argued that the lipid enrichment provided more energy for swimming and hunting. Similarly, Bransden et al. (2005) found that *Latris lineata* larvae fed low dietary 22:6n-3 displayed erratic swimming behavior as compared with those fed on higher 22:6n-3 levels.

10.6 Concluding remarks: applying feeding behavior knowledge to larval rearing technology

In spite of the species-specific differences in development, a series of general rules and trends can be considered. During larval development and growth, the mouth gape increases, allowing the consumption of relatively larger prey; swimming capacity also improves, and consequently, predatory ability, allowing more efficient energy expenditure in hunting. Thus, the growing larvae progress from passive feeders depending on prey encounter opportunity to active predators. Contrarily, other aspects may be more specific and related to the habitat, as the timing of the behavior ontogeny, preferred foraging zones within the water column, optimal illumination level, preferred hours of the day for foraging, preferred prey, and weight-specific daily ration. In the field, the particular feeding behavior of each species is adapted to an efficient exploitation of the resources offered by a given habitat. In farmed species, the environmental conditions and the feeding resources available for the larvae are different, with high larval density and low prey variability being two primary characteristics. Advances in larval rearing technology aim to satisfy the environmental and feeding requisites, trying to supply appropriate and abundant food. Some of these feeding aspects are evident and relatively easy to elucidate. Thus, searching for appropriate prey of adequate size has been a priority from the first studies on rearing fish larvae. The usual prey sequencing established during the 1970s and 1980s based on rotifers of different sizes and *Artemia* nauplii and metanauplii cover a wide range of prey size needs for many species during a large part of larval development, although not in all cases. Other prey have been tested to fit size preferences, with different results, including bivalve veligers, copepod nauplii and adults, ciliates, other rotifers, and egg and larvae of other invertebrates. The search focuses mainly on prey below 100 µm in order to advance the initiation of feeding in those species with very small mouth gapes (Wullur et al. 2009). Nevertheless, it is difficult to find other live prey organisms able to supply regularly abundant prey numbers to support large-scale production of larvae. The daily amount of live prey to be supplied and the minimum levels to prevent starvation or developmental delay have also been considered a primary issue in each new successfully reared species. Overall, the current commonly used live prey, *Brachionus* spp. and *Artemia* spp., meet well the feeding behavior of larvae (Conceição et al. 2010), except in the case of very small larvae at mouth-opening. Another relevant attribute of live prey is the nutritional quality and the extent to which it satisfies the dietary requirements of developing larvae, and this aspect is covered in other chapters of this book.

One of the objectives of the research on microdiets for replacing live prey during the larval stage is to overcome some of these constraints and provide immediate food availability at any size range required. The food particles need therefore to imitate the characteristics of prey organisms in order to be detected, ingested, and accepted at the same level. Particle size, floatability, structure stability after immersion in water, and digestibility by early larvae with poorly developed gut have been features taken into account when designing food microparticles for aquatic organisms (Langdon 2003; Yúfera et al. 2005). Most of these conditions have been achieved and fish larvae may ingest and digest microdiets from first feeding. The density of the particle mass is perhaps one feature that needs special attention. The difference in water content between live prey (>80%) and inert microparticles (<70%) has been considered one factor preventing good results with microdiets because it affects their digestibility as well as the sinking rate. As in the case of live prey, the dietary formulation of microdiets is of primary nutritional importance for larval growth. But in addition, an appropriate formulation, together with an adequate particle physical structure, is needed to allow the larvae to smell the amino acids leaked in order to identify the particle as a tempting meal (Kolkovski et al. 1997; Yúfera et al. 2002; see Chapters 12–13 for additional discussion of microdiets).

On the other hand, there are some environmental aspects that affect feeding performance on both live prey and microdiets. The influence of illumination period and intensity of ingestion has been studied in different species. While daily consumption seems to increase with increased illumination time in most species, the more favorable illumination intensity and visual feeding threshold vary among species (Ronzani Cerqueira and Chatain 1991; Downing and Litvak 2001; Puvanendran and Brown 2002; Peña et al. 2004; Carton 2005). A correlated factor with illumination intensity is the turbidity induced by the green water technique used in larval rearing protocols for many species. The effect of microalgae addition is variable, being beneficial in some species and indifferent or even prejudicial in others (Cobcroft et al. 2001; Papandroulakis et al. 2002; Carton 2005). In addition to variations in experimental protocols and illumination intensity ranges, the observed differences reflect species-specific responses that may also be changing during development. Appropriate illumination conditions for start feeding are not necessarily the optimum for the entire larval period. Circadian feeding rhythms and the influence of illumination and turbidity during the early stages still needs to be examined in detail in many farmed species. Likewise, whether continuous or discrete feeding as well as food dosage can affect feeding performance are factors that are insufficiently studied. These issues are particularly important for determining the minimum residence time of microdiets in water before being ingested. A good understanding of all these aspects is necessary to develop more efficient feeding protocols for larval fish.

Acknowledgments

I am most grateful to Joan Holt for her help in the preparation of this chapter. Acknowledgments are also due to the editorial offices of the different journal publishers for authorizing the reproduction of figures used in this chapter. This study was supported by the Plan Nacional de Investigación Desarrollo e Innovación, Spain, with the support of the European Regional Development Fund (projects Aquagenomics CSD2007-00002 and AGL2007-64450-C02-01). The publication benefits from participation in the European Cooperation in Science and Technology Program, LarvaNet COST Action FA0801.

Literature cited

Adverlund, M., and Takemura, A. 2006. The importance of chemical environmental clues for juveniles *Lethrinus nebulosus* Forsskål (Lethrinidae, Teleostei) when settling into their first benthic habitat. *Journal of Experimental Marine Biology and Ecology* 338:112–122.

Appelbaum, S., and Kamler, E. 2000. Survival, growth, metabolism and behaviour of *Clarias gariepinus* (Butchell 1822) early stages under different light conditions. *Aquacultural Engineering* 22:269–287.

Arnold, G.P., and Holford, B.P. 1990. The reactive perceptive field of larval plaice (*Pleuronectes platessa*): a three dimensional analysis of visual feeding. *Rapport et Procès-Verbaux des Réunions du Conseil International pour l'Exploration de la Mer* 191:474.

Bailey, K.M., Canino, M.F., Napp, J.M., et al. 1995. Contrasting years of prey levels and mortality of larval walleye Pollock *Theragra chalcogramma* in the western Gulf of Alaska. *Marine Ecology Progress Series* 119:11–23.

Balon, E.K. 1986. Types of feeding in the ontogeny of fishes and the life-history model. *Environmental Biology of Fishes* 16:11–24.

Barahona-Fernandes, M.H., and Conan, G. 1981. Daily food intake of reared larvae of the European sea bass (*Dicentrarchus labrax* L.). Statistical analysis and modelling. *Archivos do Museu Bocage. Serie A* 1:29–43.

Baras, E., Ndao, M., Maxi, M.Y.J., et al. 2000. Sibling cannibalism in dorada under experimental conditions. I. Ontogeny, dynamics, bioenergetics of cannibalism and prey size selectivity. *Journal of Fish Biology* 57:1001–1020.

Baskerville-Bridges, B., and Kling, L.J. 2000. Development and evaluation of microparticulate diets for early weaning of Atlantic cod *Gadus morhua* larvae. *Aquaculture Nutrition* 6:171–182.

Batty, R.S., and Blaxter, J.H.S. 1992. The effect of temperature on the burst swimming performance of fish larvae. *Journal of Experimental Biology* 170:187–201.

Bell, M.V., Batty, R.S., Dick, J.R., et al. 1995. Dietary deficiency of docosahexanoic acid impairs vision at low light intensities in juvenile herring (*Clupea harengus* L.). *Lipids* 30: 443–449.

Bengtson, D.A. 2003. Status of marine aquaculture in relation to live prey: past, present and future. In: Stottrup, J.G., and McEvoy, L.A. (eds.) *Live Feeds in Marine Aquaculture*. Blackwell Publishing, Oxford, pp. 1–13.

Bengtson, D.A., Gleason, T., and Hossain, M.A. 1999. Consumption rates of summer flounder larvae on rotifer and brine shrimp prey during larval rearing. *North American Journal of Aquaculture* 61:243–245.

Blaxter, J.H.S. 1986. Development of sense organs and behaviour in teleost larvae with special reference to feeding and predation avoidance. *Transactions of the American Fisheries Society* 115:98–114.

Bochdansky, A.B., Grønkjær, P., Pepin, P., et al. 2008. Food limitation in larval fish: ontogenic variation in feeding scope and its potential effect on survival. *Marine Ecology Progress Series* 367:239–248.

Boehlert, G.W., and Yoklavich, M.M. 1984. Carbon assimilation as a function of ingestion rate in larval pacific herring, *Clupea harengus pallasi* Valenciennes. *Journal of Experimental Marine Biology and Ecology* 79:251–262.

Boeuf, G., and Le Bail, P.Y. 1999. Does light have an influence on fish growth? *Aquaculture* 177:129–152.

Boglione, C., Giganti, M., Selmo, C., et al. 2003. Morphoecology in larval fin-fish: a new candidate species for aquaculture, *Diplodus puntazzo* (Sparidae). *Aquaculture International* 11:17–41.

Bransden, M.P., Cobcroft, J.M., Battaglene, S.C., et al. 2005. Dietary 22:6n-3 alters gut and liver structure and behaviour in larval striped trumpeter (*Latris lineata*). *Aquaculture* 248:275–285.

Browman, H.I., and O'Brien, W.J. 1992. Foraging and prey search behaviour of golden shiner (*Notemigonus crysoleucas*) larvae. *Canadian Journal of Fisheries and Aquatic Sciences* 49:913–819.

Busch, A. 1996. Transition from endogenous to exogenous nutrition: larval size parameters determining the start of external feeding and size of prey ingested by Ruegen spring herring *Clupea harengus*. *Marine Ecology Progress Series* 130:39–46.

Cañavate, J.P., Zerolo, R., and Fernández-Díaz, C. 2006. Feeding and development of Senegal sole (*Solea senegalensis*) larvae reared in different photoperiods. *Aquaculture* 258:368–377.

Carton, A.G. 2005. The impact of light intensity and algal-induced turbidity on first-feeding *Seriola lalandi* larvae. *Aquaculture Research* 36:1588–1594.

Chesney, E.J. Jr. 1989. Estimating the food requirements of striped bass larvae *Morone saxatilis*: effect of light, turbidity and turbulence. *Marine Ecology Progress Series* 53:191–200.

Chesney, E.J. 2008. Foraging behaviour of bay anchovy larvae, *Anchoa mitchilli*. *Journal of Experimental Marine Biology and Ecology* 362:117–124.

Chitty, N. 1981. Behavioral observations of feeding larvae of bay anchovy, *Anchoa mitchilli*, and bigeye anchovy, *Anchoa lamprotaenia*. *Rapports de Procès-Verbaux des Réunions du Conseil International pour l'Exploration de la Mer* 178:320–321.

Clark, D.L., Leis, J.M., Hay, A.C., et al. 2005. Swimming ontogeny of larvae of four temperate marine fishes. *Marine Ecology Progress Series* 292:287–300.

Cobcroft, J.M., and Pankhurst, P.M. 2006. Visual field in cultured striped trumpeter *Latris lineata* (Teleostei) feeding in rotifer prey. *Marine and Freshwater Behaviour and Physiology* 39:193–208.

Cobcroft, J.M., Pankhurst, P.M., Hart, P.R., et al. 2001. Effect of light intensity and algal-induced turbidity on feeding behaviour of larval striped trumpeter. *Journal of Fish Biology* 59:1181–1197.

Conceição, L.E.C., Dersjant-Li, Y., and Verret, J.A.J. 1998. Cost of growth in larval and juvenile African catfish (*Clarias gariepinus*) in relation to growth rate food intake and oxygen consumption. *Aquaculture* 161:95–106.

Conceição, L.E.C., Yúfera, M., Makridis, P., et al. 2010. Live feeds for early stages of fish rearing. *Aquaculture Research* 41:613–640.

Cox, E.S., and Pankhurst, P.M. 2000. Feeding behaviour of greenback flounder larvae, *Rhombosolea tapirina* (Günther) with different exposure histories to live prey. *Aquaculture* 183:285–297.

Cunha, I., and Planas, M. 1999. Optimal prey size for early turbot larvae (*Scophthalmus maximus* L.) based on mouth and ingested prey size. *Aquaculture* 175:103–110.

Day, O.J., Jones, D.A., and Howell, B.R. 1996. Food consumption, growth and respiration of sole, *Solea solea* (L.), during early ontogeny in a hatchery environment. *Aquaculture Research* 27:831–839.

Dempsey, C.H. 1978. Chemical stimuli as a factor in feeding and intraspecific behaviour of herring larvae. *Journal of the Marine Biological Association of the United Kingdom* 58:739–747.

Doi, M., Ohno, A., Kohno, H., et al. 1997. Development of feeding ability in red snapper *Lutjanus argentimaculatus* early larvae. *Fisheries Science* 63:845–853.

Dou, S., Seikai, T., and Tsukamoto, K. 2000. Cannibalism in Japanese flounder juveniles, *Paralichthys olivaceus*, reared under controlled conditions. *Aquaculture* 182:149–159.

Dou, S., Masuda, R., Tanaka, M., et al. 2002. Feeding resumption, morphological changes and mortality during starvation of Japanese flounder larvae. *Journal of Fish Biology* 60:1363–1380.

Døving, K.B., and Kasumyan, A.O. 2008. Chemoreception. In: Finn, R.N., and Kapoor, B.G. (eds.) *Fish Larval Physiology*. Science Publishers, Enfield, NH, pp. 331–394.

Downing, G., and Litvak, M.K. 2001. The effect of light intensity and spectrum on the incidence of first feeding by larval haddock. *Journal of Fish Biology* 59:1566–1578.

Dunbrack, R., Giguère, L.A.S., and Pierre, J.F. 2009. A comparison of gut evacuation models for larval mackerel (*Scomber scombrus*) using serial photography. *Journal of Fish Biology* 74:906–920.

Evans, B.I., and Browman, H.I. 2004. Variation in the development of fish retina. *American Fisheries Society Symposium* 40:145–166.

Fernández-Díaz, C., Pascual, E., and Yúfera, M. 1994. Feeding behaviour and prey size selection of gilthead seabream, *Sparus aurata* L., larvae fed on inert and live food. *Marine Biology* 118:323–328.

Fielder, D.S., Bardsley, W.J., Allan, G.L., et al. 2002. Effect of photoperiod on growth and survival of snapper *Pagrus auratus* larvae. *Aquaculture* 211:137–152.

Fontagné, S., Robin, J., Corraze, G., et al. 2000. Growth and survival of European sea bass (*Dicentrarchus labrax*) larvae fed from first feeding on compound diets containing medium-chain triacylglycerols. *Aquaculture* 190:261–271.

Fuiman, L.A., Smith, M.E., and Malley, V.M. 1999. Ontogeny of the routine swimming speed and startle responses in red drum, with a comparison of responses to acoustic and visual stimuli. *Journal of Fish Biology* 55A:215–226.

Fukuhara, O. 1985. Functional morphology and behaviour of early life stages of red sea bream.

Bulletin of the Japanese Society of Scientific Fisheries 51:731–743.

Geffen, A.J., van der Veer, H.W., and Nash, R.D.M. 2007. The cost of metamorphosis in flatfishes. *Journal of Sea Research* 68:35–45.

Getchis, T.S., and Bengtson, D.A. 2006. Food consumption and absorption efficiency by new settled summer flounder, *Paralichthys dentatus* (Linnaeus 1766). *Aquaculture* 257:241–248.

Gisbert, E., Conklin, D.B., and Piedrahita, R.H. 2004. Effects of delayed first feeding on the nutritional condition and mortality of California halibut larvae. *Journal of Fish Biology* 64:116–132.

Hamasaki, K., Tsurouka, K., Teruya, K., et al. 2009. Feeding habits of hatchery-reared larvae of greater amberjack *Seriola dumerili*. *Aquaculture* 288:216–225.

Helvik, J.V., Drivenes, Ø., Harboe, T., et al. 2001. Topography of different photoreceptor cell types in the larval retina of Atlantic halibut (*Hippoglossus hippoglossus*). *Journal of Experimental Biology* 204:2553–2559.

Hetch, T., and Pienaar, A.G. 1993. A review of cannibalism and its implications in fish larviculture. *Journal of the World Aquaculture Society* 24:246–261.

Higgs, D.M., and Fuiman, L.A. 1996. Ontogeny of visual and mechanosensory structure and function in Atlantic menhaden *Brevoortia tyrannus*. *Journal of Experimental Biology* 199:2619–2629.

Holt, G.J., and Holt, S.A. 2000. Vertical distribution and the role of physical processes in the feeding dynamics of two larval sciaenids *Sciaenops ocellatus* and *Cynoscion nebulosus*. *Marine Ecology Progress Series* 193:181–190.

Houde, E.D., and Schekter, R.C. 1980. Feeding by marine fish larvae: developmental and functional responses. *Environmental Biology of Fishes* 5:315–334.

Houde, E.D., and Schekter, R.C. 1983. Oxygen uptake and comparative energetics among eggs and larvae of three subtropical marine fishes. *Marine Biology* 72:283–293.

Houde, E.D., and Zastrow, C.E. 1993. Ecosystem- and taxon-specific dynamic and energetics properties of larval fish assemblages. *Bulletin of Marine Science* 53:290–335.

Hunt von Herbing, I., and Gallager, S.M. 2000. Foraging behaviour in early Atlantic cod larvae (*Gadus morhua*) feeding on a protozoa (*Balanion* sp.) and a copepod nauplius (*Pseudodiaptomus* sp.). *Marine Biology* 136:591–602.

Hunt von Herbing, I., Gallager, S.M., and Halteman, W. 2001. Metabolic cost of pursuit and attack in early larval Atlantic cod. *Marine Ecology Progress Series* 216:201–212.

Hunter, J.R. 1981. Feeding ecology and predation of marine fish larvae. In: Lasker, R. (ed.) *Marine Fish Larvae: Morphology, Ecology, and Relation to Fisheries*. Washington Sea Grant Program, Seattle, WA, pp. 33–77.

Iizawa, M. 1984. Corrélations entre la densité de proies et la quantité consommée par les larves du loup *Dicentrarchus labrax* (L.). In: Barnabé, G., and Billard, R. (eds.) *L'Aquaculture du Bar et des Sparidés*. INRA Publishers, Paris, pp. 161–173.

Iwai, T. 1980. Sensory anatomy and feeding of fish larvae. In: Bardach, J.E., Magnuson, J.J., May, R.C., et al. (eds.) *Fish Behaviour and Its Use in the Capture and Culture of Fishes*. International Center for Living Aquatic Resources Management, Manila, pp. 124–145.

James, N.C., Cowley, P.D., Whitfield, A.K., et al. 2008. Choice chamber experiments to test the attraction of postflexion *Rhabdosargus holubi* larvae to water of estuarine and riverine origin. *Estuarine and Coastal Shelf Science* 77: 143–149.

Jones, W.R., and Janssen, J. 1992. Lateral line development and feeding behaviour in mottled sculpin, *Cottus bairdi* (Scorpaeniformes: Cottidae). *Copeia* 1992:485–491.

Kasumyan, A.O., Ryg, M., and Døving, K.B. 1998. Effect of amino acids on the swimming activity of newly hatched turbot larvae (*Scophthalmus maximus*). *Marine Biology* 131:189–194.

Katavić, I., Jug-Dujaković, J., and Glamuzina, B. 1979. Cannibalism as a factor affecting the survival of intensively cultured sea bass (*Dicentrarchus labrax*) fingerlings. *Aquaculture* 77:135–143.

Kawamura, G., Tsuda, R., Kumai, H., et al. 1984. The visual cell morphology of *Pagrus major* and its adaptive changes with shift from pelagic to benthic habitats. *Bulletin of the Japanese Society of Scientific Fisheries* 50:1975–1980.

Kawamura, G., Mori, H., and Kuwahara, A. 1989. Comparison of sense organ development in wild and reared flounder *Paralichthys olivaceus* larvae. *Bulletin of the Japanese Society of Scientific Fisheries* 55:2079–2083.

Kawamura, G., Masuma, S., Tezuka, N., et al. 2003. Morphogenesis in sense organs in the bluefin tuna *Thunnus orientalis*. In: Browman, H.J., and Skiftesvik, A.B. (eds.) *The Big Fish Bang*. Proceedings of the 26th Annual Larval Fish Conference. Institute of Marine Research, Bergen, pp. 123–135.

Kelly, S.P., Larsen, S.D., Collins, P.M., et al. 2000. Quantitation of inert feed ingestion in larval silver sea bream (*Sparus sarba*) using autofluorescence of alginate-based microparticulate diets. *Fish Physiology and Biochemistry* 22:109–117.

Kentouri, M., and Divanach, P. 1982. Differences et similitudes dans la genese des comportements locomoteur et trophique des stades prelarvaires de *Sparus auratus*, *Diplodus vulgaris* et *Diplodus sargus*. *Aquaculture* 27:355–376.

Kiørboe, T., Munk, P., and Støttrup, J.G. 1985. First feeding by larval herring *Clupea harengus* L. *Dana* 5:95–107.

Knutsen, J.A. 1992. Feeding behaviour of North Sea turbot (*Scophthalmus maximus*) and Dover sole (*Solea solea*) larvae elicited by chemical stimuli. *Marine Biology* 113:543–548.

Kolkovski, S., Koven, B., and Tandler, A. 1997. The mode of action of *Artemia* in enhancing utilization of microdiet by gilthead seabream *Sparus aurata* larvae. *Aquaculture* 155: 193–205.

Koumoundouros, G., Divanach, P., and Kentouri, M. 1999. Ontogeny and allometric plasticity of *Dentex dentex* (Ostheichtyes: Sparidae) in rearing conditions. *Marine Biology* 135:561–572.

Koumoundouros, G., Ashton, C., Sfakianakis, D.G., et al. 2009. Thermally induced phenotypic plasticity of swimming performance in European seabass *Dicentrarchus labrax* juveniles. *Journal of Fish Biology* 74:1309–1322.

Kvenseth, A.M., Pittman, K., and Helvik, J.V. 1996. Eye development in Atlantic halibut (*Hippoglossus hippoglossus*): differentiation and development of the retina from early yolk sac stages through metamorphosis. *Canadian Journal of Fisheries and Aquatic Sciences* 53:2524–2532.

Lagardère, F., Amara, R., and Joussard, L. 1999. Vertical distribution and feeding activity of metamorphosing sole, *Solea solea*, before immigration to the Bay of Vilaine nursery (Northern Bay of Biscay, France). *Environmental Biology of Fishes* 256:213–228.

Langdon, C. 2003. Microparticle types for delivering nutrients to marine fish larvae. *Aquaculture* 227:259–275.

Lara, M.R. 2008. Development of the nasal olfactory organs in the larvae, settlement stages and some adults of 14 species of Caribbean reef fishes (Labridae, Scaridae, Pomacentridae). *Marine Biology* 154:51–64.

Laurence, G.C. 1977. A bioenergetic model for the analysis of feeding and survival potential of winter flounder larvae (*Pseudopleuronectes americanus*) during the period from hatching to metamorphosis. *Fishery Bulletin* 75:529–546.

Lika, K., and Papandroulakis, N. 2005. Modeling feeding processes: a test of a new model for seabream (*Sparus aurata*) larvae. *Canadian Journal of Fisheries and Aquatic Sciences* 62:425–435.

Loew, E.R., and Wahl, C.M. 2008. Photoreception. In: Finn, R.N., and Kapoor, B.G. (eds.) *Fish Larval Physiology*. Science Publishers, Enfield, NH, pp. 395–424.

Ma, A., Liu, X., Xu, Y., et al. 2006. Feeding rhythm and growth of the tongue sole, *Cynoglossus semilaevis* Günther, during its early life stages. *Aquaculture Research* 37: 586–593.

MacIntosh, K.E., and Duston, J. 2007. Effects of light intensity and eye development on prey capture by larval striped bass *Morone saxatilis*. *Journal of Fish Biology* 71:725–736.

MacKenzie, B.R., Leggett, W.C., and Peters, R.H. 1990. Estimating larval fish ingestion rates: can laboratory values by reliably extrapolated to the wild? *Marine Ecology Progress Series* 67:209–225.

MacKenzie, B.R., Ueberschär, B., Basford, D., et al. 1999. Diel variability of feeding activity in haddock (*Melanogrammus aeglefinus*) larvae in the East Shetland area, North Sea. *Marine Biology* 135:361–368.

McLaren, I.A., Avendaño, P., Taggart, C.T., et al. 1997. Feeding by larval cod in different water-masses on Western Bank, Scotian Shelf. *Fisheries Oceanography* 6:260–265.

Monk, J., Puvanendran, V., and Brown, J.A. 2006. Do different light regimes affect the foraging behaviour, growth and survival of larval cod (*Gadus morhua* L.). *Aquaculture* 257:287–293.

Mukai, Y. 2006. Role of the free neuromasts in the larval feeding of willow shiner *Gnathopogon elongates caulescens* Teleostei, Cyprinidae. *Fisheries Science* 72:705–709.

Mukai, Y., Tuzan, A.D., Lim, L.S., et al. 2008. Development of sensory organ in larvae of African catfish *Clarias gariepinus*. *Journal of Fish Biology* 73:1648–1661.

Munk, P. 1992. Foraging behaviour and prey size spectra of larval herring *Clupea harengus*. *Marine Ecology Progress Series* 80:149–158.

Munk, P., and Kiørboe, T. 1985. Feeding behaviour and swimming activity of larval herring (*Clupea harengus*) in relation to density of copepod nauplii. *Marine Ecology Progress Series* 24:15–21.

Næss, T., Germain-Henry, M., and Naas, K.E. 1995. First feeding of Atlantic halibut (*Hippoglossus hippoglossus*) using different combinations of *Artemia* and wild zooplankton. *Aquaculture* 130:235–250.

Nhu, V.C., Dierckens, K., Nguyen, T.H., et al. 2009. Can umbrella-stage *Artemia franciscana* substitute enriched rotifers for Cobia (*Rachycentron canadum*) fish larvae. *Aquaculture* 289:64–69.

Noakes, D.L.G., and Godin, J.G.J. 1988. Ontogeny of behaviour and concurrent development changes in sensory systems in teleost fishes. In: Hoar, W.S., and Randall, D.J. (eds.) *Fish Physiology, XI B. Viviparity and Posthatching Juveniles*. Academic Press, New York, pp. 345–395.

O'Brien, W.J., Evans, B.I., and Browman, H.I. 1989. Flexible search tactics end efficient foraging in saltatory searching animals. *Oecologia* 80:100–110.

O'Brien-MacDonald, K., Brown, J.A., and Parrish, C.C. 2006. Growth, behaviour and digestive enzyme activity in larval Atlantic cod, *Gadus morhua*, in relation to rotifer lipids. *ICES Journal of Marine Science* 63:275–284.

O'Connell, C.P. 1981. Development of organ systems in the northern anchovy *Engraulis mordax* and other teleosts. *American Zoologist* 21:429–446.

Olsen, A.I., Attramadal, I., Reitan, K.I., et al. 2000. Food selection and digestion characteristics of Atlantic halibut (*Hippoglossus hippoglossus*) larvae fed cultivated prey organism. *Aquaculture* 181:293–310.

Oozeki, Y., Hagiwara, A., Eda, H., et al. 1992. Development of food selectivity of striped mullet *Mugil cephalus* during larval stage. *Bulletin of the Japanese Society of Scientific Fisheries* 58:1381.

Osse, J.W.M., and van den Boogaart, J.G.M. 1999. Dynamic morphology of fish larvae, structural applications of friction forces in swimming, feeding and ventilation. *Journal of Fish Biology* 55A:156–174.

Østergaard, P., Munk, P., and Janekarn, V. 2005. Contrasting feeding patterns among species of fish larvae from the tropical Andaman Sea. *Marine Biology* 146:595–606.

Pankhurst, P.M. 2008. Mechanoreception. In: Finn, R.N., and Kapoor, B.G. (eds.) *Fish Larval Physiology*. Science Publishers, Enfield, NH, pp. 305–329.

Pankhurst, P.M., and Eagar, R. 1996. Changes in visual morphology through the life history stages of the New Zealand snapper, *Pagrus auratus*. *New Zealand Journal of Marine and Freshwater Research* 30:79–90.

Papandroulakis, N., Divanach, P., and Kentouri, M. 2002. Enhanced biological performance of intensive sea bream (*Sparus aurata*) larviculture in the presence of phytoplankton with long photophase. *Aquaculture* 204:45–63.

Papandroulakis, N., Kentouri, M., and Divanach, P. 2004. Biological performance of red porgy (*Pagrus pagrus*) larvae under intensive rearing conditions with the use of an automated feeding system. *Aquaculture International* 12:191–203.

Parazo, M.M., Avila, E.M., and Reyes, D.M. Jr. 1991. Size- and weight-dependent cannibalism in hatchery-bred sea bass (*Lates calcarifer* Bloch). *Journal of Applied Ichthyology* 7:1–7.

Parra, G., and Yúfera, M. 2000. Feeding, physiology and growth response in first-feeding gilthead seabream (*Sparus aurata* L.) larvae in relation to prey density. *Journal of Experimental Marine Biology and Ecology* 243:1–15.

Parra, G., and Yúfera, M. 2001. Comparative energetics during early development of two marine fish species, *Solea senegalensis* (Kaup) and *Sparus aurata* (L.). *Journal of Experimental Biology* 204:2175–2183.

Peña, R., Dumas, S., Saldivar-Lucio, S., et al. 2004. The effects of light intensity on the first feeding of the spotted sand bass *Paralabrax maculatofasciatus* (Steindachner) larvae. *Aquaculture Research* 35:345–349.

Polo, A., Yúfera, M., and Pascual, E. 1992. Feeding and growth of gilthead seabream (*Sparus aurata*, L.) larvae in relation to the size of the rotifer strain used as food. *Aquaculture* 103:45–54.

Puvanendran, V., and Brown, J.A. 1998. Effect of light intensity on the foraging and growth of Atlantic cod larvae: interpopulation difference. *Marine Ecology Progress Series* 167:207–214.

Puvanendran, V., and Brown, J.A. 2002. Foraging growth and survival of Atlantic cod larvae reared in different light intensities and photoperiods. *Aquaculture* 214:131–151.

Puvanendran, V., Salies, K., Laurel, B., et al. 2004. Size-dependent foraging on larval Atlantic cod (*Gadus morhua*). *Canadian Journal of Zoology* 82:1380–1389.

Reiss, C., McLaren, I., Avendaño, P., et al. 2004. Feeding ecology of silver hake larvae on the Western Bank, Scotian Shelf and comparison with Atlantic cod. *Journal of Fish Biology* 66:703–702.

Ronzani Cerqueira, V., and Chatain, B. 1991. Photoperiodic effects on the growth and feeding rhythm of the European seabass, *Dicentrarchus labrax*, larvae under intensive rearing. *European Aquaculture Society, Special Publication* 15:304–306.

Roo, F.J., Socorro, J., Izquierdo, M.S., et al. 1999. Development of red porgy *Pagrus pagrus* visual system in relation with changes in the digestive tract and larval feeding habits. *Aquaculture* 179:499–512.

Rowlands, W.L., Dickey-Collas, M., Geffen, A.J., et al. 2006. Gape morphology of cod *Gadus morhua*, L., haddock *Melanogrammus aeglefinus* (L.) and whiting *Merlangius merlangus* (L.) through metamorphosis from larvae to juveniles in the Western Irish Sea. *Journal of Fish Biology* 69:1379–1395.

Russo, T., Costa, C., and Cataudella, S. 2007. Correspondence between shape and feeding habit changes throughout ontogeny of gilthead sea bream *Sparus aurata* L., 1785. *Journal of Fish Biology* 71:629–656.

Russo, T., Boglione, C., De Marzi, P., et al. 2009. Feeding preferences of the dusky grouper (*Epinephelus marginatus*, Lowe 1834) larvae reared in semi-intensive conditions: a contribution addressing the domestication of this species. *Aquaculture* 289:289–296.

Sabate, F. de la S., Sakakura, Y., Shiozaki, M., et al. 2009. Onset and development of aggressive behaviour in the early stages of the sevenband grouper *Epinephelus septemfasciatus*. *Aquaculture* 290:97–103.

Sánchez-Amaya, M.I., Ortiz-Delgado, J.B., García-López, A., et al. 2007. Larval ontogeny of redbanded seabream *Pagrus auriga* Valenciennes, 1843 with special reference to digestive system. A histological and histochemical approach. *Aquaculture* 263:259–279.

Shand, J., Hart, N.S., Thomas, N., et al. 2002. Developmental changes in the cone visual pigments of black bream *Acanthopagrus butcheri*. *Journal of Experimental Biology* 205:3661–3667.

Shaw, G.W., Pankhurst, P.M., and Battaglene, S.C. 2006. Effects of turbidity, prey density and culture history on prey consumption by greenback flounder *Rhombosolea tapirina* larvae. *Aquaculture* 253:447–460.

Shirota, A. 1970. Studies on the mouth size of fish larvae. *Bulletin of the Japanese Society of Scientific Fisheries* 36:353–368.

Shoji, J., Maehara, T., Aoyama, M., et al. 2001. Daily ration of Japanese mackerel *Scomberomorus nophonius* larvae. *Fisheries Science* 67:238–245.

Simonsen, C.S., Munk, P., Folkvord, A., et al. 2006. Feeding ecology of Greenland halibut and sandeel larvae off West Greenland. *Marine Biology* 149:937–952.

Skiftesvik, A.B. 1992. Changes in behaviour at the onset of exogenous feeding in marine fish larvae. *Canadian Journal of Fisheries Aquatic Sciences* 49:1570–1572.

Stepien, W.P. Jr. 1976. Feeding of laboratory-reared larvae of the sea bream *Archosargus rhomboidalis* (Sparidae). *Marine Biology* 38:1–16.

Sumida, B.Y., and Moser, H.G. 1980. Food and feeding of Pacific hake larvae, *Merluccius productus*, off Southern California and Northern Baja California. *California*

Cooperative Ocean Fisheries Investigation Report 21:161–166.

Tanaka, Y., Mukai, Y., Takii, K., et al. 1991. Chemoreception and vertical movement in planktonic yolk-sac larvae of red sea bream *Pagrus major*. *Journal of Applied Ichthyology* 7:129–135.

Tandler, A., and Mason, C. 1984. The use of 14C labelled rotifers (*Brachionus plicatilis*) in the larvae of gilthead seabream (*Sparus aurata*): measurements of the effect of rotifer concentration, the lighting regime and seabream larval age on their rate of rotifer ingestion. *European Aquaculture Society, Special Publication* 8:241–259.

Teng, S.-K., El-Zahr, C., Al-Abdul-Elah, K., et al. 1999. Pilot-scale spawning and fry production of blue-fin porgy, *Sparidentex hasta* (Valenciennes), in Kuwait. *Aquaculture* 178:24–41.

Theilacker, G.H. 1987. Feeding ecology and growth energetics of larval northern anchovy, *Engraulis mordax*. *Fishery Bulletin* 85:213–228.

Villamizar, N., García-Alcazar, A., and Sánchez-Vázquez, F.J. 2009. Effect of light spectrum and photoperiod on the growth development and survival of European sea bass (*Dicentrarchus labrax*) larvae. *Aquaculture* 292:80–86.

Webb, P.W., and Weihs, D. 1986. Functional locomotor morphology of early life history stages of fishes. *Transactions of the American Fisheries Society* 115:115–127.

Wuenschel, M.J., and Werner, R.G. 2004. Consumption and gut evacuation rate of laboratory-reared spotted seatrout (Sciaenidae) larvae and juveniles. *Journal of Fish Biology* 65:723–743.

Wullur, S., Sakakura, Y., and Hagiwara, A. 2009. The minute monogont rotifer *Proales similis* de Beauchamp: culture and feeding to small mouth marine fish. *Aquaculture* 293:62–67.

Yamashita, Y., and Bailey, K.M. 1989. A laboratory study of the bioenergetics of larval walleye pollock, *Theragra chalcogramma*. *Fishery Bulletin* 87:525–536.

Yin, M.C., and Blaxter, J.H.S. 1987. Feeding ability and survival during starvation of marine fish larvae reared in the laboratory. *Journal of Experimental Marine Biology and Ecology* 105:73–83.

Yoseda, K., Yamamoto, K., Asami, K., et al. 2008. Influence of light intensity on feeding, growth and early survival of leopard coral grouper (*Plectropomus leopardus*) larvae under mass-scale rearing conditions. *Aquaculture* 279:56–62.

Yúfera, M., and Darías, M.J. 2007. The onset of feeding in marine fish larvae. 2007. *Aquaculture* 268:53–63.

Yúfera, M., and Pascual, E. 1984. La producción de organismos zooplanctónicos para la alimentación larvaria en la acuicultura marina. *Informes Técnicos del Instituto de Investigaciones Pesqueras* 119:1–27.

Yúfera, M., Polo, A., and Pascual, E. 1993a. Changes in chemical composition and biomass during the transition from endogenous to exogenous feeding of *Sparus aurata* L. (Pisces, Sparidae) larvae reared in the laboratory. *Journal of Experimental Marine Biology and Ecology* 167:149–161.

Yúfera, M., Pascual, E., Polo, A., et al. 1993b. Effect of starvation on the feeding ability of gilthead seabream (*Sparus aurata* L.) larvae at first feeding. *Journal of Experimental Marine Biology and Ecology* 169:259–272.

Yúfera, M., Fernández-Díaz, C., and Pascual, E. 1995. Feeding rates of gilthead seabream, *Sparus aurata*, larvae on microcapsules. *Aquaculture* 134:257–268.

Yúfera, M., Kolkovski, S., Fernández-Diaz, C., et al. 2002. Free amino acid leaching from a protein-walled microencapsulated diet for fish larvae. *Aquaculture* 214:273–287.

Yúfera, M., Fernández-Díaz, C., Vidaurreta, A., et al. 2004. Gastrointestinal pH and development of the acid digestion in larvae and early juveniles of *Sparus aurata* (Pisces: Teleostei). *Marine Biology* 144:863–869.

Yúfera, M., Fernández-Díaz, C., and Pascual, E. 2005. Food microparticles for larval fish prepared by internal gelation. *Aquaculture* 245:253–262.

Chapter 11

Live feeds

Gunvor Øie, Kjell Inge Reitan, Jan Ove Evjemo, Josianne Støttrup, and Yngvar Olsen

11.1 Introduction

Successful production of marine fish larvae is depended on cost-effective production of live feed organisms. To obtain a stable production of high-quality live feed requires competence in nutritional, microbiological, and zootechnical aspects. The most common zooplankton species used as live feed are different rotifers (*Brachionus* spp.) and *Artemia*. Both rotifers and *Artemia* have been produced in marine hatcheries around the world since the early 1960s. Rotifers and *Artemia* are not the natural food for marine fish larvae, but they are easy to produce in high densities and their nutritional content can be manipulated. Copepods are the natural prey for most marine fish larvae, but intensive cultivation of copepods in high densities is not a commercial industry yet. Some hatcheries harvest copepods from the sea.

In addition to different zooplankton species, microalgae are also important components in fish larval production, either directly added to the water or indirectly as food for rotifers or *Artemia*. Microalgae can be cultivated in hatcheries, as described in this chapter; it is also possible to purchase condensed microalgae or microalgae paste. This chapter will focus on aspects of biology, cultivation methodology, and nutritional value of microalgae, rotifers, *Artemia*, and copepods.

11.2 Application of microalgae in marine aquaculture

Several species of marine microalgae are used in the process of first feeding of marine fish larvae. The microalgae are used both as a feed component in cultivation and as enrichment for the zooplankton used as live feed, and they are additionally given directly to the fish larvae (Reitan et al. 1997).

Rotifers used as live feed for fish larvae are often fed and grown on microalgae during the whole cultivation process (Øie 1997). To

Larval Fish Nutrition, First Edition. Edited by G. Joan Holt.

satisfy the food requirements for rotifer culture, 5–10 times larger volumes of algal cultures are needed (Lubzens 1987). Rotifers cultivated on microalgae are considered to have higher overall quality compared with rotifers grown on formulated diets. Both the energy and protein contents of the rotifers are higher when fed with microalgae than with baker's yeast or formulated diets (Øie et al. 1994). Algae are also known to affect the microbial conditions of the tank environment and the bacteria associated with the live food (Skjermo and Vadstein 1993). Both factors may affect the microflora of the intestine of first-feeding larvae.

Due to the high cost of using microalgae as the only cultivation diet, rotifers are often fed on a variety of cheaper food sources (Olsen et al. 1993) and additionally fed with microalgae for a short period (2–24 hours) prior to their use as live food for the larvae in order to improve their nutritional quality (Lubzens 1987; Hayashi et al. 1993). During the period of feeding with microalgae, rotifers will obtain nutritional value and the bacteria composition that is structured by the microalgae species used.

Microalgae are also frequently added directly to the larval tanks together with the live feed. This addition is frequently termed the "green water technique." Microalgae are available as food not only for the zooplankton in the fish tank but also for the fish larvae. Some marine fish larvae (e.g., Atlantic cod, halibut, and turbot) ingest microalgae during their early life stages (Reitan et al. 1994a, 1998). This ingestion is significantly higher than can be explained by passive uptake through the process of drinking. The use of green water (e.g., using the microalgae *Isochrysis galbana* and *Tertaselmis* sp.) in first-feeding turbot (Figure 11.1) has, in many experiments, resulted in better survival of larvae at the end of the first-feeding period than without using microalgae. The positive effect of adding microalgae can result from a generally enhanced nutritional value of the live feed as a result of their feeding on microalgae, together with a positive direct nutritional effect of microalgae on the fish larvae. It may also be a result of improved microbial conditions for fish larvae when algae are added (Reitan et al. 1997). Additionally, it has also been suggested that the addition of algae to the water will change the light milieu of the tank in a way that positively affects the feeding behavior of the larvae during the early feeding period (Naas et al. 1992). All these factors may work independently or in combination with each other, and thus explain the

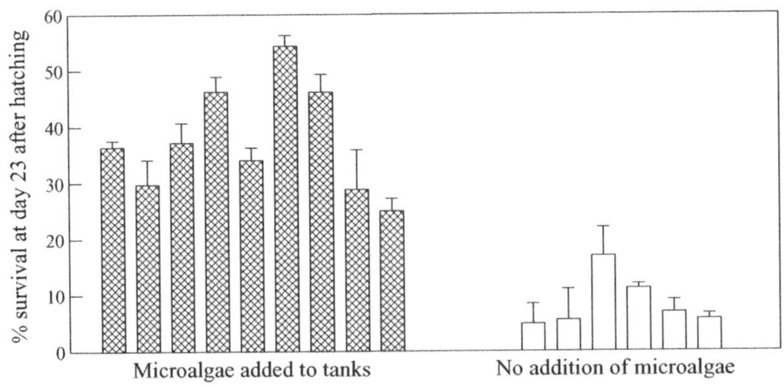

Figure 11.1 Survival of turbot larvae at 23 days after hatching when fed on rotifers and *Artemia*, with and without addition of microalgae. Shaded bars represent larval groups that received microalgae (*Isochrysis galbana* and *Tertaselmis* sp.) along with the live feed, and open bars represent larval groups that did not receive microalgae. (Unpublished data.)

observed improved feeding behavior as well as the higher survival and growth rate of the larvae in green water culture.

11.2.1 Algae species

Several microalgae are cultivated for use in marine aquaculture. Important species are *Isochrysis galbana* (different strains) and *Pavlova lutheri* in the class Prymnesiophyceae; *Nannochlorosis* sp. in Estigmatophyceae; *Tetraselmis* spp. in Prasinophyceae; and *Rhodomonas* sp. and *Chryptomonas* sp. belonging to Chryptophyceae. In addition, species of Bacillariophyceae (diatoms) such as *Chaetoceros* spp., *Skeletonema costatum*, and *Thalassiosira* spp. are frequently cultivated for use in bivalve cultivation. Characteristics of species of microalgae that are cultivated are ease of culture, neutral buoyancy, and optimal size (2–15 μm).

11.2.2 Cultivation

Microalgae are unicellular organisms that have been cultivated under laboratory conditions since the last part of the 1800s (Andersen 2005). Large-scale cultivation was tested during the first part of the 20th century. For use in marine aquaculture, microalgae are cultivated with the use of light energy, carbon dioxide (CO_2), and inorganic nutrients. If light energy is the only energy source for microalgae, an intensity of 70–300 μE/m^2/s is normally recommended. Microalgae can be cultivated in natural light in special greenhouses or under direct sunlight, but in many cases the microalgae are cultivated in indoor systems in marine hatcheries, illuminated with artificial light. In dense cultures of large volumes, the algae cells will tend to shade the light for the other cells, and as a result the highest light intensity will be at the surface of the culture, whereas most of the cells experience very low light intensities. As a result of this, at high light intensity (i.e., direct sunlight) the culture may experience photoinhibition at the surface and photolimitation in the center of the culture.

In order to avoid culture breakdown or contamination of other algae species or microzooplankton, the seawater for the algae cultures needs to be disinfected by using different procedures such as microfiltration (0.5 μm), UV treatment, ozonation, chlorination, or high-temperature treatments. The seawater used for the algae growth media is enriched with mineral nutrients balanced to fit the requirements of the algae species (Guillard 1975). There are several formulations of growth media for algae, and there are also commercially available mixtures of cultivation medium (Andersen 2005).

Temperature and salinity will also affect the growth and culture of microalgae. Increasing the temperature up to a certain level provides increased growth rate of the algae; however, it is advisable to keep culture temperatures relatively low (20–25°C). Cultures are very often aerated in order to keep the cultures well mixed and to prevent sedimentation of the algae cells. With mixing, all the cells in the culture will be equally exposed to light and nutrients, and thermal stratification (e.g., in outdoor cultures) will be avoided. Photosynthesis of microalgae produces oxygen, and it is important for the algae to remove this oxygen from the culture because O_2 inhibits the growth of microalgae. Air bubbling will also (see above) improve gas exchange between the culture and the air. Carbon dioxide is added to the air, as the CO_2 is the carbon source for the microalgae. Correct addition of carbon dioxide will also contribute to the pH of the culture within the optimal range for algae.

Microalgae grow by normal cell division and cultures supplied with excess resources will normally exhibit an exponential increase in cell numbers (Andersen 2005). After inoculation, the cultures need to acclimate to the new culture condition and the microalgae

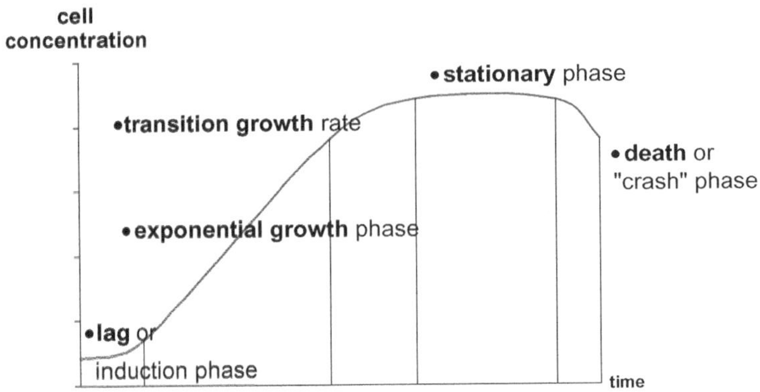

Figure 11.2 Normal growth curve for microalgae, following lag, exponential, transition, stationary, and death phases.

culture will have an induction phase, or a so-called lag phase (Figure 11.2). When the cells have become acclimatized to conditions of excess resources, they grow and divide faster and the culture will follow an exponential growth phase. During this exponential growth phase, the cell density increases, while the nutrient in the growth medium will be exhausted and the increased density of algae cells will increase self-shading of the culture. The culture will then come into a phase of declining growth rate before it continues into the so-called stationary phase. In the stationary phase, the net increase in cell numbers is zero, meaning that the net specific growth rate is zero. After some time, depending on the species, the culture will continue into the death phase, where mortality rate exceeds growth rate.

11.2.3 Production systems

Several methods for large-scale cultivation of microalgae are developed for both indoor and outdoor cultivation systems, including ponds and tanks. Typical indoor cultivation systems are standing plastic or glass tubes and closed photobioreactors. The methods developed are often adapted to the microalgae species cultivated and dependent on the volume required. However, for all types of production systems, it is important to ensure an adequate supply of light to the cultures. In most large-scale systems, the light is often the first limiting factor.

Figure 11.3 Microalgal cultures at a marine hatchery in Norway. (Photo from Scalpro, Norway.)

Microalgae are grown in systems of different volumes, from relatively small (some mL) to larger open culture volume (m^3). Typical cultures used in many types of hatcheries are standing cylindrical tanks or tubes, with volumes of 100–500 L (Figure 11.3). These are often illuminated by standing lamps of light tubes placed around the tank. In cases where the need for algae is not so large, it is appropriate to cultivate algae in glass bottles of 10–20 L. When the need for microalgae is larger, bigger cultivation volumes are needed. Typical large-scale systems are rectangular shallow tanks illuminated from the surface.

During the last decades, hatcheries with high requirement for microalgae produce it by photobioreactors built up by horizontal or vertical tubes (Figure 11.4). The benefit of such reactors is that they have a very large surface in relation to the volume, with good supply of light energy to the cultures. Such cultures may therefore achieve higher biomass production yield than the traditional standing tube cultures. Such tube photobioreactors can be placed standing, angled, or lying on the ground, depending on the direction to the light source.

Common microalgae production strategies use either batch cultures or continuous cultures. A batch culture follows a typical growth curve, starting with an inoculation phase, an exponential growth phase, and further into a stationary phase (see above), where the carrying capacity of the culture is reached (Figure 11.2). Batch cultures are started at a low density and are harvested when the desired density is achieved. Continuous cultures mean that the cultures are harvested, more or less, continuously. There are several types of continuous cultures, and the most common is the chemostat or semicontinuous culture. In both, a specific volume is harvested each day, and the harvested volume is replaced by new growth media. A chemostat culture is continuously diluted (harvested) and will have stable conditions (light, nutritional state) over time and the produced biomass and composition of algae can be predicted.

Figure 11.4 Photobioreactor for cultivation of microalgae at IGV GmbH, Germany. (Photo from IGV GmbH, Germany).

11.2.4 Biochemical composition

Microalgae are the natural diet for the natural prey of marine fish larvae. They have high contents of essential n-3 fatty acids, especially eicosapenatenoic acid (EPA) and docosahexaenoic acid (DHA), and therefore have been included as a feed component in marine fry production. Microalgae synthesize their lipid *de novo*, and several microalgae contain large amounts of polyunsaturated n-3 fatty acids (Reitan et al. 1994c). The lipid content varies among species and is affected by the growth condition; however, there is obvious similarity in the different algal classes (Table 11.1). Diatoms have high contents of 14:0,

Table 11.1 Dominant fatty acids in several classes of microalgae.

Algae group	Dominant fatty acids	References
Bacillariophyceae	14:0, 16:0, 16:1, 18:1n-9, 16-PUFA, 20:5n-3	1, 2, 4, 6, 7
Prymnesiophyceae	14:0, 16:0, 16:1 or 18:1, 18:4n-3 or 20:5n-3, 22:6n-3	1, 3, 6, 7
Chlorophyceae and Prasinophyceae	High variation, 18:1n-9, 18-PUFA, 20:5n-3	1, 3, 5
Cryptophyceae	18:3n-3, 18:4n-3, 20:5n-3, 22:6n-3	6
Dinophyceae	16:0, 18:4n-3, 20:5n-3, 22:6n-3	5, 7

1: Ackman et al. (1968); 2: Chuecas and Riely (1969); 3: Ben-Amotz et al. (1985); 4: Mortensen et al. (1988); 5: Olsen (1990); 6: Volkman et al. (1989); 7: Reitan et al. (1994c).

16:0, 16:1, and 20:5 n-3 fatty acids. Species in the class Prymnesiophyceae have high contents of 14:0, 16:0, 16:1 or 18:1, and 22:6 n-3, and some species of this group have, in addition, high levels of 18:4n-3 or 20:5n-3. The green algae (Chlorophyceae) have a greater variation among species in the dominant fatty acids, but contain several 18-chain polyunsaturated fatty acids. Dinoflagellates tend to have very high concentrations of 22:6n-3 and slightly lower contents of 16:0, 18:4n-3, and 20:5n-3.

The lipid content and fatty acid composition of microalgae are also, to some extent, influenced by the culture conditions, including temperature, light environment, and nutritional state of the algae. Growth rate of the cultures decreases as the density of the cultures increases, and the growth rate can be limited by nutrients or light, resulting from increased cell density. Changing growth conditions of the algal cells will affect their chemical composition, in particular the content of lipid and fatty acids. Nutrient limitation (both phosphorus and nitrogen) results in increased lipid content in most of the lipid-rich species. Lipid accumulation is mainly a result of ongoing lipid synthesis combined with reduced cell division and protein synthesis due to reduced availability of nutrients. The increase in lipid content with nutrient limitation reflects an increased fraction of neutral lipid, such as triglycerides and hydrocarbons. The microalgae species that do not accumulate lipids often have lower lipid content and accumulate photosynthesis products as carbohydrates rather than as lipids. However, there are several species that accumulate both lipids and carbohydrates under nutrient limitation.

The carbohydrate content (e.g., glucan) varies significantly among microalgae species and growth conditions (Størseth et al. 2006). Carbohydrate content, especially β-1,3-glucan content, has received increased attention because these β-glucans have been shown to have a positive immunostimulating effect on fish larvae (Skjermo et al. 2006)

Most vitamins are found in the algae species used in marine aquaculture, with the content of vitamins C and E being particularly high. In general, the vitamin content in microalgae is high compared with that found in other food products and higher than the reported requirements for fish.

11.3 Rotifers in aquaculture

Rotifers in the genus *Brachionus* have been used in aquaculture since 1960 (Ito 1960; reviewed by Nagata and Hirata 1986) and are still very important as live feed in the production of fish larvae and crustaceans in most parts of the world. Rotifers are used because they are easy to cultivate in high densities all year round, their size fits the mouth of many species of fish larvae, and their swimming speeds are slow so they are easily captured by fish larvae. Rotifers tolerate pronounced variations in environmental conditions such as salinity, pH, and temperature, and their nutritional value is easy to manipulate and can be quite well adapted to the requirements of each species of interest (Øie et al. 1997; Reitan 1993). The first weeks of exogenous feeding are pivotal in the production of marine fish, and rotifers are the most important live feed for both fish larvae and crustacean species in aquaculture (Figure 11.5).

11.3.1 General biology

The phylum Rotifera is divided into two superclasses named Seisona and Eurotatoria, and more than 2,000 different rotifer species are registered. The *Brachionus* complex is one of the most studied groups of rotifers, and their taxonomy is still under constant review. In aquaculture, different species from the genus *Brachionus* are used in larviculture. Genetic analysis has indicated that *Brachionus plicatilis* is not a single species, but is in fact,

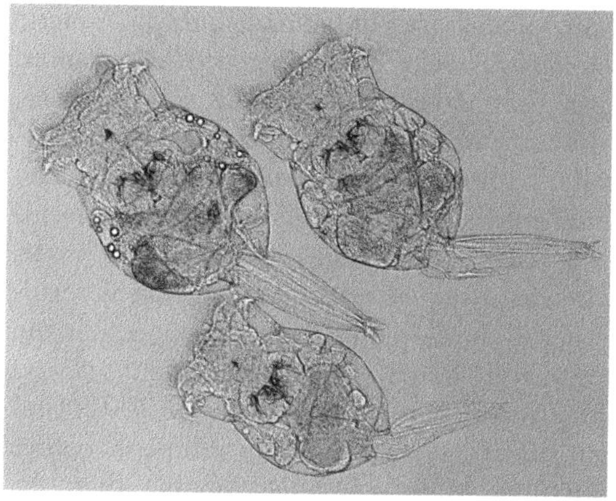

Figure 11.5 Picture of rotifers (*Brachionus plicatilis* [Nevada]). (Photo by Tora Bardal.)

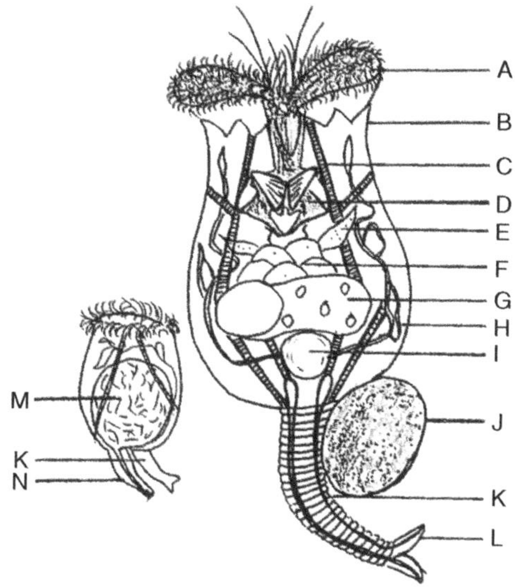

Figure 11.6 Drawing of *Brachionus plicatilis*. Male (left), and female with one egg (right). A: corona; B: lorica; C: muscle; D: mastax; E: gastric gland; F: stomach; G: ovary; H: kidneys; I: bladder; J: egg; K: foot; L: toe; M: prostate; N: penis.

a complex of at least 14 putative species (Ciros-Pérez et al. 2001; Gómez et al. 2002; Suatoni et al. 2006). Genetic analysis has revolutionized the taxonomic classification of *Brachionus* species during the last decade.

Identifying the species used in experiments and hatcheries is a crucial step toward a better understanding of rotifer biology and a more scientific approach to larval rearing. *Brachionus plicatilis* was the subject of approximately 750 peer-reviewed articles between 1950 and 2000 (Yúfera 2001). Unfortunately, for the majority of these articles, it is unknown which of the 14 putative species were used. This limits the utilization of this large accumulation of knowledge (Suatoni et al. 2006), which is especially important since *Brachionus* spp. rotifers exhibit different growth rates (Kostopoulou and Vadstein 2007), lorica lengths, optimum for salinity and temperature, swimming activity, size preferences for food particles (Vadstein et al. 1993; Baer et al. 2008), and biochemical composition.

11.3.2 Morphology

Rotifers are small aquatic invertebrates, and most species live in freshwater. *Brachionus* species have some distinctive features, including the ciliated corona and mastax. The anterior end of the rotifer carries the corona ("head"), which is composed of two concentric ciliary crowns (Figure 11.6). This rotary organ mediates swimming and sweeping of food particles toward the mouth. The body is covered by a lorica, consisting of keratin-like proteins, and different shapes of their anterior spins can be used in species identification. Rotifers also have a foot with two toes that can secrete a sticky substance that enables them to attach to a surface. They are able to contract the corona and the foot into the lorica by longitudinal muscles. The movement is also facilitated by contractions of muscles inside the rotifers. Below the corona, rotifers have a single large cerebral ganglion, and paired neurons proceed from the simple brain to the foot, branching off to various organs. Rotifers also have sensory cells divided into mechanoreceptors, chemoreceptors, and photoreceptors in the head region (Lubzens and Zmora 2003).

11.3.3 Feeding and digestion

The ciliated corona is important in food capture. Particles from the water enter the mouth-opening and pass into the pharynx. The pharynx houses the mastax, where the food particles are crushed by the masticatory organs called trophi, probably with the aid of enzymes. The hard parts of the trophi are important in the taxonomic characterization of rotifers. The crushed food particles enter the alimentary tract together with gastric juices from large gastric glands, and extracellular digestion occurs within the stomach lumen. Several digestive enzymes have been reported from *Brachionus plicatilis* (Kühlea and Kleinow 1985). The stomach leads into an intestine and a cloaca, which also voids fluids from the bladder and eggs from the oviduct. Rotifers do not have a respiratory system or a circulatory system, and body fluids are located in the pseudocoelom (Lubzens and Zmora 2003). Rotifers exchange gasses and dispose nitrogenous waste by diffusion through the body surface. *Brachionus plicatilis* is an osmoconformer, and will adjust its body fluid osmolarity to the external concentrations (Epp and Winston 1977).

11.3.4 Reproduction

During favorable cultivation conditions for rotifers, the individuals are normally females and the population increases by diploid parthenogenesis. Diploid females produce diploid eggs known as amictic eggs, which will hatch and develop into a diploid female. Sexual reproduction may occur under special circumstances, and haploid males will develop. The mechanism behind the change from asexual to sexual reproduction is not well understood, but laboratory studies found nutritional, density, salinity, and genetic factors (Pourriot and Snell 1983; Serra and King 1999; Ricci 2001). Diploid females can either be amictic or mictic, and morphologically, they are indistinguishable. Amictic females produce parthenogenetically diploid eggs that develop mitotically into females, while mictic females produce parthenogenetically haploid eggs by meiosis. If a mictic female does not mate and the egg is not fertilized, the haploid eggs form into a male, whereas a mated mictic female that is fertilized will develop diploid resting eggs. Resting eggs will hatch after a dormant period (Hagiwara 1996; Lubzens et al. 2001) and can survive several years in sediments before hatching. Males are much smaller than the females, and their movement is much faster.

11.3.5 Cultivation of rotifers

Rotifers can be cultivated in a wide range of production systems. In hatcheries around the world, there are round, square, cylindrical, conical, or flat bottom tanks, with volume ranging from 100 to 300,000 L. In European hatcheries, fiberglass conical tanks (2–3 m^3) are most often used. The tanks are aerated with an air-stone, and oxygen is added if the density exceeds 500 rotifer/mL. The rotifers are often fed refrigerated and condensed freshwater *Chlorella* to maintain a high population density. Other types of feed for rotifers include live algae, baker's yeast, formulated diets, and emulsified oils.

11.3.6 Cultivation methods

The different methods used for rotifer production include batch cultivation, continuous cultivation, and combinations of these two.

Batch cultivation is production in a closed culture system supplied with air, oxygen, and feed. The cultures are completely harvested at a certain development stage (2–5 days) or in the late phase of growth. Continuous cultivation is production in an open culture system that is supplied with necessary resources and harvested regularly by replacing a fixed volume of the culture by seawater once daily (semicontinuous) or continuously.

Both methods and appropriate combinations are feasible for production. The priorities of the producer regarding rotifer quality, scale of production, costs, risks, and hatchery routines may affect the choice of method. Adequate knowledge of rotifer biology, their nutritional and environmental requirements, and traits of the specific cultivation system are important for sustainable and safe cultivation of rotifers.

11.3.7 Evaluating the state of cultures

The egg ratio (eggs per rotifer) is an important indicator of the state of the rotifer culture. Egg ratio is determined as the number of eggs divided by the number of females in the sample, and can easily be counted using a microscope. The value will reflect the expected growth and production of the culture for the next day. The egg ratio, like the growth rate, generally depends on feed quality and quantity, oxygen level, ammonia level, pH, temperature, and salinity.

11.3.8 Water quality in rotifer cultures

Rotifer cultures have to be evaluated daily to ensure a predictable and stable supply of high-quality rotifers for larval production in hatcheries. A common problem in rotifer cultivation is overfeeding, which in turn may cause oxygen deficiency and high ammonia concentrations. The optimal level for ammonia is <1 mg/L, and an acceptable range for ammonia and nitrate is 6–10 mg/L (Lubzens and Zmora 2003). In water, ammonia occurs in two forms, represented as NH_4^+ (ionized ammonia) and NH_3 (unionized ammonia). This is important to know because NH_3 is toxic to rotifers. Water temperature, salinity, and pH will affect which form of ammonia is predominant at any given time in the production system. The optimal pH in rotifer cultures is 7.5–8.5 (Hirano 1987). To secure mixing and an adequate aeration in rotifer cultures, air or oxygen supply is required. The oxygen level should be maintained above 4 ppm.

11.3.9 Enrichment and nutritional quality

Rotifers feed on a wide range of food particles, including bacteria, protozoa, microalgae, and dead organic materials within a certain size range, and it is therefore important to consider the size of the particles, both for cultivation and enrichment. *Brachionus plicatilis* is able to consume larger particles than *Brachionus rotundiformis*. Efficient cultivation feeds must also cover the nutritional demands of the rotifers and secure proper hygienic conditions in the production tanks. Many species of microalgae are excellent food for rotifers, and today, concentrated microalgae cultures or algal pastes are commercially available. The use of microalgae as feed often results in a higher egg ratio and growth rate, in addition to a lower bacterial content than with yeast and formulated diets. A relatively cheap cultivation diet is live baker's yeast together with a small amount of microalgae (5–10%). These cultivation diets do not cover the nutritional needs of fish larvae; thus rotifers need short-term enrichments before being used as live feed for fish larvae.

In short-term enrichment, it is important to use products containing high n-3 highly unsaturated fatty acids (HUFA). Rotifers are fed the enrichment diet for a short period (2–24 hours) to change the nutritional content in the rotifer to near that of the enrichment diet. Lipid and fatty acid content after short-term enrichment will be influenced by the nutritional content in the cultivation diet, the ambient conditions, the rotifer strain, and particularly, by the properties of the enrichment diet.

11.3.9.1 Protein

Protein is a major component of the biomass, and the content will, to some extent, reflect the energy level of the organism. The protein content per individual rotifer is a dynamic variable that is related to food availability and specific growth rate. In the literature, the protein content range from 28–63% of dry matter, this large variation in reported levels may reflect both variations in methods and variations in rotifer protein content. Nitrogen-based estimates of protein derived using the common conversion factor of 6.25 mg protein/mg N yields overestimates, whereas amino acid–based estimates are likely to be slightly underestimated. A conversion factor of 4.2 mg protein/mg N has been established through testing and analysis of amino acids (Lie et al. 1997; Øie et al. 1997). The protein content per individual rotifer is related to the food source and specific growth rate of the rotifer culture. A common range of variation in protein level in *Brachionus plicatilis* (Nevada strain) is 100–200 ng protein per individual, with the lower level representative of starved rotifers and the higher of well-fed rotifers. Protein per dry matter appears to be less variable because protein makes up a major fraction of the dry matter. Even though the protein content of individual rotifers is variable, the amino acid profiles (% distribution) of rotifers are unaffected by food ration or type of food provided to the rotifers (Makridis and Olsen 1999).

11.3.9.2 Lipids

The lipid content and fatty acid composition of rotifer tissues are mainly a function of the composition of dietary lipids and their inherent metabolic (Øie et al. 1997) and genetic traits. *Brachionus plicatilis* is presumably evolutionarily adapted to warm water (Olsen 1999), and its requirements for n-3 fatty acids are low. Dietary lipids will then affect the composition of both its triglycerides and its phospholipids. This means that diet composition is paramount, whereas the metabolic activity becomes less important, but still significant. This makes *Brachionus plicatilis* well suited for fatty acid manipulation; fatty acid composition of the triglycerides becomes equal to that of the dietary lipids. The rotifer content of essential n-3 fatty acids can be completely controlled by the selection of the dietary lipid source used during cultivation. When the percent n-3 fatty acid composition of the dietary lipids is known, the rotifer composition may be estimated as follows (Olsen 2004):

$$\text{Rotifer}_{\%\text{n-3}} = 0.86 \times \text{Feed}_{\%\text{n-3}} \quad (11.1)$$

$$\text{Rotifer}_{\%\text{EPA}} = 0.81 \times \text{Feed}_{\%\text{EPA}} \quad (11.2)$$

$$\text{Rotifer}_{\%\text{DHA}} = 0.72 \times \text{Feed}_{\%\text{DHA}} \quad (11.3)$$

11.3.9.3 Carbohydrates

The carbohydrate content in rotifers ranges from 11 to 27% of dry weight (DW), and is composed of 61–80% glucose, 9–18% ribose, and 0.8–7.0% galactose, mannose, deoxyglucose, fucose, and xylose (Whyte and Nagata 1990; Nagata and Whyte 1992).

11.3.9.4 Minerals

The mineral levels in rotifers, except calcium, magnesium, and iron, are generally lower than in copepods (see Chapter 4). Magnesium and selenium are also lower in rotifers than the predicted requirements of larger fish given by the National Research Council (NRC 1993). In an experiment where rotifers were enriched with iodine and selenium up to levels found in copepods, there was a 32% increase in survival of cod larvae (Hamre et al. 2008a). However, the nutrient levels in copepods vary and the actual requirements of the larvae may be lower than the levels found in some copepods.

11.3.9.5 Vitamins

The actual requirements for vitamins are not known for most marine fish larva species (see Chapter 4). Several nutritional studies have found a direct effect of several vitamins in chondrogenic and osteogenic development during the early life stages of marine fish species. It is suggested that rotifers should be enriched with vitamins C, E, A, and B_1 (thiamine), but the actual requirements are not known. Experiments with higher levels of vitamin A than those in commercial emulsions for rotifer enrichment led to skeletal deformities in gilthead sea bream larvae (Fernández et al. 2008). Today, hatcheries normally use microalgae as feed in rotifer cultivation since rotifers fed on algae contain sufficient amounts of water-soluble vitamins to meet the nutritional requirements of fish larvae (Lie et al. 1997). Rotifers cultivated on *Chlorella* contained 2,300 μg AA/g DW (Merchie et al. 1995). Fat-soluble vitamins (A, D, and E) are shown to increase rotifer reproduction.

11.3.9.6 Stability of nutritional value

The choice of cultivation and enrichment diets will influence the bacterial content in the culture. To decrease the bacterial level, it is important to have good washing procedures. The water should be at the same temperature and salinity as the culture. After washing, it is possible to cool down the rotifers to stabilize their nutritional value. Storage at 5–10°C results in a 10% decrease in lipid levels, while storage at 20°C results in a 40% decrease (Olsen et al. 1993)

11.4 *Artemia* in aquaculture

Brine shrimp (*Artemia* spp.) are used in marine aquaculture worldwide. Annually, more than 2,000 metric tons of dry cysts are used for cultivation of fish, crustacean, and shellfish larva. Although *Artemia* is not a part of the natural diet of marine larvae, it has been favored due to its convenience for use and high nutritional value (Léger and Sorgeloos 1992). Dormant cysts of *Artemia* can be stored for long periods in cans and then used as live food. The cysts require only 24 hours of incubation, prior to hatching, followed by a 24-hour feeding period (enrichment) with diets high in n-3 HUFA. This makes the whole production process simpler and less labor intensive as compared with other live food organisms available for land-based production of larval fish and shrimp species.

With the introduction of fish and shrimp aquaculture, the demand for *Artemia* cyst has gradually increased, and today cysts are harvested from more than 200 locations. Two commercial sources of *Artemia* cysts were available in the beginning, the coastal salt works in San Francisco Bay (California, United States) and Great Salt Lake (Utah, United States). More then 90% of the supply of cyst was at times dependent on the source in Great Salt Lake. In the 1990s, new sites were found in Iran, China, and Russia (Lavens and Sorgeloos 1996), but most of the cyst market is still supplied from Great Salt Lake. Harvested cysts are dehydrated, dried, and packed in oxygen-free condition, and stored in boxes for sale. For the hatcheries, the cysts represent great advantages since the eggs can be stored for several years.

Today, the technology for intensive *Artemia* production is well established and forms the basis for adapting *Artemia* as live food organisms into marine aquaculture. This work led by Professor Patrick Sorgeloos and his colleagues at Ghent University, Belgium (see manuals by Sorgeloos et al. 1986; Lavens and Sorgeloos 1996) has been of major importance for the aquaculture industry. Their pioneering work, which started in the early 1970s, focused on the use of brine shrimp *Artemia* as a food source for fish and shellfish larvae. Later it appeared that the nutritional value of *Artemia* was not optimal, especially for marine organisms, since the nutritional variation among strains and within batches of each strain was found to be suboptimal for most marine larvae. By introducing emulsified lipid diets rich in n-3 HUFA, as food for *Artemia* nauplii, their nutritional quality was improved. Application of this method of bio-encapsulation, also called *Artemia* enrichment or boosting, has been very important and improved the larviculture production for most marine species. For many species of fish and crustacean larvae, this is important not only to survival, growth, and successful metamorphosis but also to their quality, for example, reduced incidence of malformations, improved pigmentation, and stress resistance.

Improvements in culturing biology and strain characterization, nutritional manipulation, improved techniques for enrichment, and new enrichment products have optimized the nutritional value of *Artemia* and led to widespread use of this live food organism in hatcheries throughout the world. *Artemia* are used for both warm- and cold-water species. The application and use of nauplii in hatcheries had improved over the years and become more effective. This is seen in a reduction in the required amount of cysts per unit fish or shrimp produced. For example, the sea bass and sea bream production in the Mediterranean Sea has been reduced from 150 kg cysts per 1 million fry produced to a required amount of cysts today of only 90 kg for sea bass and 70 kg for sea bream (Dhont and Van Stappen 2003).

11.4.1 General biology of *Artemia*

The brine shrimp *Artemia* (class: Crustacea; subclass: Branchiopoda; order: Anostraca; family: Artemiidae; and genus: *Artemia*) is found in more than 500 saline lakes throughout the world (Sorgeloos et al. 1986). This includes different climatic zones in both coastal and inland areas (Lavens and Sorgeloos 1996). Among these, *Artemia* populations are found in both zygogenetic and parthenogenetic forms. The genus *Artemia* is a complex group of sibling and subspecies maintained mainly by reproductive isolation. This has resulted in numerous geographic strains or genetically different populations within the same sibling species (Dhont and Van Stappen 2003).

Through their physiological adaptation to saline biotopes, *Artemia* spp. have a novel defense mechanism against predation in that they are found only in habitats where potential predators cannot survive (salinity 70 ppt and higher). *Artemia* can live in natural seawater, but without any behavioral or physical defense mechanism, *Artemia* are easy prey for most predators.

11.4.1.1 Morphology and life cycle

Newly hatched *Artemia* nauplii (Instar I) are between 400- and 500-μm long and a dark brownish or orange color, which is related to the yolk. The nauplii do not ingest food particles before entering the Instar II stage, 8–10 hours after hatching. The antenna (Figure 11.7a) is used for locomotion and filtering food particles (1–50 μm). Small particles (bacteria and algal cells) are captured by the filtering setae, transferred toward the labrum, and handled by the mandibles before they are ingested.

The nauplii grow through a number of molts, develop compound eyes and one pair of thoracopods in each thoracal segment, the antenna gradually lose their locomotory function (Figure 11.7b), and it undergoes sexual differentiation. The males can be easily recognized by the graspers and the females by the uterus situated behind the 11th pair of thoracopodes. Adult *Artemia* are 8–12-mm long, covered with a thin exoskeleton of chitin. Under optimal conditions, it takes newly hatched *Artemia* 12–14 days to reach the adult stage.

The reproduction cycle of *Artemia* is affected by environmental conditions, for

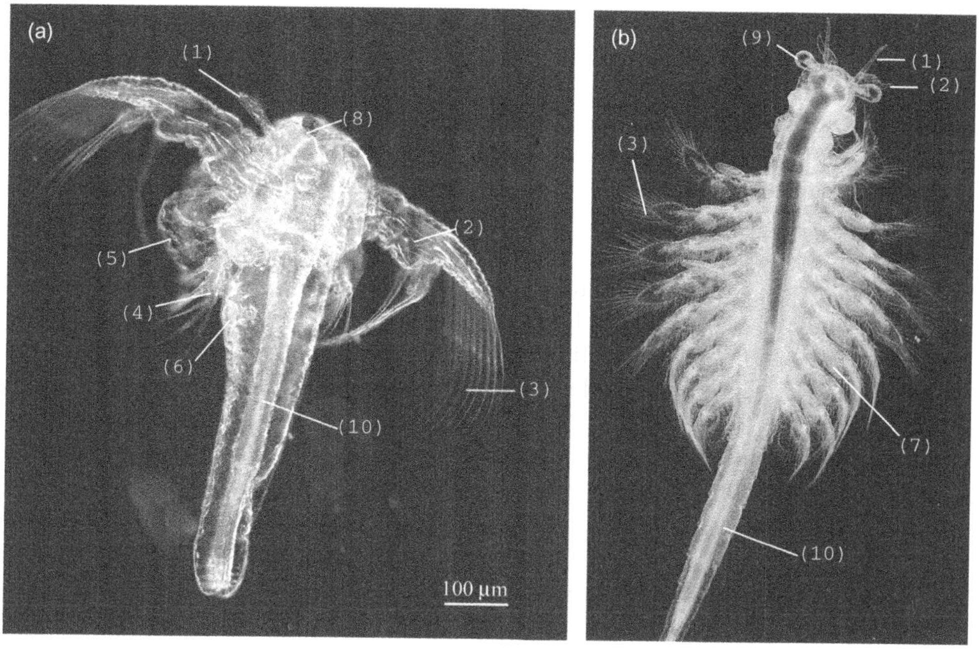

Figure 11.7 *Artemia franciscana*: (a) Instar II and (b) postlarval stage. (1) antennula; (2) antenna; (3) filtering setae; (4) mandible; (5) labrum; (6) thoracal segment; (7) thoracopode (11 pairs); (8) nauplius eye; (9) complex eye; (10) digestive tract. (Photo by T. Bardal, Norwegian University of Science and Technology [NTNU].)

example, females can switch reproductive modes from one ovulation to the next. Under optimal conditions, each female produces free-swimming nauplii (ovoviviparous reproduction) at a rate of 300 nauplii every 4 days. Under suboptimal or extreme conditions (high salinity or low oxygen level), the embryos develop in the uterus up to the gastrula stage and become surrounded by a thick shell (chorion), enter a state of dormancy (diapause), and are then released by the female (oviparous reproduction). The dormant cysts usually float and accumulate in the seashore where they dry and can be stored until the conditions become favorable. The cysts can remain dormant for many years as long as they are kept dry and oxygen free.

11.4.1.2 Feeding and ingestion

Artemia are continuous, nonselective, obligate phagotrophic filter-feeders (Provasoli and Shiraishi 1959; Reeve, 1963; D'Agostino 1980); they start to ingest food at the Instar II stage (metanauplius I) using the larval antenna (Barlow and Sleigh 1980). During postembryonic development, the feeding function is gradually taken over by the multifunctional phyllopods (Schrehardt 1987), which are fully developed at the preadult stage (Blanchard 1987).

How filter-feeders respond to variable food concentrations is normally described by a functional response curve (Holling 1966), defined as the relationship between the prey density and the average numbers of prey killed per unit of time by a single predator (Abrams 1982). For different stages of *Artemia*, the ingestion rate increases as a function of increasing food concentration up to a saturation level where the animals reach the maximum ingestion rate (I_{max}). Thereafter, the ingestion rate is high and constant even if the food concentration is further increased

(Evjemo et al. 2000). The incipient limiting concentration (ILC), defined as the food concentration where the animals reach I_{max} (McMahon and Rigler 1965), is around 5–7 mg C/L, and the feeding response of *Artemia* can be described by a sigmoidal type 3 functional response curve (Holling 1966; Evjemo et al. 2000). This shows that *Artemia* are adapted to relatively high concentrations of food compared with many other planktonic crustaceans, such as copepods and cladoceran, where the I_{max} is in the order of 10–100 times lower.

It is of major importance to consider the feeding behavior of *Artemia* when they are produced in hatcheries. Overfeeding will always stimulate microbial growth in a culture, and because *Artemia* are nonselective filter-feeders, bacteria will be ingested and transferred to the fish, shrimp, or shellfish larvae. A good strategy is to feed *Artemia* at a rate close to, or slightly higher, than the ILC. This means that *Artemia* can ingest food at the maximum rate and at the same time avoid overfeeding and high bacterial loads in the cultures.

11.4.2 *Artemia* production

Artemia are mostly used as freshly hatched nauplii or enriched nauplii fed emulsified lipid diets or algae for a period of 24 hours. Juvenile and preadult/adult *Artemia* have also been used for first feeding of some species such as halibut (Olsen et al. 1999), shrimp (Dhert et al. 1993), or lobster larvae (Evjemo, unpublished data). Biomass from harvested *Artemia* (frozen or freeze-dried) is also used for feeding several species of penaeid shrimp even though fresh live forms have the highest nutritional value.

11.4.2.1 Cultivation methods

Although intensive production of *Artemia* appears to be simple, several factors are critical for successful hatching, good survival, low bacterial load, and high nutrition quality of the *Artemia* before the nauplii/metanauplii can be fed to fish and shrimp larvae.

The whole process includes four steps: decapsulation, hatching, enrichment/growth, and harvesting. During this process, *Artemia* can be cultivated in a variety of tanks as long as aeration ensures proper oxygenation and adequate mixing of food particles and nauplii. The hatching procedure has been improved through decapsulation of the cysts, a process whereby the chorion that encysts the embryo is removed by exposure to a hypochlorite solution (Bruggeman et al. 1980). This step improves the hatching of cysts, eliminates hatching debris, and reduces the bacterial load in the *Artemia* culture.

The optimum hatching conditions are as follows: temperatures between 25 and 28°C, salinity of 15–35 g/L, pH of 8–8.3, oxygen levels higher than 2.5 mg O_2/L, and illumination of 2,000 lux. The maximum density of cysts in the hatching tank should not exceed 2 g/L. After hatching, the nauplii must be concentrated, washed, and transferred to enrichment tanks under the same physical conditions as described above. The density of nauplii during enrichment should be in the range of 200–300 ind/mL.

11.4.3 Enrichment and nutritional quality

11.4.3.1 Feeding and enrichment

Since *Artemia* were introduced into marine larviculture, much work has focused on the challenge of supplementing n-3 HUFA to larvae through live food organisms. Enrichment with emulsified lipid diets is by far the most important technique to increase the nutritional value of *Artemia* and thereby provide a high-quality live food organism for fish and shellfish larvae. Incorporation of n-3 HUFA in *Artemia* is obtained by different enrichment procedures (Watanabe et al. 1978, 1983; Léger et al. 1986; Lavens et al. 1989), and considerable improvements have been achieved in the HUFA content of *Artemia*, in particular the content of DHA and EPA (Watanabe et al. 1983; Dhert et al. 1993; Navarro et al. 1993; Evjemo et al. 1997, 2001). Not only do the amounts of these essential fatty acids affect the fish larvae, but also the DHA/EPA ratio is found to be important for pigmentation of flatfish (Rainuzzo 1993; Reitan et al. 1994a, 1994b).

Newly hatched nauplii ingest small oil particles in the enrichment diets and show a linear relationship between the DHA content in *Artemia* and the duration of the enrichment period up to 24 hours (Evjemo et al. 1997; Evjemo 2001). Various enrichment diets, with both high and low DHA contents, have confirmed these results, and the variable DHA content in *Artemia* is a reflection of the contents in the diets (Evjemo 2001). Enrichment can also increase the total lipid content of the nauplii, which might increase from 18% of DW in newly hatched nauplii to more than 25% of DW after enrichment. Most of these diets contain only lipids, and will cause high mortality in the *Artemia* cultures if the enrichment period exceeds 36 hours.

11.4.3.2 Lipids

It is generally believed that copepods, which constitute a major part of the natural diet of marine larvae, can meet nutritional requirements, particularly with regard to the requirements of n-3 HUFA (Shields et al. 1999; Evjemo et al. 2003). The lipid content and the fatty acid composition of cultivated live food organisms (*Artemia* and rotifers) and copepods show major differences. This is mainly related to total lipid content and the content of n-3 HUFA (Evjemo and Olsen 1997). Newly hatched *Artemia* contain low and variable contents of n-3 HUFA (Léger et al. 1986), but through normal enrichment procedures the content of DHA might reach 18–22% of total fatty acids and a DHA/EPA ratio close to 2. The total content of n-3 HUFA might increase from less than 10% before enrichment to more than 35% (of total fatty acids) after enrichment (Evjemo et al. 1997; Evjemo et al. 2003), which is close to the corresponding values for marine copepods (Evjemo and Olsen 1997; Evjemo 2001).

11.4.3.3 Proteins

In newly hatched *Artemia*, the protein content constitutes 41% of DW. Following enrichment with emulsified lipids, the protein content usually decreases to 32–35% of DW (Evjemo 2001) because most of the enrichment diets contain only lipids. It seems, however, that both nauplii and adults (protein content 44–51% of DW) contain sufficient levels of the 10 amino acids that are considered essential for fish larvae (Fyhn et al. 1993; Helland et al. 1999). The protein and amino acid profiles show less variation between strains and live stages of *Artemia* than the lipid and fatty acid profiles (Dhont and Van Stappen 2003).

11.4.3.4 Vitamins and minerals

Artemia appear to cover the minimal dietary requirements of fish larvae and have sufficient amounts of vitamins (van Stappen et al. 1996; see also Chapter 4). With the exception of ascorbic acid and thiamine, Mæland et al. (2000) has shown that *Artemia* nauplii contained higher vitamin levels than natural zooplankton. Through normal enrichment procedures, vitamins can efficiently be transferred into *Artemia*. After enrichment, it appears that some components (ascorbic acid and B_6) are selectively catabolized at a higher rate than others (Olsen et al. 2000). On the other hand, the thiamine content remains constant during starving conditions. Generally, it seems that the variation in vitamins is comparable with that of n-3 HUFA, with clear effects of selective retention of some components, and selective and faster catabolism for others.

Seawater is considered to be a sufficient source of minerals for most marine organisms. The mineral content in *Artemia* seems to be sufficient for most marine larvae except for selenium (van Stappen et al. 1996).

11.4.3.5 Stability of nutritional value post enrichment

The main challenge in *Artemia* production is to maintain high DHA levels after enrichment because DHA is selectively catabolized at a rate significantly higher than other fatty acids (Evjemo et al. 1997, 2001). Both during and after enrichment, DHA is retroconverted into EPA (Navarro et al. 1999) and there is a significant reduction in DHA content in the nauplii by 12 hours post enrichment (Evjemo et al. 1997). It is important to consider that even short delays, or storage periods, between enrichment and larval consumption might reduce the nutritional value of the nauplii. This specific loss rate (SLR, per day) of components (or nutrients) is highly dependent on the temperature and can be calculated by the equation $SLR = (At + B)$. The regression coefficients (A and B) are obtained by linear regression analysis and are valid for temperatures (t) in the range 5–30°C (Table 11.2). By calculating the SLR for total protein and lipid content, total non-n-3 HUFA fatty acids, EPA, and DHA, it is clear that the DHA content is reduced several times faster than both EPA and non-n-3 HUFA fatty acids. The protein content is only slightly reduced (Table 11.2) and indicates that *Artemia* use fatty acids rather than proteins in their metabolism during starving conditions. Using the SLR, the loss rate (LR, as %/day) can be calculated from the equation $\%\ LR = (e^{SLR} - 1)\,100$.

11.5 Copepods in aquaculture

Free-living marine copepods are natural prey for most marine fish larvae, abundant in, and distributed throughout, estuarine, coastal, and open ocean systems. More than 11,500 species of copepods have been classified (Humes 1994). In nature, these small crustaceans constitute the vital link from primary producers to fish. Comprehensive reviews available include a review on copepod biology and ecology (Dussart and Defaye 2001; Sazhina 2006), copepod evolution (Huys and Boxshall 1991), biology of calanoid species (Mauchline 1998), and culture of copepods (Støttrup 2003). In aquaculture, the number of species that have been cultured as prey for fish larvae are limited, and these species generally belong to the order Calanoida, although a number of harpacticoid and cyclopoid species has also been cultured. Lists of species of copepods used for intensive or extensive cultures are provided by Støttrup (2003).

Table 11.2 Coefficients (±SE) for calculating the specific loss rate (SLR, per day) of different nutrients in *Artemia franciscana* post enrichment.

	A	*B*	r^2
Protein content	−0.015 ± 0.002[a]	−0.144 ± 0.044[a]	0.96
Total lipid content	−0.021 ± 0.004[a]	−0.002 ± 0.082[a]	0.88
Non-n-3 HUFA	−0.038 ± 0.003[a]	−0.057 ± 0.065[a]	0.97
EPA	−0.028 ± 0.005[a]	−0.171 ± 0.091[a]	0.88
DHA	−0.089 ± 0.007[a]	−0.298 ± 0.145[a]	0.97

[a] Regression coefficients significant from 0 ($P < 0.05$).
Data from Evjemo et al. (2001).

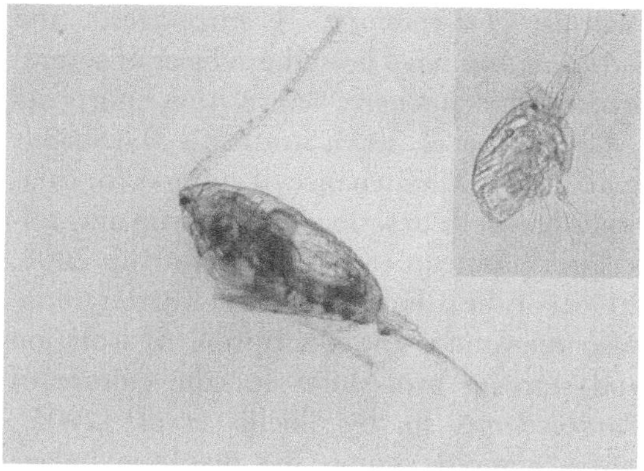

Figure 11.8 Picture of *Acartia tonsa* adult and nauplius. (Photo by Sune R. Sørensen, Technical University of Denmark, National Institute of Aquatic Resources [DTU Aqua].)

11.5.1 Copepod biology

Like other crustaceans, copepods have an exoskeleton composed primarily of chitin. Planktonic species have generally cylindrical bodies with a narrow abdomen and thorax (Figure 11.8). Benthic or surface-living forms may have broader or dorsoventrally compressed abdomen. The head is fused with the thorax and bears anteriorly a typical median naupliar eye and a conspicuous set of antennae. The thorax also bears the various appendages used for feeding or locomotion. The genital opening is usually located in the first segment of the abdomen, while the last segment bears the rami and attached setae. Most copepods reproduce sexually, and a sac containing viable sperm called the spermatophore is produced by the male and deposited near the genital aperture of the female. Eggs are either shed directly into the water column (broadcasters) or produced in one or two egg sacs that remain attached to the female (Mauchline 1998). Each egg sac may contain up to a couple of hundred eggs, depending on species and condition. The nauplii hatch directly from the egg sac whereupon the egg sac is discarded and a new one produced.

The nauplii have oval-shaped, dorsoventrally compressed, unsegmented bodies and a naupliar eye (Figure 11.8). The nauplii molt through six stages. The first three stages have generally few appendages, while there is a progressive development of appendages and setae in the posterior end of the body during the last three stages (Hicks and Coull 1983; Dussart and Defaye 2001). The last, sixth stage molts into a copepodite resembling the adult form. Copepodites also undergo six molts resulting in the final mature stage, where reproduction can take place.

The calanoids are generally planktonic species feeding on phytoplankton, which they filter from the surrounding water. Some species are able to switch to a predatory mode and feed on nauplii of their own or of other species (Daan et al. 1988; Kiørboe et al. 1996). Copepods belonging to the order Harpacticoida are generally benthic species grazing on a variety of food sources including bacteria and macro- and microalgae (Støttrup 2003). Cyclopoid copepods are omnivorous, with a variety of feeding modes from planktonic filter-feeders to detrivorous benthic grazers and can switch between herbivorous and carnivorous feeding.

Together with the egg and naupliar stages, copepods provide a broad range of prey sizes for larval fish. Copepod eggs range from 70 to around 800 µm in diameter (Mauchline 1998). The naupliar and adult stages together provide size ranges from around 100 µm to 10 mm in length, although the majority of adults are 1–2 mm in size (Mauchline 1998; Støttrup 2003).

11.5.2 Cultivation of copepods

Copepods have been successfully cultivated in outdoor extensive systems (see Svåsand et al. 1998; Engell-Sørensen et al. 2004) and in intensive systems (see review by Støttrup 2003; Lee et al. 2005).

11.5.2.1 Extensive cultivation

Mixed copepod cultures have been obtained from outdoor ponds or enclosed lagoons using fertilizers to enhance phytoplankton production, yielding densities of 10–30 ind/L. Seawater is pumped into the ponds through filters (20–40 μm), which ensure removal of predators but allow phytoplankton to enter the system. Through selective addition of nutrients, production of a particular phytoplankton is favored, such as the proliferation of diatoms, which has been considered more favorable for copepod production in open ponds than small flagellates (Naas et al. 1991). The species produced in such systems are primarily calanoids, but other copepods such as *Oithona* sp. and *Tisbe* sp. also occur (van der Meeren and Naas 1997). They can be collected with plankton nets to selectively sieve nauplii (80–250 μm), copepodite stages (80–350 μm), or primarily adult stages (250–600 μm) (Støttrup 2003).

Pond cultures are used in commercial systems in Denmark for rearing mixed copepods for feeding to first-feeding larvae of flatfish species or cod (e.g., Engell-Sørensen et al. 2004). In Norway, enclosed lagoons are used to extensively culture cod on natural zooplankton. Production is limited to summer months in extensive systems, which is probably why few hatcheries rely on this method for production. It is also difficult to control which species are being produced in highest numbers, making it difficult to optimize prey size or quality.

11.5.2.2 Intensive systems

Due to the diversity of habitats inhabited by copepods in nature, different systems have been developed for intensive cultivation of copepods (see reviews by Støttrup 2003, 2006; Lee et al. 2005). Calanoids are largely planktonic and occur in abundance in coastal areas. They are easily collected by nets pulled through the water. Species of the genera *Acartia*, *Eurytemora*, *Centropages*, and *Parvocalanus* have been the subject of several studies for intensive cultivation purposes (McKinnon et al. 2003; Støttrup 2003; Shields et al. 2005). Maintenance of cultures for calanoids has been described by Støttrup and colleagues (Støttrup et al. 1986; Støttrup 2003, 2006) for the calanoid copepod *Acartia tonsa*. Also relevant is the description of isolation and start-up procedures for the culture of *Parvocalanus* sp. by Shields et al. (2005). Most calanoids require live planktonic algae as food, the ambient density increasing 10-fold with decreasing algal size (Støttrup 2003, 2006). For example, it was recommended to use around 10^5 cells/mL when feeding copepods small algae (<5 μm) such as *Isochrysis galbana* or *Nannochloropsis oculata*, 10^4 cells/mL for larger algae such as *Rhodomonas salina*, and 10^3 cells/mL when algae >12 μm such as *Ditylum brightwelli*, *Thalassiosira weissflogii*, or *Crypthecodinium* sp. were used (Muller-Fuega et al. 2003; Støttrup 2003, 2006). Concurrently, it was important to maintain copepod densities to ensure that the phytoplankton was eaten and to minimize sedimentation and subsequent decomposition of algal matter on the bottom of the cultivation tanks. Optimal densities vary with species, from 100/L for *Acartia tonsa* (Støttrup et al. 1986) to 2,000/L for *Eurytemora affinis* (Turk et al. 1982).

Copepods in intensive cultures may be fed a single algal species. For example, an average birth rate of 31 eggs/female/day was obtained from *Parvocalanus crassirostris* females fed the dinoflagellate *Heterocapsa niei* (McKinnon et al. 2003). The copepods may alternatively be fed a mixture of algae. It really depends on finding the correct combination of cell size and nutritional value of the microalgal species used for feeding the copepods. Copepod fecundity is impacted by the available dietary lipids, and the provision of algae with high levels of essential fatty acids, such as EPA (20:5n-3) and DHA (22:6n-3), has been shown to be important (Støttrup and Jensen

1990; Lacoste et al. 2001). Shields et al. (2005) demonstrated improved fecundity in *Parvocalanus* sp. when fed a mixture of *Chaetoceros* sp. and *Isochrysis* sp. This is probably due to the mixture of algae providing a more balanced lipid distribution because *Chaetoceros* sp. contains low DHA but high EPA levels, whereas *Isochrysis* sp. has a high level of DHA but is low in EPA (Muller-Fuega et al. 2003). Fed this complementary combination, *Parvocalanus* sp. kept in 400-L culture systems produced 3,750 nauplii/L culture daily. Generally, however, production is lower; for example, the daily production was 444/L in a 1,000-L batch culture of a tropical *Acartia* species (Schipp et al. 1999) and 878/L for *Gladioferens imparipes* in 500-L batch cultures (Payne and Rippingale 2000b).

Calanoids are primarily broadcast spawners, and this type of reproduction has led to the development of a system whereby the sedimented eggs are collected daily from the tank bottom (Støttrup et al. 1986). Eggs are separated by filtration (45-μm mesh size) from the siphoned bottom detritus and put into separate containers for hatching. They can then either be fed directly to the fish larvae or allowed to grow for a specific time to be fed at various sizes to the growing fish larvae. This daily removal of up to 10% of the tank volume from the bottom to harvest eggs has several advantages, as listed in Støttrup (2006), and includes maintaining tank hygiene, provision of eggs free from debris, separation of eggs from the culture system reducing egg cannibalism within the system, ability to target prey size for feeding, and ability to store nondiapaused eggs for later use. Eggs from *Acartia tonsa* can be stored in anoxic conditions in a refrigerator at 4°C for a period of up to 6 months, with a linear decrease in hatching of roughly 4% every 20 days from 85% fresh hatch (Peck and Holste 2006).

The production method for copepods bearing eggs sacs has to be modified relative to methods for broadcast spawners. Often batch systems are implemented whereby the whole culture is filtered daily through different mesh sizes that separate the naupliar stages from the adult stages and the detritus from the naupliar stages. Most copepods seem able to tolerate daily filtration. Several modifications of this batch method exist, including automated daily collection of nauplii (Payne and Rippingale 2000a), but production output was well below the potential and further work was needed to increase the efficiency.

Benthic species generally require a large surface area, but this can be obtained within a small laboratory space as was demonstrated in the intensive batch system developed by Støttrup and Norsker (1997). Harpacticoids are highly productive and can attain high densities, exceeding 40,000/L in culture, and species of the genera *Tisbe*, *Tigriopus*, and *Euterpina* have been most commonly targeted for culture (Støttrup 2003). Several species have planktonic, free-swimming nauplii, which make them ideal for culture and their nauplii ideal as prey for marine fish larvae. Older stages are, however, generally not ideal as prey for fish larvae due to their benthic lifestyles. Harpacticoid species are generally relatively robust for culture (Støttrup and Norsker 1997; Cutts 2002). From batch cultures in small flat trays (40 × 60 cm), 100,000–170,000 nauplii/L were harvested daily. One problem encountered was the separation of nauplii from debris from such systems (Støttrup 2006). On the other hand, harpacticoids can be fed inert feeds and their culture is not dependent on a supply of fresh microalgae.

Several cyclopoid species of the genera *Apocyclops* have been cultured with success (Lipman 2001; Su et al. 2005). *Apocyclops royi* has a short generation time, developing to maturity within 4–5 days and is very ideal for culturing (Su et al. 2005). As for harpacticoids, these species can be fed inert feeds, although most researchers use phytoplankton or a mixture of phytoplankton and yeast

(Phelps et al. 2005). Cyclopoids can be cultured at similar high densities as harpacticoids (James and Al-Kars 1986; Phelps et al. 2005) and their planktonic nauplii are easily available for first-feeding fish larvae. Similar to harpacticoids, cyclopoids produce egg sacs, and culture methods developed for these species involve the harvest of nauplii from culture tanks. Productivity levels are similar to those obtained in harpacticoids (Lipman 2001; Støttrup 2003, 2006; Phelps et al. 2005). The recommended adult densities for culturing *Apocyclops* sp. was 2,000/L, providing an average daily harvest of around 8 nauplii/female (Phelps et al. 2005). Separation of nauplii from debris is less of a problem as long as planktonic feed is provided.

11.5.3 Enrichment and nutritional quality

Copepods are nutritionally adequate for marine fish larvae and do not generally need to be enriched before their use as live feed. Indeed, the biochemical content of copepods is being explored in detail to provide information for improving enrichment emulsions, with the aim to mimic the biochemical content of copepods in traditional live feed through enrichment (van der Meeren et al. 2008).

Copepods have generally lower total lipid and neutral lipid contents as compared with traditional cultured live prey (rotifers and *Artemia*), but have a higher proportion of more easily digestible phospholipids (Evjemo et al. 2003; van der Meeren et al. 2008). Phospholipids in adult mixed copepods can constitute up to 57% of total lipid and up to 63% in nauplii (van der Meeren et al. 2008; Figure 11.2). Cultured copepods maintain their relatively high phospholipid content even through several hundred generations in culture. In cultured *Acartia tonsa* adults and newly hatched nauplii fed *Rhodomonas baltica*, phospholipids constituted 58 and 50% of total lipids, respectively (Støttrup et al. 1999; Figure 11.9). Phospholipids also facilitate digestion of other lipids (Koven et al. 1993), but it has, to date, proven difficult to enhance phospholipid content in *Artemia* and rotifers. Furthermore, marine fish larvae require high levels of essential fatty acids such

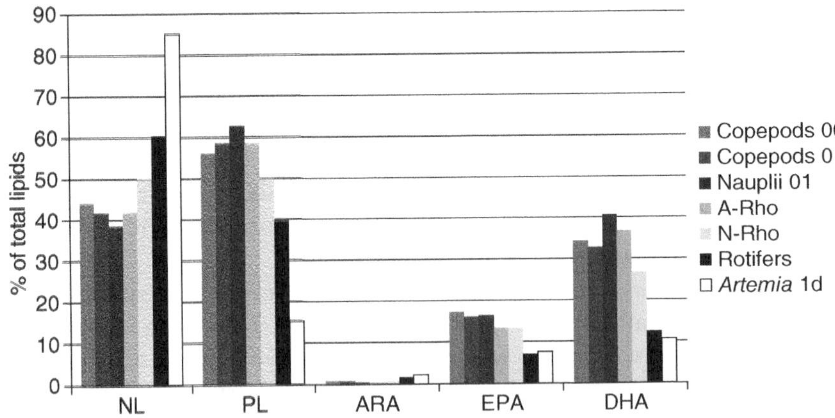

Figure 11.9 Contents of neutral lipids (NL), phospholipids (PL), arachidonic acid (ARA), eicosapentaenoic acid (EPA), and docosahexaenoic acid (DHA) as percentage of total lipids in mixed species of adult copepods sampled from outdoor ponds in 2000 (Copepods 00) and 2001 (Copepods 01) and copepod nauplii in 2001 (Nauplii 01); in laboratory-cultured *Acartia tonsa* adults fed *Rhodomonas baltica* (A-Rho) and nauplii (N-Rho), as well as in rotifers and 1-day-old *Artemia* nauplii (*Artemia* 1d). Data calculated from van der Meeren et al. (2008) and taken from Støttrup et al. (1999).

as DHA (22:6n-3), EPA (20:5n-3), and arachidonic acid (ARA, 20:4n-6). It may therefore be easier for larvae to assimilate DHA from copepods than from *Artemia* nauplii, where these essential fatty acids are mainly available as neutral lipids (McEvoy et al. 1998).

Documented higher growth rates, higher survival, and better pigmentation in larval and juvenile fish have been attributed to the content and ratio of the essential fatty acids DHA, EPA, and ARA in copepods (Bell et al. 1995a, 1995b; Sargent et al. 1997; Estevez et al. 1999). Calanoids require these essential long-chain unsaturated fatty acids and may procure them through their microalgal diet. On the other hand, the harpacticoids *Tisbe holothuriae* and *Nitocra lacustris* were shown to have the necessary desaturase and elongase enzymes necessary for bioconversion of shorter-chain fatty acids to EPA, DHA, and ARA (Norsker and Støttrup 1994; Nanton and Castell 1998; Rhodes and Boyd 2005). The mixture of copepods tested by van der Meeren et al. (2008) contained high levels of both EPA and DHA (Figure 11.9), and the ratio of EPA to ARA was also high in copepods. The ratio of DHA to EPA in calanoids and harpacticoids, whether harvested from natural systems or cultured, is generally >2 (McEvoy et al. 1998; Nanton and Castell 1998; Støttrup et al. 1999). High EPA:ARA ratio is important for successful development of pigmentation during metamorphosis in turbot (Estevez et al. 1999), halibut (McEvoy et al. 1998), yellowtail flounder (Copeman et al. 2002), and Senegalese sole (Villalta et al. 2005). Similar to wild copepods, cultured copepods also contain high ratios of EPA to ARA. EPA:ARA ratios in a mixture of copepods harvested from natural systems was 23–25 (van der Meeren et al. 2008). Adults and nauplii of *Acartia tonsa* fed *Rhodomonas baltica* had EPA-to-ARA ratios of 27 and 24, respectively (Støttrup et al. 1999).

The high levels of HUFA increase the need to protect against free radicals from lipid peroxidation. Copepods provide ample protection through the extremely high levels of astaxanthin, which are almost absent in rotifers and absent in *Artemia* (van der Meeren et al. 2008). These pigments provide natural antioxidants for fish and are also important precursors for vitamin A, although this may be provided through the high levels of another pigment, canthaxanthin, in *Artemia* (van der Meeren et al. 2008). However, Rønnestad et al. (1998) found that halibut larvae fed 14 days on *Artemia* nauplii contained 50–80% lower vitamin A (retinol and retinal) than larvae fed copepods, suggesting that the fish larvae may be incapable of assimilating or converting canthaxanthin in *Artemia*. The antioxidative capacity of astaxanthin in copepods is further enhanced by the high levels of ascorbic acid, which were double those in rotifers but similar to levels in *Artemia* nauplii (van der Meeren et al. 2008). An interesting feature was the low level of vitamin E in copepods compared with traditional live prey. This is interesting because it may not be necessary to provide the high levels of vitamin E in enrichment emulsions for traditional live prey.

Copepods have generally high total protein content (302–443 μg/mg DW), free amino acid content (56–86 μg/mg DW), and essential amino acids (van der Meeren et al. 2008). Free amino acids serve as an energy substrate and sustain protein synthesis in fish larvae, which have a high demand for energy to support high daily growth rates that may exceed 20% in weight. Indeed high initial daily weight gains were observed in cod fed natural zooplankton, achieving growth rates close to optimal when sufficient feed was provided and far superior weight gain than so far achieved for cod larvae fed rotifers (Busch et al. 2010). Higher growth rates achieved during the larval stage by feeding cod with live zooplankton as compared with rotifers persisted beyond the larval stage through the juvenile stage (Koedijk 2009). The early dietary effect not only resulted in improved growth during the juvenile phase but also improved the adaptive response to poor rearing environment such as high ambient ammonia and low oxygen levels (Koedijk 2009).

Micronutrients in copepods may also be different from those found in rotifers or *Artemia* nauplii, but little information is available on larval fish requirements. Levels of most trace elements are higher in copepods (Hamre et al. 2008b). Other factors may also play a role in confirming the superior nutritional value of copepods. Larval fish have limited proteolytic capacity, but digestive capacity may be aided by autolysis of ingested copepods (Fyhn et al. 1993; Lacoste et al. 2001). Also, zooplankton provide a wide range of sizes relative to, for example, rotifers, enabling the fish larvae to forage optimally, and this may explain the results obtained by Koedijk (2009) and Busch et al. 2010 for cod fed zooplankton relative to cod fed rotifers.

Literature cited

Abrams, P.A. 1982. Functional responses of optimal foragers. *The American Naturalist* 120:382–390.

Ackman, R.G., Tocher, C.S., and McLachland, J. 1968. Marine phytoplankton fatty acids. *Journal of the Fisheries Research Board of Canada* 25:1603–1620.

Andersen, R.A. 2005. *Algal Culturing Techniques*. Elsevier Academic Press, Burlington, MA.

Baer, A., Langdon, C., Mills, S., Schulz, C., and Hamre, K. 2008. Particle size preference, gut filling and evacuation rates of the rotifer *Brachionus* "Cayman" using polystyrene latex beads. *Aquaculture* 282(1–4):75–82.

Barlow, D.I., and Sleigh, M.A. 1980. The propulsion and use of water current for swimming and feeding in larval and adult *Artemia*. In: Persoone, G., Sorgeloos, P., Roels, O.A., and Jaspers, E. (eds.) *The Brine Shrimp Artemia, Vol. 1, Morphology, Genetics, Radiobiology, Toxicology*. Universa Press, Wetteren, Belgium, pp. 61–73.

Bell, J.G., Batty, R.S., Dick, J.R., Fretwell, K., Navarro, J.C., and Sargent, J.R. 1995a. Dietary deficiency of docosahexaenoic acid impairs vision at low light intensities in juvenile herring (*Clupea harengus* L.). *Lipids* 30:443–449.

Bell, J.G., Castell, J.D., Tocher, D.R., MacDonald, F.M., and Sargent, J.R. 1995b. Effects of different dietary arachidonic acid:docosahexaenoic acid ratios on phospholipid fatty acid compositions and prostaglandin production in juvenile turbot (*Scophthalmus maximus*). *Fish Physiology and Biochemistry* 14:139–151.

Ben-Amotz, A., Tornabene, T.G., and Thomas, W.H. 1985. Chemical profile of selected species of microalgae with emphasis on lipids. *Journal of Phycology* 21:72–81.

Blanchard, C.E. 1987. A scanning electron-microscope study of the development of the phyllopods in *Artemia*. In: Sorgeloos, P., Bengtson, D.A., Decleir, W., and Jaspers, E. (eds.) *Artemia Research and Its Application, Vol. 1, Morphology, Systematics, Genetics, Radiobiology, Ecotoxicology*. Universa Press, Wetteren, Belgium, pp. 5–32.

Bruggeman, E., Sorgeloos, P., and Vanhaecke, P. 1980. Improvements in the decapsulation technique of *Artemia* cysts. In: Persoone, G., Sorgeloos, P., Roels, O., and Jaspers, E. (eds.) *The Brine Shrimp Artemia, Vol. 3, Ecology, Culturing, Use in Aquaculture*. Universa Press, Wetteren, Belgium, pp. 261–269.

Busch, K.E.T., Falk-Petersen, I.-B., Peruzzi, S., Rist, N.A., and Hamre, K. 2010. Natural zooplankton as larval feed in intensive rearing systems for juvenile production of Atlantic cod (*Gadus morhua* L.). *Aquaculture Research* 41:1727–1740.

Chuecas, L., and Riley, J.P. 1969. Component fatty acids of the total lipids of some marine phytoplankton. *Journal of the Marine Biological Association of the United Kingdom* 49: 97–116.

Ciros-Pérez, J., Gómez, A., and Serra, M. 2001. On the taxonomy of three sympatric sibling species of the *Brachionus plicatilis* (Rotifera) complex from Spain, with the description of *B. ibericus* n. sp. *Journal of Plankton Research* 23:1311–1328.

Copeman, L.A., Parrish, C.C., Brown, J.A., and Harel, M. 2002. Effects of docosahexaenoic, eicosapentaenoic and arachidonic acids on the early growth, survival, lipid composition and pigmentation of yellowtail flounder (*Limanda ferruginea*): a live food enrichment experiment. *Aquaculture* 210:285–304.

Cutts, C.J. 2002. Culture of harpacticoid copepods: potential as live feed for rearing marine fish. *Advances in Marine Biology* 44:295–316.

Daan, R., Gonzalez, S.R., and Klein Breteler, W.C.M. 1988. Cannibalism in omnivorous calanoid copepods. *Marine Ecology Progress Series* 47:45–54.

D'Agostino, A.S. 1980. The vital requirements of *Artemia*, physiology and nutrition. In: Persoone, G., Soorgeloos, P., Roels, O., and Jaspers, E. (eds.) *The Brine Shrimp Artemia, Vol. 2, Physiology, Biochemistry, Molecular Biology*. Universa Press, Wetteren, Belgium, pp. 55–82.

Dhert, P., Sorgeloos, P., and Devresse, B. 1993. Contribution towards a specific DHA enrichment in the live food Brachionus plicatilis and *Artemia* sp. In: Reinertsen, H., Dahle, L.H., Jørgensen, L., and Tvinnerheim, K. (eds.) *Proceedings from the International Conference on Fish Farming Technology*, Trondheim, Norway. A.A. Balkema Publishers, Rotterdam, pp. 109–115.

Dhont, J., and Van Stappen, G. 2003. Biology, tank production and nutrition value of *Artemia*. In: Støttrup, J.G., and McEvoy, L. (eds.) *Live Feeds in Marine Aquaculture*. Blackwell Publishing, Oxford, pp. 65–111.

Dussart, B.H., and Defaye, D. (eds.) 2001. *Introduction to the Copepoda*. Backhuys, Leiden.

Engell-Sørensen, K., Støttrup, J.G., and Holmstrup, M. 2004. Rearing of flounder (*Platichthys flesus*) juveniles in semi-extensive systems. *Aquaculture* 230:475–491.

Epp, R.W., and Winston, P.W. 1977. Osmotic regulation in the brackish-water rotifer *Brachionus plicatilis* (Muller). *The Journal of Experimental Biology* 68:151–156.

Estevez, A., McEvoy, L.A., Bell, J.G., and Sargent, J.R. 1999. Growth, survival, lipid composition and pigmentation of turbot (*Scophthalmus maximus*) larvae fed live-prey enriched in arachidonic and eicosapentaenoic acids. *Aquaculture* 180:321–343.

Evjemo, J.O. 2001. Production and nutritional adaptation of the brine shrimp *Artemia* sp. as a live food organism for larvae of marine cold water fish species. PhD thesis, Norwegian University of Science and Technology (NTNU), Department of Zoology, Trondheim, Norway.

Evjemo, J.O., and Olsen, Y. 1997. Lipid and fatty acid content in cultivated live feed organisms compared to marine copepods. *Hydrobiologia* 358:159–162.

Evjemo, J.O., Coutteau, P., Olsen, Y., and Sorgeloos, P. 1997. The stability of decosahexaenoic acid in two *Artemia* species following enrichment and subsequent starvation. *Aquaculture* 155:135–148.

Evjemo, J.O., Vadstein, O., and Olsen, Y. 2000. Feeding and assimilation kinetics of *Artemia franciscana* given different concentrations of algae (T-iso). *Marine Biology* 136:1099–1109.

Evjemo, J.O., Danielsen, T.L., and Olsen, Y. 2001. Stability of lipid, protein and n-3 fatty acids in enriched *Artemia franciscana* starved at different temperatures. *Aquaculture* 193:65–80.

Evjemo, J.O., Reitan, K.I., and Olsen, Y. 2003. Copepods as live food organisms in the rearing of Atlantic halibut (*Hippoglossus hippoglossus* L.) with special emphasis on nutritional value. *Aquaculture* 227:191–210.

Fernández, I., Hontoria, F., Ortiz-Delgado, J.B., Kotzamanis, Y., Estévez, A., Zambonino-Infante, J.L., and Gisbert, E. 2008. Larval performance and skeletal deformities in farmed gilthead sea bream (*Sparus aurata*) fed with graded levels of vitamin A enriched rotifers (*Brachionus plicatilis*). *Aquaculture* 283(1–4):102–115.

Fyhn, H.J., Finn, R.N., Helland, S., Rønnestad, I., and Lømsland, E.R. 1993. Nutritional value of phyto- and zooplankton as live food for marine fish. In: Reinertsen, H., Dahle, L.A., Jørgensen, L., and Tvinnereim, K. (eds.) *Proceedings from the International Conference on Fish Farming Technology*, Trondheim. Norway, August 1993. A.A. Balkema Publishers, Rotterdam, pp.121–126.

Gómez, A., Serra, M., Carvalho, G.R., and Lunt, D.H. 2002. Speciation in ancient cryptic species complexes: evidence from the molecular phylogeny of *Brachionus plicatilis* (Rotifera). *Evolution* 56:1431–1444.

Guillard, R.R.L. 1975. Culture of phytoplankton for feeding marine invertebrates. In: Smith, W.L., and Chanley, M.H. (eds.) *Culture of Marine Invertebrate Animals*. Plenum Publishing, New York, pp. 19–60.

Hagiwara, A. 1996. Appearance of floating resting eggs in the rotifers *Brachionus plicatilis* and *B. rotundiformis*. *Bulletin of the Faculty of Fisheries, Nagasaki University* 77:111–115.

Hamre, K., Mollan, T.A., Sæle, Ø., and Erstad, B. 2008a. Rotifers enriched with iodine and selenium increase survival in Atlantic cod (*Gadus morhua*) larvae. *Aquaculture* 284:190–195.

Hamre, K., Srivastava, A., Rønnestad, I., Mangor-Jensen, A., and Stoss, J. 2008b. Several micronutrients in the rotifer *Brachionus plicatilis* may be limiting for growth, survival and normal development of cod larvae. *Aquaculture Nutrition* 14:51–60.

Hayashi, M., Toda, K., Yoneji, T., Sato, O., and Kitaoka, S. 1993. Dietary value of rotifers and *Artemia* enriched with *Euglena gracilis* for red sea bream. *Nippon Suisan Gakkaishi* 59:1051–1058.

Helland, S., Triantaphyllidis, G.V., Fyhn, H.J., Evjen, M.S., Lavens, P., and Sorgeloos, P. 1999. Modulation of the free amino acid pool and protein content in populations of the brine shrimp *Artemia*. *Marine Biology* 137:1005–1016.

Hicks, G.R.F., and Coull, B.C. 1983. The ecology of marine meiobenthic harpacticoid copepods. *Oceanography and Marine Biology Annual Review* 21:67–175.

Hirano, K. 1987. Studies on the culture of the rotifer (*Brachionus plicatilis* O.F. Muller). *Bulletin of the Faculty of Agriculture, Miyazaki University* 34:57–122.

Holling, C.S. 1966. The functional response of invertebrate predators to prey density. *Memoirs of the Entomological Society of Canada* 48:1–85.

Humes, A.G. 1994. How many copepods. *Hydrobiologia* 292/293:1–7.

Huys, R., and Boxshall, G.A. (eds.) 1991. *Copepod Evolution*. The Ray Society, London.

Ito, T. 1960. On the culture of the mixohaline rotifer *Brachionus plicatilis* O.F. Muller, in sea water. *Report of the Faculty of Fisheries, Prefectural University of Mie* 3:708–740.

James, C.M., and Al-Kars, A.M. 1986. Studies on the production of planktonic copepods for aquaculture. In: Schriever, G., Schminke, H.K., and Shih, C.-T. (eds.) *Proceedings of the Second International Conference on Copepoda*, Ottawa, Canada, August 13–17, 1984, pp. 333–340.

Kiørboe, T., Saiz, E., and Viitasalo, M. 1996. Prey switching behaviour in the planktonic copepod *Acartia tonsa*. *Marine Ecology Progress Series* 143:65–75.

Koedijk, R.M. 2009. Phenotypic plasticity influenced by diet during early development. A comparative study on Atlantic cod. PhD thesis, University of Bergen, Norway.

Kostopoulou, V., and Vadstein, O. 2007. Growth performance of the rotifers *Brachionus plicatilis*, B. "Nevada" and B. "Cayman" under different food concentrations. *Aquaculture* 273:449–458.

Koven, W.M., Tandler, A., Sklan, D., and Kissil, G.W. 1993. The association of eicosapentaenoic and docosahexaenoic acids in the main phospholipids of different-age *Sparus aurata* larvae with growth. *Aquaculture* 116:71–82.

Kühlea, K., and Kleinow, W. 1985. Measurements of hydrolytic enzymes in homogenates from *Brachionus plicatilis* (Rotifera). *Comparative Biochemistry and Physiology. Part B, Comparative Biochemistry* 81:437–442.

Lacoste, A., Poulet, S.A., Cueff, A., Kattner, G., Ianora, A., and Laabir, M. 2001. New evidence of the copepod maternal food effects on reproduction. *Journal of Experimental Marine Biology and Ecology* 259:85–107.

Lavens, P., and Sorgeloos, P. 1996. Manual on the production and use of live food for aquaculture. FAO Fisheries Technical Paper No. 361, pp. 45–295.

Lavens, P., Léger, P., and Sorgeloos, P. 1989. Manipulation of the fatty acid profile in *Artemia* offspring produced in intensive culture systems. In: De Paw, N., Jaspers, E., Ackefors, H., and Wilkins, N. (eds.) *Aquaculture and Biotechnology in Progress*. European Aquaculture Society, Bredne, pp. 731–739.

Lee, C.-S., O'Bryen, P.J., and Marcus, N.H. 2005. *Copepods in Aquaculture*. Blackwell Publishing, Oxford.

Léger, P., and Sorgeloos, P. 1992. Optimized feeding regimes in shrimp hatcheries. In: Fast, A.W., and Lester, L.J. (eds.) *Culture of Marine Shrimp: Principles and Practices*. Elsevier Science Publishers, New York, pp. 225–244.

Léger, P., Bengtson, D.A., Simpson, K.L., and Sorgeloos, P. 1986. The use and nutritional value of *Artemia* as a food source. *Marine Biology Annual Review* 24:521–623.

Lie, O., Haaland, H., and Hemre, G.-I. 1997. Nutritional composition of rotifers following a change in diet from yeast emulsified oil to microalgae. *Aquaculture International* 5:427–438.

Lipman, E.E. 2001. Production of the copepod *Apocyclops panamensis* under hatchery conditions. MSc thesis, Auburn University, Alabama.

Lubzens, E. 1987. Raising rotifers for use in aquaculture. *Hydrobiologia* 147:245–255.

Lubzens, E., and Zmora, O. 2003. Production and nutritional value of rotifers. In: Stottrup, J.C., and McEvoy, L.A. (eds.) *Live Feeds in Marine Aquaculture*. Blackwell, Oxford, pp. 17–64.

Lubzens, E., Zmora, O., and Barr, Y. 2001. Biotechnology and aquaculture of rotifers. *Hydrobiologia* 446/447:337–353.

Mæland, A., Ronnestad, I., Fyhn, H.J., Berg, L., and Waagebo, R. 2000. Water-soluble vitamins in natural plankton (copepods) during two consecutive spring blooms compared to vitamins in *Artemia franciscana* nauplii and metanauplii. *Marine Biology* 136:765–772.

Makridis, P., and Olsen, Y. 1999. Protein depletion of the rotifer *Brachionus plicatilis* during starvation. *Aquaculture* 174:343–353.

Mauchline, J. (ed.) 1998. *Advances in Marine Biology. The Biology of Calanoid Copepods*. Academic Press, New York.

McEvoy, L., Næss, T., Bell, J.G., and Lie, O. 1998. Lipid and fatty acid composition of normal and malpigmented Atlantic halibut (*Hippoglossus hippoglossus*) fed enriched *Artemia*: a comparison with fry fed wild copepods. *Aquaculture* 163:235–248.

McKinnon, A.D., Duggan, S., Nichols, P.D., Rimmer, M.A., Semmens, G., and Robino, B. 2003. The potential of tropical paracalanid copepods as live feeds in aquaculture. *Aquaculture* 223:89–106.

McMahon, J.W., and Rigler, F.H. 1963. Mechanisms regulating the feeding rate of *Daphnia manga* Straus. *Canadian Journal of Zoology* 41: 321–332.

van der Meeren, T., and Naas, K.E. 1997. Development of rearing techniques using large enclosed ecosystems in the mass production of marine fish fry. *Reviews in Fisheries Science* 5:367–390.

van der Meeren, T., Olsen, R.E., Hamre, K., and Fyhn, H.J. 2008. Biochemical composition of copepods for evaluation of feed quality in production of juvenile marine fish. *Aquaculture* 274:375–397.

Merchie, G., Lavens, P., Dhert, P., Dehasque, M., Nelis, H., Leenheer, A.D., and Sorgeloos, P. 1995. Variation of ascorbic acid content in different live food organisms. *Aquaculture* 134:325–337.

Mortensen, S.H., Børsheim, K.Y., Rainuzzo, J.R., and Knutsen, G. 1988. Fatty acid and elemental composition of the marine diatom *Chaetoceros gracilis* Schütt. Effects of silicate deprivation, temperature and light intensity. *Journal of Experimental Marine Biology and Ecology* 122:173–185.

Muller-Fuega, A., Moal, J., and Kaas, R. 2003. The microalgae of aquaculture. In: Støttrup, J.G., and McEvoy, L.A. (eds.) *Live Feeds in Marine Aquaculture*. Blackwell Science, Oxford, UK, pp. 206–252.

Naas, K.E., van der Meeren, T., and Aksnes, D.L. 1991. Plankton succession and responses to manipulations in a marine basin for larval rearing. *Marine Ecology Progress Series* 74:161–173.

Naas, K.E., Næss, T., and Harboe, T. 1992. Enhanced first feeding of halibut larvae (*Hippoglossus hippoglossus* L.) in green water. *Aquaculture* 105:143–156.

Nagata, W.D., and Hirata, H. 1986. Mariculture in Japan: past, present and future perspectives. *Mini Review and Data File of Fisheries Research* 4:1–38.

Nagata, W.D., and Whyte, J.N.C. 1992. Effect of yeast and algal diets on the growth and biochemical composition of the rotifer *Brachionus plicatilis* (Muller) in culture. *Aquaculture Fish Management* 23:12–21.

Nanton, D.A., and Castell, J.D. 1998. The effects of dietary fatty acids on the fatty acid composition of the harpacticoid copepod, *Tisbe* sp., for use as a live food for marine fish larvae. *Aquaculture* 163:251–261.

Navarro, J.C., Batty, R.S., Bell, M.V., and Sargent, J.R. 1993. Effects of two *Artemia* diets with different content of polyunsaturated fatty acids on the lipid composition of larvae of Atlantic herring (*Clupea harengus*). *Journal of Fish Biology* 43:503–515.

Navarro, J.C., Henderson, R.J., McEvoy, L.A., Bell, M.V., and Amat, F. 1999. Lipid conversions during enrichment of *Artemia*. *Aquaculture* 174:155–166.

Norsker, N.H., and Støttrup, J.G. 1994. The importance of dietary HUFA's for fecundity and HUFA content in the harpacticoid, *Tisbe holothuriae* Humes. *Aquaculture* 125:155–166.

NRC. 1993. *Nutrient Requirements of Fish*. National Research Council, National Academy Press, Washington, DC.

Øie, G. 1997. Evaluation of rotifer (*Brachionus plicatilis*) quality in early first feeding of turbot (*Scophthalmus maximus* L) larvae. Dr. thesis. NTNU, Norway.

Øie, G., Reitan, K.I., and Olsen, Y. 1994. Comparison of rotifer culture quality with yeast plus oil and algal-based cultivation diets. *Aquaculture International* 2:225–238.

Øie, G., Makridis, P., Reitan, K.I., and Olsen, Y. 1997. Protein and carbon utilization of rotifers (*Brachionus plicatilis*) in first feeding of turbot larvae (*Scophthalmus maximus* L.). *Aquaculture* 153:103–122.

Olsen, Y., Evjemo, J.O., and Olsen, A. 1999. Status of the cultivation technology for production of Atlantic halibut (*Hippoglossus hippoglossus*) juveniles in Norway/Europe. *Aquaculture* 176:3–13.

Olsen, Y. 1990. Cultivated microalgae as a source of omega-3 fatty acids. Proceedings of the International Symposium on Fish Lipids and Their Influence on Human Health, Svanøy, August 9–11, 1989, Svanøy Foundation, Report No. 9.

Olsen, Y. 1999. Lipids and essential fatty acids in aquatic food webs: what can freshwater ecologists learn from mariculture? In: Arts, M.T., and Wainman, B.C. (eds.) *Lipids in Freshwater Ecosystem*. Springer, New York, pp. 161–202.

Olsen, Y. 2004. Live food technology of cold-water marine fish larvae. In: Moksness, E., Kjørsvik, E., and Olsen, Y. (eds.) *Culture of Cold-Water Marine Fish*. Blackwell Publishing, Oxford, pp. 73–111.

Olsen, Y., Rainuzzo, J.R., Reitan, K.I., and Vadstein, O. 1993. Manipulation of lipids and ω3 fatty acids in *Brachionus plicatilis*. In: Reinertsen, H., Dahle, L.A., Jørgensen, L., and Tvinnereim, K. (eds.) *Proceedings of the First International Conference on Fish Farming Technology*, Trondheim, Norway, August 9–12, 1993. A.A. Balkema, Rotterdam, pp. 101–108.

Olsen, A.I., Mæland, A., Waagebø, R., Olsen, Y. 2000. Effect of algal addition on stability of fatty acids and some water-soluble vitamins in juvenile *Artemia franciscana*. *Aquaculture Nutrition* 6(4):263–273.

Payne, M.F., and Rippingale, R.J. 2000a. Rearing West Australian seahorse, *Hippocampus subelongatus* juveniles on copepod nauplii and enriched *Artemia*. *Aquaculture* 188:353–361.

Payne, M.F., and Rippingale, R.J. 2000b. Intensive cultivation of the calanoid copepod *Gladioferens imparipes*. *Aquaculture* 201:329–342.

Peck, M.A., and Holste, L. 2006. Effects of salinity, photoperiod and adult stocking density on egg production and egg hatching success in *Acartia tonsa* (Calanoida: Copepoda): optimizing intensive cultures. *Aquaculture* 255:341–350.

Phelps, R.P., Sumiarsa, G.S., Lipman, E.E., Lan, H.-P., Moss, K.K., and Davis, A.D. 2005. Intensive and extensive production techniques to provide copepod nauplii for feeding larval red snapper *Lutjanus campechanus*. In: Lee, C.-S., Bryen, P.J.O., and Marcus, N.H. (eds.) *Copepods in Aquaculture*. Blackwell Publishing Ltd., Oxford, UK, pp. 151–168.

Pourriot, R., and Snell, T.W. 1983. Resting eggs in rotifers. *Hydrobiologia* 104:213–224.

Provaoli, L., and Shiraishi, K. 1959. Axenic cultivation of the brine shrimp *Artemia salina*. *Biological Bulletin of the Marine Biological Laboratory (Woods Hole)* 117:347–355.

Rainuzzo, J.R. 1993. Lipids in early stages of marine fish. PhD thesis, Norwegian University of Science and Technology (NTNU), Trondheim, Norway.

Reeve, M.R. 1963. The filter-feeding in *Artemia* in suspensions of various particles. *The Journal of Experimental Biology* 40:207–214.

Reitan, K.I. 1993. Nutritional effects of algal addition in first feeding of turbot (*Scophthalmus maximus* L) larvae. *Aquaculture* 118: 257–275.

Reitan, K.I., Bolla, S., and Olsen, Y. 1994a. A study of the mechanism of algal uptake in yolk-sac larvae of Atlantic halibut (*Hippoglossus hippoglossus* L). *Journal of Fish Physiology* 44:303–310.

Reitan, K.I., Rainuzzo, J.R., and Olsen, Y. 1994b. Influence of lipid composition of live feed on growth, survival and pigmentation of turbot larvae. *Aquaculture International* 2:33–48.

Reitan, K.I., Rainuzzo, J.R., and Olsen, Y. 1994c. Effect of nutrient limitation on fatty acid and lipid content of marine microalgae. *Journal of Phycology* 30:972–979.

Reitan, K.I., Rainuzzo, J.R., Øie, G., and Olsen, Y. 1997. A review of the nutritional effects of algae in marine fish larvae. *Aquaculture* 155:207–221.

Reitan, K.I., Natvik, C., and Vadstein, O. 1998. Uptake of micro algae and bacteria in turbot (*Scophthalmus maximus* L.) larvae. *Journal of Fish Biology* 53:1145–1154.

Rhodes, A., and Boyd, L. 2005. Formulated feeds for harpacticoid copepods: implications for population growth and fatty acid composition. In: Lee, C.-S., Bryen, P.J.O., and Marcus, N.H. (eds.) *Copepods in Aquaculture*. Blackwell Publishing Ltd, Oxford, UK, pp. 61–73.

Ricci, C. 2001. Dormancy pattern in rotifers. *Hydrobiologia* 446/447:1–11.

Rønnestad, I., Helland, S., and Lie, Ø. 1998. Feeding *Artemia* to larvae of Atlantic halibut (*Hippoglossus hippoglossus* L.) results in lower larval vitamin A content compared with feeding copepods. *Aquaculture* 165:159–164.

Sargent, J.R., McEvoy, L.A., and Bell, J.G. 1997. Requirements, presentation and sources of polyunsaturated fatty acids in marine fish larval feeds. *Aquaculture* 155:117–127.

Sazhina, L.I. (ed.) 2006. *Breeding, Growth Rates, and Production of Marine Copepods*. Universities Press, Andhra Pradesh, India.

Schipp, G.R., Bosmans, J.M.P., and Marshall, A.J. 1999. A method for hatchery cultivation of tropical calanoid copepods, *Acartia* spp. *Aquaculture* 174:81–88.

Schrehardt, A. 1987. A scanning electron-microscope study of the post-embryonic development of *Artemia*. In: Soorgeloos, P., Bengtson, D.A., Decleir, W., and Jaspers, E. (eds.) *Artemia Research and Its Applications, Vol. 1, Morphology, Genetics, Strain Characterization, Toxicology*. Universa Press, Wetteren, Belgium, pp. 5–31.

Serra, M., and King, C.E. 1999. Optimal rates of bisexual reproduction in cyclic parthenogens with density-dependent growth. *The Journal of Evolutionary Biology* 12:263–271.

Shields, R.J., Bell, J.G., Luizi, F.S., Gara, B., Bromage, N.R., and Sargent, J.R. 1999. Natural copepods are superior to enriched *Artemia* nauplii as feed for larvae (*Hippoglossus hippoglossus*) in terms of survival, pigmentation and retinal morphology: relation to dietary essential fatty acids. *Journal of Nutrition* 129(6):1186–1194.

Shields, R.J., Kotani, T., Molnar, A., Marion, K., Kobashigawa, J., and Tang, L. 2005. Intensive cultivation of a subtropical paracalanid copepod, *Parvocalanus* sp., as prey for small marine fish larvae. In: Lee, C.-S., Bryen, P.J.O., and Marcus, N.H. (eds.) *Copepods in Aquaculture*. Blackwell Publishing Ltd, Oxford, UK, pp. 209–223.

Skjermo, J., and Vadstein, O. 1993. The effect of microalgae on skin and gut bacterial flora of halibut larvae. In: Reinertsen, H., Dahle, L.A., Jørgensen, L., and Tvinnereim, K. (eds.) *Fish Farming Technology*. A.A. Balkema Publishers, Rotterdam, pp. 61–67.

Skjermo, J., Størseth, T.R., Hansen, K., Handå, A., and Øie, G. 2006. Evaluation of β-(1→3, 1→6)-glucans and High-M alginate used as immunostimulatory dietary supplement during first feeding and weaning of Atlantic cod (*Gadus morhua* L.). *Aquaculture* 261: 1088–1101.

Sorgeloos, P., Lavens, P., Léger, P., Tackaert, P., and Versichele, D. 1986. Manual for the culture and use of brine shrimp *Artemia* in aquaculture. Manual prepared for the Belgian Administration for Development Cooperation and the Food and Agriculture Organization of the United Nations. Artemia Reference Center, Faculty of Agriculture, State University of Ghent, Belgium.

van Stappen, G., Merchie, G., Dhont, J., Lavens, P., Baert, P., Bosteels, T., and Sorgeloos, P. 1996. Artemia. In: Lavens, P., and Sorgeloos, P. (eds.) *Manual on the Production and Use of Live Food for Aquaculture*. Food and Agriculture Organization of the United Nations, pp. 79–136.

Størseth, T.R., Kirkvold, S., Skjermo, J., and Reitan, K.I. 2006. A branched beta-D-(1→3,1→6)-glucan from the marine diatom *Chaetoceros debilis* (Bacillariophyceae) characterized by NMR. *Carbohydrate Research* 341:2108–2114.

Støttrup, J.G. 2003. Production and nutritional value of copepods. In: Støttrup, J.G., and McEvoy, L.A. (eds.) *Live Feeds in Marine Aquaculture*. Blackwell Science, Oxford, UK, pp. 145–205.

Støttrup, J.G. 2006. A review on the status and progress in rearing copepods for marine larviculture. Advantages and disadvantages among calanoid, harpacticoid and cyclopoids copepods. *Avences en Nutrición Acuícola VIII*.

Støttrup, J.G., and Jensen, J. 1990. Influence of algal diet on feeding and egg production of the calanoid copepod *Acartia tonsa* Dana. *Journal of Experimental Marine Biology and Ecology* 141:87–105.

Støttrup, J.G., and Norsker, N.H. 1997. Production and use of copepods in marine fish larviculture. *Aquaculture* 155:231–247.

Støttrup, J.G., Richardsen, K., Kirkegaard, E., and Pihl, N.J. 1986. The cultivation of *Acartia tonsa* Dana for use as live food source for marine fish larvae. *Aquaculture* 52:87–96.

Støttrup, J.G., Bell, J.G., and Sargent, J.R. 1999. The fate of lipids during development and cold-storage of eggs in the calanoid copepod *Acartia tonsa* Dana, and in response to different algal diets. *Aquaculture* 176:257–269.

Su, H.-M., Cheng, S.-H., Chen, T.-I., and Su, M.-S. 2005. Culture of copepods and application to marine finfish larval rearing in Taiwan. In: Lee, C.-S., O'Bryen, P.J., and Marcus, N.H. (eds.) *Copepods in Aquaculture*. Blackwell Publishing Ltd, Oxford, UK, pp. 183–194.

Suatoni, E., Vicario, S., Rice, R., Snell, T., and Caccone, A. 2006. An analysis of species boundaries and biogeographic patterns in a cryptic species complex: the rotifer—*Brachionus plicatilis*. *Molecular Phylogenetics and Evolution* 41:86–98.

Svåsand, T., Kristiansen, T.S., Pedersen, T., Salvanes, A.G.V., Engelsen, R., and Nødtvedt, M. 1998. *Havbeite med Torsk—Artsrapport*. Norges Forskningsråd, Oslo, Norway.

Turk, P.E., Krejci, M.E., and Yang, W.T. 1982. A laboratory method for the culture of *Acartia tonsa* (Crustacea: Copepoda) using rice bran. *Journal of Agriculture and Aquatic Sciences* 3:25–27.

Vadstein, O., Øie, G., and Olsen, Y. 1993. Size dependent feeding by the rotifer *Brachionus plicatilis*. *Hydrobiologia* 255/256:261–267.

Villalta, M., Estevez, M.P., and Bransden, M.P. 2005. Arachidonic acid enriched live prey induces albinism in Senegalese sole (*Solea senegalensis*) larvae. *Aquaculture* 245: 193–209.

Volkman, J.K., Jeffrey, S.W., Nichols, P.D., Rodgers, G.I., and Garland, C.D. 1989. Fatty acid and lipid composition of 10 species of microalgae used in mariculture. *Journal of Experimental Marine Biology and Ecology* 128:219–240.

Watanabe, T., Oowa, T.F., Kitajima, C., and Fujita, S. 1978. Nutritional quality of the brine shrimp, *Artemia salina*, as live feed from view point of essential fatty acids for fish. *Bulletin of the Japanese Society of Scientific Fisheries* 44:1115–1121.

Watanabe, T., Kitajima, C., and Fujita, S. 1983. Nutritional values of live feed organisms in Japan for mass propagation of fish: a review. *Aquaculture* 34:115–143.

Whyte, J.N.C., and Nagata, W.D. 1990. Carbohydrate and fatty acid composition of rotifer, *Brachionus plicatilis*, fed monospecific diets of yeast and phytoplankton. *Aquaculture* 89:263–368.

Yúfera, M. 2001. Studies on *Brachionus* (Rotifera): an example of interaction between fundamental and applied research. *Hydrobiologia* 446/447: 383–392.

Chapter 12

Microparticulate diets: technology

Chris Langdon and Rick Barrows

12.1 Introduction

Culture of the larval stages of many fish species is still dependent on expensive live feeds, such as rotifers and *Artemia*, despite several decades of research (Liao et al. 2001; Langdon 2003; Støttrup and McEvoy 2003). Production of live feeds can represent a significant proportion of the running costs of commercial hatcheries (Person Le Ruyet et al. 1993). Typically, the early larval stages of marine fish species are fed on live prey, such as rotifers or *Artemia* nauplii, and the larvae are weaned onto artificial diets at a later stage of development. However, the nutritional quality of these live feeds is often found to be inferior to that of natural prey, such as copepods, and mortalities of hatchery-reared larvae are often reported to be high (Shields et al. 1999; Evjemo et al. 2003; Drillet et al. 2006; Hamre 2006; Rajkumar and Kumaraguru vasagam 2006).

Microparticulate diets for fish larvae are typically in the 25–250-μm size range, and the optimal particle size varies depending on developmental stage and species (Dabrowski and Bardega 1984a; Fernández-Díaz et al. 1994). High surface-to-volume ratios of micron-sized particles result in high rates of diffusional exchange of nutrients and gases between particles and the surrounding medium. Furthermore, settlement (Backhurst and Harker 1988), clumping, and adherence of particles onto surfaces can result in substantial losses of feed (Figure 12.1).

One of the greatest challenges in designing artificial microparticles to deliver diets to fish larvae is to obtain a digestible particle type that also retains water-soluble nutrients (Figure 12.1). Unlike the feeding appendages of many crustacean larvae, fish larvae typically do not possess mechanisms to physically break down captured particles; therefore, their breakdown depends on the action of digestive enzymes coupled with the mechanical action of peristalsis.

The lack of a fully developed digestive system in the early developmental stages of many fish larval species (Munilla-Moran et al. 1990; Kjørsvik et al. 1991; Hamlin et al.

Larval Fish Nutrition, First Edition. Edited by G. Joan Holt.

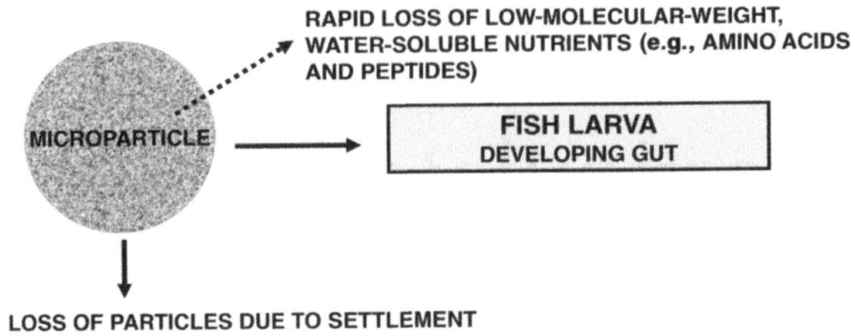

Figure 12.1 Challenges in developing microparticles to deliver nutrients to fish larvae are due to rapid leakage losses of low-molecular-weight, water-soluble nutrients, immature development of the larva's digestive system, and settlement of particles from suspension.

2000; Kolkovski 2001) suggests that nutrients within the food particle must be in a form that is readily digestible (Rust et al. 1993; Rust 1995; Cahu and Zambonino Infante 2001; see Chapter 1). Studies indicate that larvae benefit from additions of partially hydrolyzed, water-soluble dietary proteins to their diets (Cahu and Zambonino Infante 1994; Zambonino Infante et al. 1997; Takeuchi et al. 2003; see Chapter 3).

Increasing water solubility of dietary ingredients also increases potential rates of losses of these nutrients through diffusion. Most dietary low-molecular-weight, water-soluble nutrients, such as amino acids and water-soluble vitamins, are rapidly lost from microbound and cross-linked protein walled capsules after only a few minutes of suspension in water (López-Alvarado et al. 1994; Baskerville-Bridges and Kling 2000b). Water-soluble proteins and hydrolysates can also be rapidly lost from these particle types; for example, Chui and Wan (1997) reported that about 55% of incorporated trypsin was lost from alginate beads after only 10 minutes of aqueous suspension. Losses of nutrients into the culture medium typically result in increases in bacterial concentrations and probability of disease outbreaks (Muir and Sutton 1994; Alabi et al. 1999). On the other hand, loss of dietary water-soluble components might be desirable if they stimulate feeding when dissolved in the culture water by affecting "taste" and "smell" (Kolkovski et al. 1997; Cahu et al. 1998; Lazo et al. 2000; Nikolaeva and Kasumyan 2000; Koven et al. 2001; Lindsay and Vogt 2004).

In this chapter, we review technologies that have been used to prepare microparticles (Figure 12.2) for rearing fish larvae or for enrichment of their prey species, updating and broadening the scope of earlier reviews by Cahu and Zambonino Infante (2001), Langdon (2003), and Kvåle et al. (2007). We will not discuss diet formulation or nutritional requirements as these topics are addressed by others in this publication.

12.2 Delivery of high-molecular-weight proteins, carbohydrates, and water-insoluble nutrients

12.2.1 Microbound particles

The most common particle type used in feeding experiments with fish larvae are microbound particles in which dietary ingredients are bound together in a polymer matrix. These particles do not possess walls and, technically, should not be designated as microcapsules. Formation of microbound particles typically requires activating or gelling a mixture of ingredients, followed by drying.

Microbound particles can be further characterized as crumbled particles or shaped par-

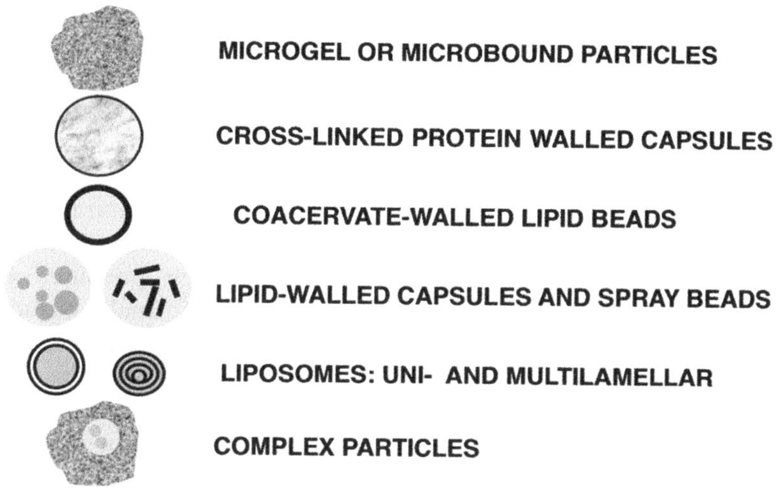

Figure 12.2 Microparticle types for delivery of diets to fish larvae.

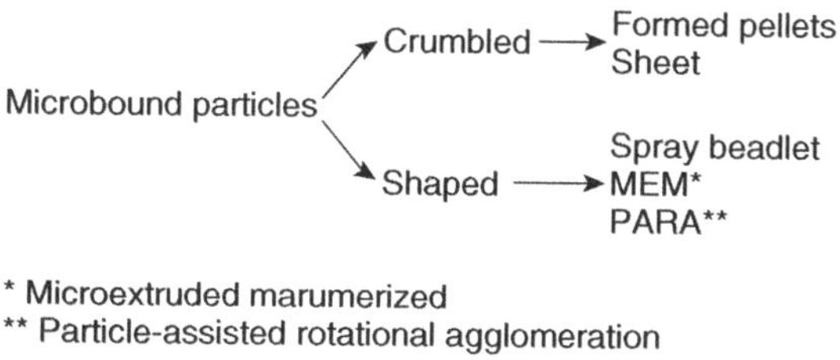

Figure 12.3 Classification of different microbound particle types.

ticles (Figure 12.3). If large particles are made in the initial gelling or activation step, either in the form of pellets or thin sheets of dietary material, then further grinding and sieving is required to obtain particles in the desired size range. Grinding typically results in irregular-shaped particles and dietary ingredients will not necessarily be completely embedded in the binder, resulting in loss of integrity when these particles are added to aqueous suspension. Whether the ragged, irregular shapes of these particles affects consumption by larvae can be debated, but the resulting increase in surface area-to-volume ratios will adversely affect the leaching rates of water-soluble nutrients.

Binding agents used to make microbound particles include hydrocolloids, such as alginate, carrageenan, and carboxymethylcellulose; starches; chitosan; gelatin; fish proteins; and zein. The choice of binding agent depends on the desired physical characteristics, such as particle integrity and leakage rates, as well as the biological responses of fish larvae, such as palatability and digestibility (see Langdon 2003 for a review). One binder may not be optimal for all species; for example, Teshima et al. (1982) reported that carrageenan-bound particles were optimal for the ayu *Plecoglossus altivelis* whereas zein-bound diets were optimal for the red sea bream *Chrysophrys major*. Person Le Ruyet et al. (1993) successfully weaned sea bass (*Dicentrarchus labrax*) larvae on diets bound with alginate or zein. In contrast, Partridge and Southgate (1999) reported that alginate- and zein-bound particles were poorly digested by barramundi (*Lates calcarifer*) larvae, and they recommended gelatin or carrageenan as binders instead. Carrageenan was not considered suitable as a binder for larvae of white sturgeon due to poor digestibility (Gawlicka et al. 1996). Guthrie et al. (2000) also reported that carrageenan was poorly accepted by first-feeding walleye (*Stizostedion vitreum*) larvae.

Takeuchi et al. (2003) described the preparation of microparticulate diets bound together with the addition of 25% w/w calcium salt of fish oil. The diet included peptides, amino acids, and water-soluble vitamins. No leakage data were presented, but larvae of the Japanese flounder *Paralichthys olivaceus* utilized the diets and overall larval performance was better than that of larvae fed on a commercial microbound diet.

Shaped microbound particles differ from crumbled particles in that they are formed as micron-sized particles that require no additional grinding (Figure 12.3). Shaped types include microextruded marumerized (MEM) particles, particle-assisted rotational agglomerated (PARA) particles (Barrows et al. 1993; Barrows and Lellis 2000), and spray beads (Önal and Langdon 2000).

MEM particles have been produced in the pharmaceutical industry for many decades, and recently have been used to produce experimental and commercial larval feeds. Commercial manufacture systems are currently being used to produce specialty, ornamental, and salmonid feeds. Two modifications of the typical feed production methodologies are needed for extrusion of particles in the size range needed for larval feeds. Extremely finely ground ingredients are required to produce all types of larval feeds, but it is particularly critical with the MEM process to prevent die blockage. It is generally recommended to have all ingredients smaller than 20% of the die opening. The best grinding method is dictated by the physical characteristic of the ingredients, and may include hammer mills, ball mills, pin mills, and homogenizers. The moisture, starch, and fat content, as well as other undefined characteristics affect how well an ingredient will be ground in a certain type of grinder. For high-fat, low-moisture ingredients, the air-swept pulverizer has been proven to be very effective on a large scale but costly in terms of energy demands.

In addition to finely ground ingredients, alternative extruder designs are required. Recent advancements in extruder design have resulted in a variety of machines capable of producing particles less than 300 μm. The primary constraint to extruding such a small particle is that the thickness of the die plate/screen is always related to the diameter of the particle hole. In traditional, axial discharge extruder designs with a flat plate/screen, the high pressure generated from making such a small extrudate results in damage to the die. Radial discharge extruders allow for increased surface area of the die since the die holes wrap around the screws, resulting in less pressure on the screen. Other extruder designs such as the twin-dome extruder also reduce pressure at the die screen and allow for production of these very small particles. With both radial discharge and twin-dome extruders, however, a cutter head cannot be effectively installed so an additional piece of equipment is needed to break the long extrudates into usable particles. A spheronizer (originally termed a marumerizer) consists of a cylindrical chamber with a rotating bottom plate. The plate is grooved to break the noodles, reshaping and compacting the particles. The spheronizer is equipped with a variable speed motor to allow a range of energies to be delivered to the extrudates. Some diet formulations will result in very strong noodles, requiring a coarse plate and high rotational velocities to adequately reshape the particle. Other diet formulations result in very weak noodles, and lower speeds and a fine or flat plate is used in the bottom of the spheronizer.

Moisture is added to the mixture prior to extrusion and can vary from 20 to 48% w/w depending on the nature of the ingredients (Barrows and Lellis 2006). If moisture levels are too low, either weak noodles (limited binding) or extruder plugging will result. If moisture levels are too high, the material will reagglomerate in the spheronizer, resulting in the formation of very large particles. The centrifugal force in the spheronizer causes internal moisture of each particle to migrate to the surface of the particle, causing an increase in apparent moisture level.

MEM is a high-pressure agglomeration method due primarily to extrusion and the action of the spheronizer, producing a dense, fast-sinking particle. When formulation, moisture levels, and extruder and spheronizer speeds are optimized, the MEM method produces practically no fines and does not require sieving. As a result, the diets must be extruded through different size dies if a specific range of particle sizes is desired.

The PARA method was developed to increase the yield of particles less than 400 μm (Barrows and Rust 2000). This method differs from MEM because only a cylindrical shaper is used and the process does not involve an extruder (Barrows and Hardy 2001). The ingredients are mixed and moisture added as with the MEM process, but slightly higher moisture levels are required. The wet mixture of ingredients is then added to the spheronizer with a charge of 3.0-mm inert particles (75% weight of beads/weight of wet mash). The rotation of the spheronizer imparts energy to the inert particles, which in turn transfer energy to the mash, producing roughly spheroid particles. High rotational velocity is most effective, yet PARA is a low-pressure agglomeration method. Contact time in the spheronizer can vary from 15 seconds up 1 minute, but 30 seconds is usually adequate to form particles (Barrows and Lellis 2006). Particles and inert beads are discharged from the spheronizer and sieved through a 2.0-mm screen to remove the beads. The PARA process produces a wide range of particle sizes, but manipulation of moisture levels and spheronization time can alter the particle size distribution. Increasing moisture levels along with increasing spheronization time will result in larger particles. The PARA process is also more dependent on adequate binding capacity from the diet formulation than the MEM method since high pressure is not used to agglomerate the particles. PARA particles have a lower sinking rate than MEM particles, which can be beneficial for slow-feeding larvae of some species, such as halibut (Kvåle et al. 2007).

Shaped microbound particles can also be formed by spraying a mixture of diet and binder into a gelling solution or drying chamber where particles are formed on contact; for example, alginate- and carboxymethylcellulose-bound particles can be prepared by spraying a mixture of dietary components and binder into a solution of calcium salt. The resulting particles will vary in size and must be sieved to obtain appropriate size categories. The spray method results in lower-density particles with a slower sinking rate than MEM particles. Furthermore, binding occurs through the action of specific binders and not through the action of dietary components, resulting in more flexibility in diet formulation than possible for the PARA method.

A disadvantage of spraying dietary mixtures into an aqueous medium is that leaching of water-soluble dietary ingredients will likely occur, changing the dietary composition of the particles. The problem can be addressed by spraying a mixture of binder and diet into heated air to evaporate the solvent and deposit the binder material around the dietary particles (Figure 12.4; Önal and Langdon 2004b). Another approach is to form microbound particles by gelling the binder and dietary mixture after emulsification in oil, then separating the particles from the oil by centrifugation or filtration (Vandenberg and De La Noüe 2001; Yúfera et al. 2005); however, effective removal of oil from the gelled particles may require aqueous washing that could result in nutrient leakage.

Microbound particles are not effective in retaining low-molecular-weight, water-soluble nutrients, such as amino acids, with high losses occurring within a few minutes of suspending microparticles in water (see review by Langdon 2003; Table 12.1). Losses of high-molecular-weight, water-soluble dietary

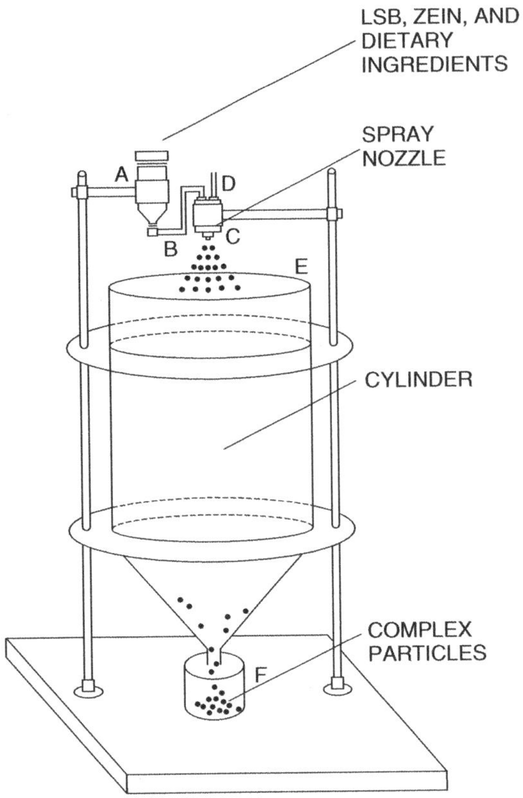

Figure 12.4 Spray apparatus for preparing zein-bound microparticles (after Önal and Langdon 2004b). LSB = lipid spray beads.

components, such as soluble proteins and peptides, may also be rapid and substantial; for example, Tanaka et al. (1984) reported 50% loss of α-lactoalbumin (molecular weight less than 2×10^4 Da) from alginate beads after less than 15 minutes of suspension. Similarly, Chui and Wan (1997) reported that up to 55% trypsin was lost from alginate beads after only 10 minutes of aqueous suspension. Kvåle et al. (2006) reported high losses of algal protein, hydrolysates, and serine from heat-coagulated and agglomerated particles. Leaching rates increased with decreasing particle size and molecular weight of the core material.

Coating microbound particles can reduce leakage rates; for example, Anas et al. (2008) coated starch-microbound particles with chitosan to reduce leakage of amino acids. Hurtado-López and Murdan (2005) reported good retention of ovalbumin by zein-polyvinylpyrrolidone microbeads in aqueous suspension. Cara et al. (2007) described a casein-protamine microparticle prepared by complex coacervation for feeding to early fish larvae. This particle type exhibited lower protein leakage rates than both alginate beads prepared by ionic gelation and cross-linked protein walled capsules.

Despite poor retention of water-soluble nutrients by microbound particles, researchers have reported success on weaning larvae onto microbound diets after an initial period of feeding on live feeds or a combination of live feeds and microbound diets (e.g., Kanazawa et al. 1989; Holt 1993; Hart and Purser 1996; Kolkovski et al. 1997; Cahu et al. 1999; Cañavate and Fernández-Díaz 1999; Lazo et al. 2000; Baskerville-Bridges and Kling 2000a; Hamre et al. 2001). In contrast to marine fish, larvae of several freshwater fish species have been successfully reared entirely on microbound diets; for example, larvae of coregonid (*Coreogonus schinzi*; Dabrowski et al. 1984b), goldfish (*Carassius auratus*; Szlaminska et al. 1993), pike perch (*Sander lucioperca*; Ostaszewska et al. 2005), and zebrafish (*Danio rerio*; Carvalho et al. 2004).

In practice, high concentrations of water-soluble nutrients added to microbound diets likely compensate for rapid leakage rates as long as larvae ingest food particles soon after the particles are added to the culture medium. Larvae may also acquire leached nutrients by drinking the culture medium. Frequent exchanges of culture medium remove leached nutrients and reduce the potential for development of high bacterial concentrations. The economic cost of wasted leached nutrients may be acceptable given the small amounts of feed required for larval production; however, the costs of poor larval performance due to inadequate delivery of essential nutrients may pose a serious problem for commercial fish hatcheries.

Table 12.1 Comparison of microparticle types for delivery of nutrients to fish larvae.

Microparticle type	Advantages	Disadvantages
Microbound particles	• Inexpensive • Easy to produce • Binders can be nutritionally inert	• Poor retention of LMWS nutrients and possibly water-soluble proteins
Cross-linked protein walled capsules	• Possible to modify capsule wall properties • Digestible for some species of fish larvae	• Expensive • Use of organic solvents • Poor retention of LMWS nutrients
Lipid-walled capsules and lipid spray beads	• Inexpensive • Easy to produce • Better retention of LMWS nutrients compared with microbound particles and cross-linked protein walled capsules	• Hard-lipid particles not digestible by most species of fish larvae and depends on mechanical breakdown • Possible oxidation of unsaturated lipids during preparation and storage
Liposomes	• Better retention of LMWS nutrients compared with microbound particles and cross-linked protein walled capsules • Digestible • Phospholipid wall material may contribute to larval nutrition	• Expensive • Use of organic solvents • Preparation involves several steps • Possible oxidation of unsaturated lipids during preparation and storage

LMWS = low-molecular-weight, water-soluble.

12.2.2 Cross-linked protein walled capsules

The cross-linked protein walled capsule type originated from earlier work of Chang (1964), who described the preparation of nylon-protein walled, semipermeable microcapsules for encapsulation of enzymes for treatment of human kidney diseases. Jones et al. (1974) applied the technique for delivery of dietary nutrients to marine suspension-feeders. Subsequently, researchers eliminated the need to incorporate nylon in the wall by simply cross-linking dietary protein with various cross-linking agents to form capsule walls (Jones 1980; Hayworth 1983; Langdon 1989).

The capsules are prepared by first forming an emulsion of an aqueous protein solution and other dietary ingredients in an organic solvent, such as cyclohexane. A cross-linking agent is then added, such as sebacoyl chloride, to form the wall. Capsules are washed by repeated centrifugation or filtered to remove the organic solvent and excess cross-linking agent. Capsules can be freeze-dried and stored as a powder. The method is dependent on the use of organic solvents that are potentially toxic to fish larvae, unless they can be completely removed during the preparation process. The use of organic solvents also increases the cost of preparing this capsule type (Table 12.1).

The permeability of the capsule wall depends on the choice of cross-linking agents and process of wall formation (Langdon and DeBevoise 1990). Low-molecular-weight dietary ingredients, such as amino acids, are rapidly lost from these capsules in aqueous suspension (Önal and Langdon 2000; Aragão et al. 2007), and loss of dietary proteins can also be significant (Alabi et al. 1999). Kvåle et al. (2006) reported that losses of algal protein, hydrolysate, and serine were lower from cross-linked protein walled capsules than from either protein-bound or agglomerated particles; however, it is possible that losses of these water-soluble materials occurred during their preparation as a result of aqueous washing (Yúfera et al. 2002). Although washed capsules retained water-soluble nutrients better than protein-bound or agglomerated particles, Atlantic cod (*Gadus morhua*) larvae preferred to ingest these microbound particle types compared with capsules (Kvåle et al. 2006).

Attempts to reduce leakage rates by coating capsules with lipid have shown limited success (Langdon and DeBevoise 1990). Alternatively, lipid-walled beads containing low-molecular-weight, water-soluble nutrients have been incorporated in cross-linked protein walled capsules in order to reduce nutrient leakage rates (Chu and Ozkizilcik 1999).

Digestion of this capsule type is not always evident in early larvae (Kanazawa et al. 1982; Teshima et al. 1982; Walford et al. 1991), but modifications of cross-linking agents and preparation process can improve their digestibilities (Fernández-Díaz and Yúfera 1995). Cross-linked protein walled capsules have been used in experimental feeding studies with larvae for a wide range of species (Dover sole *Solea solea*, Appelbaum 1985; sea bass *Lates calcarifer*, Walford et al. 1991; gilthead sea bream *Sparus aurata*, Fernández-Díaz et al. 1994; striped bass *Morone saxatilis*, Ozkizilcik and Chu 1996; sea bream *Sparus aurata*, Yúfera et al. 1999, 2000; zebrafish *Brachydanio rerio*, Önal and Langdon 2000; and sea bream *Sparus aurata*, Aragão et al. 2007).

12.3 Delivery of lipids

12.3.1 Emulsions

Lipid emulsions can be readily prepared by vigorously mixing the lipid with a solution of an emulsifying agent, such as gelatin or gum acacia. The resulting emulsion can be directly fed to fish larvae or used to enrich live prey with lipids, such as essential fatty acids and fat-soluble vitamins. There are many reports of enriching live prey with eicosapentaenoic acid (EPA; 20:5n-3) and docosahexaenonic acid (DHA; 22:6n-3) to improve the nutritional value of prey species for fish larvae (e.g., Southgate and Kavanagh 1999; Sorgeloos et al. 2001; Lie et al. 2006; Srivastava et al. 2006; see Chapter 2).

Stabilization of lipid emulsions is desirable for long-term storage. One approach is to simply store the emulsion; however, coalescence of lipid droplets can occur over time. Addition of surfactants, such as phospholipids or nontoxic food-grade agents, such as sorbitan monostearate, can help stabilize emulsions. Alternatively, emulsions can be spray-dried and stored as a powder (Jin et al. 2007).

Another approach in stabilizing lipid emulsions is to coat lipid droplets with a coacervate wall, such as gelatin-acacia (e.g., Langdon and Waldock 1981; Ozkizilcik and Chu 1994a). The method involves emulsifying the lipid in an aqueous solution of one of the wall-forming polymers, then adding the other oppositely charged copolymer followed by adjustment of pH or a salting-out process to form coacervate walls surrounding lipid droplets. The walls may be further strengthened with a cross-linking agent, such as glutaraldehyde. Encapsulated lipid droplets can then be concentrated and washed by centrifugation or spray-dried for long-term storage.

High surface area-to-volume ratios provide conditions for rapid diffusion of oxygen into microparticles (Table 12.1). Oxidation of labile lipids, such as polyunsaturated fatty acids, can be avoided by storing the lipid products under nitrogen in the dark at low temperature (Kolanowski et al. 2004). Addition of antioxidants, such as vitamin E, may also be necessary to limit losses; however, high levels of peroxidation products associated with emulsions of unsaturated fatty acids have been reported even with additions of antioxidants (McEvoy et al. 1996; Monroig et al. 2007b). Bustos et al. (2003) prepared chitosan-coated lipid droplets of krill oil to reduce oxidative losses of n-3 polyunsaturated fatty acids and carotenoids in microparticles designed for feeding to fish larvae; however, the coating had little beneficial effect, and astaxanthin losses were as rapid as with noncoated krill oil emulsions.

12.4 Delivery of low-molecular-weight, water-soluble nutrients

Most available artificial diets for fish larvae are made up of insoluble dietary components bound together with insoluble binders; however, early larvae may require dietary supplies of soluble proteins, low-molecular-weight peptides, and amino acids that are more easily digested and absorbed than proteins (Rust et al. 1993; Cahu et al. 1999; Rønnestad et al. 1999, 2003; Cahu and Zambonino Infante 2001; Rojas-García and Rønnestad 2003; Kvåle et al. 2007; see Chapter 3). Soluble proteins are present at higher concentrations in rotifers (Srivastava et al. 2006) and *Artemia* (Kvåle et al. 2002) than in most artificial diets and free amino acids are present at concentrations of up to 6.7% dry weight in live prey (Helland et al. 2003a, 2003b, 2003c; Aragão et al. 2004) but are rapidly lost from most microparticle types suspended in water.

Apart from peptides and amino acids, Hamre et al. (2008) reported that unenriched rotifers (*Brachionus* sp.) may not meet the nutritional requirements of marine fish larvae for some minerals and water-soluble vitamins (see Chapter 4). As discussed above, microbound and protein walled microcapsules do not effectively deliver these low-molecular-weight, water-soluble nutrients to larvae due to rapid leakage losses, and other particle types need to be used to effectively deliver these nutrients to fish larvae and their prey.

12.4.1 Lipid-walled capsules and lipid spray beads

Both lipid-walled capsules and lipid spray beads can be prepared using a spray method in which the core material is first emulsified in melted lipid (Hincal and Kas 1994; Önal and Langdon 2004a). The emulsion is then sprayed into a chilled container to harden the lipid particles. The product can be collected as a powder and stored at low temperature under nitrogen. Tripalmitin, triolein, and fish oils have been used in wall formation of lipid beads for use in aquaculture (Villamar and Langdon 1993; López-Alvarado et al. 1994; Ozkizilcik and Chu 1996; Buchal and Langdon 1998; Chu and Ozkizilcik 1999; Baskerville-Bridges and Kling 2000b; Önal and Langdon 2004b). Phospholipids (Nordgreen et al. 2007) and waxes (Önal and Langdon 2004a; Langdon et al. 2008) have also been used to prepare lipid spray beads for enrichment of *Artemia* as a feed for fish larvae.

Önal and Langdon (2004a) reported that larvae of glowlight tetra (*Hemigrammus erythrozonus*) and zebrafish (*Brachydanio rerio*) could break down lipid spray beads prepared with methyl palmitate (melting point 28–29°C), and clownfish larvae (*Amphiprion percula*) could break down beads prepared with menhaden stearine that were soft at culture temperatures (24–26°C). In contrast, beads prepared with lipids of high melting point, such as tripalmitin (melting point 65°C), are not readily digested by fish larvae (López-Alvarado et al. 1994). *Artemia* and larvae of other crustacean species possess piercing and grinding mouth parts that may physically rupture the lipid matrix to release the core material (Villamar and Langdon 1993; Langdon et al. 2008).

More than 90% of the dry weight of lipid spray beads can be represented by lipid (Langdon et al. 2007) that could potentially be utilized as a dietary lipid source, creating an imbalance in the nutrition of the target organism (Brinkmeyer and Holt 1995). Delivery of adequate amounts of trace nutrients may be possible by adding small amounts of the lipid particles to the diet without significantly affecting its overall lipid composition. Alternatively, the use of waxes or other inert hydrophobic materials in bead preparation may eliminate or reduce the contribution of the matrix material to the nutrition of the target organism. However, Langdon et al. (2008) reported that substantial losses of labile micronutrients, such as vitamins A, E, and C, thiamine, and selenium, occurred during preparation and storage of lipid spray beads even though antioxidants were added to the lipid mixture and the beads were stored under nitrogen at −80°C.

Incorporation of small amounts of lipid-walled capsules or lipid spray beads containing essential micronutrients in other particle types that contain bulk nutrients (proteins, etc.) may allow delivery of a complete diet to the target organism (Figure 12.5). Delivery of combinations of particle types for delivery of nutrients to marine suspension-feeders in a "complex" particle (Villamar and Langdon 1993; Chu and Ozkizilcik 1999; Baskerville-Bridges and Kling 2000b; Önal and Langdon 2005; Langdon et al. 2007) is an advantage in that the target organism is not provided with an opportunity to selectively capture different particle types.

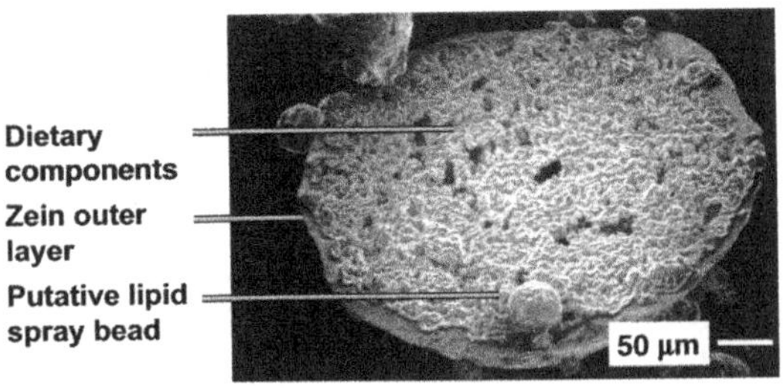

Figure 12.5 Scanning electron microscope image of a zein-bound complex microparticle (after Önal and Langdon 2005).

12.4.2 Liposomes

Liposomes have been developed for many uses, including pharmacological and medical applications (Basu and Basu 2002). There has been considerable interest in applying this technology to deliver water-soluble nutrients to marine suspension-feeders (see review by Coutteau and Sorgeloos 1997). Liposomes are primarily prepared with phospholipids that form a bilayer or bilayers surrounding aqueous solutions of core materials. The phospholipid bilayers reduce leakage of core material from the liposomes.

There is a large body of literature that describes different methods of preparation (Gregoriadis 1983). Preparation method determines the size and properties of the liposomes. Size can vary from less than 1 µm to over 10 µm (Kulkarni et al. 1995). Methods for preparing freeze-dried liposomes have been described (e.g., Miyajima 1997) that allow long-term storage. Liposomes prepared with phospholipids that contain unsaturated fatty acids are susceptible to oxidation and should be stored under conditions to reduce losses of unsaturated lipids and labile nutrients (Monroig et al. 2007b).

Several researchers have used liposomes to enrich *Artemia* with water-soluble nutrients to improve its nutritional value for fish larvae (Hontoria et al. 1994; Ozkizilcik and Chu 1994b; Koven et al. 1999; Tonheim et al. 2000; Monroig et al. 2003, 2006, 2007a. Monroig et al. (2003) reported that large, unilamellar (LUV) liposomes showed better retention of a water-soluble tracer (carboxyfluorescein) than multilamellar liposomes under conditions used for enrichment of *Artemia*.

Barr and Helland (2007) described a simplified method for mass production of large (2–8 µm) liposomes suitable for delivery of free amino acids to filter-feeding organisms. The method was based on reverse-phase evaporation of chloroform from a phospholipid-aqueous emulsion followed by hydration of the empty liposomes with an amino acid solution. The authors reported that 46.3% of the initial amino acid solution was encapsulated and that 50% of the dry weight of the liposomes was made up of amino acids. After removing nonencapsulated amino acids by washing, liposomes retained 90.9% of amino acids after 2 hours suspension in seawater. This liposome type was successfully used to enrich the free amino acid content of rotifers, *Brachionus* sp., and *Artemia* (Barr et al. 2005; Saavedra et al. 2008, 2009). The wall material will likely contribute to the nutrition of the target organism because phospholipids are more digestible than many other lipid types.

Fish larvae may have difficulty capturing small-sized liposomes, except as a consequence of larvae drinking the culture medium or filtering liposomes from suspension (Van der Meeren 1991; Reiten et al. 1994, 1997). Aggregates of liposomes commonly form in seawater suspensions and can be of sufficient size for active capture by fish larvae (Koven et al. 1999). Alternatively, it may be necessary to incorporate liposomes in larger-sized particles, such as microbound particles or cross-linked protein walled capsules, to produce "complex" microparticle types (Ozkizilcik and Chu 1996; Figures. 12.2 and 12.5).

One of the disadvantages of liposomes for large-scale use in commercial aquaculture is that the method of preparation is often complex, involving several different steps and the use of organic solvents. Nonetheless, liposomes are a useful experimental tool and have been successfully used to enrich prey organisms to study the nutritional requirements of suspension-feeders

12.5 Conclusions

Despite several decades of research, there is a lack of understanding of the nutritional requirements of marine and freshwater suspension-feeders, including the larvae of many fish species. One of the main reasons for this lack of understanding is due to technical difficulties in delivering nutrients in micron-sized particles that are digestible and retain essential low-molecular-weight, water-soluble nutrients. Leakage losses and consequent fouling of the culture medium are much more significant problems for micron-sized particles required by larvae of many species of fish, compared with leakage from larger-sized dietary particles used in the fingerling and growout stages of production.

The development and application of various microencapsulation technologies offer the opportunity to develop feed particles that are capable of retaining low-molecular-weight, water-soluble nutrients and providing them to fish larvae in a digestible form, possibly in combination with other particle types that deliver bulk nutrients. Until this goal is achieved, aquaculture of the larval stages of many fish species will be dependent on the use of live prey species that are expensive and often not reliable in supply and quality. Development of microparticulate diets for rearing fish larvae is an important goal in establishing commercial aquaculture of many marine fish species.

Literature cited

Alabi, A.O., Cob, Z.C., Jones, D.A., and Latchford, J.W. 1999. Influence of algal exudates and bacteria on growth and survival of white shrimp larvae fed entirely on microencapsulated diets. *Aquaculture International* 7:137–158.

Anas, A., Philip, R., and Singh, I.S.B. 2008. Chitosan as a wall material for a microencapsulated delivery system for *Macrobrachium rosenbergii* (de Man) larvae. *Aquaculture Research* 39:885–890.

Appelbaum, S. 1985. Rearing of the Dover sole, *Solea solea* (L.), through its larval stages using artificial diets. *Aquaculture* 49:209–221.

Aragão, C., Conceição, L.E.C., Dinis, M.T., and Fyhn, H.J. 2004. Amino acid pools of rotifers and *Artemia* under different conditions: nutritional implications for fish larvae. *Aquaculture* 234:429–445.

Aragão, C., Conceição, L.E.C., Lacuisse, M., Yúfera, M., and Dinis, M.T. 2007. Do dietary amino acid profiles affect performance of gilthead seabream? *Aquatic Living Resources* 20:155–161.

Backhurst, J.R., and Harker, J.H. 1988. The settling rates of larval feeds. *Aquacultural Engineering* 7:363–366.

Barr, Y., and Helland, S. 2007. A simple method for mass-production of liposomes, in particular large liposomes, suitable for delivery of free amino acids to filter feeding zooplankton. *Journal of Liposome Research* 17:79–88.

Barr, Y., Terjesen, B.F., and Helland, S. 2005. Intensive short-term enrichment—an efficient method for enhancing phospholipids and free amino acids in live feeds. In: Hendry, C.I., Van Stappen, G., Willie, M., Sorgeloos, P. (eds.) *Larvi 2005—Fish and Shellfish Larviculture Symposium*, Special Publication No. 36. European Aquaculture Society, Oostende, Belgium.

Barrows, F.T., and Hardy, R.W. 2001. Nutrition and feeding. In: Wedemeyer, G. (ed.) *Fish Hatchery Management*, 2nd edition. John Wiley & Sons, Inc., NewYork, pp. 483–558.

Barrows, F.T., and Lellis, W.A. 2000. Microbound feeds. In: Stickney, R.R. (ed.) *Encyclopedia of Aquaculture*. John Wiley & Sons, Inc., New York, pp. 525–528.

Barrows, F.Y., and Lellis, W.A. 2006. Effect of diet processing method and ingredient substitution on feed characteristics and survival of larval walleye, *Sander vitreus*. *Journal of the World Aquaculture Society* 37:154–160.

Barrows, F.T., and Rust, M.B. 2000. Larval feeding—fish. In: Stickney, R.R. (ed.) *Encyclopedia of Aquaculture*. John Wiley & Sons, Inc., New York, pp. 465–469.

Barrows, F.T., Zitzow, R.E., and Kindschi, G.A. 1993. Effects of surface water spray, diet, and phase feeding on swim bladder inflation, survival, and cost of production of intensively reared larval walleyes. *The Progressive Fish Culturist* 55:224–228.

Baskerville-Bridges, B., and Kling, L.J. 2000a. Early weaning of Atlantic cod (*Gadus morhua*) larvae onto a microparticulate diet. *Aquaculture* 189:109–117.

Baskerville-Bridges, B., and Kling, L.J. 2000b. Development and evaluation of microparticulate diets for early weaning of Atlantic cod (*Gadus morhua*) larvae. *Aquaculture Nutrition* 6:171–182.

Basu, S.C., and Basu, M. 2002. Liposome methods and protocols. In: Basu, S.C., and Basu, M. (eds.) *Methods in Molecular Biology*, Vol. 199. Humana Press, Totowa, NJ.

Brinkmeyer, R.L., and Holt, G.J. 1995. Response of red drum larvae to graded levels of menhaden oil in semipurified microparticulate diets. *The Progressive Fish Culturist* 57:30–36.

Buchal, M., and Langdon, C. 1998. Evaluation of lipid spray beads for the delivery of water-soluble materials to a marine suspension-feeder, the Manila clam *Tapes philippinarum* (Deshayes 1853). *Aquaculture Nutrition* 4:265–284.

Bustos, R., Romo, L., Yáñez, K., Díaz, G., and Romo, C. 2003. Oxidative stability of carotenoid pigments and polyunsaturated fatty acids in microparticulate diets containing krill oil for nutrition of marine fish larvae. *Journal of Food Engineering* 56:289–293.

Cahu, C.L., and Zambonino Infante, J.L. 1994. Maturation of the pancreatic and intestinal digestive functions in sea bass (*Dicentrarchus labrax*): effect of weaning with different protein sources. *Fish Physiology and Biochemistry* 14:431–437.

Cahu, C.L., and Zambonino Infante, J.L. 2001. Substitution of live food by formulated diets in marine fish larvae. *Aquaculture* 200:161–180.

Cahu, C.L., Zambonino Infante, J.L., Péres, A., Quazuguel, P., and Le Gall, M.M. 1998. Algal addition in sea bass (*Dicentrarchus labrax*) larvae rearing: effect on digestive enzymes. *Aquaculture* 161:479–489.

Cahu, C.L., Zambonino Infante, J.L., Quazuguel, P., and Le Gall, M.M. 1999. Protein hydrolysate vs. fish meal in compound diets for 10-day old sea bass larvae *Dicentrarchus labrax* larvae. *Aquaculture* 171:109–119.

Cañavate, J.P., and Fernández-Díaz, C. 1999. Influence of co-feeding larvae with live and inert diets on weaning the sole *Solea senegalensis* onto commercial dry feeds. *Aquaculture* 174:255–263.

Cara, B., Moyano, F.J., Gander, B., and Yúfera, M. 2007. Development of a novel casein-protamine based microparticle for early feeding of fish larvae: *in vitro* evaluation. *Journal of Microencapsulation* 24(6):505–514.

Carvalho, A.P., Sá, R., Oliva-Teles, A., and Bergot, P. 2004. Solubility and peptide profile affect the utilization of dietary protein by common carp (*Cyprinus carpio*) during early larval stages. *Aquaculture* 234:319–333.

Chang, T.M.S. 1964. Semipermeable microcapsules. *Science* 146(3643):523–525.

Chu, F.L., and Ozkizilcik, S. 1999. Acceptability of complex microencapsulated diets by striped bass (*Morone saxatilis*) larvae. *Journal of Experimental Marine Biology and Ecology* 237:1–9.

Chui, W.K., and Wan, L.S.C. 1997. Prolonged retention of cross-linked trypsin in calcium alginate microspheres. *Journal of Microencapsulation* 14:51–61.

Coutteau, P., and Sorgeloos, P. 1997. Manipulation of dietary lipids, fatty acids and vitamins in zooplankton cultures. *Freshwater Biology* 38:501–512.

Dabrowski, C., Charlton, N., Bergot, P., and Kaushik, S. 1984b. Rearing of coregonid (*Coregonus schinzi palea* cuv. et val.) larvae using dry and live food. I. Preliminary data. *Aquaculture* 41(1):11–20.

Dabrowski, K., and Bardega, R. 1984. Mouth size and predicted preferences of larvae of three cyprinid fish species. *Aquaculture* 40:41–46.

Drillet, G., Jorgensen, N.O.G., Sørensen, T.F., Ramløv, H., and Hansen, B.W. 2006. Biochemical and technical observations supporting the use of copepods as live feed organisms in marine larviculture. *Aquaculture Research* 37:756–772.

Evjemo, J.O., Reitan, K.I., and Olsen, Y. 2003. Copepods as live food organisms in the larval rearing of halibut larvae (*Hippoglossus hippoglossus*) with special emphasis on the nutritional value. *Aquaculture* 227:191–210.

Fernández-Díaz, C., and Yúfera, M. 1995. Capacity of gilthead seabream, *Sparus aurata* L., larvae to break down dietary microcapsules. *Aquaculture* 134:269–278.

Fernández-Díaz, C., Pascual, E., and Yúfera, M. 1994. Feeding behavior and prey size selection of gilthead seabream, *Sparus aurata*, larvae fed on inert and live food. *Marine Biology* 118:323–328.

Gawlicka, A., McLaughlin, L., Hung, S.S.O., and de la Noüe, J. 1996. Limitations of carrageenan microbound diets for feeding white sturgeon, *Acipenser transmontanus*, larvae. *Aquaculture* 141:245–265.

Gregoriadis, G. 1983. *Preparation of Liposomes, Vol. 1, Liposome Technology*. CRC Press, Boca Raton, FL.

Guthrie, K.M., Rust, M.B., Langdon, C.J., and Barrows, F.T. 2000. Acceptability of various microparticulate diets to first feeding walleye (*Stizostedion vitreum*) larvae. *Aquaculture Nutrition* 6:153–158.

Hamlin, H.J., Hunt von Herbing, I., and Kling, L.J. 2000. Histological and morphological evaluations of the digestive tract and associated organs of haddock throughout post-hatching ontogeny. *Journal of Fish Biology* 57:716–732.

Hamre, K. 2006. Nutrition in cod (*Gadus morhua*) larvae and juveniles. *ICES Journal of Marine Science* 63:267–274.

Hamre, K., Naess, T., Espe, M., Holm, J.C., and Lie, Ø. 2001. A formulated diet for Atlantic halibut (*Hippoglossus hippoglossus*, L.) larvae. *Aquaculture Nutrition* 17:123–132.

Hamre, K., Srivastava, A., Rønnestad, I., Mangor-Jensen, A., and Stoss, J. 2008. Several micronutrients in the rotifer *Brachionus* sp. may not fulfill the nutritional requirements of marine fish larvae. *Aquaculture Nutrition* 14:51–60.

Hart, G.J., and Purser, G.J. 1996. Weaning of hatchery-reared greenback flounder (*Rhombosolea tapirina* Gunter) from live to artificial diets: effect of age and duration of the changeover period. *Aquaculture* 145:171–181.

Hayworth, L.W. 1983. Microencapsulation process. UK Patent 2103568.

Helland, S., Terjesen, B.F., and Berg, L. 2003a. Free amino acid and protein content in the planktonic copepod *Temora longicornis* compared to *Artemia franciscana*. *Aquaculture* 215:1213–1218.

Helland, S., Nejstgaard, J.C., Fyhn, H.J., Egge, J.K., and Båmstedt, U. 2003b. Effects of starvation, season, diet on the free amino acid and protein content of *Calanus finmarchicus* females. *Marine Biology* 143:297–306.

Helland, S., Nejstgaard, J.C., Humlen, R., Fyhn, H.J., and Båmstedt, U. 2003c. Effects of season and maternal food on *Calanus finmarchicus* reproduction, with emphasis on the free amino acids. *Marine Biology* 142:1141–1151.

Hincal, A.A., and Kas, H.S. 1994. Preparation of micropellets by spray congealing. *Drugs and the Pharmaceutical Sciences* 65:17–34.

Holt, G.J. 1993. Feeding larval red drum on microparticulate diets in a close recirculation water system. *Journal of the World Aquaculture Society* 24:225–230.

Hontoria, F., Crowe, J.H., Crowe, L.M., and Amat, F. 1994. Potential use of liposomes in larviculture as a delivery system through *Artemia* nauplii. *Aquaculture* 127:255–264.

Hurtado-López, P., and Murdan, S. 2005. Formation and characterization of zein microspheres as delivery vehicles. *Journal of Drug Delivery Science and Technology* 15:267–272.

Jin, Y., Pierre, C., Zhang, W., Diepen, C.C., Curtis, J., and Barrow, C. 2007. Microencapsulation of marine lipids as a vehicle for functional food delivery. In: Barrow, C., and Shahidi, F. (eds.) *Marine Nutraceuticals and Functional Foods*. CRC Press, Boca Raton, FL, pp. 115–149.

Jones, D.A. 1980. Microencapsulation process. UK Patent 2040863.

Jones, D.A., Munford, J.G., and Gabbott, P.A. 1974. Microcapsules as artificial food particles for aquatic filter feeders. *Nature* 247:233–235.

Kanazawa, A., Teshima, S., Inamori, S., Sumida, S., and Iwashita, T. 1982. Rearing larval red sea bream and ayu with artificial diets. *Memoirs of the Faculty of Fisheries, Kagoshima University* 31:185–193.

Kanazawa, A., Koshio, S., and Teshima, S. 1989. Growth and survival of larval red sea bream *Pagrus major* and Japanese flounder *Paralichthys olivaceus* fed microbound diets. *Journal of the World Aquaculture Society* 2:31–37.

Kjørsvik, E., van der Meeren, T., Kryvi, H., Arnfinnson, J., and Kvenseth, P.G. 1991. Early development of the digestive tract of cod larvae, *Gadus morhua* L., during start-feeding and starvation. *Journal of Fish Biology* 38:1–15.

Kolanowski, W., Laufnberg, G., and Kunz, B. 2004. Fish oil stabilization by microencapsulation with modified cellulose. *International Journal of Food Science and Nutrition* 55(4):333–343.

Kolkovski, S. 2001. Digestive enzymes in fish larvae and juveniles—implications and applications to formulated diets. *Aquaculture* 200:181–201.

Kolkovski, S., Koven, W., and Tandler, A. 1997. The mode of action of *Artemia* in enhancing utilization of microdiet by gilthead seabream *Sparus aurata* larvae. *Aquaculture* 155:193–205.

Koven, W., Barr, Y., Hadas, E., Ben-Atia, I., Chen, Y., Weiss, R., and Tandler, A. 1999. The potential of liposomes as a nutrient supplement in first-feeding marine fish larvae. *Aquaculture Nutrition* 5:251–256.

Koven, W., Kolkovski, S., Hadas, E., Gamsiz, K., and Tandler, A. 2001. Advances in the development of microdiets for gilthead seabream, *Sparus aurata*: a review. *Aquaculture* 194:107–121.

Kulkarni, S.B., Betageri, G.V., and Singh, M. 1995. Factors affecting microencapsulation of drugs in liposomes. *Journal of Microencapsulation* 12:229–246.

Kvåle, A., Harboe, T., Espe, M., Næss, T., and Hamre, K. 2002. Effect of pre-digested protein on growth and survival of Atlantic halibut larvae (*Hippoglossus hippoglossus* L.). *Aquaculture Research* 33:311–321.

Kvåle, A., Yúfera, M., Nygård, E., Aursland, K., Harboe, T., and Hamre, K. 2006. Leaching properties of three different microparticulate diets and preference for the diets in cod (*Gadus morhua* L.) larvae. *Aquaculture* 251:402–415.

Kvåle, A., Nordgreen, A., Tonheim, S.K., and Hamre, K. 2007. The problem of meeting dietary protein requirements in intensive aquaculture of marine fish larvae, with emphasis on Atlantic halibut (*Hippoglossus hippoglossus* L.). *Aquaculture Nutrition* 13:170–185.

Langdon, C.J. 1989. Preparation and evaluation of protein microcapsules for a marine suspension-feeder, the Pacific oyster *Crassostrea gigas*. *Marine Biology* 101:217–224.

Langdon, C.J. 2003. Microparticle types for delivery nutrients to marine fish larvae. *Aquaculture* 227:259–275.

Langdon, C.J., and DeBevoise, A.E. 1990. Effect of microcapsule type on delivery of dietary protein to a marine suspension-feeder, the oyster *Crassostrea gigas*. *Marine Biology* 105:437–443.

Langdon, C.J., and Waldock, M.J. 1981. The effect of algal and artificial diets on the growth and fatty acid composition of *Crassostrea gigas* spat. *Journal of the Marine Biological Association U.K.* 61:431–448.

Langdon, C.J., Clack, B., and Önal, U. 2007. Complex microparticles for delivery of low-molecular-weight, water-soluble nutrients and pharmaceuticals to marine fish larvae. *Aquaculture* 268:143–148.

Langdon, C.J., Nordgreen, A., Hawkyard, M., and Hamre, K. 2008. Evaluation of wax spray beads for delivery of low-molecular-weight, water-soluble nutrients and antibiotics to *Artemia*. *Aquaculture* 284:151–158.

Lazo, J.P., Dinis, M.T., Holt, G.J., Faulk, C., and Arnold, C.R. 2000. Co-feeding microparticulate diets with algae: towards eliminating the need of zooplankton at first feeding in larval red drum (*Sciaenops ocellatus*). *Aquaculture* 188:339–351.

Liao, I.C., Su, H.M., and Chang, E.Y. 2001. Techniques in finfish larviculture in Taiwan. *Aquaculture* 200:1–31.

Lie, Ø., Haaland, H., Hemre, G.-I., Maage, A., Lied, E., Rosenlund, G., Sandes, K., and Olsen, Y. 2006. Nutritional composition of rotifers following a change in diet from emulsified oil to microalgae. *Aquaculture International* 5:427–438.

Lindsay, S.M., and Vogt, R.G. 2004. Behavioral responses of newly hatched zebrafish (*Danio rerio*) to amino acid chemostimulants. *Chemical Senses* 29:93–100.

López-Alvarado, J., Langdon, C.J., Teshima, S., and Kanazawa, A. 1994. Effects of coating and encapsulation of crystalline amino acids on leaching in larval feeds. *Aquaculture* 122:335–346.

McEvoy, L.A., Navarro, J.C., Hontoria, F., Amat, F., and Sargent, J.R. 1996. Two novel *Artemia* enrichment diets containing polar lipid. *Aquaculture* 144:339–352.

Miyajima, K. 1997. Role of saccharides for the freeze-thawing and freeze drying of liposome. *Advanced Drug Delivery Reviews* 24:151–159.

Monroig, Ó., Navarro, J.C., Amat, F., González, P., Amat, F., and Hontoria, F. 2003. Enrichment of *Artemia* nauplii in PUFA, phospholipids, and water-soluble nutrients using liposomes. *Aquaculture International* 11:151–161.

Monroig, Ó., Navarro, J., Amat, F., González, P., Bermejo, A., and Hontoria, F. 2006. Enrichment of *Artemia* nauplii in essential fatty acids with different types of liposomes and their use in the rearing of gilthead sea bream (*Sparus aurata*) larvae. *Aquaculture* 251:491–508.

Monroig, O., Navarro, J.C., Amat, F., and Hontoria, F. 2007a. Enrichment of *Artemia* nauplii in vitamin A, vitamin C and methionine using liposomes. *Aquaculture* 269:504–513.

Monroig, Ó., Navarro, J.C., Amat, F., González, P., and Hontoria, F. 2007b. Oxidative stability and changes in the particle size of liposomes used in the *Artemia* enrichment. *Aquaculture* 266:200–210.

Muir, P.R., and Sutton, D.C. 1994. Bacterial degradation of microencapsulated foods used in larval culture. *Journal of the World Aquaculture Society* 25:371–378.

Munilla-Moran, R., Stark, J.R., and Barbour, A. 1990. The role of exogenous enzymes in digestion in cultured turbot larvae *Scophthalmus maximus* L. *Aquaculture* 88:337–350.

Nikolaeva, E.V., and Kasumyan, A.O. 2000. Comparative analysis of the taste preferences and behavioral responses to gustatory stimuli in females and males of the guppy *Poecilia reticulata*. *Journal of Ichthyology* 40:479–484.

Nordgreen, A., Hamre, K., and Langdon, C. 2007. Development of lipid microbeads for delivery of lipid and water-soluble materials to *Artemia*. *Aquaculture* 273:614–623.

Önal, U., and Langdon, C.J. 2000. Characterization of two microparticle types for delivery of food to altricial fish larvae. *Aquaculture Nutrition* 6:159–170.

Önal, U., and Langdon, C.J. 2004a. Lipid spray beads for delivery of riboflavin to first-feeding fish larvae. *Aquaculture* 233:477–493.

Önal, U., and Langdon, C.J. 2004b. Characterization of lipid spray beads for delivery of glycine and tyrosine to early marine fish larvae. *Aquaculture* 233:495–511.

Önal, U., and Langdon, C.L. 2005. Development and characterization of complex particles for delivery of amino acids to early marine fish larvae. *Marine Biology* 146:1031–1038.

Ostaszewska, T., Dabrowski, C., Czuminska, K., Olech, W., and Olejniczak, M. 2005. Rearing pike-perch larvae using formulated diets—first success with starter feeds. *Aquaculture Research* 36:1167–1176.

Ozkizilcik, S., and Chu, F.E. 1994a. Evaluation of omega-3 fatty acid enrichment of *Artemia* nauplii as food for striped bass *Morone saxatilis* Walbaum larvae. *Journal of the World Aquaculture Society* 25:147–154.

Ozkizilcik, S., and Chu, F.E. 1994b. Uptake and metabolism of liposomes by *Artemia* nauplii. *Aquaculture* 128:131–141.

Ozkizilcik, S., and Chu, F.E. 1996. Preparation and characterization of a complex microencapsulated diet for striped bass *Morone saxatilis* larvae. *Journal of Microencapsulation* 13:331–343.

Partridge, G.J., and Southgate, P.C. 1999. The effect of binder composition on ingestion and assimilation of microbound diets (MBD) by barramundi *Lates calcarifer* Bloch larvae. *Aquaculture Research* 30:879–886.

Person Le Ruyet, J.P., Alexander, J.C., Thebaud, L., and Mugnier, C. 1993. Marine fish larvae feeding: formulated diets or live prey? *Journal of the World Aquaculture Society* 24:211–224.

Rajkumar, M., and Kumaraguru vasagam, K.P. 2006. Suitability of the copepod, *Acartia clausi* as a live feed for seabass larvae (*Lates calcarifer* Bloch): compared to traditional live-food organisms with special emphasis on the nutritional value. *Aquaculture* 261:649–658.

Reiten, K.I., Bolla, S., and Olsen, Y. 1994. A study of the mechanism of algal uptake in yolk-sac larvae of Atlantic halibut (*Hippoglossus hippoglossus*). *Journal of Fish Biology* 44:303–310.

Reiten, K.I., Rainuzzo, J.R., Øie, G., and Olsen, Y. 1997. A review of the nutritional effects of algae in marine fish larvae. *Aquaculture* 155:207–221.

Rojas-García, C.R., and Rønnestad, I. 2003. Assimilation of dietary free amino acids, peptides and protein in post-larval Atlantic halibut (*Hippoglossus hippoglossus*). *Marine Biology* 142:801–808.

Rønnestad, I., Thorsen, S.K., and Finn, R.N. 1999. Fish larval nutrition: a review of recent advances in the roles of amino acids. *Aquaculture* 177:201–216.

Rønnestad, I., Tonheim, S.K., Fyhn, H.J., Rojas-García, C.R., Kamisaka, Y., Koven, W., Finn, R.N., Terjesen, B.F., Barr, Y., and Conceição, L.E.C. 2003. The supply of amino acids during early feeding stages of marine fish larvae: a review of recent findings. *Aquaculture* 227:147–164.

Rust, M. 1995. Quantitative aspects of nutrient assimilation in six species of fish larvae. PhD dissertation, University of Washington, School of Fisheries, Seattle, WA.

Rust, M.B., Hardy, R.W., and Stickney, R.R. 1993. A new method for force-feeding larval fish. *Aquaculture* 116:341–351.

Saavedra, M., Conceição, L.E.C., Helland, S., Pousão, P., and Dinis, M.T. 2008. Effect of lysine and tyrosine supplementation in the amino acid metabolism of *Diplodus sargus* larvae fed rotifers. *Aquaculture* 284:180–184.

Saavedra, M., Barr, Y., Pousão-Ferreira, P., Helland, S., Yúfera, M., Dinis, M.T., and Conceição, L.E.C. 2009. Supplementation of tryptophan and lysine in *Diplodus sargus* larval diet: effects on growth and skeletal deformities. *Aquaculture Research* 40:1191–1201.

Shields, R.J., Bell, J.G., Luizi, F.S., Gara, B., Bromage, N.R., and Sargent, J.R. 1999. Natural copepods are superior to enriched *Artemia* nauplii as feed for halibut larvae (*Hippoglossus hippoglossus*) in terms of survival, pigmentation and retinal morphology: relation to dietary essential fatty acids. *Journal of Nutrition* 129:1186–1194.

Sorgeloos, P., Dhert, P., and Candreva, P. 2001. Use of brine shrimp, *Artemia* spp., in marine fish larviculture. *Aquaculture* 200:147–159.

Southgate, P.C., and Kavanagh, K. 1999. The effect of dietary n-3 highly unsaturated fatty acids on growth, survival and biochemical composition of the coral reef damselfish *Acanthochromis polyacanthus*. *Aquatic Living Resources* 12(1):31–36.

Srivastava, A., Hamre, K., Stoss, J., Chakrabarti, R., and Tonheim, S. 2006. Protein content and amino acid composition of the live feed rotifer (*Brachionus plicatilis*): with emphasis on the water soluble fraction. *Aquaculture* 254:534–543.

Støttrup, J.G., and McEvoy, L.A. 2003. *Live Feeds in Marine Aquaculture*. Blackwell Press, Oxford, UK.

Szlaminska, M., Escaffe, A.M., Charlon, N., and Bergot, P. 1993. Preliminary data on semi synthetic diets for goldfish (*Carassius auratus* L.) larvae. In: Kaushik, S.J., and Luquet, P. (eds.) *Fish Nutrition in Practice*, Les Colloques 61. INRA Editions, Paris, pp. 607–612.

Takeuchi, T., Wang, Q., Furuita, H., Hirota, T., Ishida, S., and Hayasawa, H. 2003. Development of microparticulate diets for Japanese flounder *Paralichthys olivaceus* larvae. *Fisheries Science* 69:547–554.

Tanaka, H., Matsumura, M., and Veliky, I.A. 1984. Diffusion characteristics of substrates in Ca-alginate gel beads. *Biotechnology Bioengineering* 26:53–58.

Teshima, S., Kanazawa, A., and Sakamoto, M. 1982. Microparticulate diets for the larvae of aquatic animals. *Mini Review and Data File of Fisheries Research, Kagoshima University* 2:67–86.

Tonheim, S.K., Koven, W., and Rønnestad, I. 2000. Enrichment of *Artemia* with free methionine. *Aquaculture* 190:223–235.

Van der Meeren, T. 1991. Algae as first food for cod larvae, *Gadus morhua* L.: filter feeding or ingestion by accident? *Journal of Fish Biology* 39:225–237.

Vandenberg, G.W., and De La Noüe, J. 2001. Evaluation of protein release from chitosan-alginate microcapsules produced using external or internal gelation. *Journal of Microencapsulation* 18:433–441.

Villamar, D.F., and Langdon, C.J. 1993. Delivery of dietary components to larval shrimp (*Penaeus vannamei*) by means of complex microcapsules. *Marine Biology* 115:635–642.

Walford, J., Lim, T.M., and Lam, T.J. 1991. Replacing live foods with microencapsulated diets in the rearing of seabass (*Lates calcarifer*) larvae: do the larvae ingest and digest protein-membrane microcapsules? *Aquaculture* 92:225–235.

Yúfera, M., Pascual, E., and Fernández-Díaz, C. 1999. A highly efficient microencapsulated food for rearing early larvae of marine fish. *Aquaculture* 177:249–256.

Yúfera, M., Fernández-Díaz, C., Pascula, E., Sarasquete, M.C., Moyano, F.J., Díaz, M., Alarcon, F.J., García-Gallego, M., and Parra, G. 2000. Towards an inert diet for first-feeding gilthead seabream *Sparus aurata* L. larvae. *Aquaculture Nutrition* 6:143–152.

Yúfera, M., Kolkovski, S., Fernández-Díaz, C., and Dabrowski, K. 2002. Free amino acid leaching from a protein walled microencapsulated diet for fish larvae. *Aquaculture* 214:273–287.

Yúfera, M., Fernández-Díaz, C., and Pascual, E. 2005. Food microparticles for larval fish prepared by internal gelation. *Aquaculture* 248:253–262.

Zambonino Infante, J.L., Cahu, C.L., and Péres, A. 1997. Partial substitution of di- and tripeptides in sea bass diets improves *Dicentrarchus labrax* larval development. *Journal of Nutrition* 127:608–614.

Chapter 13

Microparticulate diets: testing and evaluating success

G. Joan Holt, Ken A. Webb, and Michael B. Rust

13.1 Introduction

The advent of intensive and semi-intensive larviculture brought with it the need for more efficient ways to feed marine fish larvae. While the successful production of larval fish depends on a variety of concerns such as environmental conditions and broodstock quality, the availability of a suitable food supply is the most important factor. The larvae of most commercially important marine fish are altricial and, at least early in their development, typically have a very limited ability to capture and ingest prey. Suitable prey items must be small enough for the larvae to ingest and nutritious enough to provide sufficient nutrients and energy for growth. Historically, marine fish larvae were reared in extensive systems such as nursery ponds or nursery pens to take advantage of natural zooplankton availability. Until the latter part of the 20th century, this normally meant that larviculture was either extremely seasonal or possible only in limited geographic regions. This was because wild zooplankton had to be available within the culture system itself or available for harvest from the wild and transferred into the culture system.

Following the development of rotifer (*Brachionus* spp.) and *Artemia* spp. feeding techniques, larviculture of marine fish experienced a dramatic expansion. Utilization of rotifers and *Artemia* gave culturists a source of prey that was not dependent on environmental and seasonal effects. The identification of thousands of rotifer species and dozens of *Artemia* species has allowed for customization of prey size tailored to match the ability of the larvae to consume them, limited only by the availability of a given species of prey item. Development of commercial enrichment products for both rotifers and *Artemia* has also allowed for some modification of nutritional values such as protein and fatty acid content to better match the needs of the larvae. Despite these advances, tremendous efforts have gone into developing artificial microdiets (also referred to as microparticulate diets) for use in larviculture over the past 30 or so years. This is due to the fact that

Larval Fish Nutrition, First Edition. Edited by G. Joan Holt.

while rotifer and *Artemia* feeding protocols have significant advantages when compared with wild zooplankton, they also have severe limitations. The development of artificial feeds capable of replacing live zooplankton would be a tremendous benefit to larviculture in the same way that the development of complete diets for juvenile fish has been a boon to aquaculture as a whole. In this chapter, we will discuss the theoretical advantages of artificial diets as well as the current and future roles and research needs for microdiets in larviculture.

13.2 The need for microparticulate diets

There are many advantages to replacing live prey with microparticulate diets. First and foremost, microdiets have a uniform nutritional composition, which can be tailored to match the needs of a given species or adjusted for research purposes. Current methods of enrichment for live prey items have come a long way in their ability to modify nutritional composition, but enrichment of live prey items will always be limited by the capacity of the organism to ingest the enrichment and by the alteration of the enrichment by metabolism in the prey or limited to the undigested material in the prey's digestive system. Outside of certain conditions that will be discussed in this chapter, the nutrient composition of artificial diets is potentially constant and predictable. Second, microdiets are consistently available year-round without having to dedicate culture space and time to maintain stocks. Sufficient feed can be procured in advance of a culture season and stored so that it is available as needed. The same is true for *Artemia*, which are typically sold as cysts and are stable for months; however, many fish larvae are incapable of preying on *Artemia* nauplii at first feeding and must be provided with rotifers. While it is feasible to raise rotifers from diapaused eggs (Gilbert and Schröder 2004) that can be obtained and stored, obtaining sufficient eggs to meet the needs of even a relatively small hatchery would be prohibitively expensive. And third, even though feed costs remain a significant fraction of the total cost of production, once larvae can be weaned to artificial diets, the feed costs per individual decline dramatically. Current microdiets are extremely expensive when compared with juvenile growout feeds, often costing in excess of 10 times more per kilogram, but are still much cheaper than growing and maintaining live prey for feeding larvae. Despite these advantages, there is no commercially available microdiet that can completely replace live prey in the culture of marine fish larvae, and commercial hatcheries remain dependent on the supply of live feed.

From a larval fish nutrition perspective, live feeds are not suitable for determining most nutrient requirements for larval fish using the traditional dose–response feeding trials that have been the backbone of nutritional research for larger fish and all other domestic animals. Until a suitable diet is found that allows for feeding graded levels of a nutrient in a predictable manner with high survival, further nutritional work with larvae will be difficult at best.

This is not to say that there have not been a number of successes. For instance, larvae of the European sea bass (*Dicentrarchus labrax*) have been raised with some success on a microdiet composed mainly of yeast and fish hydrolysate (Cahu et al. 1998; Cahu and Infante 2001), and red drum (*Sciaenops ocellatus*) larvae have been reared successfully on a commercial microdiet with only 5 days of cofeeding rotifers and microdiet (Holt 1993; Figure 13.1) or solely on microdiet and algae (Lazo et al. 2000). These cases mark the high point of microdiet development in marine fish though it typically comes with both reduced growth and survival when compared with live prey (the sea bass) or without completely replacing live prey (the red drum).

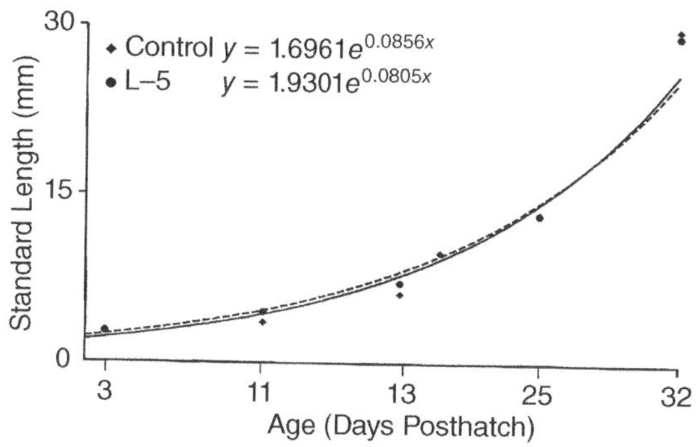

Figure 13.1 Exponential growth of red drum (*Sciaenops ocellatus*) larvae raised for 1 month on a combination of live food (rotifers grown on a mixture of *Isochrysis galbana* and *Chlorella* sp.) for 5 days and microdiet thereafter (L-5; dashed line), and controls fed live prey (rotifers and *Artemia*) throughout (solid line) (from Holt 1993).

Perhaps the ultimate question in working with larval diets is, "Why do fish survive and grow better on live prey than on artificial diets?" Even rearing marine fish larvae using live diets is not straightforward, and the issue becomes even more complex when dealing with microparticulate diets. The diagram in Figure 13.2 illustrates some of the many factors that can result in poor performance in larvae in intensive culture systems no matter what diet they are fed. For each of these factors, there is likely an optimal and/or minimal value or condition that can make the difference between high survival and high mortality. While some of these issues are common to both live and microparticulate diets, others are unique to one type of diet or the other. Many of the issues are related to ingestion (availability, appearance, long- and short-range attraction, and texture), digestion (metabolism, digestion, and assimilation), and nutrient sufficiency (maternal stores, metabolism, nutrient requirements, leaching, and formulation) of the microdiets compared with live feeds. The figure is not meant to be exhaustive of factors impacting larval

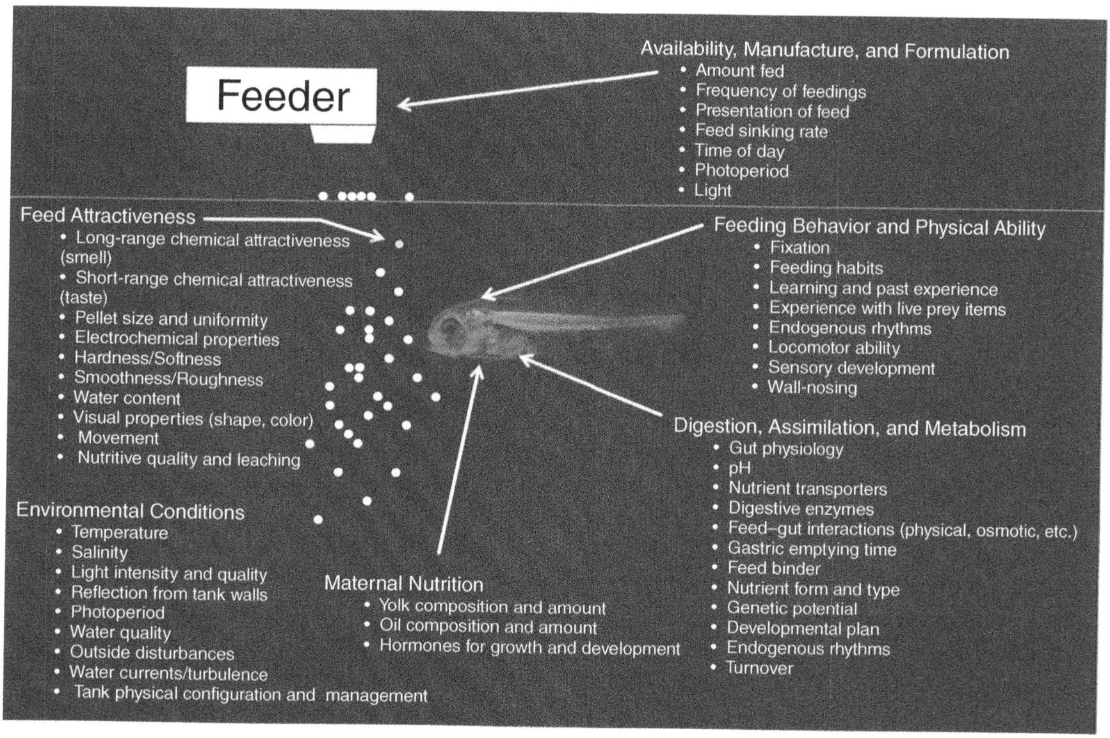

Figure 13.2 Conceptual diagram of some of the factors that can impact the success of microparticulate diets used for feeding larval fish. Optimizing the major factors for each species of interest may improve the results with the currently available microparticulate diets.

nutrition, just suggestive that there are many. Broadly speaking, the factors illustrated in Figure 13.2 can be lumped into three categories: (1) factors that impact ingestion of the diet; (2) factors that impact digestion, assimilation, and metabolism of the diet; and (3) factors that impact both.

Larvae must eat the feed before digesting and assimilating the nutrients it contains, so maximizing ingestion is the first hurdle, and one that should be obtainable with the current tools we have available. Studies showing poor survival and growth on microparticulate diets often attribute poor performance to the diet or the larvae's inability to digest and utilize nutrients in the diet without first demonstrating that the larvae are actually consuming the diets at levels that could support growth and survival. Furthermore, some studies have declared that larvae will not consume microparticulate diets without first testing the variety of conditions that might need to be adjusted and optimized to allow the larvae to consume microparticulate diets. Many of the items in Figure 13.2 can be optimized using consumption data alone, and this needs to be done before addressing the nutritional quality of a microparticulate diet or establishing the minimum age at which a larva will consume microparticulate diets.

13.3 Live versus microparticulate feeds

In addition to the physical and chemical differences between live prey and artificial diets, an important consideration is how the larvae respond to those diets. Successful feeding depends on the characteristics of the larvae such as locomotor abilities, sensory development, discrimination or feeding preference, and experience. Most marine larvae begin feeding at a time when their motor and sensory skills are undergoing development and they have limited ability to find prey (see Chapter 10 for more details). Thus, food needs to be presented so as to increase encounter rates during initial feeding. As larvae grow and develop, this becomes less of a problem. The larvae of some fishes are voracious and will consume almost anything they can fit in their mouths, but others can be incredibly selective predators. Some species may inherently have narrow prey specificity; for example, scombrid larvae have precocious development of piscivory, with Spanish mackerel (*Scomberomorus niphonius*) feeding on fish larvae in the first-feeding stage (Kaji et al. 1996). Feeding success is not only related to ontogenetic development, but food preference and experience are important as well. Some larvae may need experience. They may not recognize microparticles as food but if it is cofed along with live prey they will often ingest microdiets. Feeding success is also influenced by the characteristics of the culture system and the diet itself.

13.3.1 Culture systems

To obtain high consumption rates with microparticulate diets, the entire culture system used for microparticulate diets may need to be fundamentally different from that used for live feeds (Barrows and Rust 2000) Because live feeds reside in the water column and do not appreciably degrade the water quality in the tank, typical intensive systems have low flow (perhaps 1–2 volumes per day—to retain the expensive live feeds and "green water"), have low or variable turbulence, and can be fed relatively infrequently (on the order of hours between feedings). Because microparticulate diets often sink out of the water column in a matter of minutes and then pile up on the bottom and decay, they degrade the water quality. Using the same culture system as used with live diets can result in very low opportunity for the larvae to feed and poor water quality. Feeding microparticulate diets is most successful when high water turnover rates (perhaps 1 volume per hour), frequent

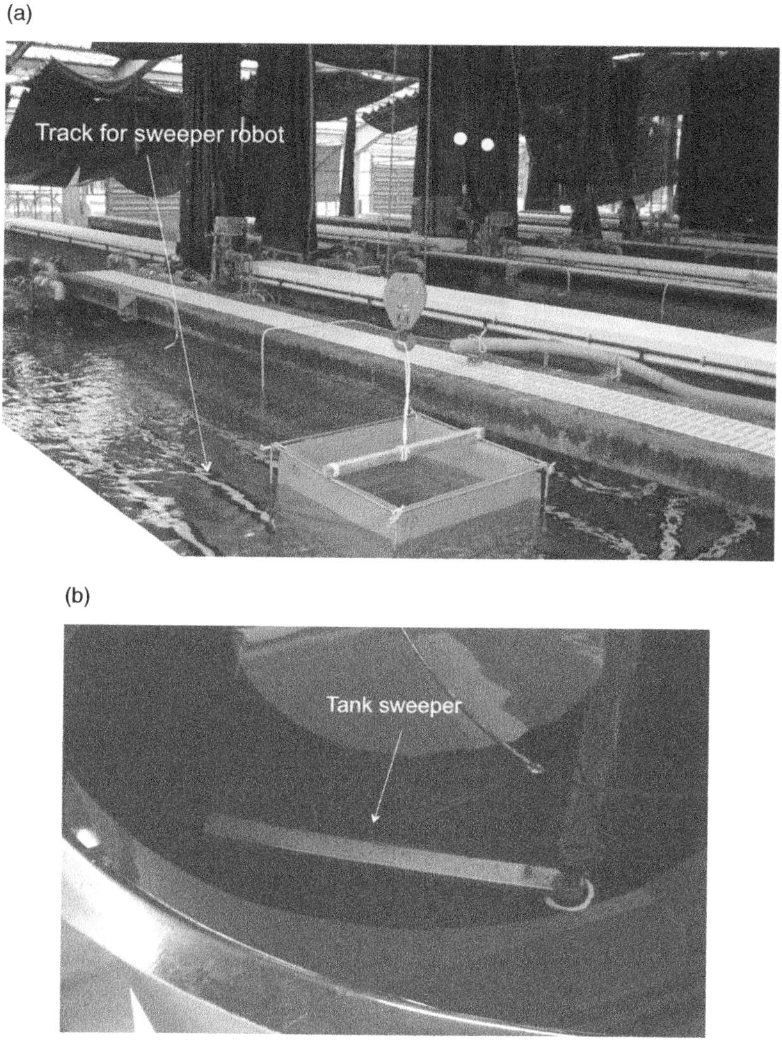

Figure 13.3 Automated tank bottom cleaners for larval rearing tanks. (a) The photo shows a larval rearing tank that has just been filled prior to stocking black rockfish (*Sebastes schlegeli*) at the Miyako Research Station in Japan. Note the white track on the bottom of the tank. It is used to optically guide a small cleaning robot around the bottom of the tank. The robot is similar to sweepers commonly used for swimming pools. (b) An aluminum arm with a rubber knife on the bottom (taken from a truck windshield wiper) makes a slow rotation around the bottom of this larval rearing tank at the Austevoll Aquaculture Research Station in Norway. A small motor is mounted under the tank to provide movement to the arm. Concentrating the waste makes cleaning easier. (Photos by Michael Rust.)

feedings (perhaps every 10–30 minutes), and frequent regular cleaning of the bottom are employed. Devices to continuously clean the bottom of tanks such as shown in Figure 13.3 would be preferred. Water inflow structures that increase flow without increasing local turbulence also require careful consideration and design.

13.3.2 Physical properties

Physical properties of the microdiet in the rearing system that are important for increasing its detection by and attractiveness to fish larvae include size, shape, color, movement, smell/taste, and texture. Guthrie et al. (2000)

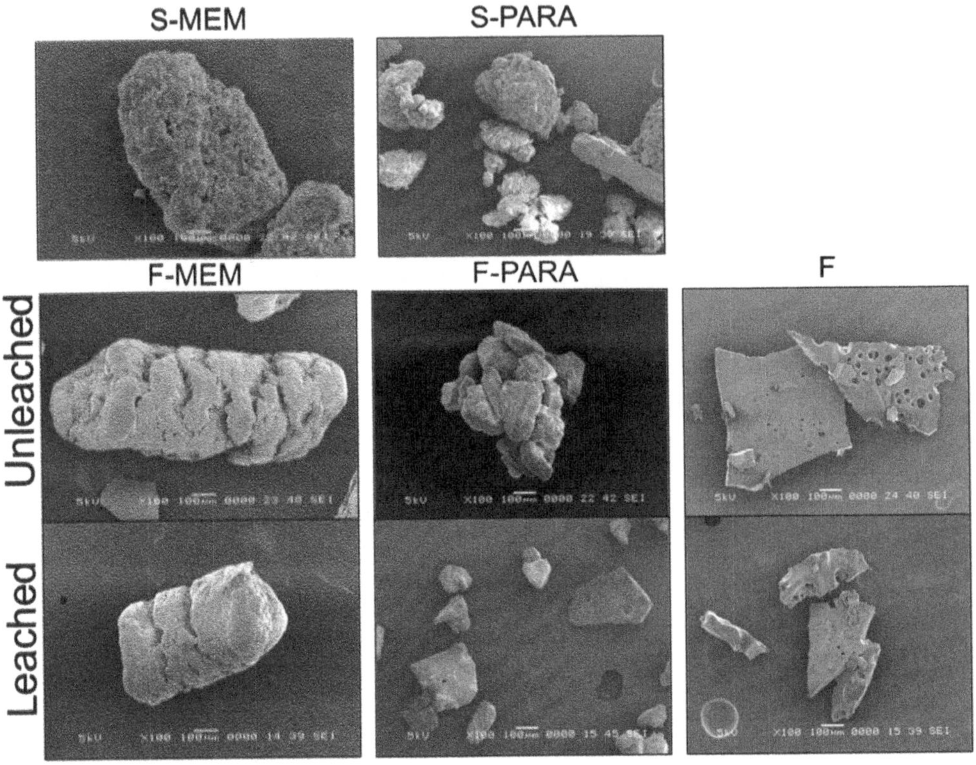

Figure 13.4 SEM photographs of microparticulate diets made using different processes. All diets were processed by three methods: flaking (F), microextrusion followed by marumerization (MEM), and particle-assisted rotational agglomeration (PARA). Each of these three diets had a unique formulation. Additional two diets were made by taking the flake diet and further processing it using the MEM (F-MEM) or PARA (F-PARA) methods. The F, F-PARA, and F-MEM diets had the same formulation and only differed by postflaking processing method. Each diet was screened to the 400–700-μm range. The lower set of SEM photos is of the F-MEM, F-PARA, and F microparticles after leaching in freshwater at room temperature for 15 minutes. F was made in Michael Rust's lab at NOAA's Northwest Fisheries Science Center, and the MEM and PARA processes were performed in Dr. Rick Barrows's lab at the United States Department of Agriculture, Agriculture Research Service (USDA/ARS) feeds lab in Bozeman, MT. (Photos by Carla Stehr and Emily Hart, National Oceanic and Atmospheric Administration [NOAA].)

demonstrated different consumption rates by larval walleye (*Stizostedion vitreum*) fed diets with a similar formulation but different methods of binding and manufacture. Methods to manufacture different microparticulate diets result in very different particle types that vary in physical properties. Figure 13.4 illustrates particle types made by three different methods: flakes, microextrusion followed by marumerization (MEM), and particle-assisted rotational agglomeration (PARA), and two additional combinations of these methods. The details of manufacture of these diets are found in Chapter 12. Even when formulated with the same ingredients (all the "F" diets), these different particle types display different densities, sinking rates, leaching curves, textures, and other physical properties (Table 13.1). To get the maximum feeding rates from larvae fed these different diets may require different feeding conditions.

13.3.2.1 Size and color

Particle size is particularly important in first-feeding larvae and needs to be adjusted for

Table 13.1 Composition, leaching, and sinking characteristics of the five diets shown in Figure 13.4.

Feed	Moisture (% as fed)	*Proximate composition (% dry weight)*			*Fatty acid concentrations (mg fatty acid/100 mg total fatty acids)*			Protein leaching, $t_{1/2}$/% lost	Sinking rate
		Protein	Lipid	Ash	ARA (n-6)	EPA (n-3)	DHA (n-3)	([seconds]/[%])	(cm/s)
S-MEM	5.7	53.8	30.5	8.4	0.4	7.5	5.7	98/23.0	0.84±0.010[a]
S-PARA	5.3	54.2	29.7	8.9	0.4	7.4	5.6	94/30.2	0.34±0.001[c]
F	9.0	75.9	10.9	7.3	1.0	6.0	15.2	20/44.3	0.33±0.001[c]
F-MEM	13.3	75.6	13.4	7.5	1.0	5.3	13.0	50/41.8	0.78±0.060[a]
F-PARA	11.0	75.6	12.6	7.3	1.0	6.0	15.2	33/47.2	0.61±0.001[b]

There were significant differences in sinking rate among diets where the superscript letters are different.

different ontogenetic stages of the target species (see prey selection in Chapter 10). Diet color can vary depending on diet formulation, the water and light conditions, and the background color of the rearing system. Often larvae are reared in "green water" culture that is either a phytoplankton bloom or algae added to the system (Tredici et al. 2009). Among the many advantages of green water culture, one is its important benefit for visually feeding larvae: the algae may increase contrast so that larvae can better see the particles. Lighting and water depth are crucial in establishing shading and visual contrast for proper feeding and should be established for each species and rearing condition. Reddish brown- to orange-colored diets appear overall most attractive to red drum and cobia (*Rachycentron canadum*) larvae (Holt, unpublished data) in either clear water or green water culture.

13.3.2.2 Movement or behavior

Live prey swim in the water column with movements that capture the attention of visually feeding predators such as fish larvae. Likewise, microdiets may need to remain in the water and have motion that is fast enough to draw attention but slow enough so the larva can catch it. This is accomplished through adjusting the density of the particle, and water and/or air movement in the rearing system. At least some, perhaps most, larvae will feed on particles moving with the currents as evidenced by *Artemia* cysts, small bubbles, bits of plastic, tree pollen, and fish eggs that have been reported in larval fish guts. This suggests that endogenous movement of microparticles is not necessary. But since many larvae feed near or just below the surface, microparticles need to remain slightly negatively buoyant and in the water column long enough for the larvae to find, catch, and ingest them. Particles that float are generally not acceptable to most fish larvae since they would need to break the surface tension to feed. Eventually, microparticulate diets generally sink to the bottom of the tank. Since few species feed on the bottom, especially in their early stages, the amount and timing of feeding needs to be adjusted for this loss. In practice, automatic feeders that will dispense microdiets on schedule into a properly turbulent system are the most successful (Figures 13.5 and 13.6).

13.3.2.3 Chemical attractiveness

Fish live in an enriched chemical environment with highly developed chemosensory systems and are especially attracted to amino acids, organic acids, nucleotides, and bile salts (Hara 1994). Live prey, especially copepods,

Figure 13.5 Larval fish rearing tanks with automatic feeding system and water movement in the University of Texas Marine Science Institute, Coastal Conservation Association of Texas (UTMSI CCA-Texas) Larviculture Lab in Port Aransas, TX. Six 350-L circular tanks in a 4,000-L rectangular tank (1.2 width × 2.4 length × 0.5 height, in meters) with temperature control, biofiltration, sand filtration, aeration, and pumped return water. (Photo by Avier Montalvo, UTMSI.)

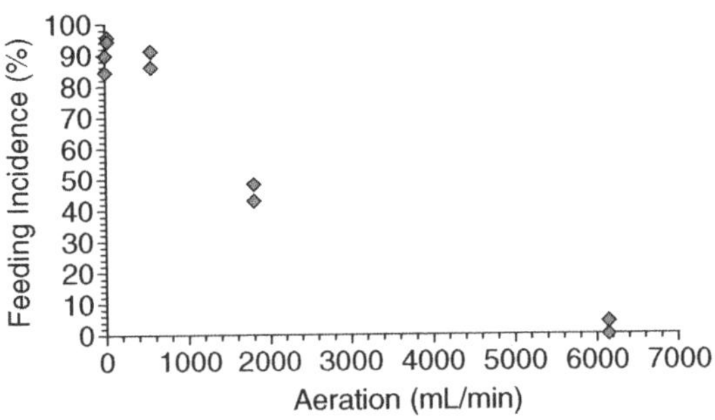

Figure 13.6 Effect of aeration-induced turbulence on feeding incidence on rotifers in first-feeding larval Pacific cod, *Gadus macrocephalus* (Rust, unpublished data). The diamonds are the percent feeding incidence of replicate tanks of fish at the given turbulence level.

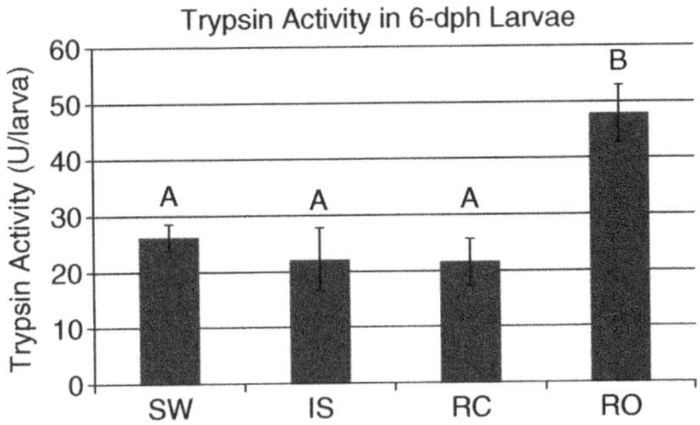

Figure 13.7 Trypsin responses (and standard error) of 6-dph red drum larvae exposed to four treatments for 2 h: SW = saltwater control, IS = *Isochrysis galbana* homogenate, RC = rotifer culture water, RO = rotifer homogenate. Columns with different letters indicate significant differences ($\alpha \leq 0.05$) among treatments (from Webb 2008).

contain relatively large amounts of free amino acids (FAA) including glycine, arginine, and betaine shown to be strong inducers of feeding behavior in larval fish (Kolkovski et al. 2009). Soluble components of rotifers have also been shown to elicit cholecystokinin (CCK) and trypsin responses in larvae, preparing the primitive gut for digestion of the perceived prey (Webb 2008; Figure 13.7). Attractants in microparticulate diets may be important for increasing detection of particles in the rearing environment, but taste and texture also likely determine final acceptance of the diet by the larvae. For example, although some cobia larvae accept commercial microdiets in the first 2–3 weeks of feeding, most reject the particles even if they take them into their mouth (Holt et al. 2007). Texture of dry microdiets may be an issue; soft diets may be more readily ingested by larvae. Future research should focus on enhancing the attractiveness and palatability of microdiets.

13.3.3 Leaching

Another difference between live and microparticulate diets is in the degree of leaching of small-molecular-weight water-soluble (SMWS) nutrients. There is good evidence that these types of compounds could be very important in the nutrition of fish larvae (see Chapters 3 and 12). Live feeds use energy to maintain these compounds in their tissues while they are alive, but then lose them as soon as homeostasis stops when they die. This makes a pretty sophisticated delivery system as the SMWS nutrients are retained while the feed is swimming in the water column but likely released quickly by osmotic pressure as soon as they die in the digestive

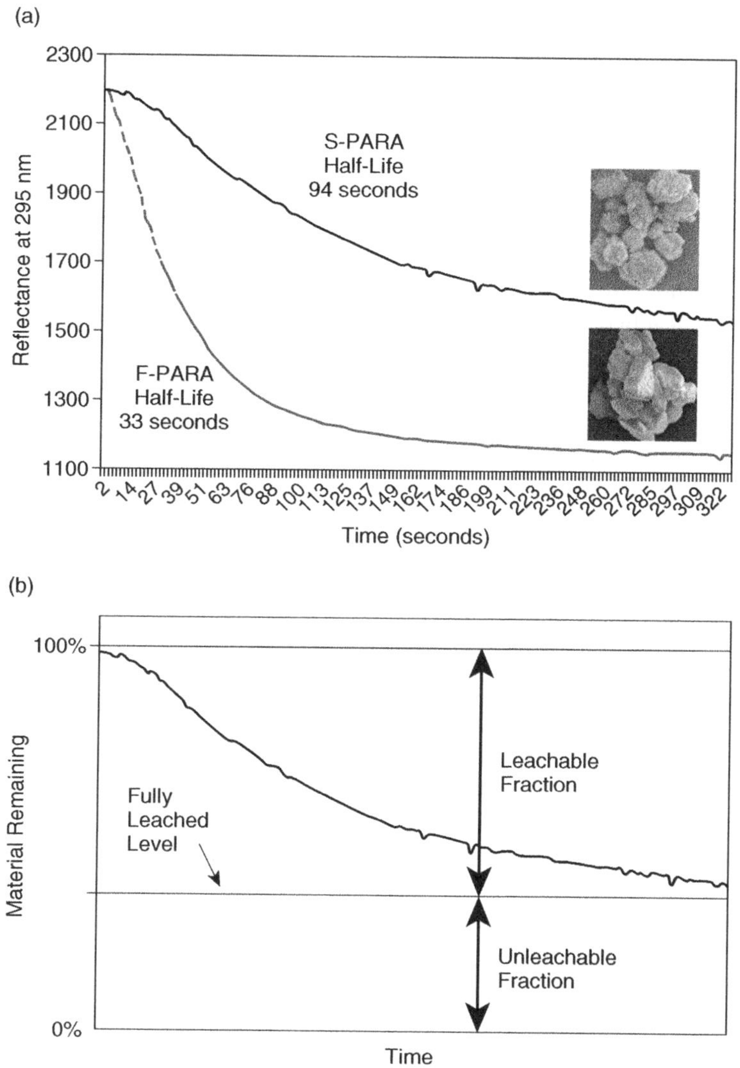

Figure 13.8 (a) Leaching curves for the F-PARA and S-PARA diets shown in Figure 13.4 using the method of Nicklason and Johnson (2008). The conjugated double bond in aromatic amino acids such as tryptophan and phenylalanine absorbs light at 295 nm. The signal from a spectrometer set to measure the signal at this wavelength is used as a proxy for soluble protein loss from microdiets. Fitting a decay curve to the data provides a calculation of the half-life of the leachable material. (b) An idealized leaching curve showing real-time leaching, percent leached, and percent unleached concepts.

tract of the larvae. So far, this has been difficult to replicate with a microparticulate diet; however, by understanding and adjusting the leaching of SMWS nutrients from microparticles and knowing the time that a larva has to feed on a microparticle (which is a function of the sinking rate) before it sinks out of the water column, it may be possible to provide the same effect.

Leaching has been studied by numerous methods and numerous authors and is discussed in Chapter 12. A typical leaching curve produced by the methods used in Nicklason and Johnson (2008) is shown in Figure 13.8a. In this case, the curve shows the difference between diets of different sizes but otherwise formulated and manufactured the same way. As expected, the smaller particle leaches faster due to the higher surface area-to-volum ratio. From the decay curve, a half-life can be computed to compare among diets or to compare with the calculated time the particle would be available to the larvae (sinking rate × tank depth, for example). Table 13.1 provides half-life and percent leached data for the five diets shown in Figure 13.4. Note for the F diets that the processing did impact half-life but not the total percent leached.

Additional aspects of leaching that may need to be determined for each diet type and size are the maximum total percent of mass leached and the composition of the leached and unleached materials (Figure 13.8b). This can be done by leaching a known dry weight of particles for a period of time long enough for the leaching to be "complete" (hours) and then filtering out the leached microparticles, drying and weighing to determine the amount leached (Table 13.1). It would also be important to know the composition of the leachate and the leached particles relative to the starting material so the compounds that are lost quickly from a particle can be added at higher levels to accommodate leaching, or put into a special inclusion particle as a part of a complex microparticle to reduce leaching (see Chapter 12 for more information and definition of terms). While total mass may be a good general indication of the leaching characteristics of a diet, individual essential nutrients may also need to be determined (see review by Langdon 2003, and Chapter 12).

It is interesting to note that the physical appearance of the microparticle before and after leaching may not change much, as shown in Figure 13.4 (with the exception of the F-PARA). Based on looks alone, a fish larva may feed on particles that may be fully leached.

13.3.4 Sinking rate

Along with leaching, the sinking rate determines the actual composition of what the larvae eat once the diet is introduced to the tank. Slower sinking diets have the advantage of being more available to the larvae for longer periods of time, are presumably easier to catch, and result in less buildup on the tank bottom. One disadvantage is that diets that are too light will float on the tank water surface until they have become sufficiently hydrated to break the surface tension and sink. Diets floating on the surface have already started to leach so the nutritional quality could already be compromised before they are even available to the larvae. Perhaps there is a need for a feeding system that disperses microparticulate diets underwater? In addition, the longer the time the particle is available to the fish larvae, the more leaching has occurred. Various methods have been used to record sinking times, but the simplest is to simply time the sinking of particles through a known distance in a column of still water. Table 13.1 shows the sinking times for the diets shown in Figure 13.4.

13.4 Quantifying larval responses to live and microparticulate diets

While it seems intuitive that once prey is captured it will be consumed, this is not necessarily true. Quantifying the ingestion of microparticulate diets by larvae is not a trivial matter due to the sizes of the animals and the feed involved, and thus specialized techniques have been developed to accomplish this task. In addition, once the diet is captured and retained by the larva, it must be successfully digested (see review by Rust 2002). Altricial larvae begin feeding before the digestive system has fully matured, and the formulation of microparticulate diets must take this into account (see Chapter 1). A number of techniques, from the very simple to the extremely complex, have been developed to quantify this aspect of larval nutrition as well.

Table 13.2 Tools available for fish nutrition in larval and juvenile fish.

Need	Juvenile fish	Larval fish
Determining ingestion and consumption	Various methods to directly measure fish consumption using • Recording consumption directly • Markers • Fish behavior	Acceptability trials using • Feeding incidence • Gut contents counting • Image analysis • Markers • Fish behavior
Determining nutrient requirements	Dietary dose–response trials using defined diets and various end points such as • Growth • Survival • Disease resistance • Reproductive performance • Feed efficiency • Others	Limited dose–response information from limited manipulation of the composition of live diets—mostly fatty acids Infer from composition of eggs, larvae, and/or key prey species End points often compromised by high mortality rates in "control" treatments
Determine digestibility	Well-developed universal *in vivo* method to determine apparent digestibility	A variety of methods, mostly qualitative and based on observation; quantitative methods are rare and usually require advanced chemical analysis *In vitro* methods Use of stable isotopes and other tracers Tube feeding studies

13.4.1 What we can learn from juvenile fish nutrition

In contrast to larval fish nutrition, huge advances in the nutrition of juveniles have been made over the previous decade. This can be attributed to the well-honed and powerful tools available to researchers for juvenile fish. Table 13.2 illustrates some of the more important ones for juvenile fish nutrition and the current counterparts used in larval fish nutrition. The only area where good tools exist for larval fish is in determining consumption. It is still difficult, at best, to determine nutrient requirements and digestibility for fish larvae.

The use of semipurified and practical diet formulations where a single nutrient can be adjusted in a dose–response manner and fed to replicated groups of juveniles is a powerful tool to determine nutrient requirements and improve diets. The larger particle size and typically voracious feeding response of juveniles renders consideration of leaching losses almost moot. Pair feeding and apparent satiation feeding allows researchers to be able to tease out differences in diet performance due to nutritional quality, diet palatability, or a combination of both. End points of survival and growth are typically high and can be measured directly. While portions of this approach could be used for larval fish, there first needs to be a diet that can be manipulated to provide single nutrient doses, and next, survival of the best treatments needs to be much greater than what are currently observed.

Likewise, the use of indigestible markers and fecal collection make it possible to routinely determine the digestibility of a diet in juvenile fish. This can be done on a dry matter basis, a nutrient class basis (protein, lipid, carbohydrate), an energy basis, or even an individual nutrient or compound basis (e.g., lysine, individual vitamins and minerals, phospholipids, fatty acids). The practical difficulty of collecting and analyzing the microgram quantities of feces produced by larval fish makes this approach challenging, but not impossible, with fish larvae (Johnson et al. 2009).

13.4.2 Tools available for testing microparticulate diets in larvae

Glencross et al. (2007) considered five components that need to be assessed for ingredients used in feeds for aquatic animals. Their list can easily apply to microparticle diets in general as well as the feedstuffs used to make them. The five areas are (1) ingredient characterization, which they define as composition, variability in composition, source, and other defining qualities; (2) ingredient digestibility; (3) ingredient palatability, which we expand to all factors that lead to consumption or acceptance of a diet; (4) nutrient utilization, which we refer to as bioavailability; and (5) ingredient functionality, which relates to the physical and chemical properties an ingredient imparts to a pellet. Elements 2, 3, and 4 on this list will be discussed, starting with palatability. As has been previously discussed, (6) nutrient leaching should be added to this list for microparticulate feeds.

13.4.2.1 Measuring consumption

Acceptability tests measure the amount of a given feed a larva actually ingests and/or consumes. Obviously, this is an important quality in a diet and these methods can be used to optimize the diet, to optimize the culture system, or to develop feeding stimulants and palatability enhancers (Figure 13.2). Since studies of consumption can proceed without the need for a nutritionally complete diet, but larval nutrition cannot proceed until consumption is demonstrated, these studies are a logical first step toward developing high-quality microparticulate diets for fish larvae. Not until routinely high consumption rates and levels are obtained can the nutritional value of a diet be convincingly tested.

There are at least three methods to measure consumption. The first method simply involves dissection and counting the particles or prey items in a larval fish gut following a feeding event. This method has been widely used but is time-consuming to do, and sometimes it is difficult to determine from samples that have been digested. The second approach uses image analysis to take advantage of the transparent gut of most larval fish (Rust and Barrows 1998; Guthrie et al. 2000). The third method incorporates an indigestible marker (a rare earth metal oxide) into the feed, which is later quantified on a spectrophotometer. Cook et al. (2008) demonstrated that dysprosium oxide (Dy_2O_3), ytterbium oxide (Yb_2O_3), yttrium oxide (Y_2O_3), and lanthanum oxide (L_2O_3) could be used to label and enumerate live prey and microparticulate diets ingested by marine fish larvae.

For each of these methods, the basic design is the same. Each diet, a live feed control and/or a starvation control, is applied to replicate tanks of larvae and feeding is allowed to take place for a short period of time. All larvae from each tank are harvested from the treatment tank at the same time and feeding incidence (percent of fish with food in the gut) and gut fullness are measured. Gut fullness can be measured by any of the three methods outlined in the previous paragraph: (1) counting; (2) determination of the cross-sectional optical area of the bolus by image analysis (Rust and Barrows 1998); or (3) for diets marked with rare earth metals, larvae are simply harvested and counted into vials, digested, and read on an emission or mass spectrophotometer (Cook et al. 2008). Data from this last method are given as milligrams of diet ingested per fish.

This type of test is good for optimizing factors that lead to maximal ingestion of diets.

Figure 13.9 Photo showing larval greenling (*Hexagrammos stelleri*) in a C-shaped attack position (lower middle). Note that the larva in the upper right is swimming normally. (Photo by Michael Rust.)

It is also important to show that diets are actually consumed by the larvae and to what extent they are consumed. Diet preference studies using a common garden design are also possible using multiple metal oxide markers, with the methods in Cook et al. (2008), assuming the markers do not impart any selective attributes to the diets. The quantification of the amount of feed ingested allows calculation of the dry weight of material actually delivered to the larval gut. When Guthrie et al. (2000) made this calculation, they found a couple of microparticulate diets that delivered more nutrients to the gut than did *Artemia* even though consumption was less.

An additional method that might greatly simplify these types of studies, though quantification of the material delivered to the gut is not possible, is the use of videography and image analysis to quantify larval feeding behavior. Fish larvae commonly use a sigmoidal attack posture (Munk and Kiorbe 1985). After spotting a prey, the larva curls its body into a C shape or S shape (Figure 13.9). The actual feeding attack takes place when the larva springs from this posture, straightens its body, and lunges at the prey. Such changes in posture can be quantified using image analysis (Puvanendran and Brown 1999). This approach was used by Rabe and Brown (2000) to study a pulse feeding strategy for yellowtail flounder.

Whatever short-term method is used, it is important to carry these conditions out to demonstrate that the short-term measurements used such as feeding incidence relate to long-term outcomes such as survival and growth.

13.4.2.2 Digestibility

There is currently no single easily applied method to determine nutrient digestibility in fish larvae. This sort of assay is fundamental to nutrition studies in larger fish and all types of domestic animals (see Glencross et al. 2007 for a recent review). Broadly speaking, digestibility measures the percent of a diet, nutrient, or other compound in a diet that is retained by the animal consuming the diet. Thus, it largely deals with a measure of the digestion and assimilation of the diet over the time period of one or several meals. Bioavailability is a related concept; however, it includes metabolic processes. Bioavailability is a measure of the percentage of a diet, nutrient, or other compound in a diet that is incorporated into the consuming animal's tissue.

Two methods have been used to determine digestibility in fish larvae. The first method is an *in vitro* method where larval digestive enzymes are extracted and mixed with the diet of interest and the resulting hydrolysis is measured (Dimes et al. 1994). The second method is an *in vivo* approach where an indigestible marker (typically a rare earth oxide, but other things have been used) is added to the diet and the ratio of the concentration of the marker to the concentration of the nutrient of interest is determined in the feed and feces of the consuming animal to compute the percentage of a nutrient retained (Windell 1978; Glencross et al. 2007). This second process is referred to as "apparent digestibility."

The *in vitro* method has been used widely for various types of feeds and feedstuff (Dimes et al. 1994). This approach has been adapted to larval fish (e.g., Tonheim et al. 2007) and can be a useful tool for determining what feed ingredients and manufacturing methods might work for fish larvae. For example, Alarcón et al. (1999) used this method to determine that cuttlefish and casein were hydrolyzed more completely than ovalbumin by larval sea bream enzymes. The practical implication is to utilize the more digestible products as feedstuffs in larval microparticulate diets. Still lacking is a study that validates any larval *in vitro* method with an *in vivo* method.

To determine the *in vivo* apparent digestibility of a nutrient (or energy) in a diet, the resulting ratios of nutrient to marker in the feed and feces are converted into an apparent digestibility coefficient (ADC) using the following formula:

$$ADC = 100 \times \left[\frac{[\text{Marker}]_{\text{feed}} / [\text{Protein}]_{\text{feed}}}{[\text{Marker}]_{\text{feces}} / [\text{Protein}]_{\text{feces}}} \right]$$

Johnson et al. (2009) adapted the ADC method to (8-week-old) cod larvae by working out four issues that made this method difficult for larval fish. First, they developed a method to collect feces from larval fish. Then the authors used the approach of Cook et al. (2008) to mark *Artemia* and a microdiet with an inert marker. Third, they developed a method to quantify the marker in the microgram quantities of larval fish feces. Finally, they refined methods to quantify the protein in microgram quantities of larval fish feces. They were not able to work with small first-feeding cod, primarily because of the limitations of the protein analysis method used. There are methods out there that could be used for smaller samples; however, they were not available to this group when the work was done. Interestingly, Johnson et al. (2009) found that the microparticulate diet was more digestible than *Artemia* even when accounting for leaching.

Tracers may also have a larger role to play in larval fish nutrition research (see review by Conceição et al. 2007), especially to determine nutrient fates. The use of stable isotopes as tracers has been used to determine the bioavailability of nutrients from diets fed to fish larvae (Schlechtriem et al. 2004). Further development of these methods may be promising as the analysis is inexpensive and relatively straightforward and the stable isotope ratio of microparticulate diets can easily be manipulated.

13.4.2.3 Feeding trials

Once it is determined that a diet will be consumed by the larvae and the diet itself has been well defined and characterized, feeding trials can be conducted to determine when that microparticulate diet can support growth and survival. Diets should be tested against live feed and starvation controls. Two basic trials can be used to improve microparticulate diets. The first trial is designed to determine the youngest stage during ontogeny that a given microparticulate diet can support growth and survival in the species of interest. The second trial starts feeding microparticulate diets at the ontological point determined from the first trial, using adjustments to the microparticulate diet to improve growth and survival from that point on. After one or more of the second type of trial demonstrates significant improvement in the microparticulate diet, the first trial can be repeated with the best diets from the second trial(s). As diets are refined, it will be necessary to repeat both types of growth trials so that the earliest starting point for microparticulate diets can be continuously pushed back toward first feeding by continuously improving microparticulate diets.

An alternative and complementary approach to the above is to determine dependent variable(s) that can be traced to survival and growth but occurs before mortality happens and can be easily and precisely measured. Numerous examples of such end points include stress tests, biochemical or molecular assays, ammonia excretion rate, oxygen consumption, carbon dioxide production, RNA/DNA, changes in physiology, and behavior indices.

13.5 Measuring success

The ultimate goal is to develop microdiets that provide all the nutrients required for optimum development and growth of fish larvae. Although research in this area has been underway for many years, the goal has not been reached for most marine species and live prey remains a critical part of most rearing protocols. Larvae may appear to feed on microdiets but active ingestion needs to be separated from accidental or passive ingestion. Cofeeding is common and often increases growth and influences larval feeding on microdiets. But we define successful weaning as larvae feeding on microdiet alone with no live prey, growing and surviving as well as or better than controls fed live prey (usually rotifers and *Artemia*). Weaning might occur after a period of days to weeks of cofeeding or, in rare cases, at first feeding. Some examples of various approaches to weaning are presented.

13.5.1 Red drum

The weaning of red drum larvae (*Sciaenops ocellatus*) from a very early stage (8 days posthatch [dph]) has been achieved using a commercial microdiet (Kyowa Fry Feed, BioKyowa, Inc., Chesterfield, MO) in a closed recirculating system (Holt 1993; Figure 13.1). Microdiets (less than 0.2 mm) are readily accepted by first-feeding red drum larvae and a standard feeding protocol was developed using the microparticulate diet and live enriched rotifers for 5 days (3–7 dph), followed by the microparticulate diet only. Larvae grow as well on this diet as they do on live prey (enriched rotifers and *Artemia*), with the added benefit of avoiding cannibalism that often occurs during weaning from *Artemia*. Further studies have been carried out to eliminate rotifers so that fish production would be independent of live prey production. A remarkable finding was that good growth and survival of red drum larvae on this microdiet could be realized with only algae (*Isochrysis galbana*) added to the tanks but not rotifers (Lazo et al. 2000). Follow-up studies showed that red drum can be raised successfully to the juvenile stage solely on a microparticulate diet supplemented with algae in the rearing tank (Holt 2002; Figure 13.10). Although growth is not as fast as when larvae are fed rotifers for 5 days, survival is high and the elimination of rotifers represents significant cost savings. We are investigating these findings further to understand why the addition of algae promotes increased survival and growth when larvae are fed exclusively on microdiets and whether other additives to the tank would act similarly. The overall goal is to enhance growth, development, and survival of red drum larvae while eliminating the need for zooplankton prey.

13.5.2 Sea bream and sea bass

Gilthead sea bream *Sparus aurata* and European sea bass *Dicentrarchus labrax* are species of great value to the marine aquaculture industry in Europe. Both sea bass and sea bream will accept microdiets at first feeding, and investigations are focused on improving the diets and protocols to increase growth and survival. Sea bream larvae that are fed only microcapsules from first feeding have low survival (11%), but this is improved significantly (42%) with cofeeding with small amounts of live prey (5% of the total food supplied on a dry weight basis) (Fernández-Diaz and Yufera 1997). In other studies, sea bream larvae fed rotifers from 4–7 dph and

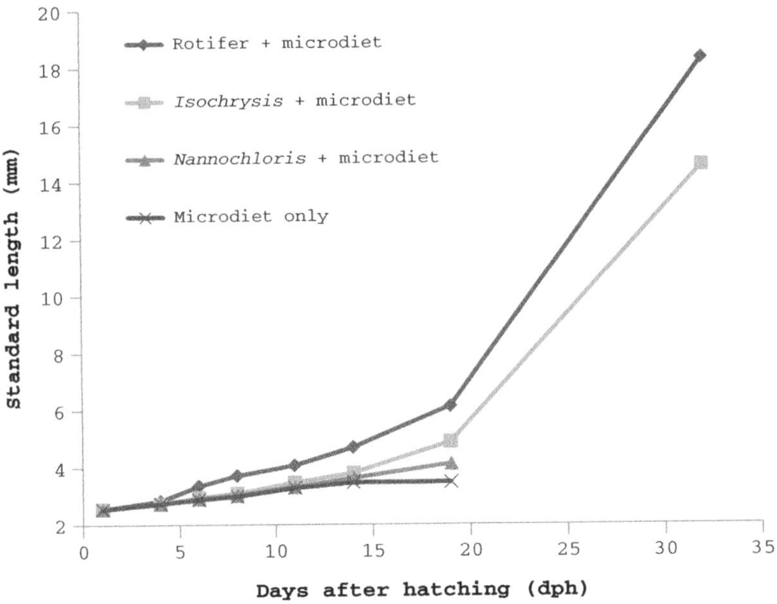

Figure 13.10 Growth of red drum (*Sciaenops ocellatus*) larvae on a commercial microdiet from first feeding (3 dph) through metamorphosis. The four treatments were (1) the control with rotifers for 5 days only (5/mL; 3–7 dph) + microdiet; (2) *Isochrysis* (40,000/mL added daily) + microdiet; (3) *Nannochloris* (40,000/mL) + microdiet; and (4) microdiet only. (Modified from Holt 2002.)

then soft gelatin microcapsules showed comparable growth and survival at 15 dph as the live-fed controls (Yúfera et al. 1999), but further studies are needed to determine if comparable growth continues throughout the larval period. These authors have shown that sea bream larvae prefer live prey and suggest that cofeeding small amounts of live food with a properly designed diet would improve their performance.

An important step in developing an acceptable diet is to assure the larvae will consume the food. Ingestion rates of microdiets by sea bream larvae are reported to increase when diets are supplemented with fractions from *Artemia*, FAA, or phosphatidylcholine (PC), or if FAA are added to the water (Koven et al. 2001). The authors suggested that these nutrient factors found in live prey not only stimulate feeding but also digestion, perhaps through an endocrine response, and enhance assimilation by increasing lipid transport. Microdiets that incorporate these and similar findings would maximize their utilization by marine fish larvae.

Sea bass larvae have been raised solely on microdiets composed of yeast, fish protein hydrolysate, fish oil, and lecithin (Cahu et al. 1998). These authors reported that 35% of sea bass larvae survived to day 28 exclusively on microdiet from first feeding but with significantly reduced survival and growth compared with larvae fed live prey. In another study, sea bass raised from mouth-opening on a microdiet to which live yeast was added had significantly higher survival rates than larvae on live feed, as well as lower incidence of spinal malformations and precocious gut maturation; growth, however, was significantly less (50%) than the control larvae fed live prey (Tovar-Ramírez et al. 2004). Research is continuing on the development of a nutritionally complete diet for sea bass, but most commercial hatcheries remain dependent on the production of rotifers and *Artemia* to grow the larvae.

In sea bream, sea bass, and other species such as Senegalese sole *Solea senegalensis* (Yúfera et al. 2005) and red sea bream *Pagrus major* (Takeuchi 2001), complete replacement of live prey with microdiets results in poor survival and growth, but weaning is possible after a few days (3–7) of feeding on live prey. This suggests that microparticle diets could replace much of the live prey used in commercial hatcheries if the diets contained the appropriate nutrients for those species. Further work is ongoing to address this issue, but research is also focused on identifying the characteristics of live prey that are important for first-feeding larvae to further reduce the live feed requirement.

13.5.3 Cobia

The weaning of cobia, like many other marine fish including southern flounder *Paralichthys lethostigma* (Faulk et al. 2007a; Faulk and Holt 2009), has only been achieved after they have metamorphosed to the juvenile stage with a fully developed stomach. Some cobia larvae begin feeding on microdiets at ~13 dph (9–10-mm standard length [SL]) and all are weaned by 24 dph (16–18-mm SL) well after stomach development (Faulk et al. 2007b). The presence of gastric glands and positive periodic acid–Schiff (PAS) staining of surface mucous cells in the stomach together with an increase in pancreatic enzyme activity 8–12 dph (5.7–8.1 mm) suggests that cobia larvae would be capable of digesting microdiets by 14 dph (Faulk et al. 2007a). But when trials have been run using several commercial microdiets, weaning at 13, 15, and 17 dph resulted in very low survival rates to metamorphosis (1–2% vs. 18–24% controls; unpublished data). Cobia have a very fast growth rate and possess a differentiated and effective digestive system early in development, yet efforts to wean cobia larvae onto commercial feeds have been met with limited success. Even though cobia larvae appear to be capable of weaning earlier, many do not ingest the microdiets that are offered, and those that do often reject the particles after taking them into their mouth. These studies suggest that cobia could be weaned earlier if offered a microparticulate diet with enhanced attractiveness and/or appropriate texture.

13.6 Future needs

There are many questions remaining to be answered about larval fish nutrition. For example, even if microdiets are ingested by larvae, they likely act very differently in the gut than live prey. For example, they may not stimulate digestive enzyme secretion nor reside in the gut long enough to be digested sufficiently to provide the needed nutrients; they may not have the nutrients in the appropriate form; and they may damage the epithelium of the gut by drawing water into the dry diet. Many of these issues were addressed during a workshop held in Bergen, Norway, in 2007 to discuss the status of knowledge on marine larval fish nutrition. The topics covered in the workshop included the nutritional requirements of marine fish larvae; digestive function and gut ontogeny; feed ingestion and assimilation; feed technology; larviculture and feeding strategies; practical hatchery operation; and broodstock nutrition. At the end of the presentations, the attendees discussed the deficiencies in larval nutrition information and how these could be resolved to advance weaning and early use of microparticulate diets. Recommendations from the group are summarized below under two categories: what do we need to know about larval fish nutrition and how do we get there.

13.6.1 What do we need to know?

- The absolute requirements and optimal ranges of concentrations for most nutrients for most species

- How the context of nutritional studies (genetics, physiology, ontogeny, hormones, behavior, and culture conditions) affect the results
- How environmental factors affect feeding, including time of day, frequency, light, turbulence, fish density, and turbidity
- Broodstock nutritional requirements and maternal contributions to the eggs/larvae
- The hormonal control of digestion and appetite in fish larvae
- The mechanisms that regulate gut transit and retention
- The osmotic effects of feeds on the gut and absorption processes
- Role of microbes and probiotics in feeding

13.6.2 How do we get there?

- Apply new tools to mechanistic understanding, such as using the zebrafish model and gene silencing, microarrays, and gene sequencing
- Design and develop diets for both broodstock and larvae that can be manipulated to test specific ingredients or forms of nutrients
- Develop defined diets that overcome the technical problems involved in buoyancy and nutrient leaching
- Apply new technologies to diet development such as nanotechnology (nutrient molecular caging), electrochemical signaling, encapsulation, and slow-/controlled-release technologies
- Use standardized methods for evaluating protein, lipid, and starch digestibility so they are comparable over species
- Develop open formulated, reference standard diets and/or enrichments for scientific studies with
  - Standard ingredients
  - High-quality oil (to control oxidation)
  - High-quality fish meal and other ingredients (from fresh fillets)
  - Standardized feed shape and particle size
  - Inert marker for determining ingestion rate
  - Standard production methods that are available for easy sharing
- Design and use standardized procedures for evaluating microparticulate diets and communicating results
- Increase international collaborations, exchange of information, and personnel/student exchange

These recommendations are still timely and, if implemented, would quickly advance our knowledge of larval fish nutrition. Coupled with the suggested innovations in microdiets, this would accelerate the replacement of live prey with microparticulate diets.

Literature cited

Alarcón, F.J., Moyano, F.J., Díaz, M., et al. 1999. Optimization of the protein fraction of microcapsules used in feeding of marine fish larvae using *in vitro* digestibility techniques. *Aquaculture Nutrition* 5:107–113.

Barrows, F.T., and Rust, M.B. 2000. Larval feeding—fish. In: Stickney, R.R. (ed.) *Encyclopedia of Aquaculture*. John Wiley & Sons, New York.

Cahu, C., and Infante, J.Z. 2001. Substitution of live food by formulated diets in marine fish larvae. *Aquaculture* 200:161–180.

Cahu, C., Zambonino Infante, J., Escaffre, A.M., et al. 1998. Preliminary results on sea bass (*Dicentrarchus labrax*) larvae rearing with compound diet from first feeding. Comparison with carp (*Cyprinus carpio*) larvae. *Aquaculture* 169:1–7.

Conceição, L.E.C., Morais, S., and Rønnestad, I. 2007. Tracers in fish larvae nutrition: a review of methods and applications. *Aquaculture* 267:62–75.

Cook, M.A., Johnson, R.B., Nicklason, P., et al. 2008. Marking live feeds with inert metal oxides for fish larvae feeding and nutrition studies. *Aquaculture Research* 39:347–353.

Dimes, L.E., Haard, N.F., Dong, F.M., et al. 1994. Estimation of protein digestibility. II. *In vitro* assay of protein in salmonid feeds. *Comparative Biochemistry and Physiology. Part A, Physiology* 108:363–370.

Faulk, C.K., and Holt, G.J. 2009. Early weaning of southern flounder, *Paralichthys lethostigma*, larvae and ontogeny of selected digestive enzymes. *Aquaculture* 296:213–218.

Faulk, C.K., Benninghoff, A.D., and Holt, G.J. 2007a. Ontogeny of the gastrointestinal tract and selected digestive enzymes in cobia *Rachycentron canadum* (L.). *Journal of Fish Biology* 70:567–583.

Faulk, C.K., Kaiser, J.B., and Holt, G.J. 2007b. Growth and survival of larval and juvenile cobia *Rachycentron canadum* in a recirculating raceway system. *Aquaculture* 270:149–157.

Fernández-Diaz, C., and Yufera, M. 1997. Detecting growth in gilthead seabream, *Sparus aurata* L., larvae fed microcapsules. *Aquaculture* 153:93–102.

Gilbert, J.J., and Schröder, T. 2004. Rotifers from diapausing, fertilized eggs: unique features and emergence. *Limnology and Oceanography* 49:1341–1354.

Glencross, B.D., Booth, M., and Allan, G.L. 2007. A feed is only as good as its ingredients—a review of ingredient evaluation strategies for aquaculture feeds. *Aquaculture Nutrition* 13:17–34.

Guthrie, K.M., Rust, M.B., Langdon, C.J., et al. 2000. Acceptability of various microparticulate diets to first-feeding walleye *Stizostedion vitreum* larvae. *Aquaculture Nutrition* 6: 153–158.

Hara, T.J. 1994. Olfaction and gustation in fish: an overview. *Acta Physiologica Scandinavica* 152:207–217.

Holt, G.J. 1993. Feeding larval red drum on microparticulate diets in a closed recirculating water system. *Journal of the World Aquaculture Society* 24:225–230.

Holt, G.J. 2002. Ecophysiology, growth and development of larvae and juveniles for aquaculture. *Fisheries Science* 68(Suppl. 1):867–871.

Holt, G.J., Kaiser, J.B., and Faulk, C.K. 2007. Advances in cobia *Rachycentron canadum* research in Texas. In: Liao, I.C., and Leaño, E.M. (eds.) *Cobia Aquaculture: Research, Development and Commercial Production.* Asian Fisheries Society, Manila, Philippines; World Aquaculture Society, Baton Rouge, LA; The Fisheries Society of Taiwan, Keelung, Taiwan; and National Taiwan Ocean University, Keelung, Taiwan, pp. 45–56.

Johnson, R.B., Cook, M.A., Nicklason, P.M., et al. 2009. Determination of apparent protein digestibility of live *Artemia* and a microparticulate diet in 8-week-old Atlantic cod *Gadus morhua* larvae. *Aquaculture* 288:290–298.

Kaji, T., Tanaka, M., Takahashi, Y., et al. 1996. Preliminary observations on development of Pacific bluefin tuna *Thunnus thynnus* (Scombridae) larvae reared in the laboratory, with special reference to the digestive system. *Marine and Freshwater Research* 47:261–269.

Kolkovski, S., Lazo, J.P., and Izquierdo, M. 2009. Fish larvae nutrition and diet: new developments. In: Burnell, G., and Allan, G. (eds.) *New Technologies in Aquaculture.* CRC Press, Boca Raton, FL, pp. 315–360.

Koven, W., Kolkovski, S., Hadasa, E., et al. 2001. Advances in the development of microdiets for gilthead seabream, *Sparus aurata*: a review. *Aquaculture* 194:107–121.

Langdon, C. 2003. Microparticle types for delivering nutrients to marine fish larvae. *Aquaculture* 227:259–275.

Lazo, J.P., Dinis, M.T., Holt, G.J., et al. 2000. Co-feeding microparticulate diets with algae: toward eliminating the need of zooplankton at first feeding in larval red drum (*Sciaenops ocellatus*). *Aquaculture* 188:339–351.

Munk, P., and Kiorbe, T. 1985. Feeding behaviour and swimming activity of larval herring (*Clupea harengus*) in relation to density of copepod nauplii. *Marine Ecology Progress Series* 24:15–21.

Nicklason, P.M., and Johnson, R.B. 2008. Real-time measurement of protein leaching from micro-particulate larval fish feeds. *Aquaculture Research* 39:1793–1798.

Puvanendran, V., and Brown, J.A. 1999. Foraging, growth and survival of Atlantic cod larvae reared in different prey concentrations. *Aquaculture* 175:77–92.

Rabe, J., and Brown, J.A. 2000. A pulse feeding strategy for rearing larval fish: an experiment with yellowtail flounder. *Aquaculture* 191: 289–302.

Rust, M.B. 2002. Nutritional physiology. In: Halver, J.E., and Hardy, R.W. (eds.) *Fish Nutrition.* Academic Press, Amsterdam, the Netherlands, pp. 367–452.

Rust, M.B., and Barrows, F.T. 1998. An image analysis approach to determine microparticulate feed acceptability with larval fish. In: Howell, W.H., Keller, B.J., Park, P.K., et al. (eds.) *Nutrition and Technical Development of Aquaculture*. New Hampshire Sea Grant Publication, Durham, NH, pp. 193–198.

Schlechtriem, C., Focken, U., and Becker, K. 2004. Stable isotopes as a tool for nutrient assimilation studies on larval fish feeding on live food. *Aquatic Ecology* 38:93–100.

Takeuchi, T. 2001. A review of development for early life stages of marine finfish in Japan. *Aquaculture* 200:203–222.

Tonheim, S.K., Nordgreen, A., Hogoy, I., et al. 2007. *In vitro* digestibility of water-soluble and water-insoluble protein fractions of some common fish larval feeds and feed ingredients. *Aquaculture* 262:426–435.

Tovar-Ramírez, D., Zambonino Infante, J., Cahu, C., et al. 2004. Influence of dietary live yeast on European sea bass (*Dicentrarchus labrax*) larval development. *Aquaculture* 234:415–427.

Tredici, M.R., Bondi, N., Ponis, E., et al. 2009. Advances in microalgae culture for aquaculture feeds and other uses. In: Burnell, G., and Allan, G. (eds.) *New Technologies in Aquaculture*. CRC Press, Boca Raton, FL.

Webb, K.A. 2008. *Cholecystokinin and the Ontogeny of Digestion in the Red Drum (Sciaenops ocellatus)*. Marine Science, University of Texas at Austin, Port Aransas, TX.

Windell, J.T. 1978. Estimating food consumption rates of fish populations. In: Bagenal, T. (ed.) *Methods for Assessment of Fish Production in Freshwaters*. Blackwell Scientific, Oxford, UK, pp. 227–254.

Yúfera, M., Pascual, E., and Fernández-Díaz, C. 1999. A highly efficient microencapsulated food for rearing early larvae of marine fish. *Aquaculture* 177:249–256.

Yúfera, M., Fernández-Díaz, C., and Pascual, E. 2005. Food microparticles for larval fish prepared by internal gelation. *Aquaculture* 248:253–262.

Chapter 14

Methods for assessing embryonic and larval growth in fish

Rafael Perez-Dominguez and Ralf Dahm

14.1 Introduction

Previous chapters have introduced the current understanding of nutritional requirements and digestive function in larval fishes. This chapter covers methodologies that have been used to gather the experimental evidence behind such knowledge. Nearly all experimental procedures use larval growth (or a proxy measure) as the performance estimate for the evaluation of causal relationships among nutritional variables. The simple rationale behind this overwhelming choice is that more appropriate formulations or feeding regimes will support optimal growth rates and, conversely, deficient ones will be less successful for supporting normal growth.

There are numerous approaches to the assessment of growth and these include an extremely large list of methodological approaches and diversity of protocols. The aim of this work is to introduce and provide a brief account of the rationale behind the main methodologies but leave the technical aspects of the assays to the methods section in the references cited here and in previous chapters.

For those concerned with the rearing of larval fishes, it is important to consider the dual nature of growth during early life: (1) the rate increase in size (length or biomass), and (2) the sequential acquisition of new or improved organs and structures leading to physiological or behavioral refinements. Developmental changes are extreme in altricial larvae, but they are also present in precocial types[1] (Balon 1981). A direct consequence of the interplay between development and growth is the constant alterations in body proportions during ontogeny.[2] This type of growth is referred to as allometric, and it is

[1]Altricial larvae are those larvae that hatch after a short incubation period and at very small size and primitive stage of development, completing most of their development relying on exogenous food sources. Precocial larvae, on the contrary, have large yolk, hatch after longer incubation times, and at larger size and advanced stage of development.

[2]Ontogeny is the gradual change from a simple to a more complex form during development.

Larval Fish Nutrition, First Edition. Edited by G. Joan Holt.

fundamentally different from the isometric-type growth of most fishes after metamorphosis. Therefore, growth is not a unique or constant process. There are precise growth schedules particular to each species and life stage. These are hardwired in the developmental programming of each species and under strict regulation (Balon 1981; Blaxter 1988; Mommsen 2001).

Most nutritional studies evaluate the direct effect of diet formulations or feeding regimes on the size progression and survival of larvae. However, it has been long realized that the potential for growth is strongly influenced by previous growth history and other inherited biological factors, some of which are sensitive to nutritional manipulation. For example, the amount and quality of yolk, and hence future larval performance, is determined by the nutrition and size of the mother (see Chapter 5). Likewise, larval nutritional condition can be a useful predictor of performance in nutritional bioassays since good condition is a prerequisite for exponential growth. Biochemical indices of condition, particularly nucleic acid (RNA : DNA) ratios, have been used widely in ecological and nutritional studies. It is now technically possible to trace these physiological responses as well as developmental progress down to their molecular onset through gene expression assays. Recent studies in the zebrafish model have demonstrated the sensitivity and speed of this approach in developmental studies and its potential for nutritional assessments.

14.1.1 Experimental methods and growth assessment

14.1.1.1 Tools for assessment: growth metrics

There is a wide range of growth metrics reporting on different structural, behavioral, and/or physiological quality elements of larval fish. Some will show a significant response to a particular nutritional experiment, while others will lack or have a delayed response, rendering them ineffective for any meaningful assessment. Variations in response arise due, in part, to differences in absorption rates and metabolic pathways of different nutrients (vitamins, fats, amino acids, etc.) and to the sensitivity[3] of the metric chosen. For example, in Japanese flounder (*Paralichthys olivaceus*) larvae, the inclusion level of vitamin A in live prey has direct effects on growth, but it also has a delayed detrimental effect on ossification, inducing skeletal deformities (Dedi et al. 1995). It is clear that in this example, two independent metrics, size progression and incidence of deformities, will be necessary for a complete assessment and that the administration of the test diet needs to cover the sensitive period where ossification is affected by toxic levels of vitamin A. Therefore, the choice of metrics has to be decided based on the nature of the expected response and timescale of the sensitive period.

It is difficult to overstress the importance of the choice of metrics as this underpins the entire experimental design, affecting data analysis and hypothesis testing. Hence, it is advisable to undertake a careful search for previous published examples as a first step into the design of any experiment. Further, a pilot test trial is a very desirable exercise prior to deciding on the final design. Test trials provide priceless help identifying problems and highlighting the modifications needed to improve the quality of the data; some of the factors determining a good experimental design will be covered next.

14.1.1.2 Experimental design

The experimental design is the way evidence is collected and how the relevance of outcomes

[3] The term sensitivity is used to express the minimum time duration of a certain treatment necessary to induce a measurable change in a given metric.

will be decided. It uses well-defined experimental methods and protocols aiming to produce the strongest evidence possible with which to challenge a working hypothesis. Probably, the single most important consideration while deciding any experimental approach is the robustness of the response to the experimental treatment as this influences the size of the experiment. That is, experiments with greater variability in response will need more replicates and vice versa. Likewise, more stringent hypothesis testing, that is, experiments aiming to detect increasingly smaller effects, will require more replication (increasing size) or more sensitive metrics. Given an adequate level of power[4] in the experimental design, even statistically nonsignificant results are meaningful because there is enough confidence to conclude that there is no effect and this is just as important as finding that there is an effect (Ruxton and Colegrave 2006).

There are necessary compromises to be made, and the size of experiments cannot increase indefinitely. There may be practical reasons or ethical considerations (limited number of tanks, technical and analytical bottlenecks, limit on experimental animals, etc.) imposing limitations that create a situation where a test is run at suboptimal power. Then, trials should be repeated several times and preferably conducted on spawns from different sources to eliminate problems with spurious tests. Repeating an experiment is in any case advisable to gain confidence in the generality of the results and eliminate criticism of pseudoreplication.[5]

14.1.1.3 Sampling procedures

The design of the sampling procedure must be ethical and cause as little distress as possible to the experimental subjects. Distress can be minimized with the use of anesthesia; tricaine methylsulfonate (MS 222) and clove oil have been employed with great success. Lowering the temperature may also lessen the detrimental effect of handling larvae. When required, animals must be sacrificed quickly by a method that ensures complete termination of coordinated neural activity. Most journals and funding organizations now have strict ethical guidelines with respect to animal experimentation, including sampling procedures. Nonstandard procedures require evidence that recommended methods will negatively affect the reliability of the data and justification that no alternative approaches are available.

Morphometric measurements such as length or weight are possible on anesthetized larvae and are best if obtained from fresh specimens because preservation method can severely affect larval fish measurements (Hjorleifsson and Kleinmacphee 1992). However, in many instances, fresh processing is not possible and the samples have to be preserved for later biometric evaluation or further processing (i.e., histological and biochemical techniques). There is, in general, a preferred sample preservation method for each analytical technique. The most widely used fixative in larval fish for biometrical measurements is 4–5% buffered formosaline solution (pH 8.0–9.0; 0.9–3.0% NaCl). Larvae are normally fixed whole for a relatively short time (i.e., hours) since their small size allows an adequate fixative penetration in a short period of time; overfixing can lead to artifacts. Simple calibration exercises can

[4]The power of a statistical test is the probability of rejecting a null hypothesis when it is false. It is normally set at 80%. Power is negatively affected by variability in the replicate scores and by the size of the treatment effect meaningful to detect. Power analysis is available in most statistical packages.

[5]Pseudoreplication is defined as the use of inferential statistics to test for treatment effects, with data from experiments where either treatments are not replicated or replicates are not statistically independent, and therefore uncontrolled for random effects (Hurlbert 1984). Nutritional experiments conducted once on a single batch of larvae, or evaluated using several related samples drawn from the same experimental unit, are often criticized as pseudoreplicated, limiting the value of the trials and the generality of the results.

be implemented to bring measurements on preserved individuals to their fresh equivalents. In any case, to ensure precision, any biometric assessment should be conducted after the fixation is complete and no further changes in relative body proportions are expected (Hay 1982; Fey 1999).

Protocols for histological techniques often require specific fixing cocktails or freshly made reagents such as Boiun's solution, paraformaldehyde, or glutaraldehyde in different concentrations and reaction buffers (Takashima and Hibiya 1982; Genten et al. 2009). Great sources of best performing protocols are dedicated technical reviews, research articles, and manufacturers of analytical equipment and reagents. The researcher must ensure compatibility with all assessment metrics and allow for duplicate sample sets in the experimental design if necessary.

14.1.2 Factors not related to nutrition

Experimental conditions must be reasonably controlled and treatments assigned randomly to the experimental tanks. Uncontrolled effects such as variable light intensity, temperature, water quality, aeration strength, food delivery, sampling, and other physical or husbandry factors may induce tank effects, adding undesirable experimental variability or bias affecting the ability to detect true treatment effects[6] (Ruxton and Colegrave 2006; see also Chapters 10 and 13 for potential uncontrolled effects due to larval behavior and husbandry procedures).

Individual response to any experimental manipulation always varies across subjects. This is known as the between-subject variation. The individual scores will therefore not all be precisely the same, resulting in a more or less clustered distribution around the mean treatment effect. With too few subjects, the variability among individual scores will make the sample distribution somewhat uncertain, affecting the precision in the estimation of the treatment effects. What size sample is reasonable? Each metric has a characteristic degree of variability in the scores that is a priori not precisely known. But, for almost all practical applications, it can be estimated by calculating the standard deviation of the sampling distribution for the relevant parameter. This statistic, known as the standard error (SE), can be used to calculate confidence intervals when the sample size becomes large (i.e., >25). These intervals can then be used to decide the sample size at which the random sampling error (noise) is expected to be less than the smallest detectable effect set for the experiment.

But, even in carefully planned and executed designs, there is the possibility for artifacts, leading to false conclusions. Effective and timely delivery of nutrients continues to pose real challenges in altricial larvae, especially during the first-feeding stage where nutritional manipulations are difficult and often confounded by the presence of maternally derived nutrients or the need to use live prey (see Chapters 5 and 11). But, even in larvae fully weaned onto formulated feeds, variation in feeding rates is an ever-present confounding factor during nutritional trials. A diet formulation can be grossly misjudged if it is not ingested or if nutrients leach out from food particles before they are ingested (see Chapter 13). Overlooking these potential problems in the design stage may result in inconclusive outcomes.

Finally, under conditions where large variations in size develop between individuals, the assessment of nutritional outcomes is likely to be confounded by developmental-related effects, that is, when the outcome of the test is influenced by the size of the larvae

[6] Statistical tests to decide the relevance of experimental outcomes will not be discussed here.

(Fuiman et al. 1998). In this case, it is not always possible, or practical, to draw a strictly random sample; thus, a stratified sampling design targeting animals in the same ontogenetic state may be a better strategy. Some artifacts in growth rate assessment may also result from size-specific mortality; for example, smaller larvae may not ingest a large food particle after a diet shift and starve. The disappearance of smaller larvae can give the false impression of sudden increases in overall growth rates under the new diet. Similar artifacts happen when smaller larvae are cannibalized by older siblings. In all these situations, a sampling criterion can be implemented to reduce variability and increase the chances of detecting true treatment effects.

Many of the potential problems are difficult to predict, and a small pilot trial is always a good insurance policy. It can provide invaluable information not only to get accurate estimates of replicate variation or appropriateness of growth metrics, and statistical methods, but also to prevent pitfalls and improve the overall quality of the study.

(a)

(b)

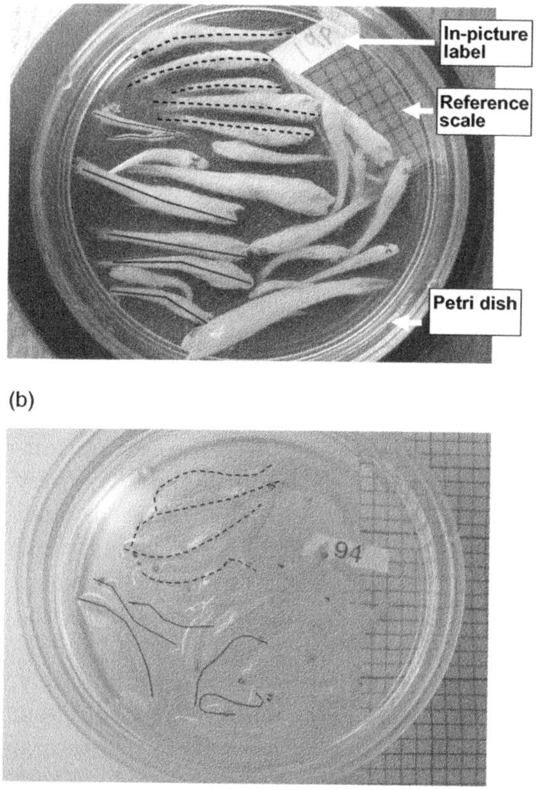

Figure 14.1 Digital photographs of (a) sand goby (*Pomatoschistus minutus*) and (b) sprat (*Sprattus sprattus*) late larvae during measuring using image processing software. Segmented lines (total length) and solid lines (standard length) are translated into length measurements using the in-picture reference scale. The photographs were taken unaided with a standard digital camera in macro mode. Smaller larval stages require the use of suitable scopes and adaptor photographic tubes or lenses for image capture.

14.2 Direct growth assessment

In this section, the different ways of modeling growth as a direct change in size or biomass will be presented along with simple mathematical expressions to calculate growth rates. In addition, growth evaluation based on quantum increments in the complexity of larval structures (ontogeny) will be introduced to measure the developmental pace during the larval period and transformation into juveniles.

14.2.1 Somatic growth

14.2.1.1 Length

Growth can be most simply calculated by measuring the change in length (L) with time.

In larvae, length is usually derived from measurements along the midline of the body, from the tip of the snout to the end of the notochord or the caudal peduncle. This distance is known as standard length (SL). In some instances, length measurements are taken to the end of the caudal fin rays (total length, TL), although it is generally not advisable because these rays are easily damaged or broken (Figure 14.1).

Some other measurements between different body structures have been used alone or in the form of ratios but rather than providing a growth estimate they are used to estimate nutritional condition (see Ferron and Leggett 1994 for a review of early studies). These altered body ratios are most apparent after severe nutritional impairment and are therefore less sensitive than simple growth rate estimations.

Length can be easily obtained from digitalized pictures using image processing software packages.[7] This technique is very accurate and has virtually replaced older direct measuring techniques such as those using an eyepiece's micrometer or a camera lucida. The basic setup requires a dissecting scope fitted with a digital camera and a reference scale of an appropriate size. Alternatively, a video camera with an image capture system can be used. In any case, there are several commercial options on the market along with free software packages[8] with very similar functions and overall performance.

The speed of the size change, that is, growth rate, can be obtained from two discrete length measurements, L_0 and L_t, taken at time t_0 (start of experiment) and at time t_t (after a period of time, t). The difference in size, $\Delta L = L_t - L_0$, is termed *absolute growth*, and is under most circumstances greater than 0. In practice and for comparative purposes, growth is normally expressed as the rate of size change over a defined time interval, that is, *absolute growth rate* (AGR). This is calculated by expressing the absolute growth (ΔL) as the proportional growth increment by unit of time:

$$\mathrm{AGR} = \Delta L/(t - t_0)$$

AGR has velocity units and can be interpreted as the speed at which larval fishes increase their size. Calculated in this way, AGR only reflects the speed of the size change that has occurred during a specific time interval but does not provide any indication of the relative magnitude of the change with respect to the size of the larvae.

Since larvae grow in size over one or two orders of magnitude before metamorphosis, it is often useful to represent the growth rate as a relative measure in proportion to the initial size of the larva. The *proportional growth rate* (g) does this by dividing the AGR by the initial length of the larva (L_0):

$$g = \mathrm{AGR}/L_0$$

The units are the inverse of time and can be interpreted as the number of body length increments attained per unit of time. The proportional growth rate can also be expressed as a percentage increase ($g_{\%} = g \times 100$) for easier interpretation.

These calculations imply that the rate of growth remains constant within the period between the two assessments (i.e., linear growth model). This is a reasonable assumption for most species over short time intervals. However, over longer time periods or during periods where growth is accelerating, the calculation of specific growth rates using simple linear models does not provide adequate results (Figure 14.2). Under these circumstances, a better fit results from using exponential growth models.

$$L_t = L_0 \cdot e^{G(t-t_0)}$$

or in the case of two measurements:

$$G = (\ln L_t - \ln L_0)/(t - t_0),$$

[7] Similar techniques can be used to take measurements on histological preparations. Histological techniques will further require a high-power compound microscope fitted with an image documentation system.

[8] ImageJ, available from http://rsbweb.nih.gov/ij/index.html, is a free image processing software with a wide range of functions that include the calculation of distances on digitalized pictures.

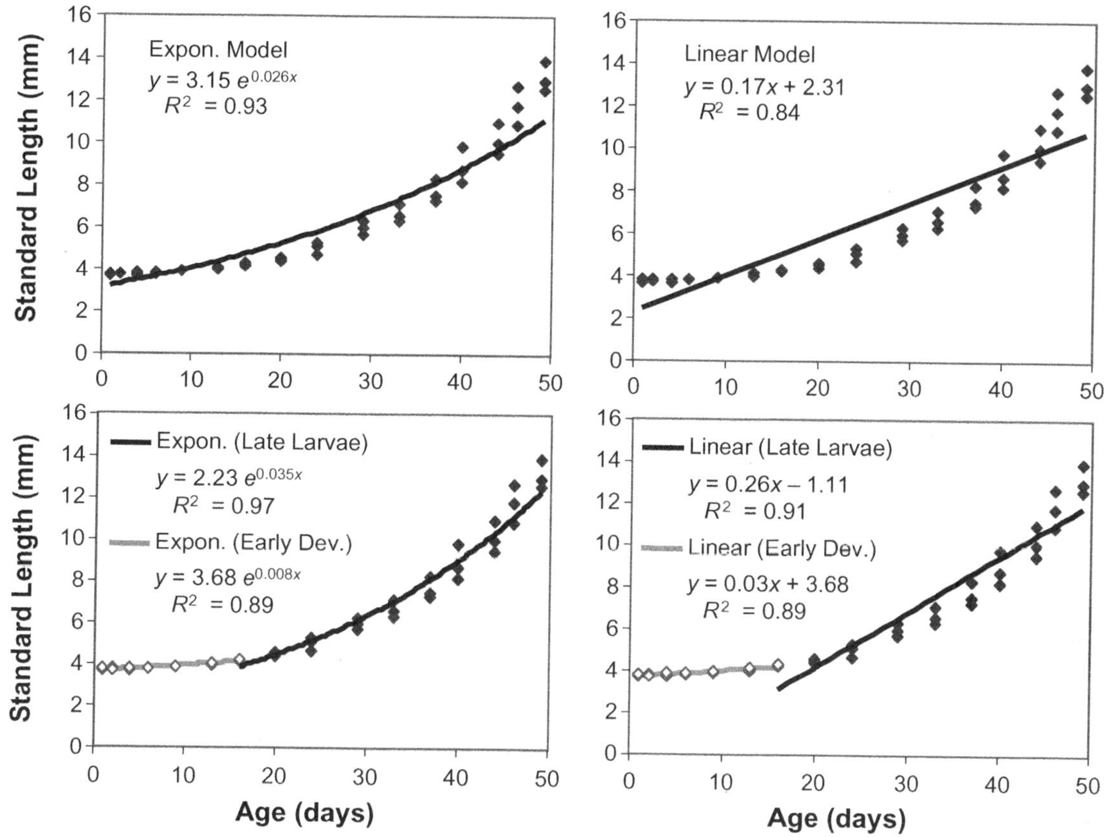

Figure 14.2 European flounder (*Platichthys flesus*) larval growth trial showing different growth models fitted to the same data. Growth models and R^2 are given in the figures. The trial was conducted in triplicate, and 25 individuals were sampled at regular intervals throughout the experiment. Data points in the figures represent the estimated average length from each replicate tank and sampling date (age). The data points used for early larvae growth regressions are indicated by open diamonds.

where the coefficient of growth (G) indicates the instantaneous growth rate. Note that G has the same units as the proportional growth rate g (i.e., inverse of time). G indicates the shape of the growth curve representing the proportional change in length per day or any other unit of time.

Linear or exponential models will provide a reasonable estimation of growth for comparing larval performance under different treatments. However, they should not be used to forecast growth into the juvenile phase or when extreme allometric changes are suspected. Both models assume a constant growth rate, which is not an accurate representation of the growth dynamic under these circumstances.

While the measurement of length is straightforward, there are some different ways to compute the timeline over which any growth assessment is made. Days are the most frequently used time unit as these are a reasonable fraction of the duration of the larval phase for most species, normally 25–60 days. Also, it is conceptually easy to understand as days are equivalent to age in hatchery-reared fish. However, when temperature cannot be precisely controlled, the use of a fixed unit of time may induce large bias.[9] This problem is common in rearing trials conducted outdoors, when treatments have to be run at different times or when comparisons across species with different optimal temperatures are needed. Under these circumstances, it is advantageous to use a modified timeline that introduces a temperature weighting factor to the day unit. This new "relative-time" unit is known as *degree-day* (DD) and is commonly used in ecological studies. There are several ways to calculate the weighting factor, but all are based on a summary statistic (mean, median, maximal, etc.) reporting on the temperature dose experienced within 1 day (or any other length of time), and then totaling the results in a cumulative index.[10] Growth rates calculated on DD timelines are often better than simple day computations (Neuheimer and Taggart 2007). It must be noted that DD is not applicable outside the range for which growth rate is known to be proportional to the temperature dose. It is common for growth rates to fall outside that range when ambient temperatures approach the species biological zero or the upper temperature tolerance limit. Therefore, these extreme temperature ranges should be avoided if the thermal weighting is to be derived from a simple linear relationship.

14.2.1.2 Weight

Weight is the simplest way of depicting growth as a change in biomass. It is often a more robust measure than length, partially due to the allometric growth of early larvae. Weight is measured gravimetrically with a calibrated microbalance. Weight of fresh or preserved specimens is termed *wet weight* (WW). More precise biomass measurements are obtained using oven- or freeze-dried[11] specimens, termed *dry weight* (DW). These measurements are, on occasion, corrected for inorganic content and are reported as ash-free dry weight (AFDW). The correction is done by burning[12] the sample (or a representative fraction) and subtracting the estimated weight of the ash residue from the original DW. Weight estimates (WW, DW, and AFDW) can be modeled directly using the same growth equations given for length in the previous section.

Many individual weight determinations, especially DW, conducted on the smallest larvae require very precise balances and are somewhat laborious to obtain. This level of accuracy is sometimes not possible and pooled measurements are necessary. Since length is by comparison easily gathered, it is often used as a proxy measurement for biomass. Usually, there is a reasonably close relationship between length and weight in larval fish; in the general case, the relation is that of a power function

$$W = aL^b \text{ or } \ln W = \ln a + b(\ln L)$$

where W is weight, L is length, *parameter* a is a simple scaling factor (intercept), and b is the power coefficient determining the function's rate of increase. The different

[9] Temperature in poikilotherms controls the rate of biochemical reactions, and hence the rate of growth. Developmental rates of embryos are particularly sensitive to temperature effects; for example, incubation times are substantially reduced at higher temperatures.

[10] A further refinement is to use the range of temperature above the minimum temperature threshold for growth. This related index is termed *effective degree-day* (EDD), and it is calculated by subtracting the temperature of no growth (i.e., biological zero of the species) from all temperature readings before calculating the degree-day.

[11] Samples are usually dried in a thermostatic oven at constant temperature (<60°C) for 24–48 hours or until they achieve constant weight. Similar results are attained by freeze-drying or using desiccating substances (i.e., silica gel or NaOH) to a constant weight.

[12] Samples are combusted in a muffle furnace at 400–450°C for 8–12 hours. In larval fish without calcified structures, the amount of ash residue is a very small proportion of the total weight.

parameters are estimated empirically, which is normally done by taking large numbers of larvae across the size interval of interest, measuring their length and weight, and then determining the precise nature of the relationship. Under isometric growth, *b* is equal to 3, while under allometric growth, it requires empirical calculation. The power coefficient is widely used in adult fish as an index of condition termed *Fulton's condition factor* (*K*). This is not generally applicable to larval studies unless properly calibrated because changes in *K* could be related to developmental rather than nutritional effects (Ferron and Leggett 1994).

14.2.1.3 Protein content

The greatest fraction of biomass in fish is made up of protein. Changes in protein content or protein synthesis rate can therefore be used to derive growth rate estimates. Two procedures are currently in use.

Biochemical measurement. There are a number of analytical methods that can be applied to the scale of larval fishes. Among the most widely used are the Bradford assay, the bicinchoninic acid method (BCA), and the Lowry assay. All three methods are colorimetric protein assay protocols based on a reagent that undergoes a color shift proportional to the amount of soluble protein in the sample. Alternatively, determinations can be attained using protein's intrinsic UV absorbance (280 nm) or fluorometrically with the use of specific protein-binding fluorochromes. Originally run in test tubes, these assays have now been miniaturized to high-throughput microplate formats. Bradford and BCA assays are readily available in kit form from several manufacturers and are within the technical capabilities of a standard biochemical laboratory. The most sensitive of these assays can measure as little as 5–10 µg protein per well (microplate format), which allows individual measurements of eggs and newly hatched larvae in duplicate. However, most determinations are normally done on pooled samples of 200–800 µg of DW to ensure enough tissue for protein and additional biochemical measurements from the same sample.[13] Properly calibrated, protein content can provide accurate biomass estimations even from preserved larvae and is a very sensitive alternative to DW and ultramicrobalances.

Raw protein concentration values derived from biochemical techniques are normally expressed as absolute protein weight content per larvae (Wp). As with length or weight, changes in Wp are then used to compute the *instantaneous coefficient of protein growth* (Gp; or a percent change $Gp_{\%} = Gp \times 100$).

Tracer studies. An alternative way to calculate protein growth is by using radiolabeled protein precursors such as small proteins, polypeptides, or amino acids (Houlihan et al. 1995; Conceição et al. 2007). In this technique, a marked protein precursor, that is, tracer, is incorporated in the test diet and the mixture fed to the larvae. Due to inconsistent feeding rates and leaching of tracers even from microencapsulated or coated formulations (see Chapter 12), most experimental designs for larval fish incorporate the tracer directly into the rearing water (usually a flooding dose of tritiated phenylalanine) from where it is absorbed by the larvae. Alternatively, larvae can be force-fed a precise amount of a test mixture containing the tracer using microinjection protocols (see Chapter 3). It is important to note that experiments using radioactive materials are only possible in strictly controlled setups and at small scale requiring dedicated licensed facilities. A non-radioactive alternative is the use of stable isotopes (e.g., ^{15}N) incorporated into diets,

[13] Often protein content is used to standardize the proportion and ratios of enzymatic activities and other biochemical compounds that require larger sample sizes for analysis.

although this is a less sensitive method (Owen et al. 1999; Conceição et al. 2007)

Under normal conditions, the rate of protein synthesis or protein growth is assumed to be proportional to the rate at which the tracer shows up bound into the protein fraction. This assumption holds well in larval fish over short time intervals since protein turnover rates are normally constant over this time frame. The specific radioactivity in the tissues (disintegrations per minute [dpm]) is translated into protein synthesis rates using time-dependent tracer incorporation equations:

$$\text{Tracer incorporation (mol/day)} = \frac{\text{Tissue dose (dpm/day)}}{\text{Tracer SA (dpm/mol)}}$$

where Tracer SA (specific activity) is a measure of the amount of radioactivity per unit amount of tracer substance and "day" represents days or any unit of time. This approach assumes that the rate of incorporation of radiolabel into body protein remains constant over the test period and there is no, or negligible, contribution of endogenous unmarked tracer substance. However, these assumptions require previous validation (Houlihan et al. 1995; Conceição et al. 2007). Protein synthesis is commonly expressed in relative units or fractional rates of protein synthesis that represent the percentage of protein synthesized per day relative to the protein content of the larvae (measured using biochemical techniques. This is, in concept, homologous to the proportional growth rate introduced earlier in this section and is the most sensitive measurement of growth rate over very short time intervals (hours).

14.2.1.4 Energy content

Combusted organic matter releases heat, termed heat of combustion or calorific value, which is proportional to the amount and composition of the biological material. This property can be used to derive a direct energy-based quantification of biomass (see Chapter 8) useful for estimating growth rates. Many studies, however, do not directly measure this parameter but estimate energy content (*Ce*) indirectly by measuring total protein, lipids, and carbohydrates (i.e., proximate analysis) and adding up their known energy equivalents. An alternate proxy approach is the use of CHNS elemental analyzers.[14] Again, the theoretical energy equivalent of the sample is computed from the C, H, O, and S fractions (the O fraction is inferred after the sample weight is corrected for ash content) (Kamler et al. 1994). Elemental analysis is potentially a more accurate proxy method but requires a more expensive setup. Proximate analysis is nevertheless a good approximation due to the uniformity of the energy yield per unit weight of energy-rich chemical species (Domalski 1972; Craig et al. 1978).

The reference methodological procedure to calculate the heat of combustion or calorific value of materials is known as *bomb calorimetry*. This method uses special insulated vessels immersed in water baths where the sample is burned as fuel under a pressurized oxygen atmosphere. The heat released results in a proportional temperature rise in the water bath, which is a direct measure of the energy contained in the sample. The energy released (*Ce*) is expressed as joules per larva; this unit measure can be used to model growth in the exact same way as we have seen for length, weight, and protein content.

The smaller bomb calorimeter systems available have a sensitivity ranging from 50 to 1,200 calories that will correspond to 25–200 mg of wet larval tissue, normally requir-

[14] CHNS analyzers measure the percentage composition of the main elements, carbon (C), hydrogen (H), nitrogen (N), and sulfur (S), in dry organic matter.

ing the pooling of several larvae into one sample. Less demanding in terms of sample needs is the determination of the calorific value by wet oxidation with a mixture of potassium dichromate in sulfuric acid (Craig et al. 1978).

14.2.2 Developmental progress

As we have seen in the previous section, ontogeny can be followed by measuring changes in size through different physical measures. During early ontogeny and along with the change in size, primitive structures and simple behaviors mature into complex and more elaborated ones, ultimately giving rise to a miniature adultlike fish. Thus, the timing and speed of developmental events can, in its own right, provide an independent evaluation of growth.

Tests reporting on developmental progress may exploit simple qualitative presence–absence morphological criteria or be more elaborate and include a quantitative estimation of a dynamic biochemical feature or gene expression. As important as the precise quantification of diagnostic features is the definition of a timeline for the evaluation of experimental manipulations. The natural choice in experimental studies is age; however, alternative timescales are possible using DD. Moreover, studies reporting on the ontogeny of larval fish have found that size effectively integrates the flow of time into one convenient measure of "biological age" (Fuiman et al. 1998; Neuman and Able 2002) that combines all past environmental experiences, including nutrition. The use of size as a benchmark to evaluate developmental progress is uncommon in nutritional studies but, given that development often correlates better with size than age, this is a method that should deserve more attention in the future (Fernández-Díaz et al. 2001).

Modeling the developmental progress of larval fish using size is similar in concept to the use of time units. In the simplest case, a valid analysis is possible by examining a representative sample and giving two alternate scores: zero to all individuals still at an undeveloped stage (or absence of a test feature), and one to those subjects that have advanced to the next stage (or display the test feature). The distribution of individual scores with respect to larval size can be drawn and fitted to a cumulative normal distribution. The mean value of the corresponding normal distribution will be the average size at which the stage transition takes place. The approach is effective for separating size effects from developmental effects, with the advantage of allowing precise statistical comparison using hypothesis testing (Figure 14.3).

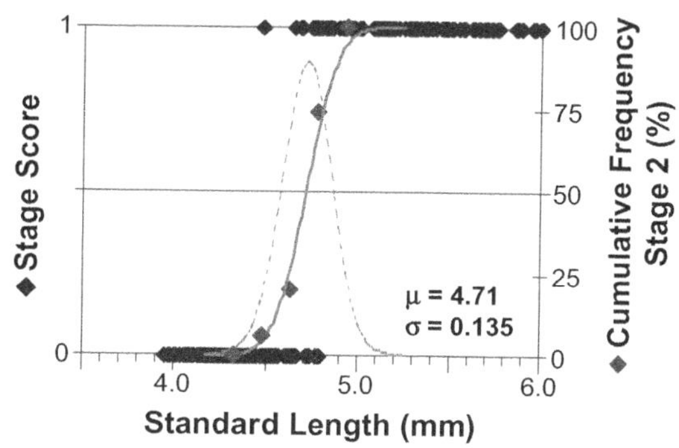

Figure 14.3 Progression of European flounder (*Platichthys flesus*) with fully developed loop (stage 1) in the digestive tract with respect to body size (standard length, mm). A cumulative (normal) probability curve (solid line) was fitted to the cumulative frequency distribution (0.15-mm bin size) of larvae reaching stage I (developed gut loop) using nonlinear regression techniques (r^2 = 0.96, n = 5). The corresponding normal probability density curve is also shown (solid line) (mean and standard deviation are given in the figure). The center of the distribution corresponds to the mean size at which European flounder made the transition to the late larval gut morphology.

14.2.2.1 Morphological features

Direct assessment

The simplest way to compare nutritional manipulations is by reporting the percentage of larvae having completed the development of a certain feature, or group of features, for a certain age group (hatching is day 1 in the life of a larva) (Yúfera et al. 2005). For example, one can report the percentage of larvae that have fully absorbed their yolk sacs by day 4, have pigmented eyes by day 3, or are at the postflexion stage by day 26. A different approach to the same assessment is to measure the time in days required for half of the larvae to acquire a determined structural feature (or group of features, i.e., developmental stage). Following the same example, one will report the average number of days to complete yolk absorption or eye pigmentation or onset of acid secretions in the stomach. Most studies base the assessments on characters that clearly mark the start of the juvenile phase, hence the end of the larval period (i.e., the position of the eye in metamorphosing flatfishes; Fernández-Díaz et al. 2001).

Along with the external changes that signal the end of the larval phase, there have been nutritional studies that assess the functional changes in internal organs and cellular components. Morphological changes of the digestive system, for example, are often visible through the larval body and correlate with digestive function (see Chapter 1, also Section 14.4). Observations on fresh specimens under regular transmission microscopy can be greatly enhanced using phase contrast, dark field, or differential interference contrast (DIC) microscopes, or by using contrast-enhancing dyes or vital stains such as methylene blue, pH indicators, and other water-soluble stains, including fluorescent dyes (Lessard et al. 1996; Lin and Hwang 2004). Fluorescent dyes require epifluorescent modules and suitable filters or confocal microscopes[15] for visualization.

Histological protocols

Much precise insight into the developmental progress of different organs and nutritional effects has been possible using histological preparations for conventional light microscopy (cellular and histological assessment) or electron microscopy (ultrastructural assessment) (McManls and Mowry 1960). This is a much more time- and labor-intensive methodology requiring sectioning and staining of tissue samples. All protocols start with frozen preparations or fixed tissue embedded in paraffin or special resins. These are sectioned in thin slices and attached to glass slides and treated with dyes. Sectioning ranges from a few micrometers (cell structure) to nanometers (organelle ultrastructure) and must be conducted to a thickness compatible with the scale of the histological target under assessment. The various dyes react differently with different cell or organelle chemical components, giving a characteristic appearance to the preparation and a means to identify first activation and function of larval organs, thus providing a qualitative reference of developmental progress.

A wide range of staining methodologies is available (Takashima and Hibiya 1982; Genten et al. 2009). The simplest and more widely used protocols use colored chemical agents for the light microscope and salts con-

[15] Confocal laser scanning microscopy (CLSM) and other laser-based scanning techniques are alternatives to epifluorescence, allowing enhanced resolution, higher contrast, and more suitable depth of field without the use of histological procedures. In confocal microscopy, out-of-focus emitted fluorescent light is rejected by a simple pinhole placed in front of the detector, resulting in a virtual optical section of the specimen. CLSM allows the construction of three-dimensional (3D) stacks and the capture of time-lapsed clips in live specimens. CLSM also can take advantage of natural autofluorescence in tissues or fixatives for general structural observation, providing real-time observations, exceptional contrast, and details previously not accessible in whole-mounted specimens. These are, however, not generally used in nutrition studies primarily due to cost of the equipment.

taining heavy metals for the electron microscope. There are generally termed *histological* protocols, which are simple to conduct but limited in terms of tissue specificity. Hematoxylin-eosin stain is the most common general stain for light microscopy. Increased staining specificity is possible with stains specific for certain macromolecules. Examples are the periodic acid–Schiff (PAS) stain specific for carbohydrates or Sudan-based stains for lipids (Takashima and Hibiya 1982). The specificity can be increased further by using colored substrates for specific enzymatic reactions taking place within cells or tissues. Termed collectively as *histochemical* protocols, they are useful as they reveal early phases in the differentiation of tissue types within organs often before positive identification with general stains. Digestive and metabolic enzymes are common targets in histochemical preparations for which protocols are readily available (Govoni et al. 1986; Gawlicka et al. 1995; Genten et al. 2009).

A difficulty of histochemical staining protocols is the need for endogenous enzymatic activity linked to the target feature. This limitation can be overcome using *immunohistochemical* protocols. In these, the specificity resides in the use of specific antibody- and enzyme-based labeling methods. Horseradish peroxidase and alkaline phosphatase are the most widely used enzyme labels (Hsu et al. 1981; Genten et al. 2009). This approach gives tremendous flexibility and there are multiple labeling kits available commercially. Moreover, all procedures can be carried out in a standard histological laboratory setup.

Immunohistochemical techniques have been expanded with the use antibodies conjugated with fluorescent tags for direct tissue visualization. These immunofluorescence protocols may be used to track different features simultaneously by using several antibodies conjugated to differently colored fluorochromes (Zaidi et al. 2000). But they also have been limited in their application due to autofluorescence of tissues (Baschong et al. 2001). However, the most serious limitation of this, and all antibody-based methods, is the need for specific antibodies, which are expensive or may not be commercially available and are laborious to produce and validate (Kurokawa and Suzuki 1998).

In situ hybridization (ISH) is an alternative "staining" method available in histological protocols. The specificity is given by the use of a short DNA or RNA probe that is complementary in sequence to a target messenger RNA (mRNA). The probe is either radioactive or has an attached tag (enzymatic, fluorescent, or luminescent) to allow its detection. The advantage of ISH is the relative ease of producing the probe compared with specific antibodies. Also, as gene expression is an early step in the functional activation of cells and tissues, it has been used to map the first activation of enzyme or hormone pathways (Kurokawa et al. 2004, 2005). It is important to notice that mRNA only provides an indirect indication of protein activity (i.e., phenotype), therefore any functional interpretation must be done with extreme caution as protein transcription rates, posttranslational modifications, and effects of modulating factors are not known. But, for the simple assessment of developmental events, it is very useful. The last section in this chapter includes references to developmental assessments using multilabel immunofluorescence and ISH protocols in the zebrafish.

14.2.2.2 Biochemical features

Biochemical tests are, in concept, similar to histochemical methods but they are conducted on tissue homogenates and provide precise quantitative estimates. Many have been used as proxy measures of digestive and metabolic capacity. For example, the digestive potential of first-feeding larvae has been evaluated by studying the onset and the variation of digestive enzymes (pancreatic and intestinal) and the response of these enzymes to nutritional manipulations (see Chapter 1).

Most biochemical analyses require the samples to be frozen. Samples for the analysis of proteins or lipids, for example, can be stored for short periods of time in conventional freezers (–20°C). However, different assays will require lower storage temperatures, with those measuring enzyme activities requiring –80°C storage. These samples must be collected in appropriate cryotubes, especially if liquid nitrogen is used.

SDS-PAGE[16] zymograms of larval extracts have been used to identify specific proteases based on size and particular inhibition profile (see Chapter 1). In this technique, the protease pool is size-fractioned and immobilized in a polyacrylamide gel before the substrate is applied. This is similar in concept to histochemical protocols as the substrate–enzyme interaction is conducted on a solid matrix. Similarly, it does not allow easy quantification of enzymes.

Enzymatic activities

Enzyme determination assays vary, with the most common methods utilizing incubations of larvae homogenate with suitable soluble substrates. During the incubation, the active enzyme in the homogenate hydrolyzes the substrate at a speed that is proportional to the level of enzyme present, allowing quantification.

These simple assays are well established and, in principle, easily accessible to any laboratory (Table 14.1). Some assays use specific substrates, allowing quantification of particular enzymes, but most commonly they use generic substrates (casein, hemoglobin, etc.) and narrow pH ranges to provide optimal conditions for the target enzyme activity, and hence the best sensitivity and specificity of the assay. Assay buffers, therefore, range in pH and, along with the incubation temperature and time, must be carefully chosen while validating the assays.

Sample requirements for enzymatic activity quantification vary and ultimately depend on the level of the enzymes in the tissues. Homogenization is most commonly done in various formulations of phosphate buffered saline (PBS) and filtered or centrifuged to eliminate interferences in the assay. Some protocols use dissected stomachs along with accessory organs instead of whole-body homogenates, an approach that eliminates interference from, for example, nondigestive proteolytic activities.

Direct quantification of biomolecules

There are occasions where direct quantification of the target biomolecule provides the best or the only possible alternative for evaluating its presence. This is, for instance, the case of molecules without catalytic capacity such as hormones, structural proteins, and mRNA. Among the many formats for quantitative analytical chemistry, we will introduce only three main families of techniques that are frequently found in larval studies: immunoassay, gene expression analysis, and chromatography.[17]

Immunoassay. There are many different variations of immunoassays but all are based on the use of an antibody that specifically binds to a target molecule subject of the assay. Radioimmunoassay (RIA) and enzyme immunoassay (EIA) are competitive assays where a labeled synthetic version of the target molecule (tracer), along with the antibody, is added to larvae homogenates containing the

[16] Sodium dodecyl sulfate–polyacrylamide gel electrophoresis is a technique used for the separation of proteins according to size.

[17] There are other analytical methodologies, some associated with the new 'omics' disciplines, for example, stable isotope labeling by amino acids in cell culture (SILAC), time-of-flight mass spectrometer (TOFMS), or nuclear magnetic resonance (NMR) spectrometry. These have potential applications in larval nutritional studies; however, currently, they are limited to biomedical research.

Table 14.1 Summary of digestive and metabolic enzymes used in larval studies.

Enzyme	Type	Source
Acetylcholinesterase[a]	Intestinal	Gawlicka et al. (1995)
Adrenodoxin[a]	Metabolic	Sampath-Kumar et al. (1997)
Alkaline phosphatase[a]	Intestinal	Gawlicka et al. (1995); Gisbert et al. (2009)
Aminopeptidase N[a]	Intestinal	Govoni et al. (1986); Gawlicka et al. (1995); Kurokawa and Suzuki (1998); Gisbert et al. (2009)
Amylase	Pancreatic	Gisbert et al. (2009)
Aminopeptidase	Intestinal	Kurokawa and Suzuki (1998)
Chymotrypsin	Pancreatic	Govoni et al. (1986); Gisbert et al. (2009)
Cytochrome P-450_{21}[a]	Metabolic	Sampath-Kumar et al. (1997)
Citrate synthase	Metabolic	Clarke et al. (1992)
Cyclin-dependent kinase	Cell cycle	Westerman et al. (1999)
Dipeptidyl peptidase II and IV[a]	Intestinal	Gawlicka et al. (1995)
Lactase[a]	Intestinal	Gawlicka et al. (1995)
Lactate dehydrogenase	Metabolic	Clarke et al. (1992)
Leucine–alanine peptidase	Intestinal	Gisbert et al. (2009)
Lipase	Pancreatic	Govoni et al. (1986); Gisbert et al. (2009)
Maltase	Intestinal	Govoni et al. (1986); Gisbert et al. (2009)
Nonspecific esterases[a]	Intestinal	Gawlicka et al. (1995)
Ornithine decarboxylase	Metabolic	Neyfakh et al. (1983)
Pepsin	Gastric	Govoni et al. (1986); Fernández-Díaz et al. (2001)
Trypsin	Pancreatic	Govoni et al. (1986); Gisbert et al. (2009)
α-Galactosidase[a]	Intestinal	Gawlicka et al. (1995)
β-Galactosidase[a]	Intestinal	Gawlicka et al. (1995)
γ-Glutamyl transpeptidase[a]	Intestinal	Gawlicka et al. (1995)

[a] Used in enzyme histochemistry.

unlabeled endogenous target (unknowns). At equilibrium, samples with increasing amounts of endogenous target have displaced some of the labeled synthetic tracer from the antibody, giving a decreased signal in the antibody-bound fraction. This competitive displacement allows precise quantification of the endogenous target present in the homogenate. In RIA formats, the tracer is radioactive, while in EIA format, it is an enzymatic tag.

A different format is the enzyme-linked immunosorbent assay (ELISA). There is variation in ELISA protocols but the most basic consists of the immobilization of the molecule to be measured onto a solid surface (i.e., microtiter plate wells), followed by a first incubation with the specific primary antibody (specific for the target molecule) and a second incubation with a tagged antibody that latches on to the primary antibody. The tag in the second antibody is usually, as in EIA, a small enzyme able to produce a colored product. The assay is designed so that the color developed in the well is proportional to the level of enzyme in the sample, allowing precise quantification.

Immunoassays are very sensitive and can detect substances in the nanogram range with ease. The biggest problem with immunoassays is that it is hard to tell whether the standard, often a synthetic standard, binds the antibody with the same affinity and dynamics as the endogenous target. Most authors report concentration as immunoactive units instead of absolute values to circumvent this problem. In any case, valid experimental inferences can be drawn upon relative values between treatments.

Expression analysis. Quantitative or real-time polymerase chain reaction (qPCR) measures mRNA levels (gene expression) of proteins. The assay relies on the amplification of mRNA targets in special PCR machines able to measure in real time the increase in PCR product (double-strand DNA). The speed at which DNA starts to accumulate over a certain threshold is directly proportional to the initial amount of target mRNA, which can be precisely quantified by comparing the results with a series of standards. The specificity of the assay is provided by the PCR primers used in the PCR amplification process. The design of these primers and the normalization and the detection system of the PCR product are the two most critical aspects of this assay (Huggett et al. 2005; Zhong et al. 2008). Once the assay is validated, there are few complications for its implementation in a laboratory fitted with a real-time PCR machine. Samples to be used for gene expression assays must be extracted from fresh, RNA-stabilized, or flash-frozen[18] tissue always stored at the lowest temperature possible to prevent RNA degradation. Extreme care should be used for the isolation and reverse transcription of the RNA as this molecule is easily damaged.

An allied technique for the analysis of gene expression is microarray hybridization (Miller and Maclean 2008). This technique is based on the same principle as the ISH protocol described above. In the microarray approach, the mRNA expressed in the sample (copied into a labeled DNA probe) is allowed to hybridize with complementary test DNA sequences immobilized on a synthetic chip. The process results in a colored mosaic where color is present on those spots bearing sequences of genes actively expressed in the sample. The DNA chip or microarray may contain several hundred thousand test features, allowing for a broader yet relative quantification of thousands of mRNA targets at once. This technique is still considered an exploratory, discovery-based tool. However, microarray hybridization has been used with success in Atlantic halibut (*Hippoglossus hippoglossus*) to describe the activation of genes encoding several key digestive enzymes (Douglas et al. 2008), and response to dietary carbohydrates (Robison et al. 2008) and starvation (Drew et al. 2008) in zebrafish (*Danio rerio*)

Chromatography. More common analytical formats applied to larval studies are gas chromatography (GC) and high-performance liquid chromatography (HPLC). Also, chromatographic techniques are commonly used to purify nucleic acids and proteins during sample preparation steps in molecular protocols and to validate immuno- and biochemical assays, respectively. The principle of all chromatographic methods is based on the selective partition of the target biomolecule between a solid matrix and a mobile carrier substance. The carrier (gas or liquid) splits a complex mixture into its component molecules as it flows through the solid matrix packed in a chromatographic column or deposited on a glass plate. The individual component molecules are then quantitatively detected at the end of the column or on the plate. Variations in physical parameters such as chemical properties of the column and mobile phase, temperature, pH, ionic strength, as well as other parameters, are used to separate the target molecule(s) from related com-

[18] RNA-stabilized tissues are obtained using special preservative cocktails that prevent degradation of RNA molecules. Flash-freezing or quenching is usually achieved by immersing the cryotubes containing sample tissue in liquid nitrogen.

pounds. A well-designed protocol will provide information over the entire chromatographic elution, which can cover a wide range of related compounds, with great sensitivity and accuracy (Abdulkadir and Tsuchiya 2008; Lund et al. 2008)

14.2.2.3 Behavioral features

Larval feeding behavior relies on coordinated sensorial input (visual and postural perception), chemoreception, motor ability, and neural integration. Development and integration of natural behaviors requires the presence of structures and reflexes that, like any other organ or metabolic pathway, may be affected by nutritional manipulations. For example, maturation of the brain is shown to be affected by manipulations of highly unsaturated fatty acids (HUFA) levels (see Chapter 2). Thus, behavioral metrics (i.e., swimming speed, startle response, aggression, schooling, omega [Ω] posture, hunting) are used to assess performance, although there are very few larval nutritional studies that have used them (Masuda and Tsukamoto 1999; Nakayama et al. 2003). Instead, it seems common in larval studies to estimate performance through end point outcomes of behavioral skills such as feeding rates (Yúfera and Darias 2007) and percentage of settled metamorphosing larvae (Gavlik et al. 2002).

14.3 Indirect growth assessment: condition analysis

It is important to note that indirect growth assessment does not rely on a measure of size change over time. In many cases, this reduces the need for long growth trials and sampling can be reduced to a single event, limiting the size of experiments and reducing the use of experimental subjects.

Cellular and biochemical correlates of growth often produce more direct cause–effect relationships as they provide information on the physiological mechanisms mediating the effect of the nutritional factors under study (Gisbert et al. 2008). But there are disadvantages as well. Main issues are often linked to the necessary validation of the different techniques and the interpretation of the results. Evaluation techniques and assay protocols are substantially more complex, require specialized equipment, cost more, and take longer to do than simple biometric measures. We will expand on these matters as we introduce some experimental approaches in this section.

14.3.1 Biochemical composition

14.3.1.1 RNA/DNA ratios

A widely used condition estimate in larval studies is the ratio of RNA to DNA. This condition index relies on the assumption that RNA content varies according to the rate of protein synthesis. In contrast, the amount of DNA in the larvae is approximately fixed and depends mainly on body size. Thus, RNA/DNA ratios can be used to estimate the relative activation of new protein synthesis that in turn is correlated with recent growth rate (Buckley et al. 2008). RNA/protein or RNA/DW are related indices where protein or DW instead of DNA provides the necessary scaling for the size of the sample analyzed (Chicharo and Chicharo 2008). Although there are known ontogenetical changes in RNA/DNA ratios during early development, it is generally accepted that, under favorable growth conditions, the ratios will be high and, conversely, low in larvae growing under unfavorable environmental or nutritional conditions.

There are numerous assay protocols and, with proper validation, all are equally suitable and provide the required sensitivity to work with small-sized samples (Buckley et al. 1999). Nucleic acids can be determined in solution directly using UV spectrophotometers.

The analytical procedure is very simple and consists of a series of purification steps and chemical hydrolyzation of RNA first and later DNA. After the separation, RNA and DNA are measured from the absorbance of the corresponding fractions at 260 nm. Protocols based on UV absorbance require relatively large amounts of tissue homogenates, which is often a problem when working with larval fish.

More sensitive alternatives use fluorescent dyes that specifically bind nucleic acids. Fluorometric assays can provide a valid measure from as little as 10 µg DW, making possible determinations in a single larva (e.g., Clemmesen et al. 2003). The dye, typically ethidium bromide (EB),[19] is added to provide a fluorescent reading of total nucleic acids (RNA + DNA). After the RNA is eliminated with a ribonuclease[20] treatment, the fluorescence in the sample is read once more; this fluorescence is assumed to be produced by the DNA exclusively (this assumption always must be verified). RNA fluorescence is finally calculated from the difference between the first and the second reading. Due to the extreme sensitivity of fluorometric measurements, a complete validation of new assays is necessary, including tests to estimate recovery of nucleic acids and saturation or fluorescence quenching. Recent miniaturization and use of fluorescence-based microplate readers makes possible the implementation of very sensitive and high-throughput protocols. Also, the wider use of these fluorometric procedures in other molecular techniques has increased choice of fluorescent dyes with increased specificity (Mcginty et al. 2008) and made reagents and consumables more affordable.

[19]EB is a carcinogenic substance. Liquid waste and contaminated labware should be handled and disposed of in accordance with local hazardous waste regulations.

[20]Ribonucleases are enzymes that specifically degrade RNA molecules and have no effect on DNA molecules.

14.3.1.2 Cell cycle analysis

Cell division rates have been proposed as a sensitive method to estimate growth rate in larvae (Theilacker and Shen 1993). The rationale behind this novel approach is based on the assumption that growth rate during the larval stage is mainly driven by increases in cell counts. Thus, fast-growing larvae will have a larger proportion of cells undergoing the DNA duplication cycle (S phase)[21] than larvae growing at a slower rate. The proportion of dividing and nondividing cells within larval fish tissues can be precisely estimated using flow cytometric analysis (FCA). This is a very sensitive analysis mainly used in biomedical research, but it is also used in plankton research, where it is a quick alternative to manual count of phytoplankton cells in water samples. FCA measures the amount of fluorescence emitted by individual cells or nuclei. The assay requires the suspension of individual cells in a liquid medium and the use of fluorescent DNA dyes. During the analysis, a laser beam scans individual cells, inducing a fluorescent signal, which is proportional to DNA content in the nucleus. These individual readings are then used to assign the position of each scanned cell in the cell cycle. From this raw output, it is finally possible to obtain a fractional index providing a relative measure of cell division activity. Overall, the few studies using this approach suggest a valid approach that can be tailored to different tissues and suggesting FCA may be suitable as an indicator of growth and condition in fish larvae (Bromhead et al. 2000; Theilacker and Shen 2001; González-Quirós et al. 2007). A great advantage of this method is its speed and relatively simple sample processing; however, there are considerable costs associated with the FCA equipment, which is often only available at core facilities in large biomedical research centers.

[21]S phase, or "synthesis" phase, in the cell cycle is the period where the chromosomes duplicate in preparation for the final cell division into two daughter cells.

14.3.1.3 Lipid ratios

Lipids are extremely important dietary elements for normal development and growth of fish larvae, having a dual role as energy reserves and structural components of cell membranes (Rainuzzo et al. 1997; Morais et al. 2007; Cahu et al. 2009; see also Chapters 2 and 5). Triacylglycerol, for example, is an energy storage component deposited under favorable growth conditions. Other lipid classes such as cholesterol or polar lipids are essential cell membrane constituents, and thus would be expected to correlate closely with larval size and be less dependent on nutritional or physiological status. Ratios of triacylglycerol to polar lipids, cholesterol, or DW have been used in ecological studies as an indicator of recent nutrition and growth of fish larvae (Fraser et al. 1987; Hakanson 1989; Lochmann et al. 1996). These indices, however, may not be as sensitive as other measures of condition (Suthers et al. 1992) probably due to the tight energy budgets and limited energy reserves of fast-growing larvae.

14.3.2 Physiological energetics

We have just seen in the previous section that individuals with more energy reserves are likely to sustain further growth. Biochemical composition and derived energy content can and has been used to infer condition in nutritional bioassays. Further elaboration of this general idea into energy budgets has been used to compare growth physiology in fish larvae (Finn et al. 1996; Kamler 2008) and to study the effect of toxicants in small invertebrates (Maltby et al. 1990; De Coen and Janssen 1997).

14.3.2.1 Scope for growth (SFG)

SFG is the free energy available to an organism to perform vital activities, which in the larval period with no reproductive investment are limited to activity and growth (see details in Chapter 8). Formally, energetic budgets during the larval phase can be simplified in the equation

$$\mathrm{SFG} = A - (R + U),$$

where A is the energy ingested, R is the energy respired during routine metabolism, and U is the energy excreted. This model is an oversimplification, but it provides a reasonable approximation to the allocation of energy in larval fish. Growth is only possible in circumstances where there is a positive net gain of energy; that is, energy gains are greater than energy expenditures.

There are a number of methodological procedures to estimate parameters A and U. In most cases, they are no different from the calculation of energy content of larvae introduced earlier. Also, nitrogen assay (Dabrowski 1986) or tracer methods (radiotracers and rare earth oxides) are used to estimate both ingestion and excretion terms (Chapter 13; Conceição et al. 2007).

The term R is the energetic cost of all routine metabolic activities. Under aerobic conditions, metabolism can be approximated by measuring respiration rates in resting larvae. This is measured in purposely built sealed respiration chambers (respirometers) fitted with oxygen-sensing microelectrodes or sample extraction ports for oxygen determinations outside the respiration chamber. Oxygen consumption rates are expressed relative to weight and are directly translated into energy equivalents per unit of time (see references in Chapter 8). Alternatively, the energy equivalent of all routine physiological work can be measured from the total heat output produced by the resting larvae in a microcalorimeter (Finn et al. 1996; McCollum et al. 2006). This second approach directly measures total metabolic cost (aerobic and anaerobic) in energy units.

SFG determinations have been widely used in ecophysiological and pollution studies but only occasionally in nutritional studies. Practical difficulties in estimating the different terms of the energy budgets, especially in the case of larvae, are the limited access to specialized equipment and elevated costs. These issues may limit the application of SFG in larval fish studies.

14.3.2.2 Cellular energy allocation (CEA)

An alternative approach to growth assessment using physiological energetics is the CEA assay. The rationale behind CEA is much simpler than that of the SFG, as is the analysis protocol. The CEA assay uses the total energy available in the larval tissue (E_a) and energy consumption rates (E_c) to provide an indicator of metabolic condition. The E_a term is measured indirectly from the proximate composition of the larvae, while E_c is estimated *in vitro* by measuring the electron transport system (ETS) activity. The ETS assay provides cellular consumption of oxygen as measure of metabolic intensity, avoiding labor-intensive respiration rate determinations. Oxygen consumption is finally transformed into the energetic equivalent resulting from the complete oxidation of an average lipid, protein, and carbohydrate mixture. CEA has been expressed as a ratio (E_a/E_c) or energy budget ($E_a - E_c$). In any case, the CEA is analogous to a measure of standard metabolic rate where the necessary size standardization is provided by the biomass of the larvae expressed in energy equivalents.

The CEA approach has been developed and applied in toxicological studies of aquatic invertebrates (De Coen and Janssen 1997) but has only been anecdotically used in fish (Pfeiler and Govoni 1993; Rueda-Jasso et al. 2004). The little evidence available suggests that CEA may represent a cost-effective method to derive simplified energy budgets of interest for comparative purposes in larval nutrition.

14.4 The zebrafish: a model organism to study fish nutrition

In recent years, the zebrafish has become one of the most important genetic model organisms in biology. The transfer of knowledge and technologies from this species could be of great benefit when trying to understand shared processes in other fish species, including fish nutrition as well as, for example, assisting in the development of improved feeds or feeding regimes. This could happen in two ways: First, techniques, tools, and protocols developed for use with model organisms may be adapted for use with related species; second, knowledge gained from research on zebrafish may be transferred to species that are more difficult to study experimentally.

While the majority of studies employing zebrafish still focus on embryonic development, it is becoming apparent that this species also holds great potential to investigate other fields, including cell biology (Beis and Stainier 2006), genomics (Chen and Ekker 2004; Postlethwait 2007), evolution (Metscher and Ahlberg 1999; Schilling and Webb 2007), and physiology (Briggs 2002).

14.4.1 Development and anatomy of the intestinal tract

The development and morphology of the intestinal tract (Wallace and Pack 2003; Ng et al. 2005; Wallace et al. 2005a) and its accessory organs (Field et al. 2003a, 2003b; Ober et al. 2003) in zebrafish has been described in great detail. As with other aspects of embryonic and early larval development, it occurs comparatively rapidly in the zebrafish (Table 14.2; Chapter 1).

A caveat when working with the zebrafish as a model organism for studies focusing on the digestive system is that it, like other cyprinids but unlike most other teleost fish, does not have a true stomach (Harder 1975;

Table 14.2 Developmental sequence during digestive tract formation in zebrafish embryo and larvae.

Developmental event	Time (dpf)
Endodermal precursor cells aggregate in a solid band along the embryo's midline from near the prospective mouth cavity to the future anus site	1 (26 hours)
Formation of a bilayer, which gradually develops into the intestinal lumen	2 (45–50 hours)
Accessory organs, liver and pancreas, begin to form	
Enteric neurons and smooth muscle cells colonize the future intestinal tract	
Intestinal tract appears as one continuous lumen; enteroendocrine cells appear	3–4
Cells lining the lumen form a polarized (apicobasal) epithelial monolayer	
Mouth opens	4
Epithelial lining of the intestinal bulb begins to form folds; mucus-secreting goblet cells and enterocytes in the midintestine develop	
Anus opens	
Intestine shows three distinct compartments (intestinal bulb, midintestine, and posterior intestine)	
Larval intestine colonized by microorganisms (Bates et al. 2006)	
Alimentary canal subdivided (from rostral to caudal) into (1) the mouth cavity and pharynx, (2) the esophagus, (3) the intestinal bulb, (4) the midintestine, and (5) the posterior intestine with the anal opening (Field et al. 2003a; Ng et al. 2005)	5
Yolk supply exhausted; start feeding on exotrophic sources	
Midintestinal epithelium begins to fold	8
Posterior intestinal epithelium begins to fold; cell proliferation restricted to the bases of the folds	12
Goblet cells present throughout the intestine, albeit significantly less abundant in the intestinal bulb, while enterocytes remain confined to the midintestine; digestive tract fully differentiated	14

dpf = days postfertilization.

Rombout et al. 1984). This could limit the transferability of some results obtained with this species to noncyprinid species. It is, however, worth noting that the larvae of most teleosts also do not have a stomach, suggesting that the digestive system in zebrafish larva may be very similar to that of noncyprinid teleosts. The enzymes typically found in vertebrate stomachs do, however, likely exist in the zebrafish irrespective of the anatomy of its intestinal tract (Kurokawa et al. 2005). It should also be pointed out that the zebrafish intestinal tract also appears to lack Paneth cells (Ng et al. 2005).

The first sporadic activity of the larval zebrafish intestine can be observed as early as 3 days postfertilization (dpf), with a more regular pattern developing at 4 dpf (Holmberg et al. 2003), prior to the onset of feeding on exogenous food. Over the next 3 days, the pattern of contractions becomes more coordinated, with both antero- and retrograde peristaltic contraction waves occurring along the intestine, accompanied by local rectal contractions (Holmberg et al. 2003). Due to the transparency and small size of zebrafish larvae at these stages, the spatiotemporal patterns of these contractions can be observed with ease and the effects of different pharmacological treatments, feeds, or other conditions can be assessed and quantified (Berghmans et al. 2008).

14.4.2 Advantages for experimental approaches

The zebrafish is a relatively robust species that can be kept and bred with minimal effort and cost (Brand et al. 2002). When kept under laboratory conditions, zebrafish reach sexual maturity after 2–3 months, spawn throughout the year (with one female capable of producing approximately 200 eggs per week), and have a life expectancy of up to 4 years. Furthermore, their small size and social behavior allows large numbers of zebrafish to be kept in limited spaces.

Like many other larval fish, the zebrafish is largely translucent. This allows the observation of both the formation and function of internal organs, including the entire gastrointestinal tract with its associated organs.

> Fluorescently tagged or quenched phospholipids and cholesterol analogues were used (1) to assess lipid absorption in the intestinal tract of larval zebrafish, (2) to visualize phospholipase A2 (PLA2) activity with subcellular resolution, and (3) to identify mutants with impaired lipid uptake and metabolism (Farber et al. 2001). In this study, one of the reporter molecules used (PED6) was labeled with a fluorescent BODIPY conjugate attached next to the PLA2 cleavage site. Enzymatic cleavage by PLA2 changed the fluorescence of the reporter from orange to green. When swallowed by the larvae, the reporter first labeled the pharynx (in orange), was metabolized (i.e., the fluorescence switched to green), and subsequently labeled the hepatobiliary system before being secreted into the gallbladder, from where it moved to the intestinal lumen. This example demonstrates the ease with which relatively complex physiological processes can be monitored in zebrafish larvae in real time.

Although such studies are possible in other species, the degree of knowledge of the zebrafish genome and developmental pathways allows detailed mechanistic assessments of digestive and absorptive processes. Therefore, in the context of this book, the greatest potential of the zebrafish lies in the possibility of employing assays to investigate the molecular networks controlling development and function of the gastrointestinal tract and its accessory organs. This possibility has been exploited in mutagenesis screens that resulted in the identification of mutant zebrafish lines with phenotypes in various organs of the gastrointestinal tract (Chen et al. 1996; Pack et al. 1996; Stainier et al. 1996; Mohideen et al. 2003). Subsequent studies on these mutants have since yielded novel insights into gut development, identifying genes with essential or important functions in the formation and function of the digestive system in zebrafish (e.g., Mayer and Fishman 2003; Wallace et al. 2005b; Ober et al. 2006). Further, zebrafish mutant lines have been used to study the embryonic development, physiology, or predisposition to disease (Amsterdam and Hopkins 2006; Lieschke and Currie 2007), and *in vivo* effects of specific compounds, such as drugs or environmental toxins (Rubinstein 2006; Berger and Currie 2007; Barros et al. 2008; Crawford et al. 2008; Scholz et al. 2008). For example, Berghmans et al. (2008) used mutagenesis screens in a small-scale pilot study to screen compounds for their effects in a gut contraction assay.

14.4.3 Methods to study gene function

The past two decades have seen a rapid expansion of the resources, protocols, and tools available to genetically manipulate and analyze the zebrafish (Anderson and Ingham 2003; Dahm et al. 2005; Dahm and Geisler 2006). These have opened up entirely new approaches to study various aspects as well as anabolic processes, such as muscle and fat tissue biogenesis, in fish.

The zebrafish is ideally suited to study the functions of genes in fish as it is genetically more tractable than most, if not all, fish species of interest to aquaculture. For one, its genome has been sequenced and is being annotated (www.ensembl.org/Danio_rerio/Info/Index) and there are several high-resolution genetic, physical, and radiation hybrid maps of the zebrafish genome (reviewed in Dahm et al. 2005). These greatly facilitate identifying loci associated with specific traits to be used in, for example, marker-assisted breeding programs to optimize growth performance.

A number of resources to analyze gene expression on a near genome-wide scale have been generated (reviewed in Dahm and Geisler 2006). These include large libraries of sequenced expressed sequence tags (ESTs) and full-length mRNAs that are used to generate microarrays for gene expression profiling. Moreover, commercial microarrays for mRNA expression analyses are now becoming available for zebrafish. In addition to these resources, several protocols to efficiently disrupt or inactivate genes in the zebrafish or to monitor their expression *in vivo* have been established (Dahm and Geisler 2006). The more recent development of mutagenesis strategies relying on gene disruption with viruses or transposable elements greatly accelerates the identification of mutated loci in forward genetic approaches (Amsterdam and Hopkins 2006). For a list of published mutants with brief phenotypic descriptions and, where available, molecular characterizations, see Frohnhoeffer (2002) or refer to the searchable zebrafish mutant database at www.zfin.org.

More recently, reverse genetics techniques that prevent the expression of specific genes have also been established in zebrafish. These include the knockdown of genes with morpholinos, a technique that uses short, chemically modified sequence-specific antisense oligonucleotides to block the translation of endogenous mRNA transcripts. This method allows obtaining phenotypes for specific genes very rapidly (within a few hours to days) and characterizing a gene's functions in an organismic context (Nasevicius and Ekker 2000; reviewed in Dahm and Geisler 2006). Furthermore, with TILLING (from targeting induced local lesions in genomes), protocols can readily be adapted to other species in order to identify individuals with a mutation in a gene of interest. Similarly, given sufficient information on the sequence of the genes to be targeted, it may be possible to adapt a morpholino-based knockdown approach to a wider range of fish species. These approaches will greatly facilitate the identification of new genes as well as the characterization of the functions of known genes, and thus enhance our understanding of the development and function of specific organs in fish.

Finally, the generation of transgenic animals and the use of reporter gene systems is another important means of studying the functions and responses of genes *in vivo* and under different conditions as well as to generate individuals with new properties. Several methods have been established to introduce transgenes into zebrafish, several of which ensure the stable expression of transgenes over many subsequent generations or allow the expression to be switched on or off at specific time points.

As more genomic resources are being generated for aquaculture fish species, the interaction between the zebrafish and other fish research communities is likely to become more productive. In particular, research on nutrition and growth as well as stress and disease resistance in the zebrafish is expected to produce information of relevance to aquaculture fish, for instance, by allowing the development of formulated feeds.

14.4.4 Methods to study effects of intestinal microorganisms on larval gut

Next to the genes and other intrinsic factors that regulate development, morphology, and physiology of the intestinal tract, its function is also modulated by the microorganisms that colonize it. Importantly, microbes not only interact with the substances passing through the alimentary tract, for instance, making nutrients more readily available to the animal host, they also influence a number of other processes, from the proliferation and differentiation of epithelial cells and the integrity of the mucosal barrier to modulating the enteric nervous system, the immune system, and intestinal angiogenesis (for references, please see Rawls et al. 2004; Bates et al. 2006). The intestinal microbial flora thus plays a crucial and manifold role in the normal functioning of the digestive system. Surprisingly, however, large aspects of the changes that microbes elicit in their host, and the molecular mechanisms mediating these changes, are still unknown.

In this context, the recent generation of germ-free (gnotobiotic) zebrafish is of interest (Rawls et al. 2004; Bates et al. 2006). When the patterns of gene expression in these fish were analyzed with microarrays, 212 genes were found to be regulated by the presence of gut microbes (Rawls et al. 2004). These comprised genes involved in stimulating the proliferation of epithelial cells, in promoting the metabolism of nutrients, and in innate immune responses. The importance of the gut's bacterial flora for intestinal development and homeostasis as well as for fish nutrition can therefore be studied and manipulated experimentally in zebrafish.

Bates et al. (2006) found that the intestinal flora was required for normal differentiation of the gut's epithelium as evidenced, for instance, by the absence of alkaline phosphatase activity in the intestinal brush border, immature glycoconjugate expression patterns, and reduced numbers of goblet and enteroendocrine cells, indicating that the intestinal microflora plays a crucial role in the development of the intestinal tract, including cell fate specification, in zebrafish. In addition to these molecular differences, the intestines of zebrafish lacking intestinal microorganisms also displayed increased contractile movements and reduced ability to take up protein macromolecules (Bates et al. 2006). Interestingly, these phenotypes could be reversed by exposure of the larvae to the appropriate microflora.

Feeding protocols for larval and juvenile zebrafish in use in research laboratories

The protocols for the feeding of larval zebrafish kept under laboratory conditions are generally based on empirical practices developed by trial and error, not rational design. Briefly, during the first 4–5 days of development (5 dpf), zebrafish larvae are not fed but live on their supply of yolk. Starting at 5 dpf and until 10 dpf, the larvae are fed with finely ground powdered fish food, which is sprinkled onto the water surface once per day. Alternatively, some laboratories feed larvae during this period with live paramecia, which are raised in a mung bean culture. Following this period and until 15 dpf, the feed is gradually switched to live, freshly hatched *Artemia* (brine shrimp), which are provided three times per day. After 15 dpf, older *Artemia* (2–3 days) may be used to feed the larvae twice per day. This feeding regime is maintained until the zebrafish have reached adulthood (after approximately 2–4 months). At this stage, zebrafish can be fed with commercially available flake food supplemented with live or frozen *Artemia* twice per day. A number of laboratories also use the larvae of the fruit fly *Drosophila melanogaster*, a widely used model organism, to supplement the feed of adult zebrafish. See *The Zebrafish Book* (www.zfin.org) or Brand et al. (2002).

14.4.5 Concluding remarks

The potential of the zebrafish as a model organism to study aspects of fish nutrition, both in larvae and in adult individuals, has only been poorly exploited to date. Given its embryological advantages, the large number of mutants, and the genetic and genomic tools available, the zebrafish is ideally suited to study a number of different aspects of the biology of the gastrointestinal tract, especially on the molecular level. Zebrafish larvae could also serve as models to rapidly and cost-efficiently screen a large number of feed parameters and feeding protocols, and to evaluate their effects on fish growth and health.

Conversely, the zebrafish community would benefit from the development of better formulated and/or standardized feeds that would increase the success of raising zebrafish and would allow fish to be raised and kept under standardized conditions, respectively. Improved formulated feeds for zebrafish would greatly ease the workload and cost associated with maintaining healthy zebrafish colonies. This is particularly important for larval stages between developmental days 5 and 10, which are too small to feed on live *Artemia*. It might also be helpful in raising those mutant strains with notoriously low survival rates during the larval and juvenile periods to adulthood. The latter would facilitate analyses of developmental and physiological parameters that are influenced by variations in essential nutrients throughout larval development, and also at later stages of juvenile development and right through reproduction.

Acknowledgments

The authors thank Dr. Helia B. Schönthaler for critically reading the part dealing with the zebrafish as a model organism.

Literature cited

Abdulkadir, S., and Tsuchiya, M. 2008. One-step method for quantitative and qualitative analysis of fatty acids in marine animal samples. *Journal of Experimental Marine Biology and Ecology* 354(1):1–8.

Amsterdam, A., and Hopkins, N. 2006. Mutagenesis strategies in zebrafish for identifying genes involved in development and disease. *Trends in Genetics* 22:473–478.

Anderson, K.V., and Ingham, P.W. 2003. The transformation of the model organism: a decade of developmental genetics. *Nature Genetics* 33(Suppl.):285–293.

Balon, E.K. 1981. Saltatory processes and altricial to precocial forms in the ontogeny of fishes. *American Zoologist* 21:573–596.

Barros, T.P., Alderton, W.K., Reynolds, H.M., et al. 2008. Zebrafish: an emerging technology for *in vivo* pharmacological assessment to identify potential safety liabilities in early drug discovery. *British Journal of Pharmacology* 154:1400–1413.

Baschong, W., Suetterlin, R., and Laeng, R.H. 2001. Control of autofluorescence of archival formaldehyde-fixed, paraffin-embedded tissue in confocal laser scanning microscopy (CLSM). *Journal of Histochemistry and Cytochemistry* 49(12):1565–1571.

Bates, J.M., Mittge, E., Kuhlman, J., et al. 2006. Distinct signals from the microbiota promote different aspects of zebrafish gut differentiation. *Developmental Biology* 297:374–386.

Beis, D., and Stainier, D.Y. 2006. *In vivo* cell biology: following the zebrafish trend. *Trends in Cell Biology* 16:105–112.

Berger, J., and Currie, P. 2007. The role of zebrafish in chemical genetics. *Current Medicinal Chemistry* 14:2413–2420.

Berghmans, S., Butler, P., Goldsmith, P., et al. 2008. Zebrafish based assays for the assessment of cardiac, visual and gut function—potential safety screens for early drug discovery. *Journal of Pharmacological and Toxicological Methods* 58:59–68.

Blaxter, J.H.S. 1988. Pattern and variety in development. In: Hoar, W.S., and Randall, D.J. (eds.) *Fish Physiology*, Vol. 11, Part A. Academic Press, London, pp. 1–58.

Brand, M.G., Granato, M., and Nüsslein-Volhard, C.H. 2002. Keeping and raising zebrafish. In: Nusslein-Volhard, C.H., and Dahm, R. (eds.) *Zebrafish, Practical Approach Series*. Oxford University Press, Oxford, pp. 7–37.

Briggs, J.P. 2002. The zebrafish: a new model organism for integrative physiology. *American Journal of Physiology. Regulatory, Integrative and Comparative Physiology* 282:R3–R9.

Bromhead, D., Kalish, J., and Waring, P. 2000. Application of flow cytometry cell cycle analysis to the assessment of condition and growth in larvae of a freshwater teleost *Galaxias olidus*. *Canadian Journal of Fisheries and Aquatic Sciences* 57:732–741.

Buckley, L.J., Caldarone, E.M., and Ong, T. 1999. RNA–DNA ratio and other nucleic acid–based indicators for growth and condition of marine fishes. *Hydrobiologia* 401:265–277.

Buckley, L.J., Caldarone, E.M., and Clemmesen, C. 2008. Multi-species larval fish growth model based on temperature and fluorometrically derived RNA/DNA ratios: results from a meta-analysis. *Marine Ecology—Progress Series* 371:221–232.

Cahu, C.L., Gisbert, E., Villeneuve, L.A.N., et al. 2009. Influence of dietary phospholipids on early ontogenesis of fish. *Aquaculture Research* 40(9):989–999.

Chen, E., and Ekker, S.C. 2004. Zebrafish as a genomics research model. *Current Pharmaceutical Biotechnology* 5:409–413.

Chen, J.N., Haffter, P., Odenthal, J., et al. 1996. Mutations affecting the cardiovascular system and other internal organs in zebrafish. *Development* 123:293–302.

Chicharo, M.A., and Chicharo, L. 2008. RNA : DNA ratio and other nucleic acid derived indices in marine ecology. *International Journal of Molecular Sciences* 9(8):1453–1471.

Clarke, M.E., Calvi, C., Domeier, M., et al. 1992. Effects of nutrition and temperature on metabolic enzyme activities in larval and juvenile red drum, *Sciaenops ocellatus*, and lane snapper, *Lutjanus synagris*. *Marine Biology* 112:31–36.

Clemmesen, C., Bühler, V., Carvalho, G., et al. 2003. Variability in condition and growth of Atlantic cod larvae and juveniles reared in mesocosms: environmental and maternal effects. *Journal of Fish Biology* 62:706–723.

Conceição, L.E.C., Morais, S., and Rønnestad, I. 2007. Tracers in fish larvae nutrition: a review of methods and applications. *Aquaculture* 267:62–75.

Craig, J.F., Kenley, M.J., and Talling, J.F. 1978. Comparative estimations of the energy content of fish tissue from bomb calorimetry, wet oxidation and proximate analysis. *Freshwater Biology* 8:585–590.

Crawford, A.D., Esguerra, C.V., and de Witte, P.A. 2008. Fishing for drugs from nature: zebrafish as a technology platform for natural product discovery. *Planta Medica* 74:624–632.

Dabrowski, K.R. 1986. Active metabolism in larval and juvenile fish: ontogenetic changes, effect of water temperature and fasting. *Fish Physiology and Biochemistry* 1(3):125–144.

Dahm, R., and Geisler, R. 2006. Learning from small fry: the zebrafish as a genetic model organism for aquaculture fish species. *Marine Biotechnology (New York, NY)* 8:329–345.

Dahm, R., Geisler, R., and Nüsslein-Volhard, C. 2005. Zebrafish (*Danio rerio*) genome and genetics. In: Meyers, R.A. (ed.) *Encyclopedia of Molecular Cell Biology and Molecular Medicine*, 2nd edition. Wiley-VCH, Weinheim, pp. 593–626.

De Coen, W.M., and Janssen, C.R. 1997. The use of biomarkers in *Daphnia magna* toxicity testing, IV. Cellular energy allocation: a new methodology to assess the energy budget of toxicant-stressed *Daphnia* populations. *Journal of Aquatic Ecosystem Stress and Recovery* 6:43–55.

Dedi, J., Takeuchi, T., Seikai, T., et al. 1995. Hypervitaminosis and safe levels of vitamin A for larval flounder (*Paralichthys olivaceus*) fed *Artemia* nauplii. *Aquaculture* 133:135–146.

Domalski, E.S. 1972. Selected values of heats of combustion and heats of formation of organic compounds containing the elements C, H, N, O, P, and S. *Journal of Physical and Chemical Reference Data* 1:221–277.

Douglas, S.E., Knickle, L.C., Williams, J., et al. 2008. A first generation Atlantic halibut *Hippoglossus hippoglossus* (L.) microarray: application to developmental studies. *Journal of Fish Biology* 72(9):2391–2406.

Drew, R.E., Rodnick, K.J., Settles, M., et al. 2008. Effect of starvation on transcriptomes of brain and liver in adult female zebrafish (*Danio rerio*). *Physiological Genomics* 35:283–295.

Farber, S.A., Pack, M., Ho, S.Y., et al. 2001. Genetic analysis of digestive physiology using fluorescent phospholipid reporters. *Science* 292:1385–1388.

Fernández-Díaz, C., Yúfera, M., Cañavate, J.P., et al. 2001. Growth and physiological changes during metamorphosis of Senegal sole reared in the laboratory. *Journal of Fish Biology* 58:1086–1097.

Ferron, A., and Leggett, W.C. 1994. An appraisal of condition measures for marine fish larvae. In: Blaxter, J.H.S., and Southward, A.J. (eds.) *Advances in Marine Biology*. Academic Press, San Diego, CA, pp. 217–304.

Fey, D.P. 1999. Effects of preservation technique on the length of larval fish: methods of correcting estimates and their implication for studying growth rates. *Archives of Fishery and Marine Research* 47:17–29.

Field, H.A., Ober, E.A., Roeser, T., et al. 2003a. Formation of the digestive system in zebrafish. I. Liver morphogenesis. *Developmental Biology* 253:279–290.

Field, H.A., Dong, P.D., Beis, D., et al. 2003b. Formation of the digestive system in zebrafish. II. Pancreas morphogenesis. *Developmental Biology* 261:197–208.

Finn, R.N., Widdows, J., and Fyhn, H.J. 1996. Calorespirometry of developing embryos and yolk-sac larvae of turbot (*Scophthalmus maximus*). *Marine Biology* 122:157–163.

Fraser, A.J., Sargent, J.R., Gamble, J.C., et al. 1987. Lipid class and fatty acid composition as indicators of the nutritional condition of larval Atlantic herring. *American Fisheries Society Symposium* 2:129–143.

Frohnhoeffer, H.G. 2002. Table of zebrafish mutations. In: Nüsslein-Volhard, C., and Dahm, R. (eds.) *Zebrafish—A Practical Approach*. Oxford University Press, Oxford.

Fuiman, L.A., Poling, K.R., and Higgs, D.M. 1998. Quantifying developmental progress for comparative studies of larval fishes. *Copeia* 1998(5):602–611.

Gavlik, S., Albino, M., and Specker, J.L. 2002. Metamorphosis in summer flounder: manipulation of thyroid status to synchronize settling behavior, growth, and development. *Aquaculture* 203(3–4):359–373.

Gawlicka, A., Teh, S.J., Hung, S.S.O., et al. 1995. Histological and histochemical changes in the digestive tract of white sturgeon larvae during ontogeny. *Fish Physiology and Biochemistry* 14(5):357–371.

Genten, F., Terwinghe, E., and Danguy, A. 2009. *Atlas of Fish Histology*. Science Publishers Inc., Enfield, NH.

Gisbert, E., Ortiz-Delgado, J.B., and Sarasquete, C. 2008. Nutritional cellular biomarkers in early life stages of fish. *Histology and Histopathology* 23(12):1525–1539.

Gisbert, E., Giménez, G., Fernández, I., et al. 2009. Development of digestive enzymes in common dentex *Dentex dentex* during early ontogeny. *Aquaculture* 287:381–387.

González-Quirós, R., Munuera, I., and Folkvord, A. 2007. Cell cycle analysis of brain cells as a growth index in larval cod at different feeding conditions and temperatures. *Scientia Marina* 71(3):485–497.

Govoni, J.J., Boehlert, G.W., and Watanabe, Y. 1986. The physiology of digestion in fish larvae. *Environmental Biology of Fishes* 16(1–3):59–77.

Hakanson, J.L. 1989. Analysis of lipid components for determining the condition of anchovy larvae, *Engraulis mordax*. *Marine Biology* 102:143–151.

Harder, W. 1975. *Anatomy of Fishes*. Schweizerbart'sche Verlagsbuchhandlung, Stuttgart.

Hay, D.E. 1982. Fixation shrinkage of herring larvae: effects of salinity, formalin concentration, and other factors. *Canadian Journal of Fisheries and Aquatic Sciences* 39:1138–1143.

Hjorleifsson, E., and Kleinmacphee, G. 1992. Estimation of live standard length of winter flounder *Pleuronectes americanus* larvae from formalin-preserved, ethanol-preserved and frozen specimens. *Marine Ecology—Progress Series* 82(1):13–19.

Holmberg, A., Schwerte, T., Fritsche, R., et al. 2003. Ontogeny of intestinal motility in correlation to neuronal development in zebrafish embryos and larvae. *Journal of Fish Biology* 63:318–331.

Houlihan, D.F., Pedersen, B.H., Steffensen, J.F., et al. 1995. Protein-synthesis, growth and energetics in larval herring (*Clupea harengus*) at different feeding regimes. *Fish Physiology and Biochemistry* 14(3):195–208.

Hsu, S.M., Raine, L., and Fanger, H. 1981. Use of avidin-biotin-peroxidase complex (ABC) in immunoperoxidase techniques: a comparison between ABC and unlabeled antibody (PAP) procedures. *Journal of Histochemistry and Cytochemistry* 29(4):577–580.

Huggett, J., Dheda, K., Bustin, S., et al. 2005. Real-time RT-PCR normalisation; strategies and considerations. *Genes and Immunity* 6(4):279–284.

Hurlbert, S.H. 1984. Pseudoreplication and the design of ecological field experiments. *Ecological Monographs* 54(2):187–211.

Kamler, E. 2008. Resource allocation in yolk-feeding fish. *Reviews in Fish Biology and Fisheries* 18:143–200.

Kamler, E., Szlaminska, M., Kuczynski, M., et al. 1994. Temperature-induced changes of early development and yolk utilization in the African catfish *Clarias gariepinus*. *Journal of Fish Biology* 44:311–326.

Kurokawa, T., and Suzuki, T. 1998. Development of intestinal brush border aminopeptidase in the larval Japanese flounder *Paralichthys olivaceus*. *Aquaculture* 162:113–124.

Kurokawa, T., Iinuma, N., Unuma, T., et al. 2004. Development of endocrine system regulating exocrine pancreas and estimation of feeding and digestive ability in Japanese eel larvae. *Aquaculture* 234(1–4):513–525.

Kurokawa, T., Uji, S., and Suzuki, T. 2005. Identification of pepsinogen gene in the genome of stomachless fish, *Takifugu rubripes*. *Comparative Biochemistry and Physiology. Part B, Biochemistry & Molecular Biology* 140:133–140.

Lessard, E.J., Martin, M.P., and Montagnes, D.J.S. 1996. A new method for live-staining protists with DAPI and its application as a tracer of ingestion by walleye pollock (*Theragra chalcogramma* (Pallas)) larvae. *Journal of Experimental Marine Biology and Ecology* 204(1–2):43–57.

Lieschke, G.J., and Currie, P.D. 2007. Animal models of human disease: zebrafish swim into view. *Nature Reviews. Genetics* 8:353–367.

Lin, L.Y., and Hwang, P.P. 2004. Mitochondria-rich cell activity in the yolk-sac membrane of tilapia (*Oreochromis mossambicus*) larvae acclimatized to different ambient chloride levels. *Journal of Experimental Biology* 207(8):1335–1344.

Lochmann, S.E., Maillet, G.L., Taggart, C.T., et al. 1996. Effect of gut contents and lipid degradation on condition measures in larval fish. *Marine Ecology—Progress Series* 134:27–35.

Lund, I., Steenfeldt, S.J., Banta, G., et al. 2008. The influence of dietary concentrations of arachidonic acid and eicosapentaenoic acid at various stages of larval ontogeny on eye migration, pigmentation and prostaglandin content of common sole larvae (*Solea solea* L.). *Aquaculture* 276(1–4):143–153.

Maltby, L., Naylor, C., and Calow, P. 1990. Effect of stress on a freshwater benthic detritivore: scope for growth in *Gammarus pulex*. *Ecotoxicology and Environmental Safety* 19:285–291.

Masuda, R., and Tsukamoto, K. 1999. School formation and concurrent developmental changes in carangid fish with reference to dietary conditions. *Environmental Biology of Fishes* 56(1–2):243–252.

Mayer, A.N., and Fishman, M.C. 2003. Nil per os encodes a conserved RNA recognition motif protein required for morphogenesis and cytodifferentiation of digestive organs in zebrafish. *Development* 130:3917–3928.

McCollum, A.J., Geubtner, J., and von Herbing, I.H. 2006. Metabolic cost of feeding in Atlantic cod (*Gadus morhua*) larvae using microcalorimetry. *ICES Journal of Marine Science* 63(2):335–339.

Mcginty, E.L., Smith-Keune, C., and Jerry, D.R. 2008. A high through-put protocol for quantifying nucleic acids in individual microcrustaceans using new generation RNA and DNA specific dyes. *Journal of Shellfish Research* 27(2):449–455.

McManls, J.F.A., and Mowry, R.W. 1960. *Staining Methods, Histological and Histochemical*. Paul B. Hoeber Inc., New York.

Metscher, B.D., and Ahlberg, P.E. 1999. Zebrafish in context: uses of a laboratory model in comparative studies. *Developmental Biology* 210:1–14.

Miller, K.M., and Maclean, N. 2008. Teleost microarrays: development in a broad phylogenetic range reflecting diverse applications. *Journal of Fish Biology* 72(9):2039–2050.

Mohideen, M.A., Beckwith, L.G., Tsao-Wu, G.S., et al. 2003. Histology-based screen for zebrafish mutants with abnormal cell differentiation. *Developmental Dynamics* 228:414–423.

Mommsen, T.P. 2001. Paradigms of growth in fish. *Comparative Biochemistry and Physiology.*

Part B, Biochemistry & Molecular Biology 129:207–219.

Morais, S., Conceição, L.E.C., Rønnestad, I., et al. 2007. Dietary neutral lipid level and source in marine fish larvae: effects on digestive physiology and food intake. *Aquaculture* 268(1–4):106–122.

Nakayama, S., Masuda, R., Shoji, J., et al. 2003. Effect of prey items on the development of schooling behavior in chub mackerel *Scomber japonicus* in the laboratory. *Fisheries Science* 69:670–676.

Nasevicius, A., and Ekker, S.C. 2000. Effective targeted gene "knockdown" in zebrafish. *Nature Genetics* 26:216–220.

Neuheimer, A.B., and Taggart, C.T. 2007. The growing degree-day and fish size-at-age: the overlooked metric. *Canadian Journal of Fisheries and Aquatic Sciences* 64(2):375–385.

Neuman, M.J., and Able, K.W. 2002. Quantification of ontogenetic transitions during the early life of a flatfish, windowpane (*Scophthalmus aquosus*) (Pleuronectiformes, Scophthalmidae). *Copeia* 2002(3):597–609.

Neyfakh, A.A., Yarygin, K.N., and Gorgolyuk, S.I. 1983. Ornithine decarboxylase activity in embryos depends on temperature of development rather than on the stage of development. *The Biochemical Journal* 216:597–604.

Ng, A.N., de Jong-Curtain, T.A., Mawdsley, D., et al. 2005. Formation of the digestive system in zebrafish: III. Intestinal epithelium morphogenesis. *Developmental Biology* 286:114–135.

Ober, E.A., Field, H.A., and Stainier, D.Y. 2003. From endoderm formation to liver and pancreas development in zebrafish. *Mechanisms of Development* 120:5–18.

Ober, E.A., Verkade, H., Field, H.A., et al. 2006. Mesodermal Wnt2b signalling positively regulates liver specification. *Nature* 442:688–691.

Owen, S.F., McCarthy, I.D., Watt, P.W., et al. 1999. *In vivo* rates of protein synthesis in Atlantic salmon (*Salmo salar* L.) smolts determined using a stable isotope flooding dose technique. *Fish Physiology and Biochemistry* 20(1):87–94.

Pack, M., Solnica-Krezel, L., Malicki, J., et al. 1996. Mutations affecting development of zebrafish digestive organs. *Development* 123:321–328.

Pfeiler, E., and Govoni, J.J. 1993. Metabolic rates in early life history stages of elopomorph fishes. *The Biological Bulletin* 185:277–283.

Postlethwait, J.H. 2007. The zebrafish genome in context: ohnologs gone missing. *Journal of Experimental Zoology. Part B, Molecular and Developmental Evolution* 308:563–577.

Rainuzzo, J.R., Reitan, K.I., and Olsen, Y. 1997. The significance of lipids at early stages of marine fish: a review. *Aquaculture* 155: 103–115.

Rawls, J.F., Samuel, B.S., and Gordon, J.I. 2004. Gnotobiotic zebrafish reveal evolutionarily conserved responses to the gut microbiota. *Proceedings of the National Academy of Sciences of the United States of America* 101:4596–4601.

Robison, B.D., Drew, R.E., Murdoch, G.K., et al. 2008. Sexual dimorphism in hepatic gene expression and the response to dietary carbohydrate manipulation in the zebrafish (*Danio rerio*). *Comparative Biochemistry and Physiology. Part D, Genomics & Proteomics* 3(2):141–154.

Rombout, J.H., Stroband, H.W., and Taverne-Thiele, J.J. 1984. Proliferation and differentiation of intestinal epithelial cells during development of *Barbus conchonius* (Teleostei, Cyprinidae). *Cell and Tissue Research* 236:207–216.

Rubinstein, A.L. 2006. Zebrafish assays for drug toxicity screening. *Expert Opinion on Drug Metabolism & Toxicology* 2:231–240.

Rueda-Jasso, R., Conceição, L.E.C., Dias, J., et al. 2004. Effect of dietary non-protein energy levels on condition and oxidative status of Senegalese sole (*Solea senegalensis*) juveniles. *Aquaculture* 231:417–433.

Ruxton, G.D., and Colegrave, N. 2006. *Experimental Design for the Life Sciences*. Oxford University Press, Oxford.

Sampath-Kumar, R., Lee, S.T.L., Tan, C.H., et al. 1997. Biosynthesis *in vivo* and excretion of cortisol by fish larvae. *The Journal of Experimental Zoology* 277:337–344.

Schilling, T.F., and Webb, J. 2007. Considering the zebrafish in a comparative context. *Journal of Experimental Zoology. Part B, Molecular and Developmental Evolution* 308:515–522.

Scholz, S., Fischer, S., Gundel, U., et al. 2008. The zebrafish embryo model in environmental risk assessment—applications beyond acute toxicity testing. *Environmental Science and Pollution Research International* 15:394–404.

Stainier, D.Y., Fouquet, B., Chen, J.N., et al. 1996. Mutations affecting the formation and function of the cardiovascular system in the zebrafish embryo. *Development* 123:285–292.

Suthers, I.M., Fraser, A., and Frank, K.T. 1992. Comparison of lipid, otolith and morphometric condition indices of pelagic juvenile cod *Gadus morhua* from the Canadian Atlantic. *Marine Ecology—Progress Series* 84:31–40.

Takashima, F., and Hibiya, T. 1982. *An Atlas of Fish Histology, Normal and Pathological Features*, 2nd edition. Kodansha Ltd., Tokyo.

Theilacker, G.H., and Shen, W. 1993. Calibrating starvation-induced stress in larval fish using flow cytometry. *American Fisheries Society Symposium* 14:85–94.

Theilacker, G.H., and Shen, W. 2001. Evaluating growth of larval walleye pollock, *Theragra chalcogramma*, using cell cycle analysis. *Marine Biology* 138(5):897–907.

Wallace, K.N., and Pack, M. 2003. Unique and conserved aspects of gut development in zebrafish. *Developmental Biology* 255:12–29.

Wallace, K.N., Akhter, S., Smith, E.M., et al. 2005a. Intestinal growth and differentiation in zebrafish. *Mechanisms of Development* 122:157–173.

Wallace, K.N., Dolan, A.C., Seiler, C., et al. 2005b. Mutation of smooth muscle myosin causes epithelial invasion and cystic expansion of the zebrafish intestine. *Developmental Cell* 8:717–726.

Westerman, M.E., Holt, G.J., and DiMichele, L. 1999. Quantitative assay of cyclin-dependent kinase activity as a sensitive marker of cell proliferation in marine teleost larvae. *Marine Biotechnology* 1:297–310.

Yúfera, M., and Darias, M.J. 2007. The onset of exogenous feeding in marine fish larvae. *Aquaculture* 268:5363.

Yúfera, M., Fernández-Díaz, C., and Pascual, E. 2005. Food microparticles for larval fish prepared by internal gelation. *Aquaculture* 248:253–262.

Zaidi, A.U., Enomoto, H., Milbrandt, J., et al. 2000. Dual fluorescent *in situ* hybridization and immunohistochemical detection with tyramide signal amplification. *The Journal of Histochemistry and Cytochemistry* 48(10):1369–1375.

Zhong, Q.W., Zhang, Q., Wang Z., et al. 2008. Expression profiling and validation of potential reference genes during *Paralichthys olivaceus* embryogenesis. *Marine Biotechnology* 10(3): 310–318.

Index

Note: Pages in *italics* indicate figures or tables.

Printed and bound by CPI Group (UK) Ltd, Croydon, CR0 4YY
16/09/2026
14963449-0002